Spintronics

Claudia Felser · Gerhard H. Fecher
Editors

Spintronics

From Materials to Devices

Springer

Editors
Prof. Dr. Claudia Felser
Inst. Anorganische Chemie
und Analytische Chemie
Universität Mainz
Mainz, Germany

Max-Planck-Institut für
Chemische Physik fester Stoffe
Dresden, Germany

Dr. Gerhard H. Fecher
Inst. Anorganische Chemie
und Analytische Chemie
Universität Mainz
Mainz, Germany

Max-Planck-Institut für
Chemische Physik fester Stoffe
Dresden, Germany

ISBN 978-90-481-3831-9 ISBN 978-90-481-3832-6 (eBook)
DOI 10.1007/978-90-481-3832-6
Springer Dordrecht Heidelberg New York London

Library of Congress Control Number: 2013935588

Printed on acid-free paper

Springer is part of Springer Science+Business Media (www.springer.com)

Claudia Felser · Gerhard H. Fecher
Editors

Spintronics

From Materials to Devices

Springer

Editors
Prof. Dr. Claudia Felser
Inst. Anorganische Chemie
und Analytische Chemie
Universität Mainz
Mainz, Germany

Max-Planck-Institut für
Chemische Physik fester Stoffe
Dresden, Germany

Dr. Gerhard H. Fecher
Inst. Anorganische Chemie
und Analytische Chemie
Universität Mainz
Mainz, Germany

Max-Planck-Institut für
Chemische Physik fester Stoffe
Dresden, Germany

ISBN 978-90-481-3831-9 ISBN 978-90-481-3832-6 (eBook)
DOI 10.1007/978-90-481-3832-6
Springer Dordrecht Heidelberg New York London

Library of Congress Control Number: 2013935588

Printed on acid-free paper

Springer is part of Springer Science+Business Media (www.springer.com)

Foreword

Spintronics is a new field of science, developed over the past 20 years, that is focused on exploring the properties of materials in which spin-polarized currents play a key role, as well developing materials and devices for advanced nano-electronic applications. Spintronics materials have already found one important application in the form of highly sensitive magnetic field sensors in recording read heads for magnetic disk drives. The development of such "*spin-valve*" sensors was predicated on the discovery of the twin phenomena of giant magnetoresistance (GMR) and oscillatory interlayer magnetic coupling in magnetic multilayers formed from transition and noble metals in the early 1990s. More recently the same spin-engineering concepts in conjunction with the phenomenon of spin-dependent tunneling has allowed for the further development of even more sensitive recording read heads. Magnetic tunnel junction (MTJ) read heads have, since about 2007, have completely replaced sensors based on the phenomenon of GMR. Further advances in such recording read heads require new sensing materials that have a complex set of intertwined properties. Similarly, spintronics materials and phenomena, if suitably engineered, have the potential for novel solid-state memories. Magnetic tunnel junctions themselves can form a non-volatile memory cell and proposals for an MTJ based magnetic random access memory (MRAM) date from the mid 1990s. The first MTJ based MRAM was demonstrated by IBM in 1999. Today there is much excitement about the potential of MRAM for high density applications and for replacing conventional charge based memories such as DRAM, e-DRAM and even SRAM. These latter technologies may have difficulties scaling to dimensions below 10–15 nm. However, just like sensors, the success of MRAM is predicated on the discovery of new materials with the needed combination of properties for this application. For MRAM, in addition to high tunneling magnetoresistance, it must be possible to switch the magnetic state of the MTJ memory element using tiny currents passed through the MTJ that take advantage of the phenomenon of spin-momentum transfer. This requires MTJ elements in which the magnetic moment of each of the magnetic electrodes in the device are oriented perpendicular to the layers forming the tunneling junction. Moreover, these MTJ elements must be sufficiently stable against thermal fluctuations that their memory, namely the direction of magnetization of the electrode in which the data is stored, can be sustained for a decade or longer. It is challenging to find materials that can meet these stringent requirements.

This book focuses on classes of materials that exhibit half-metallicity and which have the potential for advanced sensor and memory applications. Half-metallicity is found in magnetic materials in which one of the spin-polarized bands has a band-gap at the Fermi energy and in which the other exchange-split band has a non-zero density of states at the Fermi energy. Examples include members of the perovskite and double-perovskite families, as well as CrO_2, and members of the Heusler family of compounds. The oxides typically have magnetic ordering temperatures near or below room temperature but the Heusler compounds can have Curie temperatures that exceed 1,000 K. For example, Co_2FeSi was discovered to have a Curie temperature greater than 1,120 K in 2005. Over the past decade "*simple rules*" for understanding the properties of the vast class of Heusler and half-Heusler compounds have been developed, allowing for systematic materials engineering of such materials for a variety of applications including magnetic sensors and memories. Out of the scope of this book are applications of Heuslers to, for example, thermoelectrics and solar cells. The application of Heuslers to the field of spintronics perhaps began with the inverse design of the quaternary compound, $Co_2Cr_{0.6}Fe_{0.4}Al$, by the Mainz group in 2000. Using the "*simple rules*" mentioned previously to identify potentially Heusler compounds that would not only exhibit high tunneling spin polarization by virtue of their half-metallic character, but at temperatures needed for sensor and memory applications, the Mainz group carried out density functional theoretical modeling of the electronic structure of hundreds of Heusler compounds. They calculated and then experimentally demonstrated that in $Co_2Cr_{0.6}Fe_{0.4}Al$ the Fermi energy would like in the middle of the electronic band-gap in the minority spin-polarized channel. Thus, the half-metallic nature of this material is less sensitive to thermal excitations of minority spin-polarized carriers, so ensuring near complete half-metallicity over the temperature range of applications.

This book provides an insight into our current understanding of Heusler compounds. A detailed understanding of the relationship between their crystal structure and electronic properties is discussed including the influence of atomic disorder. The current state of the art of Heusler based devices is also presented. Of particular importance are Heusler compounds including as one of their constituent elements cobalt for the magnetic electrodes in spin-valve and magnetic tunnel junction devices. These compounds have very high magnetic ordering temperatures, as mentioned above. An important recent focus of interest is the exploration of tetragonally distorted Heusler compounds, and Mn_3Ga and cousins, which display very large values of perpendicular magnetic anisotropy. Such materials have significant potential for spin transfer torque-MRAM.

The detailed understanding of the properties of complex materials such as Heusler compounds and oxides that is described in this book attests to the very bright future of half-metallic materials!

IBM Almaden Stuart S.P. Parkin

Preface

Spintronics is the child of magnetoelectronics, which was born with the discovery of the "*half-metallic ferromagnets*" in 1983 by Robert de Groot for NiMnSb and by Jürgen Kübler for Co_2MnAl and Co_2MnSn. Peculiar for this type of ferromagnets is that only electrons with one kind of spin take part in the electronic transport properties. This leads to the idea that electronic devices can be built where not the charge but the spin of the electron transports the signal and is thus free of "ohmic" energy dissipation. The exceptional property that the materials exhibit 100 % spin polarization at the Fermi energy makes them ideal candidates for spin injection devices to be used in spin electronics that is spintronics.

Spintronics is an emerging technology exploiting the spin degree of freedom and has proved to be very promising for new types of fast electronic device. Amongst the anticipated advantages of spintronics technologies, researchers have identified the non-volatile storage of data with high density and low energy consumption as particularly relevant. This monograph examines the concept of half-metallic compounds perspectives to obtain novel solutions, and discusses several oxides as well as the intermetallic Heusler compounds. Such materials can be designed and made with high spin polarization and, especially in the case of Heusler compounds, many material-related problems present in current-day 3d metal systems, can be overcome. From materials design by theoretical methods and the preparation and properties of the materials to the production of thin films and devices, this monograph provides an insight into the current research on Heusler compounds and double perovskites. It offers a general understanding of structure–property relationships, including the influence of disorder and correlations on the electronic structure and interfaces. Last not least, spintronics devices such as magnetic tunnel junctions (MTJs) and giant magnetoresistance (GMR) devices, with current perpendicular to the plane, in which Co_2 based Heusler compounds are used as new electrode materials, are also introduced.

The way from materials design to applications and devices is reviewed in 17 chapters by experts in the field. An introduction to Heusler compounds (Chaps. 1–3) and double perovskites (Chap. 4) as materials for spintronics is given in the first four chapters. In the following four chapters, the basic theoretical considerations for materials design of the half-metallicity of Heusler compounds (Chaps. 5–7) and double perovskites (Chaps. 5 and 8) are given. Next five chapters are devoted to the application of new and well established experiments to investigate the materials and devices (Chaps. 9–13). Finally, four chapters (14–17) review the successful use of the new materials in thin films and devices.

The first part is devoted to new materials from the classes of Heusler compounds and double perovskites. Chapter 1 gives an overview on functionalities and applications of Heusler compounds. The following Chap. 2 reports on new Heusler compounds and their properties starting with material design from the viewpoint of a chemist based on optimization and tuning of the Fermi energy by chemical substitution. Details of crystallographic order and disorder phenomena in Heusler compounds are discussed in Chap. 3. These phenomena are important for an understanding of the structure to property relations. Substitution effects in double perovskites are reported in Chap. 4, which investigates the influence of electron and hole doping in half-metallic Sr_2FeReO_6.

The second part reviews the design of new materials by theoretical methods. Chapter 5 deals with the theory of the half-metallic materials and reviews the half-metallic ferromagnets and ferrimagnets found in the classes of Heusler compounds and double perovskites. The special focus of this chapter is on electronic structure, nature of the magnetic moments and physics of the energy gap for a single spin channel. Chapter 6 is about correlation and chemical disorder in Heusler alloys with a special emphasis on spectroscopic studies. It is shown that effects of local electronic correlations and alloying on the properties of the Heusler compounds need an equal treatment of static and dynamic correlations. More details of the electronic structure of the half-metallic, transition metal based Heusler compounds are reviewed in Chap. 7. It reports on calculation of electronic structure and properties by means of ab-initio band structure methods for ordered, substituted and disordered Heusler compounds. The theory of the electronic structure of complex oxides is reported in Chap. 8, which discusses—based on dynamical mean field theory—the electronic structure, magnetic properties, and metal–insulator transitions in transition metal oxides.

The third part is devoted to new experimental methods and their application to the investigation of new materials. Chapter 9 is about the experimental investigation of the local structure of highly spin polarized Heusler compounds revealed by nuclear magnetic resonance spectroscopy (NMR). It shows that NMR, as a local probe, is a suitable tool to reveal structural contributions and foreign phases in spin polarized materials which are very difficult to detect with other methods. Chapter 10 reports on the investigation of new materials with high spin polarization by X-ray magnetic circular dichroism. It explains in an element specific way the spin and orbital magnetic moments of Co_2YZ (bulk and films) and confirms experimentally the tailoring of the Fermi level in quaternary substituted compounds. As a very young experimental method, hard X-ray photoelectron spectroscopy is applied in Chap. 11 to explore the electronic structure of new materials, thin films, and buried layers in MTJs. In Chap. 12, the surface electronic properties of $Co_2Cr_{1-x}Fe_xAl$ are characterized by spin resolved photo emission with low photon energies to discriminate bulk and surface states. Magneto-optical experiments and ion beam-induced modification of Heusler compounds are reported in Chap. 13 that reports on investigations of magnetic exchange stiffness, magnetic anisotropy, magnetization reversal, and magneto-optical Kerr effect in Co-based Heusler compound thin films.

The last part demonstrates the successful use of new spintronics materials in thin films and devices. Chapter 14 is devoted to a special material ($Co_2Fe(Al_{1-x}Si_x)$) and its application in spintronics. It reports on the structural and magnetic properties of epitaxial thin films and their applications to magnetic tunnel junctions (MTJs), giant magnetoresistive (GMR) devices and spin transfer magnetization switching. It reveals that the use of Heusler compounds is an effective way to reduce the switching current density in MTJs. Chapter 15 is about the transport properties of Co_2(Mn, Fe)Si thin films. Especially it is shown that the normal Hall effect undergoes a transition from a hole-like charge transport in Co_2MnSi to an electron-like transport in Co_2FeSi. Chapter 16 reports on the preparation, barrier-interface engineering, and investigation of $Co_2Cr_{1-x}Fe_xAl$ thin films for MTJs with AlO_x tunneling barriers. The tunnel magnetoresistance in tunnel junctions with Co_2MnSi as Heusler alloy electrode and MgO as barrier is investigated in Chap. 17. This completes the way from material design to successful use in spintronics devices.

Mainz, Dresden

Claudia Felser
Gerhard H. Fecher

Acknowledgements

A major part of this book is based on the interdisciplinary collaboration of groups from chemistry, material science, and physics at the universities in Kaiserslautern, Mainz (Germany), and Sendai (Japan), which was funded by the Deutsche Forschungsgemeinschaft. All project leaders of the research unit FG 559 (*New Materials with High Spin Polarization*)—H. Adrian (P2, Mainz), M. Aeschlimann (P10, Kaiserslautern), Y. Ando (P12, Sendai), M. Bauer (P10, Kaiserslautern), M. Cinchetti (P10, Kaiserslautern), N. Blümer (P6, Mainz), S.O. Demokritov (P9, Kaiserslautern), P. van Dongen (P6, Mainz), H.-J. Elmers, C. Felser (P1, P7, P8, Mainz), J. Hamrle (P9, Kaiserslautern), B. Hillebrands (P9, Kaiserslautern), G. Jakob (P2, Mainz), M. Jourdan (P11, Mainz), V. Ksenofontov (P8, Mainz), G. Schönhense (P7, Mainz), W. Tremel (P4, Mainz)—their co-workers, and PhD students are very grateful to the Deutsche Forschungsgemeinschaft DfG for the financial and administrative support granted to them through six years of very successful collaboration.

Contents

Contributors

Martin Aeschlimann Department of Physics and Research Center OPTIMAS, University of Kaiserslautern, Kaiserslautern, Germany

Yasuo Ando Department of Applied Physics, Graduate School of Engineering, Tohoku University, Sendai, Japan

Benjamin Balke Institut für Anorganische Chemie und Analytische Chemie, Johannes Gutenberg – Universität, Mainz, Germany

Irene Bonn Institut für Anorganische Chemie und Analytische Chemie, Johannes Gutenberg – Universität, Mainz, Germany

Jürgen Braun Department Chemie, Ludwig-Maximilians-Universität München, Munich, Germany

Stanislav Chadov Institut für Anorganische Chemie und Analytische Chemie, Johannes Gutenberg – Universität, Mainz, Germany; Max-Planck-Institut für Chemische Physik fester Stoffe, Dresden, Germany

Mirko Cinchetti Department of Physics and Research Center OPTIMAS, University of Kaiserslautern, Kaiserslautern, Germany

Hubert Ebert Department Chemie, Ludwig-Maximilians-Universität München, Munich, Germany

Hans-Joachim Elmers Institut für Physik, Johannes Gutenberg – Universität, Mainz, Germany

Gerhard H. Fecher Institut für Anorganische Chemie und Analytische Chemie, Johannes Gutenberg – Universität, Mainz, Germany; Max-Planck-Institut für Chemische Physik fester Stoffe, Dresden, Germany

Claudia Felser Institut für Anorganische Chemie und Analytische Chemie, Johannes Gutenberg – Universität, Mainz, Germany; Max-Planck-Institut für Chemische Physik fester Stoffe, Dresden, Germany

Oksana Gaier Institut Néel, Grenoble, France; Technische Universität Kaiserslautern and Research Center OPTIMAS, Kaiserslautern, Germany

Haitao Gao Institut für Anorganische Chemie und Analytische Chemie, Johannes Gutenberg – Universität, Mainz, Germany

Tanja Graf Institut für Anorganische Chemie und Analytische Chemie, Johannes Gutenberg – Universität, Mainz, Germany

Jaroslav Hamrle VSB – Technical University of Ostrava, Ostrava, Czech Republic; Technische Universität Kaiserslautern and Research Center OPTIMAS, Kaiserslautern, Germany

Burkard Hillebrands Technische Universität Kaiserslautern and Research Center OPTIMAS, Kaiserslautern, Germany

Kouichiro Inomata National Institute for Materials Science, Tsukuba, Japan

Gerhard Jakob Institut für Physik, Johannes Gutenberg – Universität, Mainz, Germany

Martin Jourdan Institut für Physik, Johannes Gutenberg – Universität, Mainz, Germany

Alexandra Jung Institut für Anorganische Chemie und Analytische Chemie, Johannes Gutenberg – Universität, Mainz, Germany

Michael Kallmayer Institut für Physik, Johannes Gutenberg – Universität, Mainz, Germany

Peter Klaer Institut für Physik, Johannes Gutenberg – Universität, Mainz, Germany

Jürgen T. Kohlhepp Eindhoven University of Technology, Eindhoven, The Netherlands

Vadim Ksenofontov Institut für Anorganische Chemie und Analytische Chemie, Johannes Gutenberg – Universität, Mainz, Germany

Jürgen Kübler Institut für Festkörperphysik, Technische Universität Darmstadt, Darmstadt, Germany

Alexander I. Lichtenstein Institut für Theoretische Physik, Universität Hamburg, Hamburg, Germany

Ján Minár Department Chemie, Ludwig-Maximilians-Universität München, Munich, Germany

Hiroshi Naganuma Department of Applied Physics, Graduate School of Engineering, Tohoku University, Sendai, Japan

Mikihiko Oogane Department of Applied Physics, Graduate School of Engineering, Tohoku University, Sendai, Japan

Martin Panthöfer Institut für Anorganische Chemie und Analytische Chemie, Johannes Gutenberg – Universität, Mainz, Germany

Sergey Reiman Institut für Anorganische Chemie und Analytische Chemie, Johannes Gutenberg – Universität, Mainz, Germany

Yuya Sakuraba Institute for Materials Research, Tohoku University, Sendai, Japan

Horst Schneider Institut für Physik, Johannes Gutenberg – Universität, Mainz, Germany

Hiroaki Sukegawa National Institute for Materials Science, Tsukuba, Japan

Koki Takanashi Institute for Materials Research, Tohoku University, Sendai, Japan

Wolfgang Tremel Institut für Anorganische Chemie und Analytische Chemie, Johannes Gutenberg – Universität, Mainz, Germany

Simon Trudel University of Calgary, Calgary, Canada; Technische Universität Kaiserslautern and Research Center OPTIMAS, Kaiserslautern, Germany

Sumito Tsunegi Department of Applied Physics, Graduate School of Engineering, Tohoku University, Sendai, Japan

Enrique Vilanova Vidal Institut für Physik, Johannes Gutenberg – Universität, Mainz, Germany

Georg Wolf Technische Universität Kaiserslautern and Research Center OPTIMAS, Kaiserslautern, Germany

Sabine Wurmehl Leibniz Institute for Solid State and Materials Research, Dresden, Germany

Jan-Peter Wüstenberg Department of Physics and Research Center OPTIMAS, University of Kaiserslautern, Kaiserslautern, Germany

Acronyms

2-PPE	Two photon photoemission
AAL	Around atomic limit
AF	Antiferromagnetic
AFM	Atomic force microscopy
AP	Antiparallel
APS	Appearance potential spectroscopy
ARUPS	Angular resolved ultraviolet photoelectron spectroscopy
ASA	Atomic sphere approximation
BC-MIT	Bandwidth controlled metal–insulator transition
BIS	Bremsstrahlung isochromat spectroscopy
BL	Beamline
BLS	Brillouin light scattering
CCFA	$Co_2Cr_{0.6}Fe_{0.4}Al$
CDMFT	Cluster dynamical mean field theory
CEMS	Conversion electron Mößbauer spectroscopy
CMR	Colossal magnetoresistance
COHP	Crystal orbital Hamiltonian population
CPA	Coherent potential approximation
CPP	Current perpendicular to plane
CPP-SV	Current perpendicular to plane spin valves
CPP-GMR	Current perpendicular to plane giant magnetoresistance
DE	Damon–Eshbach
DE	Double exchange
DFT	Density functional theory
DOS	Density of states
DIPS	Digital Image Processing System
DMFT	Dynamical mean field theory
DP	Double perovskite
DSC	Differential scanning calorimetry
EB	Electron beam
EB-MTJ	Magnetic tunneling junction with electron beam evaporated MgO barrier

EDX	Energy dispersive X-ray spectroscopy
EXAFS	Extended X-ray absorption fine structure spectroscopy
FC-MIT	Filling-controlled metal–insulator transition
FID	Free induction decay
FLAPW	Full potential linearized augmented plane-wave
FLEX	Local fluctuating exchange
FM	Ferromagnetic
FWHM	Full width at half maximum
GGA	Generalized gradient approximation
GMR	Giant magnetoresistance
HAADF	High-angle angular dark field
HAXPES	Hard X-ray photoelectron spectroscopy
HDD	Harddisk drive
HF	Hartree–Fock
HMF	Half-metallic ferromagnet
HRTEM	High resolution transmission electron microscope
HTS	High temperature structure
ICP	Inductively coupled plasma
ICSD	Inorganic crystal structure database
IPT	Iterated perturbation theory
JASRI	Japan synchrotron radiation research institute
KKR	Korringa–Kohn–Rostoker
LDA	Local density approximation
LDA+U	Local density approximation with added correlation potential
LDAD	Linear dichroism in the angular distribution
LDOS	Local density of states
LEED	Low energy electron diffraction
LH	Lower Hubbard
LHB	Lower Hubbard band
LMDAD	Linear magnetic dichroism in the angular distribution
LMOKE	Linear or longitudinal MOKE
LMTO	Linearized muffin-tin orbital
LNLS	Brazilian synchrotron light laboratory
LSDA	Local spin density approximation
LTS	Low temperature
LTS	Low temperature structure
MCDAD	Magnetic circular dichroism in the angular distribution
MDAD	Magnetic dichroism in the angular distribution
MIT	Metal–insulator transition
MOKE	Magneto optical Kerr effect
MR	Magnetoresistance
MRAM	Magnetic random access memory
MTJ	Magnetic tunnel junction
NCA	Non-crossing approximation
NQP	Non-quasiparticle

PDOS	Partial density of states
PMA	Perpendicular magnetic anisotropy
PSV	Pseudo spin valve
QMC	Quantum Monte Carlo
QMOKE	Quadratic magneto-optical Kerr effect
RA	Resistance area
RA_P	Resistance area for parallel magnetization configuration
RE	Rare earth
RF	Radio frequency
RHEED	Reflection high energy electron diffraction
RKKY	Rudermann–Kittel–Kasuya–Yosida
RRR	Residual resistivity ratio
RT	Room temperature
SB	Strukturberichte
SE	Superexchange
SEM	Scanning electron microscope
SIC	Self interaction correction
SIC-LDA	Self interaction corrected local density functional approximation
SOC	Spin–orbit coupling
SPEAR	Stanford synchrotron radiation laboratory
SP-MTJ	Magnetic tunneling junction with sputtered MgO barrier
SPRKKR	Spin polarized, full relativistic Korringa–Kohn–Rostocker method
SPTF	Spin polarized T-matrix plus fluctuation
SQUID	Superconducting quantum interference device
SR-PES	Spin resolved photoelectron spectroscopy
TEM	Transmission electron microscopy
TEY	Total electron yield
TM	Transmission
TMO	Transition metal oxides
TMR	Tunneling magnetoresistance
UH	Upper Hubbard
UHB	Upper Hubbard band
UPS	Ultraviolet photoelectron spectroscopy
VSM	Vibrating sample magnetometer
VB-XPS	Valence band photoemission spectra
XAFS	X-ray absorption fine structure spectroscopy
XAS	X-ray absorption spectroscopy
XMCD	X-ray magnetic circular dichroism
XPS	X-ray photoelectron spectroscopy
XRD	X-ray diffraction

Chapter 1
Heusler Compounds at a Glance

Tanja Graf and Claudia Felser

Abstract The class of Heusler compounds, including the XYZ and the X_2YZ compounds, does not only have an endless number of members, but also a vast variety of properties can be found in this class of materials, ranging from semi-conductors, half-metallic ferromagnets, superconductors, and topological insulators to shape memory alloys. With this chapter, we would like to provide an overview of Heusler compounds, focusing on basis design principles, their properties and potential applications.

1.1 Introduction

The search for new materials in the field of spintronics lead to the class of Heusler compounds about 30 years ago. They seem to be the material of choice for many applications, due to their tunable electronic structure, which makes the design of desirable properties, ranging from half-metallic ferromagnets [1, 2] over completely compensated antiferromagnets [3] to non-magnetic semiconductors [4, 5], and even superconductors [6, 7], possible. In the last few years, new areas of applications emerged, comprising environmental technologies, as thermoelectrics [8, 9] and solar cells [10]. Very recently, the discovery of topological insulators, a new quantum state of matter, opened the door to technologies of the future with unimagined possibilities for Heusler compounds [11, 12].

Heusler compounds are ternary intermetallics with the general formula XYZ (often called Half-Heusler) or X_2YZ, in which X and Y typically are transition metals and Z is a main group element. This class of materials was discovered in 1903 by Fritz Heusler [13, 14]. Their special feature is, in fact, that their properties differ completely from those of the contained elements. For instance, Cu_2MnAl is mag-

T. Graf · C. Felser
Institut für Anorganische Chemie und Analytische Chemie, Johannes Gutenberg – Universität, 55099 Mainz, Germany

C. Felser (✉)
Max-Planck-Institut für Chemische Physik fester Stoffe, 01187 Dresden, Germany
e-mail: felser@cpfs.mpg.de

C. Felser, G.H. Fecher (eds.), *Spintronics*, DOI 10.1007/978-90-481-3832-6_1,

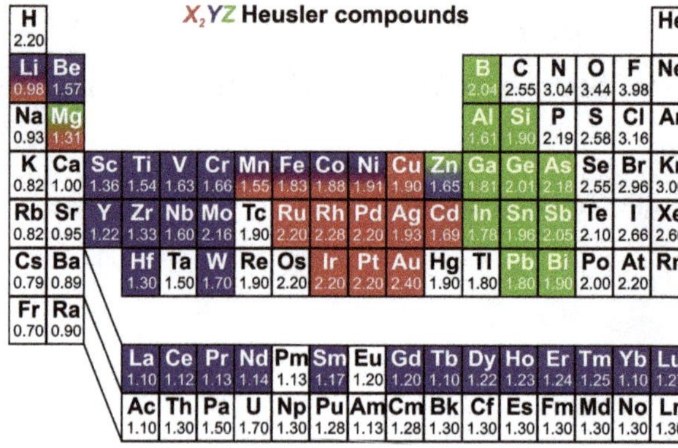

Fig. 1.1 Endless numbers of Heusler compounds can be designed. This periodic table gives an overview on possible combinations of elements. Below the element symbol, the Pauling electronegativity value is given, which helps to understand the structure and resulting properties of the compounds

netic, although neither Cu, nor Mn and Al are magnetic by themselves. Another example is TiNiSn, which is semiconducting, even though it is comprised of three metals [15]. The endless number of compounds, which can be prepared by combination of almost all elements of the periodic table, is illustrated in Fig. 1.1.

In this review article, we give an overview of the structural, electronic and magnetic properties of this exceptional class of materials. Furthermore, several areas of possible applications will be discussed.

1.2 Crystal Structure of Heusler Compounds

In the past, Heusler compounds were often understood as intermetallic alloys, although the description as an intermetallic compound is more appropriate due to their characteristic atomic order. They first attracted interest among the scientific community in 1903, when F. Heusler found that the compound Cu_2MnAl becomes ferromagnetic, although non of its constituent elements is ferromagnetic by itself [13, 14]. However, it took three decades until the crystal structure was determined to be ordered with a face centered, cubic lattice [16, 17]. Today, it is well-known that the properties of Heusler compounds depend strongly on the atomic order. Band structure calculations show that already small amounts of disorder within the distribution of the atoms on the lattice sites cause distinct changes in their electronic structure, and thus also in their magnetic and transport properties [18–20]. Therefore, a profound understanding and a careful analysis of their crystal structure is essential to comprehend the structure-to-property relation of Heusler compounds.

Figure 1.2 shows the most important crystal structure relationships concerning Heusler compounds. The zinc blende structure (ZnS) builds the basis for the understanding of Heusler compounds, since all Heusler compounds can be decomposed into a ZnS-type substructure with variations in the occupancy of the interstitial lattice sites. Due to the covalent character of its bonding interaction, this sublattice is always composed of those elements with the smallest difference in electronegativity.

Generally, two families of Heusler compounds have to be distinguished: The ternary compounds XYZ with $C1_b$ structure (often referred to as "Half"-Heusler

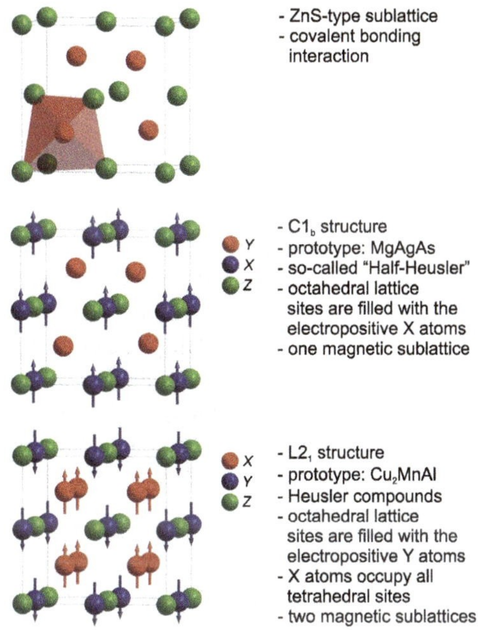

Fig. 1.2 Crystal structure of Heusler compounds: Summary of the most important characteristics

compounds, compare Fig. 1.2, middle panel) and those with the general formula X_2YZ (Fig. 1.2, bottom). A common feature of both families are four interpenetrating *fcc* sublattices. In case of X_2YZ compounds, they are fully occupied, whereas one of them is void in Half-Heusler compounds. Starting from the ZnS-type structure, the most electropositive atoms fill the octahedral vacancies, since this is electrostatically preferred. The prototype of the Half-Heusler structure is MgAgAs (space group F$\bar{4}$3m (216)), C1_b.

Heusler compounds X_2YZ crystallize in the cubic space group Fm$\bar{3}$m (225) with Cu_2MnAl (L2_1) as prototype. This structure can also be understood as a zinc blende-type sublattice, build up by one X and Z, the second X occupies in the remaining tetrahedral holes, and Y is located in the octahedral holes. Depending on the atomic arrangement, the $HgCu_2Ti$-type structure, in which one X and Y change sites, is also possible (Space group no. 216, F$\bar{4}$3m) but this "inverse" Heusler structure is less frequently observed [21].

Atomic disorder has a strong impact on the electronic structure, and thus, also on the properties of Heusler compounds. Detailed descriptions of disorder effects in Heusler compounds can be found in Chap. 2 and in Refs. [22–26]. It should be noted that the detection of anti-site disorder is not trivial, standard powder X-ray diffraction (XRD) methods are often not sufficient to fully clarify the atomic arrangement. Useful techniques to detect also small amounts of disorder include anomalous XRD investigations with synchrotron radiation [27], spin echo nuclear magnetic resonance [28–30], Mößbauer spectroscopy [31, 32], and extended X-ray-absorption fine structure (XAFS) [33, 34].

1.3 Electronic Properties

The electronic behavior of Heusler compounds can be easily understood by electron counting rules.

Half-Heusler compounds comprise the family of Nowotny–Juza phases. All compounds with a closed electron shell exhibit semiconducting behavior. For the Nowotny–Juza phases $A^I B^{II} C^V$, with A^I = Li, Cu, Ag, B^{II} = Be, Mg, Zn, Cd, and C^V = N, P, As Sb, Bi, the closed shell configuration is obtained for a valence electron number of eight, as indicated by the formal oxidation states. These compounds are well-known direct wide band gap semiconductors [35–37]. The introduction of transition metals leads to Half-Heusler compounds XYZ. Please, note that the elements are ordered according to their electronegativity, the most electropositive element is placed at the beginning of the formula, e.g. ZrNiSn, and GdAuSn [38, 39]. The electropositive X transfers its valence electrons to the more electronegative elements Y and Z. Thus, they become stable closed shell ions, i.e. a d^{10} configuration for Y and a s^2p^6 configuration for Z. This procedure requires 18 valence electrons and empties the valence atomic orbitals of X. Consequently, semiconducting behavior is found for compounds with 18 valence electrons, for instance TiCoSb, and TiNiSn [40, 41]. Changing the valence electron count to a number different from 18, should lead to the formation of a metallic and magnetic compound. Due to the fact that magnetic atoms can only be placed on one of the *fcc* sublattices, as indicated by the arrows in Fig. 1.2 (middle panel), most compounds with $N_V \neq 18$ do not crystallize with the $C1_b$ structure. Well-studied examples are FeMnSb, MnMnSb, CrMnSb, and MnCoSb, which show contain two magnetic elements, and hence, crystallize in different structure types with more than one magnetic lattice site [42–44]. The only exceptions are the MnYSb compounds with Y = Ni, Pd, and Pt, which exhibit half-metallic ferromagnetism [45–50]. For MnNiSb, the bonding can be described in terms of Mn^{3+} (d^4), Ni^0 (d^{10}), and Sb^{3-} (s^2p^6) ions [4, 51]. The electrons of the Mn^{3+} ions are strongly localized, and therefore, the compounds MnYSb are stable in the $C1_b$ structure and exhibit a magnetic moment m of 4 μ_B per formula unit.

For the Heusler compounds X_2YZ the closed shell electron configuration is obtained in case of a valence electron count of $N_V = 24$, e.g. Fe_2VAl [52]. In contrast to Half-Heusler compounds, two magnetic sublattices exist in the $L2_1$ structure as indicated in Fig. 1.2. Thus, a whole zoo of magnetic compounds with $N_V \neq 24$ is known, comprising ferromagnets, ferrimagnets and even completely compensated antiferromagnets. The magnetic properties, and half-metallic ferromagnetism in particular, are discussed in Sect. 1.4.

1.4 Magnetism in Heusler Compounds

In 1983, Kübler et al. revealed by ab initio calculations that the density of states of ferromagnetic Co_2MnSn and Co_2MnAl nearly vanishes for one spin direction at

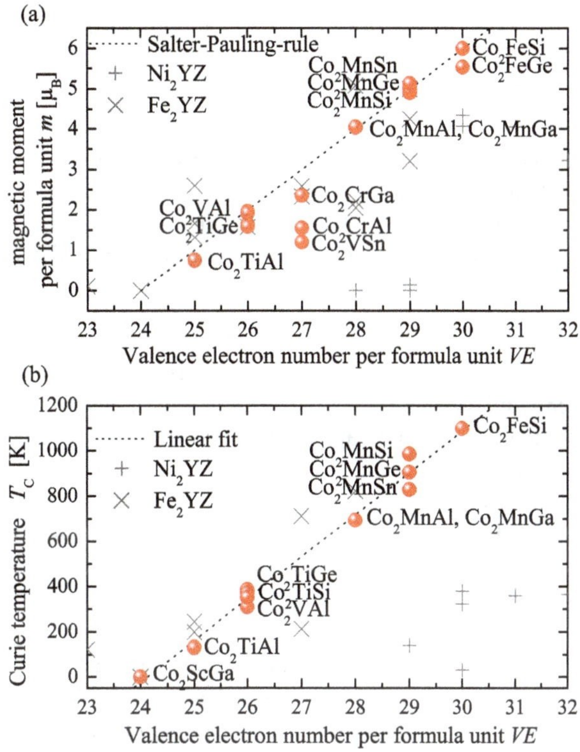

Fig. 1.3 (**a**) Magnetic moment of Heusler compounds versus the total valence electron number. Co_2-based Heusler compounds, which show half-metallic ferromagnetism, follow the Slater–Pauling rule. (**b**) The Curie temperature shows a linear dependence on the valence electron count, and therefore, also on the magnetic moment

the Fermi energy level (ϵ_F), which results in a high spin polarization. They concluded that this leads to peculiar transport properties [1]. This prediction of half-metallic ferromagnetism was extended to a large group of Co_2-based Heusler compounds [20, 53–55].

The electron spin polarization P at ϵ_F of a material is a useful measure to describe its electronic properties. P is given by

$$P = \frac{\rho_\uparrow(\epsilon_F) - \rho_\downarrow(\epsilon_F)}{\rho_\uparrow(\epsilon_F) + \rho_\downarrow(\epsilon_F)}, \tag{1.1}$$

where $\rho_\uparrow(\epsilon_F)$ and $\rho_\downarrow(\epsilon_F)$ are the spin projected density of states at ϵ_F. The arrows $\uparrow$ and $\downarrow$ denote states of opposite spin, i.e. majority and minority states, respectively. P vanishes for paramagnetic or antiferromagnetic materials, even below the magnetic transition temperature. However, it has a finite value in ferromagnetic or ferrimagnetic materials below the Curie temperature. The electrons at ϵ_F are fully spin polarized ($P = 100\ \%$), if either $\rho_\uparrow(\epsilon_F)$ or $\rho_\downarrow(\epsilon_F)$ equals zero.

For ferromagnetic materials the Slater–Pauling curve [56, 57] is a simple way to study the relation between the valence electron concentration and the magnetic moment (see Fig. 1.3). It is well known that Heusler compounds based on Co_2 follow the Slater–Pauling rule for predicting their total spin magnetic moment [58–60]. This allows for the prediction of their total spin magnetic moment. The electronic structure of these alloys exhibits a minimum in the minority density of states, in which the Fermi energy level is pinned. Therefore, the minority spin density is predefined to a finite value, and hence, the number of majority electrons increases proportional to the total number of electrons. As a direct consequence, the magnetic moment scales linearly with the number of valence electrons (for more details see references [58, 60]).

Half-metallic ferromagnets, like the Co_2-based Heusler compounds, even exhibit a gap in the minority density of states. The Slater–Pauling rule

$$m = N_V - 24, \tag{1.2}$$

with $m =$ the magnetic moment per formula unit and $N_V =$ the total valence electron number (sum of s, d electrons for the transition metals, and s, p electrons for the main group element), is strictly fulfilled. m is given by the difference between the electron counts in both spin directions, since 12 minority states, in particular, one s, three p, and eight d bands, are occupied in Heusler compounds [46, 58].

For Half-Heusler compounds with $C1_b$ structure, a similar rule was first found by Kübler et al. [46]:

$$m = N_V - 18. \tag{1.3}$$

In Fig. 1.3(a) experimental values of the magnetic moment of selected Heusler compounds are compared to the Slater–Pauling rule. All Heusler compounds are located at the *localized* part of the Slater–Pauling curve [58, 60]. In this part of the curve, predominantly materials with *bcc* or *bcc*-derived structures are found. Only the Co_2-based compounds, which are predicted to show half-metallic ferromagnetism, follow the linear trend, while the magnetic moments of many Fe_2- and Ni_2-based compounds are considerably scattered. For a detailed discussion of this context see Refs. [2, 58, 60].

Figure 1.3(b) reveals a different behavior: An approximately linear dependence is obtained for Co_2-based Heusler compounds, when plotting the Curie temperature (T_C) as function of their valence electron number (see data and references in [24, 25]). According to this plot, T_C is highest for those half-metallic compound with a high valence electron concentration, or equivalently, for those exhibiting a large magnetic moment. Co_2FeSi, in fact, is the Heusler compound with the highest magnetic moment of 5.97 μ_B at 5 K and the highest Curie temperature of 1100 K [61]. Kübler developed an ab initio estimate for the calculation of the Curie temperature of an itinerant-electron ferromagnet in the spherical approximation using the LDA scheme [62]. With this approach, a very good agreement between calculation and experiment was achieved, e.g. a T_C value of 1185 K was calculated for Co_2FeSi, which fits quite well to the experimental value of 1100 K [54, 55].

Changing the total number of valence electrons to a value below 24, leads to half-metallic ferromagnetism with a band gap in the majority channel, as calculated by the generalized gradient exchange-correlation potential method for Mn_2VAl [63]. $(Mn_{0.5}Co_{0.5})_2VAl$, which can be derived from Mn_2VAl by substitution of Co for half of the Mn atoms, is non-magnetic with 24 valence electrons.

1.5 Properties and Applications of Heusler Compounds

Co_2-based Heusler compounds were intensely studied due to their high potential as new electrode materials in spintronic devices, such as magnetic tunnel junctions (MTJs) [64–67], giant magnetoresistance (GMR) devices [68–72], and for spin injection from ferromagnetic electrodes into semiconductors [73]. The activity in this field was stimulated by Block et al., who discovered the large negative magnetoresistance (MR) at room temperature (RT) in the quaternary Heusler compound $Co_2Cr_{0.6}Fe_{0.4}Al$ (CCFA) [53, 74], which demonstrates the tunability of the spin density of states at the Fermi level by substituting constituent elements. In pressed powder compacts, which act as a series of MTJs, a MR of -30 % was reported in a small external field of 0.1 T. This triggered an enormous research activity focusing on the implementation of this material into spintronic devices [64, 75–77]. The first tunneling magnetoresistance (TMR) using a B2 sputtered CCFA electrode was reported to be 26.5 % at 5 K (16 % at RT) by Inomata et al. [64]. The incorporation of Heusler compounds into TMR devices lead to a dramatic increase in the TMR ratio in the following years as shown in Fig. 1.4. High TMR values of up to 570 % at 2 K could be reached with Co_2MnSi electrodes sandwiching a AlO_x-barrier [78]. In fact, a spin polarization of $\approx$85 % for Co_2MnSi can be deduced from these results. At RT, however, the effect is reduced considerably, which is attributed to spin-filp tunneling caused by magnetic impurities at the interface or pinholes in the tunneling barrier. A major step forward was maid by the introduction of single-crystalline MgO as a barrier material, which gives rise to a coherent tunneling mechanism [79, 80]. The TMR values at RT could be increased steadily, leading to 340 % for a CsCl-type disordered Co_2FeAl-based MTJs [81]. Since this Heusler compound does not exhibit a complete spin polarization, the high TMR value is a strong indication for pronounced coherent tunneling. A further improvement of MTJs with $Co_2FeAl_{0.5}Si_{0.5}$ electrodes fabricated by molecular beam epitaxy (MBE) was achieved and TMR ratios of 386 % at RT (832 % at 9 K) were observed [82]. The use of slightly off-stoichiometric $Co_2Mn_{1.29}Si_{1.06}$ in fully epitaxial MTJs resulted in high TMR ratios of 1135 % at 4.2 K and 236 % at RT [83].

For current perpendicular to plane (CPP) GMR devices comprised of $Co_2FeAl_{0.5}Si_{0.5}/Ag/Co_2FeAl_{0.5}Si_{0.5}$, a MR ratio of 34 % at RT (80 % at 14 K) was reported [88]. To minimize the lattice mismatch and to optimize the interface scattering properties, "all-Heusler" CPP-GMR devices with the trilayer $Co_2MnSi/Ru_2CuSn/Co_2MnSi$ were proposed. These devices yielded a MR ratio of 6.7 % at RT [72].

Thin films with perpendicular magnetic anisotropy (PMA), i.e. with the easy magnetization axis pointing perpendicular to the film surface, are strongly desired for spintronics applications to overcome the superparamagnetic limit and enhance the thermal stability of ultra-high density magnetic memory storage devices. On the other hand, materials based on $Mn_{3-x}Ga$ are promising candidates, due to the antiferromagnetic coupling of the Mn magnetic moments, which leads to a low saturation magnetization, and simultaneously, retains the high Curie temperature of [89, 90]. As an example, the successful growth of tetragonal distorted $Mn_{2.5}Ga$ thin

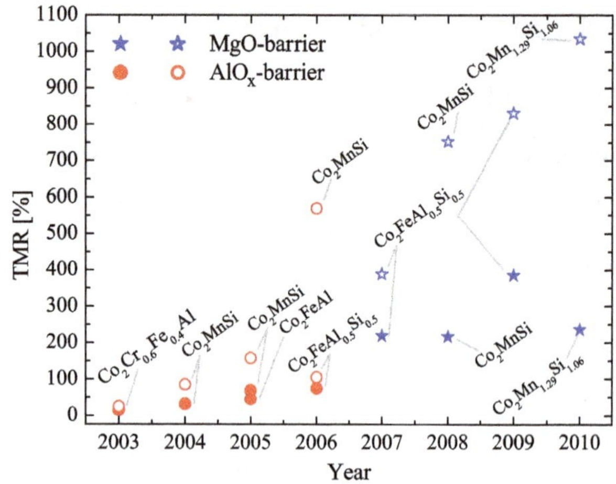

Fig. 1.4 Development of the TMR ratio for MTJs with Heusler electrodes. Open symbols denote the TMR value at low temperature, while filled symbols display the value at RT. Data taken from Refs. [64, 65, 78, 82–87]

films with the tetragonal c-axis pointing perpendicular to the surface was recently demonstrated [91]. The search for new materials with suitable properties is, in fact, a field of active research.

The most attractive properties of Half-Heusler materials for an application as thermoelectric materials are their high Seebeck coefficient S up to $\approx$300 μV K^{-1} at RT and their high electrical conductivity ($\approx$1000–10000 S cm^{-1}) [9, 92–94]. The only drawback is the relatively high thermal conductivity, which can be as high as 10 W/m K. In the past, research activities were focused on n-type TiNiSn-based compounds, but recently great progress was also made for p-type TiCoSb-based materials. Consequently, a maximum ZT value of 1.4 at 700 K was reported n-type $(Zr_{0.5}Hf_{0.5})_{0.5}Ti_{0.5}NiSn_{0.998}Sb_{0.002}$ [8]. The highest ZT value for an p-type TiCoSb-based material, reported so far, is 0.7 at 900 K for $Ti_{0.6}Hf_{0.4}Co_{0.87}Ni_{0.13}Sb$ with a power factor of 23.4 μW/cm K^2 [95]. Recently, a transition from a n-type to a p-type material was observed for $Ti_{(1-x)}M_xNiSn$ (M = Sc, V) which opens the opportunity to build thermocouples based on the same material [96].

The first superconducting Heusler compounds Pd_2RESn and Pd_2REPb (RE = rare earth) were reported by Ishikawa et al. in 1982 [97]. Up to now, several new superconductors within the family of Heusler compounds have been reported, their critical temperature, however, is too low from an applications point of view. Generally, superconductivity is often found in Heusler compounds with 27 valence electrons, which show the same finger print in their band structure, i.e. a van-Hove singularity at the L point [7, 98]. This allows for the prediction of superconductivity in new Heusler compounds. The highest transition temperature, up to now, was found in Pd_2YSn with T_c = 4.9 K [6].

The system Ni_2MnGa is intensively investigated due to its shape memory behavior. The cubic phase undergoes a ferromagnetic transition at $T_C = 376$ K [99]. Additionally, stoichiometric Ni_2MnGa undergoes a structural phase transition at $T_M = 202$ K from the high-temperature cubic $L2_1$ structure to a martensitic phase with complex tetragonal structure [99]. Due to the reversibility of this martensitic transition, a shape memory effect is observed in this system. Interestingly, both critical temperatures can be tuned by alloying. Changing the relation between T_M and T_C results in different properties, which makes these alloys promising for technological applications. Furthermore, in $Ni_{2+x}Mn_{1-x}Ga$ with $0.18 \leq x \leq 0.20$, a coupling of the magnetic and the structural transition takes place, because the transition temperatures are close to each other [100]. Consequently, it becomes possible to achieve a shape memory effect by applying a magnetic field and to induce attractive properties such as the giant magnetocaloric effect, magnetostriction, and magnetoresistance, which are important for magnetic refrigeration or magnetostrictive transducers [101–104]. For $x \geq 0.3$, T_M is higher than T_C, and therefore, the martensitic transition occurs in the paramagnetic region. Since T_M increases dramatically with increasing x, alloys with a high Ni excess can be used as high temperature shape memory alloys. Here again, the total electron count is an easy, but reliable way to qualitatively understand the relation between composition and transition temperatures. It was shown that an increasing number of valence electrons lowers T_M [105]. A profound understanding of the physical properties, however, is needed, to design new materials with predictable properties. For this purpose, first-principles calculations can give an insight into the complex relation between concentration dependent properties and the transition temperatures [106].

The recently discovered quantum spin Hall state and topological insulators are new states of quantum matter, interesting both, for fundamental condensed matter physics and material science. Recently it was demonstrated that around 50 Heusler compounds show a band inversion, similar to that of HgTe, in which the quantum spin Hall effect was observed experimentally [11, 12, 107]. The topological state in these zero-gap semiconductors can be created by applying strain or by designing an appropriate quantum well structure, similar to the case of HgTe. Many of these ternary zero-gap semiconductors (LnAuPb, LnPdBi, LnPtSb, and LnPtBi) contain the rare-earth element Ln, which can realize additional properties ranging from superconductivity (e.g. LaPtBi [108]) to magnetism (e.g. GdPtBi [109]) and heavy fermion behavior (e.g. YbPtBi [110]). These properties can open new research directions in realizing the quantized anomalous Hall effect and topological superconductors.

1.6 Conclusion and Outlook

Heusler compounds are an outstanding class of materials with a large variety of diverse properties. Their success story began 30 years ago with the prediction of half-metallic ferromagnetism in Co_2-based materials. In recent years, not only the field of spintronics jumped into Heusler materials, but also research on environmental techniques as thermoelectrics and solar cells discovered their amenities, including tunability of their properties by substitution of atoms, non-toxicity, and the low price. With the discovery of the topological state in Half-Heusler compounds, a new era dawns, and the search for new, multi-functional Heusler compounds will be a fruitful field of research in coming years.

Acknowledgements The authors are grateful to B. Balke, C.G.F. Blum, F. Casper, G.H. Fecher, V. Jung, V. Ksenofontov, J. Winterlik, and S. Wurmehl for providing data and for fruitful discussions. Financial support by the Deutsche Forschungsgemeinschaft (research unit 559) and by the Graduate School of Excellence "Material Science in Mainz" is gratefully acknowledged.

References

1. Kübler J, Williams AR, Sommers CB (1983) Phys Rev B 28:1745
2. Felser C, Fecher GH, Balke B (2007) Angew Chem, Int Ed 46:668
3. Wurmehl S, Kandpal HC, Fecher GH, Felser C (2006) J Phys Condens Matter 18:6171
4. Jung D, Koo HJ, Whangbo MH (2000) J Mol Struct, Theochem 527:113
5. Pierre J, Skolozdra RV, Tobola J, Kaprzyk S, Hordequin C, Kouacou MA, Karla I, Currat R, Lelievre-Berna E (1997) J Alloys Compd 262–263:101
6. Wernick JH, Hull GW, Geballe TH, Bernadini JE, Waszczak JV (1983) Mater Lett 2:90
7. Winterlik J, Fecher GH, Thomas A, Felser C (2009) Phys Rev B 79:064508
8. Sakurada S, Shutoh N (2005) Appl Phys Lett 86:082105
9. Uher C, Yang J, Hu S, Morelli DT, Meisner GP (1999) Phys Rev B 59:8615
10. Kieven D, Klenk R, Naghavi S, Felser C, Gruhn T (2010) Phys Rev B 81:075208
11. Chadov S, Qi X, Kübler J, Fecher GH, Felser C, Zhang SC (2010) Nat Mater 9:541
12. Lin H, Wray A, Xia Y, Xu S, Jia S, Cava RJ, Bansil A, Hasan MZ (2010) Nat Mater 9:546
13. Heusler F, Starck W, Haupt E (1903) Verh Dtsch Phys Ges 5:220
14. Heusler F (1903) Verh Dtsch Phys Ges 5:219
15. Aliev FG, Brandt NB, Moshchalkov VV, Kozyrkov VV, Skolozdra RV, Belogorokhov AI (1989) Z Phys B, Condens Matter 75:167
16. Heusler O (1934) Ann Phys 19:155
17. Bradley AJ, Rodgers JW (1934) Proc R Soc A 144:340
18. Miura Y, Nagao K, Shirai M (2004) Phys Rev B 69:144413
19. Kandpal HC, Ksenofontov V, Wojcik M, Seshadri R, Felser C (2007) J Phys D, Appl Phys 40:1587
20. Picozzi S, Continenza A, Freeman AJ (2004) Phys Rev B 69:094423
21. Puselj M, Ban Z (1969) Croat Chem Acta 41:79
22. Webster PJ (1969) Contemp Phys 10:559
23. Bacon GE, Plant JS (1971) J Phys F, Met Phys 1:524
24. Webster PJ, Ziebeck KRA (1988) In: Wijn HPJ (ed) Landolt-Börnstein – group III condensed matter, vol 19C, Springer, Heidelberg, pp 75–185
25. Ziebeck KRA, Neumann KU (2001) In: Wijn HPJ (ed) Landolt-Börnstein – group III condensed matter, vol 32C, Springer, Heidelberg, pp 64–314
26. Graf T, Casper F, Winterlik J, Balke B, Fecher GH, Felser C, Anorg Z (2009) Z Anorg Allg Chem 635:976
27. Ravel B, Cross JO, Raphael MP, Harris VG, Ramesh R, Saraf V (2002) Appl Phys Lett 81:2812
28. Wurmehl S, Kohlhepp JT, Swagten HJM, Koopmans B, Wojcik M, Balke B, Blum CGF, Ksenofontov V, Fecher GH, Felser C (2007) Appl Phys Lett 91:052506
29. Wurmehl S, Kohlhepp JT, Swagten HJM, Koopmans B, Wojcik M, Balke B, Blum CGF, Ksenofontov V, Fecher GH, Felser C (2008) J Appl Phys 103:07D706
30. Inomata K, Wojcik M, Jedryka E, Ikeda N, Tezuka N (2008) Phys Rev B 77:214425
31. Jung V, Fecher GH, Balke B, Ksenofontov V, Felser C (2009) J Phys D, Appl Phys 42:084007
32. Jung V, Balke B, Fecher GH, Felser C (2008) Appl Phys Lett 93:042507
33. Ravel B, Raphael MP, Harris VG, Huang Q (2002) Phys Rev B 65:184431
34. Balke B, Wurmehl S, Fecher GH, Felser C, Alves MCM, Bernardi F, Morais J (2007) Appl Phys Lett 90:172501

35. Juza R, Hund F (1948) Z Anorg Allg Chem 257:1
36. Nowotny H, Bachmayer K (1950) Monatsh Chem 81:488
37. Kalarasse F, Bennecer B (2006) J Phys Chem Solids 67:846
38. Hohl H, Ramirez AP, Goldmann C, Ernst G, Wolfing B, Bucher E (1999) J Phys Condens Matter 11:1697
39. Casper F, Ksenofontov V, Kandpal HC, Reiman S, Shishido T, Takahashi M, Takeda M, Felser C (2006) Z Anorg Allg Chem 632:1273
40. Xia Y, Ponnambalam V, Bhattacharya S, Pope AL, Poon SJ, Tritt TM (2001) J Phys Condens Matter 13:77
41. Tobola J, Pierre J, Kaprzyk S, Skolozdra RV, Kouacou MA (1998) J Phys Condens Matter 10:1013
42. de Groot RA (1991) Physica B 172:45
43. de Groot RA, Kraan AMvd, Buschow KHJ (1986) J Magn Magn Mater 61:61
44. Ksenofontov V, Melnyk G, Wojcik M, Wurmehl S, Kroth K, Reiman S, Blaha P, Felser C (2006) Phys Rev B 74:134426
45. de Groot RA, Mueller FM, van Engen PG, Buschow KHJ (1983) Phys Rev Lett 50:2024
46. Kübler J (1984) Physica B 127:257
47. Kulatov E, Mazin II (1990) J Phys Condens Matter 2:343
48. Youn SJ, Min BI (1995) Phys Rev B 51:10436
49. Wang X, Antropov VP, Harmon BN (1994) IEEE Trans Magn 30:4458
50. Galanakis I, Ostanin S, Alouani M, Dreysse H, Wills JM (2000) Phys Rev B 61:4093
51. Hordequin C, Lelivre-Bernab E, Pierre J (1997) Physica B 234–236:602
52. Nishino Y, Kato M, Asano S, Soda K, Hayasaki M, Mizutani U (1997) Phys Rev Lett 79:1909
53. Block T, Felser C, Jakob G, Ensling J, Mühling B, Gütlich P, Beaumont V, Studer F, Cava RJ (2003) J Solid State Chem 176:646
54. Kübler J, Fecher GH, Felser C (2007) Phys Rev B 76:024414
55. Wurmehl S, Fecher GH, Kandpal HC, Ksenofontov V, Felser C, Lin HJ, Morais J (2005) Phys Rev B 72:184434
56. Salter JC (1936) Phys Rev 49:537
57. Pauling L (1938) Phys Rev 54:899
58. Kübler J (2000) Theory of itinerant electron magnetism. Clarendon Press, Oxford
59. Galanakis I, Dederichs PH, Papanikolaou N (2002) Phys Rev B 66:174429
60. Fecher GH, Kandpal HC, Wurmehl S, Felser C, Schönhense G (2006) J Appl Phys 99:08J106
61. Wurmehl S, Fecher GH, Ksenofontov V, Casper F, Stumm U, Felser C, Lin HJ, Hwu Y (2006) J Appl Phys 99:08J103
62. Kübler J (2006) J Phys Condens Matter 18:9795
63. Weht R, Pickett WE (1999) Phys Rev B 60:13006
64. Inomata K, Okamura S, Goto R, Yezuka N (2003) Jpn J Appl Phys 42:L419
65. Kämmerer S, Thomas A, Hütten A, Reiss G (2004) Appl Phys Lett 85:79
66. Kubota H, Nakata J, Oogange M, Ando Y, Sakuma A, Miyazaki T (2004) Jpn J Appl Phys 43:L984
67. Ishikawa T, Marukame T, Kijima H, Matsuda Ki, Uemura T, Yamamoto M (2006) Appl Phys Lett 89:192505
68. Yakushiji K, Saito K, Mitani S, Takanashi K, Takahashi YK, Hono K (2006) Appl Phys Lett 88:222504
69. Furubayashi T, Kodama K, Sukegawa H, Takahashi YK, Inomata K, Hono K (2008) Appl Phys Lett 93:122507
70. Sakuraba Y, Iwase T, Mitani S, Takanashi K (2009) Appl Phys Lett 94:012511
71. Kodama K, Furubayashi T, Sukegawa H, Nakatani TM, Inomata K, Hono K (2009) J Appl Phys 105:07E905
72. Nikolaev K, Kolbo P, Pokhil T, Peng X, Chen Y, Ambrose T, Mryasov O (2009) Appl Phys Lett 94:222501

73. Dong XY, Adelmann C, Xie JQ, Palmstrøm CJ, Lou X, Strand J, Crowell PA, Barnes JP, Petford-Long AK (2005) Appl Phys Lett 86:102107
74. Felser C, Heitkamp B, Kronast F, Schmitz D, Cramm S, Dürr HA, Elmers HJ, Fecher GH, Wurmehl S, Block T, Valdaitsev D, Nepijko SA, Gloskovskii A, Jakob G, Schönhense G, Eberhardt W (2003) J Phys Condens Matter 15:7019
75. Clifford E, Venkatesan M, Gunning R, Coey JMD (2004) Solid State Commun 131:61
76. Okamura S, Goto R, Sugimoto S, Tezuka N, Inomata K (2004) J Appl Phys 96:6561
77. Conca A, Jourdan M, Adrian H (2007) J Phys D, Appl Phys 40:1534
78. Sakuraba Y, Hattori M, Oogane M, Ando Y, Kato H, Sakuma A, Miyazaki T, Kubota H (2006) Appl Phys Lett 88:192508
79. Parkin SSP, Kaiser C, Panchula A, Rice PM, Hughes B, Samant M, Yang SH (2004) Nat Mater 3:862
80. Miura Y, Uchida H, Oba Y, Nagao K, Shirai M (2007) J Phys Condens Matter 19:365228
81. Wang W, Lui E, Kodzuka M, Sukegawa H, Wojcik M, Jedryka E, Wu GH, Inomata K, Mitani S, Hono K (2010) Phys Rev B 81:140402(R)
82. Tezuka N, Ikeda N, Mitsuhashi F, Sugimoto S (2009) Appl Phys Lett 94:162504
83. Yamamoto M, Ishikawa T, Taira T, Li Gf, Matsuda Ki, Uemura T (2010) J Phys Condens Matter 22:164212
84. Okamura S, Miyazaki A, Sugimoto S, Tezuka N, Inomata K (2005) Appl Phys Lett 86:232503
85. Tezuka N, Ikeda N, Miyazaki A, Sugimoto S, Kikuchi M, Inomata K (2006) Appl Phys Lett 89:112514
86. Tezuka N, Ikeda N, Sugimoto S, Inomata K (2006) Appl Phys Lett 89:252508
87. Tsunegi S, Sakuraba Y, Oogane M, Takanashi K, Ando Y (2008) Appl Phys Lett 93:112506
88. Nakatani TM, Furubayashi T, Kasai S, Sukegawa H, Takahashi YK, Mitani S, Hono K (2010) Appl Phys Lett 96:212501
89. Winterlik J, Balke B, Fecher GH, Felser C (2008) Phys Rev B 77:054406
90. Balke B, Fecher GH, Winterlik J, Felser C (2007) Appl Phys Lett 90:152504
91. Wu F, Mizukami S, Watanabe D, Naganuma H, Oogane M, Ando Y, Miyazaki T (2009) Appl Phys Lett 95:122503
92. Kimura Y, Ueno H, Mishima Y (2009) J Electron Mater 38:934
93. Xie W, Jin Q, Tang X (2008) J Appl Phys 103:043711
94. Zhou M, Chen L, Feng C, Wang D, Li JF (2007) J Appl Phys 101:113714
95. Qui P, Huang X, Chen Y, Chen L (2009) J Appl Phys 106:103702
96. Ouardi S, Fecher GH, Balke B, Kozina X, Stryganyuk G, Felser C, Lowitzer S, Ködderitzsch D, Ebert H, Ikenaga E (2010) Phys Rev B 82:085108
97. Ishikawa M, Jorda JL, Junod A (1982) Superconductivity in d- and f-band metals. Kernforschungszentrum, Karlsruhe, Germany
98. Winterlik J, Fecher GH, Felser C, Jourdan M, Grube K, Hardy F, Löhneysen Hv, Holman KL, Cava RJ (2008) Phys Rev B 78:184506
99. Webster PJ, Ziebeck KRA, Town SL, Peak MS (1984) Philos Mag, B 49:295
100. Khovaylo VV, Buchelnikov VD, Kainuma R, Koledov VV, Ohtsuka M, Shavrov VG, Takagi T, Taskaev SV, Vasiliev AN (2005) Phys Rev B 72:224408
101. Murray SJ, Marioni M, Allen SM, O'Handley RC, Lograsso TA (2000) Appl Phys Lett 77:886
102. Sozinov A, Likhachev AA, Lanska N, Ullakko K (2002) Appl Phys Lett 80:1746
103. Marcos J, Mañosa L, Planes A, Casanova F, Batlle X, Labarta A (2003) Phys Rev B 68:094401
104. Zhou X, Li W, Kunkel HP, Williams G (2004) J Phys Condens Matter 16:L39
105. Lanska N, Söderberg O, Sozinov A, Ge Y, Ullakko K, Lindroos VK (2004) J Appl Phys 95:8074
106. Li CM, Luo HB, Hu QM, Yang R, Johansson B, Vitos L (2010) Phys Rev B 82:024201
107. König M, Wiedmann S, Brüne C, Roth A, Buhmann H, Molenkamp LW, Qi XL, Zhang SC (2007) Science 318:766

108. Goll G, Marz M, Hamann A, Tomanic T, Grube K, Yoshino T, Takabatake T (2008) Physica B 403:1065
109. Canfield PC, Thompson JD, Beyermann WP, Lacerda A, Hundley MF, Peterson E, Fisk Z, Ott HR (1991) J Appl Phys 70:5800
110. Fisk Z, Canfield PC, Beyermann WP, Thompson JD, Hundley MF, Ott HR, Felder E, Maple MB, Lopez de la Torre MA, Visani P, Seaman CL (1991) Phys Rev Lett 67:3310

Chapter 2
New Heusler Compounds and Their Properties

Benjamin Balke, Gerhard H. Fecher, and Claudia Felser

Abstract Spintronics is a multidisciplinary field and a new research area. New materials must be found for satisfying the different types of requirement. The search for stable half-metallic ferromagnets and ferromagnetic semiconductors with Curie temperatures higher than room temperature is still a challenge for solid state scientists. A general understanding of how structures are related to properties is a necessary prerequisite for material design. Computational simulations are an important tool for a rational design of new materials. The new developments in this new field are reported from the point of view of material scientists.

2.1 Introduction

A great scientific interest is attracted by materials with a complete spin polarization at the Fermi energy [1]. Such materials, being a metal for spin up and a semiconductor (or insulator) for spin down electrons, are called half-metallic ferromagnets (HMF) [2, 3]. Already in 1983 Heusler compounds have been considered potential candidates to show this property. Theoretical calculations predicted an energy gap for minority electrons for the half-Heusler compound NiMnSb [2, 4] which, however, has been controversially discussed later [5–7]. At the same time, Kübler et al. [8] recognized that the minority spin densities at the Fermi energy (ε_F) nearly vanish for Co_2MnAl and Co_2MnSn. The authors concluded that this should lead to

B. Balke · G.H. Fecher · C. Felser
Institut für Anorganische Chemie und Analytische Chemie, Johannes Gutenberg – Universität, 55099 Mainz, Germany

B. Balke
e-mail: balke@uni-mainz.de

G.H. Fecher · C. Felser (✉)
Max-Planck-Institut für Chemische Physik fester Stoffe, 01187 Dresden, Germany
e-mail: felser@cpfs.mpg.de

G.H. Fecher
e-mail: fecher@cpfs.mpg.de

C. Felser, G.H. Fecher (eds.), *Spintronics*, DOI 10.1007/978-90-481-3832-6_2,

peculiar transport properties in these Heusler compounds because only the majority density contributes to the states at ε_F.

Generally, Heusler compounds belong to a group of ternary intermetallics with the stoichiometric composition X_2YZ ordered in an $L2_1$-type structure. Many of them are ferromagnetic [9]. X and Y are transition metals and Z is usually a main group element. Y may also be replaced by a rare earth element. The $L2_1$ structure consists of a primitive *fcc* lattice with a basis of four atoms. Remarkably, the prototype Cu_2MnAl is a ferromagnet even though none of its constituents is one [10].

The quite new research area of so-called *spintronics* is a multidisciplinary field. New materials must be found for satisfying the different types of requirement. The search for stable half-metallic ferromagnets and ferromagnetic semiconductors with Curie temperatures higher than room temperature is still a challenge for solid state scientists. A general understanding of how structures are related to properties is a necessary prerequisite for material design. Computational simulations are an important tool for a rational design of new materials. The new developments in this new field are reported from the point of view of material scientists. The development of magnetic Heusler compounds specifically designed as material for spintronics applications made tremendous progress in the very recent past.

The boom about Heusler compounds for spintronics applications and with it the huge number of both, theoretical and experimental investigations (over 700 publications concerning X_2YZ Heusler compounds in the last eight–nine years) began in 2002/2003 with the observation of—at that time—large negative magnetoresistance in powder compacts of $Co_2Cr_{0.6}Fe_{0.4}Al$ [11]. A magnetoresistive effect of 30 % in a small magnetic field of 0.1 T at room temperature was found. This demonstrated for the first time the feasibility of a cheap and simple magnetic sensor based on polycrystalline, intermetallic materials. This value was later optimized by using powder compacts of $Co_2Cr_{0.6}Fe_{0.4}Al$, mixed with insulating Al_2O_3, to a maximum magnetoresistance ratio of 88 % in an induction field of 0.125 T at 295 K [12].

Based on band structure calculations and guided by striking features in the electronic structure of several magnetic Heusler compounds, $Co_2Cr_{0.6}Fe_{0.4}Al$ was chosen to obtain a half-metallic ferromagnet with a van Hove singularity in the vicinity of the Fermi energy in the majority spin channel and a gap in the minority spin channel [11]. Following this investigation, Inomata et al. [13] prepared a spin valve type tunneling junction with a $Co_2Cr_{0.6}Fe_{0.4}Al$ Heusler compound film that showed a tunneling magnetoresistance ratio of 16 % at room temperature. Afterwards, thin films of the compound $Co_2Cr_{0.6}Fe_{0.4}Al$ were successfully grown by many other groups [14–17]. A magnetoresistance ratio of 26.5 % [13] (at 5 K) and 19 % [18] (at room temperature) was found for a tunneling magnetoresistance (TMR) element of the same compound. Very recently, Marukame et al. [19] reported a TMR ratio of 74 % at 55 K for a $Co_2Cr_{0.6}Fe_{0.4}Al$-MgO-CoFe magnetic tunnel junction. A spin polarization of only less than 49 % was found for polycrystalline samples by means of Andreev reflections [20]. The observation of an incomplete spin polarization may not only be caused by the model used to interpret the data [20, 21] but also by the properties of the sample. Clifford et al. [22] reported a spin polarization of 81 % in point contacts of $Co_2Cr_{0.6}Fe_{0.4}Al$.

However, the TMR ratios found in tunneling experiments indicate a spin polarization that is still too low, and not only at room temperature. This work reports on the investigation of the reasons of this contradiction between theory and experiment and with this knowledge it was tried to overcome the problems by designing new materials suitable for spintronics applications.

2.2 Experimental Details

Polycrystalline ingots of the compounds were prepared by arc melting of stoichiometric amounts of high purity elements in an argon atmosphere to avoid oxygen contamination at a pressure of 10^{-4} mbar. Additionally, a Ti sponge was used to bind remaining oxygen. The samples were melted three times. They were turned over after each melting process to yield a homogeneous sample. The weight loss after the whole melting procedure was less than 1 %. The polycrystalline ingots were annealed in evacuated quartz tubes at 1073 K for two weeks. The crystallographic structure was investigated by X-ray powder diffraction (XRD) using excitation by Mo–K_α radiation ($\lambda = 0.7093165$ nm; Bruker, AXS D8) in reflection geometry. The experimental diffraction patterns were refined using the FULLPROF program [23]. Additional synchrotron-based anomalous XRD investigations were carried out for various samples, due to the restrictions of powder XRD using laboratory source. The anomalous XRD experiments and the temperature dependent XRD measurements were performed at the X-ray powder diffraction beamline at the bending magnet $D10$ at the Brazilian Synchrotron Light Laboratory (LNLS). For details about the characteristics of the beamline see [24]. A scanning electron microscope (SEM, Jeol JSM-6400) equipped with an energy dispersive X-ray spectroscopy (EDX) detection system (EUMEX EDX) was used to check the homogeneity and stoichiometry of the samples. The measurements were carried out at a pressure of 3×10^{-6} mbar. An acceleration voltage of 20 kV was applied and an inspection angle of 35° was set up. For the correction of the quantitative data the ZAF method was applied which relies on atomic number (Z), absorption (A) and fluorescence (F) effects. The images were acquired via the Digital Image Processing System (DIPS) and the quantitative chemical analysis was performed with the program WINEDS 4.0. Magneto-structural investigations were carried out by Mößbauer spectroscopy in transmission geometry using a constant acceleration spectrometer. For excitation, a ^{57}Co(Rh) source with a line width of 0.105 mm/s and a $Ca^{119}SnO_3$ source with a line width of 0.3 mm/s was used in the case of Fe and Sn, respectively. The spectra of powder samples were measured at 290 K. The magnetic properties were investigated by a superconducting quantum interference device (SQUID, Quantum Design MPMS-XL-5) using nearly punctual pieces of approximately 5 mg to 10 mg of the sample. The measurements of the Seebeck coefficient were carried out with a Physical Property Measurement System (Model 6000 Quantum Design) on bars of about $(2 \times 2 \times 8)$ mm^3 which were cut from the pellets and polished before the measurement. In the temperature range above 350 K the Seebeck coefficient was measured by a steady-state method using the RZ2001i measurement system from Ozawa science, Japan. For more details about the preparation, structural and magnetic properties see [25–28].

2.3 Crystal Structures of Heusler Compounds

In general, Heusler compounds crystallize in the Cu_2MnAl-type structure but in many cases certain types of disorder are observed. For a detailed description of the crystal structure as well as different types of atomic disorder see the contribution "*Crystal Structure of Heusler compounds*" by Tanja Graf and Claudia Felser. The relationship of the chemical ordering and the spin polarization is discussed there and useful experimental methods for the structural analysis of Heusler compounds are presented there.

2.4 Optimization of Co_2YZ Heusler Compounds

2.4.1 *Investigation on $Co_2Cr_{0.6}Fe_{0.4}Al$*

To understand why the spin polarization is too low, the structural and physical properties of the series $Co_2Cr_{1-x}Fe_xAl$, with varying iron concentration x, were investigated in detail. The results from X-ray diffraction (XRD) as well as an extended X-ray absorption fine structure (EXAFS) analysis and ^{57}Fe Mößbauer spectroscopy indicated structures in the bulk of the compound $Co_2Cr_{1-x}Fe_xAl$ that were different from those at the surface of the compound [29–31].

Figure 2.1 displays the EXAFS results for $Co_2Cr_{0.6}Fe_{0.4}Al$ and compares powder (higher amount of surface contribution) with bulk samples. The differences are clearly visible. The results corroborate those found from variable incidence XRD measurements, and indicate that the surface shows no ordered structure ($A2$-type) while the bulk is $B2$-like disordered. Indeed, the spectra provide no indication of the $L2_1$ structure, and this explains the reduced spin polarization.

Mößbauer spectroscopy was used to further explore the magneto-structural properties of $Co_2Cr_{0.6}Fe_{0.4}Al$. The results of both transmission and conversion electron Mößbauer spectroscopy (CEMS) were analyzed to obtain insight into both the disorder effects as well as the differences between the bulk and surface properties. It was found that rough mechanical treatment of the surfaces of bulk samples induces strong dynamic or relaxation effects. From the CEMS analysis, it was found that one of the best procedures for cleaning the surface of $Co_2Cr_{0.6}Fe_{0.4}Al$ was achieved by using grazing incidence Ar^+ ion bombardment (this was also checked for other compounds that are not shown here). Figure 2.2 shows the CEMS spectrum of ^{57}Fe enriched $Co_2Cr_{0.6}Fe_{0.4}Al$ bulk material after polishing followed by 60 min of Ar^+ ion bombardment in an UHV chamber. Even without any post-annealing for healing

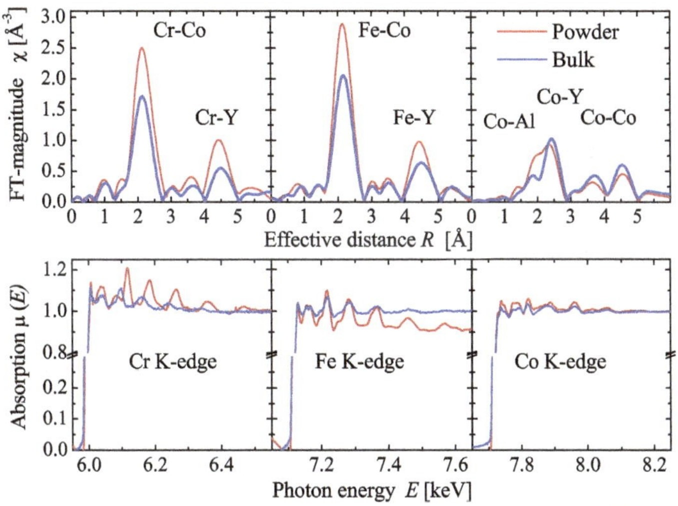

Fig. 2.1 EXAFS results of $Co_2Cr_{0.6}Fe_{0.4}Al$ comparing powder and bulk. Spectra taken at the Cr, Fe, and Co K-edges are compared for powder and bulk samples

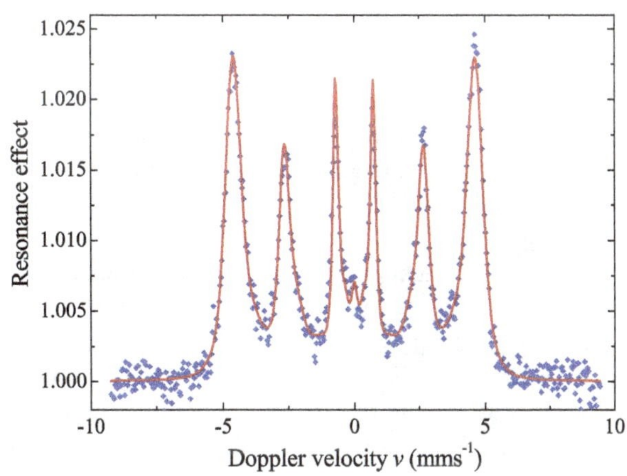

Fig. 2.2 CEMS spectrum of ^{57}Fe enriched $Co_2Cr_{0.6}Fe_{0.4}Al$ bulk material after 60 min Ar^+ ion bombardment

the surface, the spectra reveal the bulk magnetic properties as found in comparison to transmission Mößbauer experiments (please, note that, as reported in the chapter about photoemission experiments), the information depth of CEMS is comparable to that of high energy photoemission.

The studies of $Co_2Cr_{0.6}Fe_{0.4}Al$ were followed by a comprehensive characterization of other Cr-containing Heusler compounds. This is necessary for understanding the mechanism that causes a magnetic moment of Cr in that compound that is less than that found in the isoelectronic $Co_2Cr_{1-x}Fe_xGa$ compounds. Co_2CrZ Heusler compounds (Z = Al, Ga) are already known; the next compound with an element from the series of main group elements is the Co_2CrZ Heusler compound with

Z = In. This compound had not been reported before, so Co_2CrIn was synthesized and characterized [32]. In summary, this compound is $L2_1$ ordered and exhibits no antisite disorder. Co_2CrIn was found to be a ferrimagnet with a magnetic moment of 1.18 μ_B at 5 K. In addition, the hysteresis indicates a soft magnetic behavior. The value of the measured magnetic moment is not an integer, as would be expected for a half-metallic ferromagnet with four atoms in the primitive cell. Thus, unlike most of the other Co_2YZ Heusler compounds, Co_2CrIn cannot be a half-metallic ferromagnet. Finally, it should be noted that it was not possible up to now to stabilize other Co_2CrZ compounds in a pure $L2_1$ structure.

2.4.2 *Tuning the Fermi Energy in the Middle of the Gap: From Co_2FeSi to $Co_2Mn_{1-x}Fe_xSi$ and $Co_2FeAl_{1-x}Si_x$*

Due to insuperable problems such as disorder and the low magnetic moment of compounds containing Cr (and Al), we moved on to investigate new materials that clearly exhibit both the ordered $L2_1$ structure and a high Curie temperature. By inspecting the magnetic data of the known Heusler compounds in more detail, a very interesting aspect is found [33, 34]. A linear dependence is apparently obtained for Co_2-based Heusler compounds when the magnetic moments for the known, 3*d* metal-based Heusler compounds are plotted against their Curie temperatures (T_C) (see Fig. 2.3). According to this plot, T_C is highest for those half-metallic compounds that exhibit a large magnetic moment. By extrapolating from the linear dependence shown in Fig. 2.3, T_C is estimated to be above 1,000 K for compounds with 6 μ_B.

From the chemical point of view, it is clear that the exchange of Al by Si should lead to stronger covalent bonding. Indeed, Co_2FeAl does not exhibit a clean $L2_1$ structure but usually has a very high degree of $B2$-like and also $A2$-like disorder. Therefore, Co_2FeSi was revisited and its properties were analyzed in more detail. Indeed, it was found that Co_2FeSi exhibits the correct Heusler $L2_1$-type structure that is necessary for the occurrence of half-metallicity [34]. Using a Rietveld refinement, the lattice constant was determined by XRD to be 5.64 Å with an R_{Bragg} value < 5.5. A disorder between Co and Fe atoms (DO_3-type disorder) could be excluded from the XRD Rietveld refinement as well as from the EXAFS and neutron scattering data. During further structural and magnetic investigations, Mößbauer spectroscopy, EXAFS measurements, and both low and high temperature magnetometry were performed and all data analysis verified the high degree of order in Co_2FeSi, even on a short range scale, and excluded $A2$—or a $B2$-type disorder.

Low temperature magnetometry was performed to verify the estimated saturation moment. The measured magnetic moment in saturation was 5.97 μ_B at 5 K. An extrapolation to 6 μ_B per unit cell at 0 K perfectly fits the moment estimated from the Slater–Pauling rule. The experimental magnetic moment is supported by

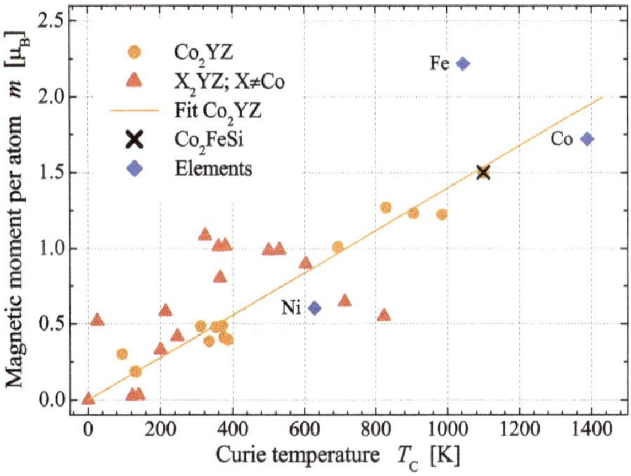

Fig. 2.3 Plot of the magnetic moment per atom of iron, nickel, cobalt, and selected Heusler compounds, as a function of the Curie temperatures. The *orange line* is a linear fit to the data for the Co_2YZ compounds

band structure calculations that indicate an HMF character with a magnetic moment of 6 μ_B. The high temperature magnetization of Co_2FeSi was measured using a vibrating sample magnetometer equipped with a high temperature stage. The ferromagnetic Curie temperature is found to be $T_C \approx 1100$ K. This value very well fits the linear behavior of T_C as a function of the magnetic moment for Co_2-based Heusler compounds (see Fig. 2.3). T_C is well below the melting point, which is determined using differential scanning calorimetry to be $T_m = 1{,}520$ K. In summary, the analysis shows that $L2_1$-ordered Co_2FeSi is a half-metallic ferromagnet that exhibits the highest values of Curie temperature and magnetic moment that have been reported for Heusler compounds.

On the other hand, next to Co_2FeSi, the Heusler alloy Co_2MnSi has attracted particular interest because it is predicted to have a large minority spin band gap of 0.4 eV and, at 985 K, has one of the highest Curie temperature, among the known Heusler compounds [35, 36]. Structural and magnetic properties of Co_2MnSi have been reported for films and single crystals [37–42]. In accordance with theoretical predictions, bulk Co_2MnSi has been stabilized in the $L2_1$ structure with a magnetization of 5 μ_B per formula unit. From tunneling magneto resistance (TMR) data with one electrode consisting of a Co_2MnSi film Schmalhorst et al. [43, 44] inferred a spin polarization of 61 % at the barrier interface. Although the desired spin polarization of 100 % was not reached, the experimental value of the spin polarization is larger than the maximum 55 % effective spin polarization of a variety of 3d-transition metal alloys in combination with Al_2O_3 barriers [45]. However, the spin polarization of photoelectrons emerging from single crystalline Co_2MnSi films grown on GaAs by pulsed laser deposition indicate a quite low spin polarization at the Fermi level of only 12 % at the free surface [42]. Wang et al. [41, 42] assumed that partial chemical disorder was responsible for this discrepancy with the theoretical predictions.

Investigations [25, 30, 33, 46] of the electronic structure of Heusler compounds indicate that on-site correlation plays an important role in these compounds and

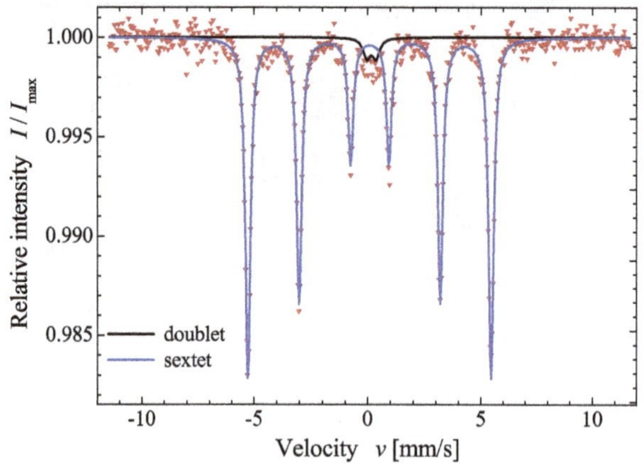

Fig. 2.4 ^{57}Fe Mößbauer spectrum of $Co_2Mn_{0.5}Fe_{0.5}Si$. The spectrum was taken at 290 K and excited by a ^{57}Co(Rh) source. *Solid lines* are results of a fit to determine the sextet and doublet contributions and to evaluate the hyperfine field

may serve to destroy the half-metallic properties of Co_2MnSi. In addition, if on-site correlation is considered in electronic structure calculations Co_2FeSi becomes a half-metallic ferromagnet with a magnetic moment of 6 μ_B.

A possible explanation is the position of the Fermi energy with respect to the minority band gap. A position close to the minority band edges (top of the minority valence band or bottom of the minority conduction band) might easily destroy the gap already at low temperatures without affecting the magnetic moment. One result from the electronic structure calculations on Co_2FeSi and Co_2MnSi is the indication of a shift of the Fermi energy from the top of the minority valence band ($Y =$ Mn) to the bottom of the minority conduction band ($Y =$ Fe) [47]. These particular positions of the minority gap with respect to the Fermi energy make both systems rather unstable with respect to their electronic and magnetic properties. Any small change in physically relevant quantities (e.g., lattice parameters) may easily serve to destroy the half-metallic character by shifting the Fermi energy outside of the minority gap. Therefore, the substitutional series of the quaternary Heusler compound $Co_2Mn_{1-x}Fe_xSi$ was synthesized and investigated both experimentally and theoretically. All samples of the substitutional series exhibit an $L2_1$ order that is independent of the Fe concentration x. The observed structural order-disorder phase transition from $L2_1 \leftrightarrow B2$ is nearly independent of x and occurs at about 1030 K. Mößbauer measurements show only a negligible paramagnetic contribution confirming the high degree of order over the whole substitutional series (see Fig. 2.4).

In agreement with the expectation from the Slater–Pauling curve, the magnetic moment increases linearly with x from 5 μ_B to 6 μ_B. True bulk sensitive, high energy photo emission bearded out the inclusion of electron–electron correlation in the calculation of the electronic structure and gave an indirect advise on the gap in the minority states. Both valence band spectra and hyperfine fields indicate an increase of the effective Coulomb exchange parameters with increasing Fe concentration. The calculated band structures suggest that the most stable compound in a half-metallic state will occur at an intermediate Fe concentration. From both the experimental and computational results it is concluded that a compound with an in-

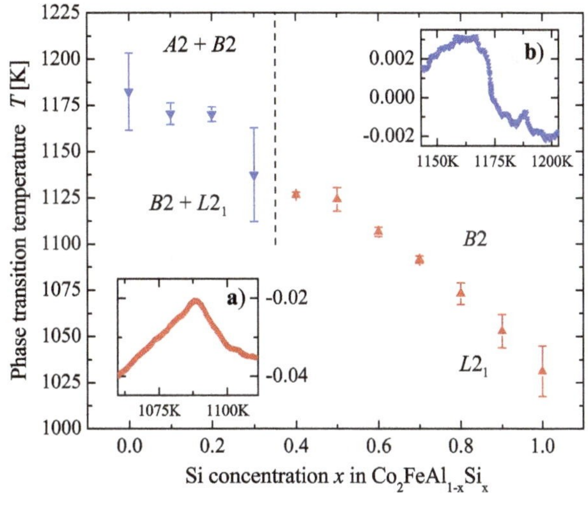

Fig. 2.5 Phase transitions in $Co_2FeAl_{1-x}Si_x$. Shown is the composition dependence of the phase transition temperature. The length of the vertical bars corresponds to the experimental hysteresis (see text). The insets (**a**) and (**b**) display typical DSC curves in high ($x = 0.7$) and low ($x = 0.1$) Si content compounds, respectively

termediate Fe concentration of about 50 % should be most stable and best suited for spintronics applications.

The end members of this series, Co_2FeSi and Co_2MnSi, have been used for fabrication of magnetic tunnel junctions [48–50]. The tunnel magnetoresistance ratios of 159 % in the Mn compound at low temperature and 41 % in the Fe compound at room temperature suggest that still an improvement with respect to the temperature stability of the TMR is necessary.

In $Co_2Mn_{1-x}Fe_xSi$ the transition metal carrying the localized moment is exchanged. This might lead to unexpected effects on the magnetic properties if the samples are not completely homogeneous. The situation is different in the isoelectronic series $Co_2FeAl_{1-x}Si_x$ where the main group element is substituted. Tezuka et al. [51, 52] reported about TMR junctions build from $Co_2FeAl_{0.5}Si_{0.5}$. The junctions exhibited TMR ratios of 76 % at 300 K and 106 % at 5 K for the $B2$ structure while the junctions with $L2_1$ structure showed 51 % and 78 % at 300 K and 5 K, respectively. The TMR ratio is 175 % at 300 K for optimized junctions with $L2_1$ structure [52] and thus larger than the ones found for pure Co_2FeAl or Co_2FeSi electrodes.

Recent ab-initio calculations [53] indicated that the compound with 50 % of the Al substituted by Si will lead to a situation where ε_F is located close to the middle of the minority gap and thus should lead to an improved temperature stability of the spin polarization, in particular if quasi particle excitations are appearing close to the band edges. The following study is devoted to the bulk properties of $Co_2FeAl_{1-x}Si_x$ in order to find the most stable compound of the series with respect to the structural and magnetic properties and thus the most promising candidate for thin film devices using them in GMR and TMR applications.

Differential scanning calorimetry (DSC) was used to find the high temperature phase transitions in the substitutional series. Figure 2.5 displays the dependence of the order-disorder transition temperature as a function of the Si concentration x. The insets display the change of the DSC signal as a function of the temperature using nominal heating rates of 15 $\mathrm{K\,min^{-1}}$. Depending on the temperature rates and the actual amount of material, a shift of the maxima is observed between heating and cooling curves (see also [47]) that is mainly due to an intrinsic hysteresis effect of the method. The length of the vertical bars in Fig. 2.5 corresponds to the range of that hysteresis. The signal at the phase transition is rather high for $x > 0.4$ (Fig. 2.5(a)), that is, for compounds exhibiting clearly the $L2_1$ structure. In that case the signal can be clearly attributed to an $L2_1 \leftrightarrow B2$ order-disorder phase transition as also observed in other quaternary Heusler compounds [47, 54]. For $x < 0.4$ the signal is much weaker.

Here, XRD indicates mainly a $B2$ structure with a very low portion of $L2_1$ and in some cases a high portion of an $A2$ structure. Therefore, the signal at low x may be attributed either to an $L2_1 \leftrightarrow B2$ or to a $B2 \leftrightarrow A2$ phase transition. Overall, Fig. 2.5 demonstrates that the structural transition temperature of the $L2_1$ to the $B2$ phase $T_t^{B2\leftrightarrow L2_1}$ decreases almost linearly with increasing Si content at least for $x > 0.4$. The results from both, XRD and DSC, demonstrate the better structural stability of the compounds with high Si content. This is expected from the stronger hybridization between Co and Si in these compounds [53]. An important detail is that for $x > 0.5$ the as-cast samples exhibit the $L2_1$ structure which make the production of thin films much easier because it may avoid extensive heat treatment to reach the correct structure of the samples.

In summary, the series $Co_2FeAl_{1-x}Si_x$ crystallizes for $x \geq 0.4$ in the $L2_1$ structure. This structure is essentially required for a high spin polarization resulting in high magnetoresistive effects. Both, structural and magnetic investigations suggest that the Si rich systems $x > 0.5$ are more stable with respect to the $L2_1$ structure, due to the stronger hybridization between Co and Si compared to Al. For $x \approx 0.5$, calculations predict that the Fermi energy is located in the middle of the gap of the minority states [53] while in Co_2FeSi ε_F is located at the bottom of the conduction band which can easily destroy the half-metallicity. This behavior will make $Co_2FeAl_{0.5}Si_{0.5}$ stable against temperature variations. From the combination of experimental (better order for high Si content) and theoretical findings (robust gap at $x \approx 0.5 \pm 0.25$) it is concluded that a compound with an intermediate Si concentration close to $x = 0.5 \ldots 0.7$ would be best suited for spintronics applications, especially for GMR and TMR applications.

Very recently, the half-metallicity of $Co_2FeAl_{0.5}Si_{0.5}$ at room temperature was demonstrated experimentally by Shan et al. [55]. Furubayashi et al. [56] reported on the structure and transport properties of current-perpendicular-to-plane spin valves (CPP-SV) with $Co_2FeAl_{0.5}Si_{0.5}$ or Co_2MnSi Heusler alloy magnetic layers and an Ag spacer layer. The CPP GMR values of 12.4 % at RT and 31 % at 12 K were measured from an optimally annealed SV with $Co_2FeAl_{0.5}Si_{0.5}$/Ag/$Co_2FeAl_{0.5}Si_{0.5}$ trilayer. On the other hand, both the upper and lower Co_2MnSi layers were $L2_1$ ordered in the SVs using Co_2MnSi/Ag/Co_2MnSi. MR ratios of 11.9 % at RT and 20 % at 10 K were obtained. This two results emphasize the high suitability of well designed Co_2-based Heusler compounds for spintronics applications.

2.4.3 Solving the Problem of Structure Determination in 3d Transition Metal-Based Heusler Compounds by Means of Anomalous X-Ray Diffraction

This section reports on the investigation of the structural properties of the 3*d* transition metal-based Heusler compounds Co_2FeZ with $Z = $ Ga, Ge, and $Ga_{0.5}Ge_{0.5}$. The high Curie temperatures and magnetic moments make them suitable candidates for spintronics applications [57, 58]. For the predicted high spin polarization and half-metallic character of those compounds it is necessary that the samples crystallize in the $L2_1$ structure. Due to the very similar scattering factors of the constituents it is not possible to distinguish the correct structure with commonly used X-ray sources, even if the compounds crystallize in the correct $L2_1$ structure.

Most of the X_2YZ Heusler compounds used in the production of TMR devices are based on 3*d* transition metals like Co for the X element and Cr, Mn, Fe, or mixtures of those for the Y element. The main group element (Z) is usually taken from the third (Al, Si) or from the fourth row (Ga, Ge) of the periodic table. The detection of the correct structure, as a necessary prerequisite for a half-metallic character, by regular X-ray diffraction (XRD) is difficult especially if all three elements in the compound are from the fourth row. In that case, the scattering factors of all three elements are very similar for excitation by standard laboratory sources (for example: Cu K_α or Mo K_α).

Various Heusler compounds appear not only in an ordered structure but also exhibit pronounced alloying. Besides the $L2_1$ structure of ordered X_2YZ compounds, the most commonly occurring types of disordered alloy have the $B2$ (CsCl-type with complete mixing of Y and Z atoms resulting in $X(Y_{0.5}Z_{0.5})$ or the $A2$ (W-type with complete mixing of all types of atom resulting in $(X_{0.5}Y_{0.25}Z_{0.25})$) structure. In particular, when all elements are taken from the fourth row of the periodic table one needs a unambiguous method to prove the $L2_1$ structure.

Ravel et al. [59] used the method of anomalous XRD to determine antisite disorder with a high precision, in particular for thin films of the Heusler compound Co_2MnGe. This technique has also been successfully used to determine the crystallographic polarity of ZnO epilayers [60], to observe an increase of the structural order parameter in Fe–Co–V soft magnetic alloy after thermal ageing [61], and to distinguish between alloying and segregation in CoCrPt films [62]. In the present work, the method of anomalous XRD is used to determine the structure in 3*d* transition metal-based Heusler compounds where the constituents have almost the same scattering factors and therefore the structure determination with commonly used X-ray sources in the laboratory failed.

The structure factors for an ordered $L2_1$ compound and disordered $B2$ and $A2$ alloys and the results for the (111) and (200) reflections are summarized in Table 2.1. (Note that the indexing of the fcc lattice was used here for all three cases, independent of the present symmetry that is for $A2$ and $B2$ different from $L2_1$.) The (111) reflex appears here only in the $L2_1$ but not in the disordered $A2$ or $B2$ structures. Therefore, the present work is concentrated on the (111) reflection. The factor for

Table 2.1 Structure factors for the ordered $L2_1$ compound and the disordered $B2$ and $A2$ alloys. Note that one has in Heusler alloys $A = X_{1/2}Y_{1/4}Z_{1/4}$ for the $A2$, or $B = Y_{1/2}Z_{1/2}$ for the $B2$ structure. The positions a, b, c, and d correspond to $(0, 0, 0)$, $(\frac{1}{2}, \frac{1}{2}, \frac{1}{2})$, $(\frac{1}{4}, \frac{1}{4}, \frac{1}{4})$, and $(\frac{3}{4}, \frac{3}{4}, \frac{3}{4})$, respectively

	a	b	c	d	f_{111}	f_{200}
$L2_1$	Z	Y	X	X	$f_Y - f_Z$	$f_Y + f_Z - 2f_X$
$B2$	B	B	X	X	0	$2(f_B - f_X)$
$A2$	A	A	A	A	0	0

the (220) reflection is in all three cases $f_{220} = 2f_X + f_Y + f_Z$. It is seen that the factor f_{111} of the $L2_1$ structure becomes very small if the Y and Z atoms behave similarly, as is the case if both atoms are from the fourth row and have nearby N, like Fe and Ga. In that case the ordered structure is almost indistinguishable from the disordered ones.

The situation changes if one approaches the photon energy close to the absorption edges of the constituents, because then the anomalous scattering factors play an important role. The complete scattering factors of the atoms with ordinal number N are described by

$$f_N = f_0(N, \theta, E) + \big(f'(N, E) + if''(N, E)\big), \tag{2.1}$$

where $f'(N, E)$ and $f''(N, E)$ are the real and imaginary part of the anomalous scattering factor. These factors do not depend on the scattering geometry but only on the photon energy (E) and the material (N).

Figure 2.6(a) displays the calculated anomalous scattering factors of Co, Fe, and Ga. For better comparison, the real part of the anomalous scattering factor $f_1 = N + f'$ is shown in the Hanke-form with the limiting value of the regular scattering factor ($\lim f_0(N, \theta, E) = N$) added.

Figure 2.6(b) displays the calculated anomalous scattering intensities for Co_2FeGa using f_1 and f_2 from Fig. 2.6(a), that is, the dependence on the scattering angle is not respected. The enhancement of the (111) reflection (Fig. 2.6(c)) indicating the $L2_1$ structure is clearly visible at energies close to the Fe K-edge.

Figures 2.6(d), (e), and (f) show the XRD data (for better comparison here only the (111) reflexes) taken at $h\nu = 7050$ eV and at the Fe K-edge ($h\nu = 7111.96$ eV) for $Co_2FeGa_{0.5}Ge_{0.5}$, Co_2FeGe, and Co_2FeGa, respectively. The enhancement of the (111) reflections when using excitation energies close to the Fe K-edge is clearly visible and fits the theory. This is clear proof of the ordered $L2_1$ for these compounds which was not possible with a standard XRD lab sources.

Before measuring the anomalous XRD data, EXAFS measurements of the samples were performed. From this experimental data the anomalous scattering factors f' were directly derived using the DIFFKK program developed by Cross et al. [63] and the excitation energies used then for the anomalous XRD were taken. Figures 2.7(a), (b), and (c) show the EXAFS data at the Fe, Co, and Ga K-edges, respectively, from Co_2FeGa and Figs. 2.7(c), (d), and (e) show the derived f's, respectively.

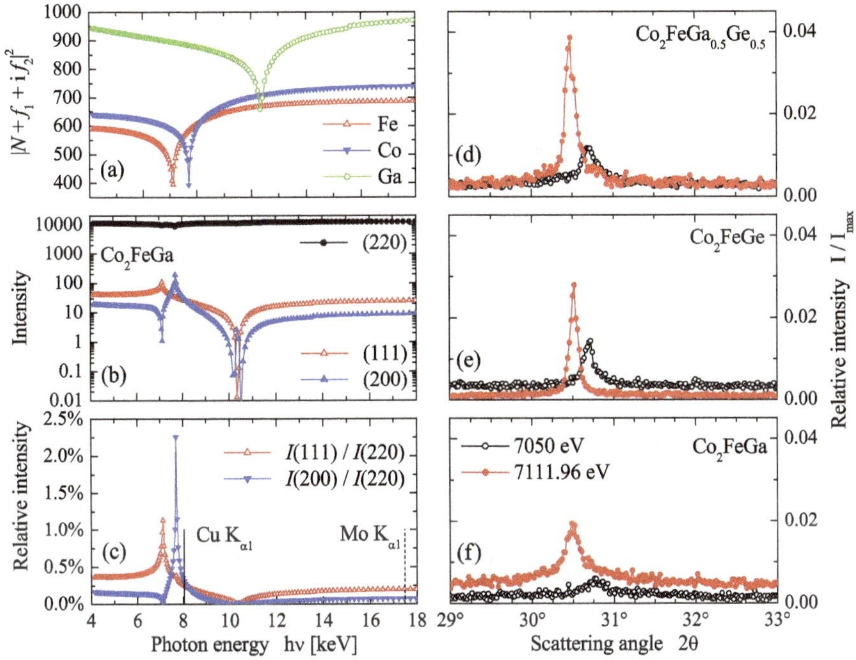

Fig. 2.6 Comparison between theory and the experiment. (**a**) shows the anomalous scattering factors of Co, Fe, and Ga. (**b**) and (**c**) show the calculated anomalous XRD intensities for Co_2FeGa. The energies for excitation by typical laboratory sources are assigned by *vertical lines* in (**c**). (**d**), (**e**), and (**f**) show the XRD data taken at $h\nu = 7050$ eV (*open circles*) and at the Fe K-edge (*closed circles*) for $Co_2FeGa_{0.5}Ge_{0.5}$, Co_2FeGe, and Co_2FeGa, respectively

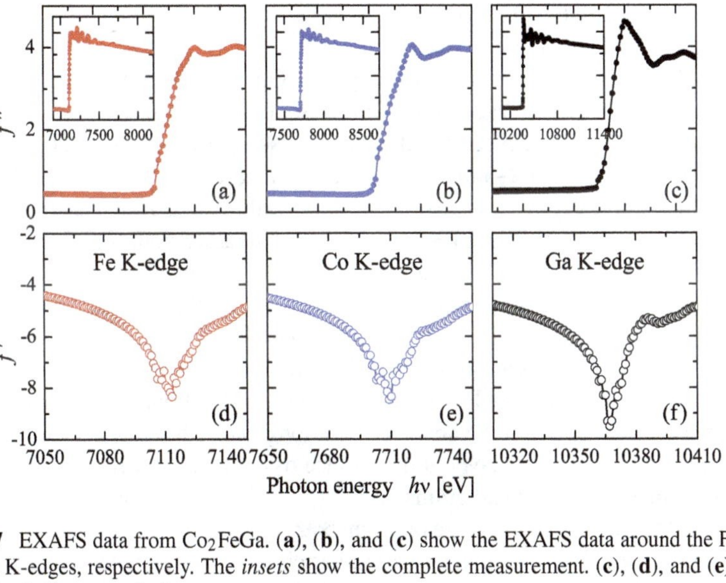

Fig. 2.7 EXAFS data from Co_2FeGa. (**a**), (**b**), and (**c**) show the EXAFS data around the Fe, Co, and Ga K-edges, respectively. The *insets* show the complete measurement. (**c**), (**d**), and (**e**) show the derived f's, respectively

In summary, it has been shown that synchrotron radiation techniques can solve the problem of crystallographic structure determination, even in the case where the scattering factors of elements in a compound are very similar in response to excitation by traditional lab sources. Knowing the anomalous scattering factors for the compounds, one can use anomalous XRD to determine the correct crystallographic structure of the sample. It was shown by other groups that this technique works quite well as well for thin film samples [59–62] so it is expected that the combination of EXAFS and anomalous XRD may also lead to a better understanding of the structure of multilayer thin films in order to improve the quality of TMR-junctions.

A large amount of other new Heusler compounds were synthesized and investigated using various different techniques to analyze their suitability as new materials for spintronics applications [64]. The prepared compounds contain ternary Cr-, Rh-, Ru-, Ni-, Pd-, and Pt-based Heusler systems as well as quaternary Co_2-based compounds and compounds of the general formula $XX'YZ$.

2.5 Highly Ordered, Half-Metallic Co_2FeSi Single Crystals

As already mentioned above, several MTJs of Co_2FeSi have successfully been fabricated, yielding a large TMR. However, in many cases the TMR is strongly temperature dependent, e.g. in Co_2FeSi and Co_2MnSi based MTJs [65–68]. According to Chioncel et al., the strong temperature dependence arises from nonquasiparticle states and their crucial contribution to the finite-temperature spin polarization [69]. In order to separate such extrinsic effects from possible intrinsic contributions to the temperature dependence of the TMR, the investigation of high quality single crystals is indispensable. Therefore, the next section reports about highly ordered, half-metallic Co_2FeSi single crystals.

Two different single crystals were prepared. One was grown by the Czochralski method using a Centor Vacuum Industries Series Crystal Puller. Metallographic investigations of polished pieces of the Czochralski grown single crystal revealed single crystalline areas of 2–3 mm width and 5–7 mm length.

The other single crystal was grown by the optical floating zone technique. Zone melting was carried out in a GERO SPO optical floating zone furnace with two 1000 W halogen lamps with the radiation focused by ellipsoidal, gold coated mirrors. To avoid oxidation the furnace was flushed with 5.0 purity argon for several hours, and during growth a gas flow rate of 300 ml/min of argon with 2 % hydrogen was maintained. The seed and feed rods were counter rotated at 45 and 15 rpm, respectively, the growth speed was set at 20 mm/h. Metallographic investigations revealed the single crystal to be about 1–3 cm long and to have nearly the width of the whole rod, except a thin layer at the surface of the freshly grown crystal, which contains some small additional grains.

Laue diffraction with a spot size of 2 mm $\times$ 2 mm was performed on the floating zone crystal with 25 kV incident copper radiation at 20 mA in reflection geometry. The resulting diffraction patterns (see Fig. 2.8) were compared with simulations

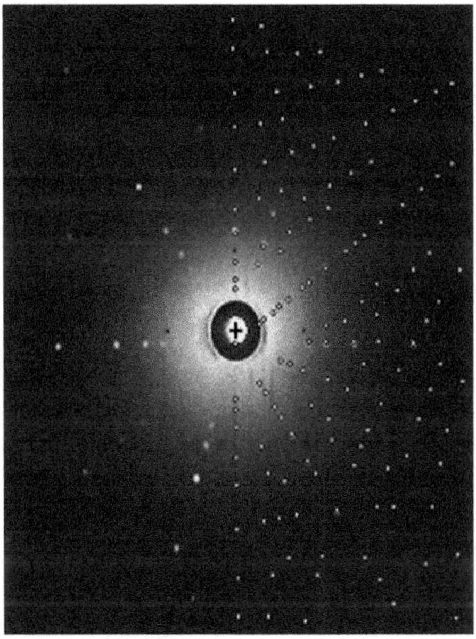

Fig. 2.8 Laue diffraction pattern of the zone molten single crystal. In addition, the simulated diffraction pattern assuming $Fm\overline{3}m$ symmetry and a lattice constant of 5.66 Å is represented by *bright dots* on the *right side* of Fig. 2.8. The perfect matching between the measured diffraction pattern and the simulation demonstrates single crystalline nature and a high degree of order of the zone melted single crystal

(bright dots on right side of Fig. 2.8). The good match verifies the single crystalline nature of the sample. High quality $L2_1$ ordering in space group $Fm\overline{3}m$ with a fitted unit cell parameter of 5.66 Å is confirmed, which is in good agreement with previous results [30]. The simulation also confirms the crystal growth along the (110) axis without twinning.

All crystals show metallic behavior in the resistivity measurements, see Fig. 2.9(a). As expected, the resistivity decreases with decreasing temperature in the temperature range between 300–50 K. Remarkably, the resistivity below 50 K is temperature independent, in agreement with the results of Ambrose et al. [70]. The residual resistivity at 2 K is 0.0155 μΩ m for the floating zone single crystal, 0.0344 μΩ m for the polycrystal and 0.134 μΩ m for the Czochralski grown single crystal, respectively. Note that all measurements of the electrical resistivity shown here are well below the Curie-temperature reported to be 1100 K for Co_2FeSi [30].

The residual resistivity ratio (RRR, here $\rho_{300\,K}/\rho_{2\,K}$) is a measure of the quality of a crystal [37]. Good ordering is inferred in both the zone melted and the polycrystal by the RRR of 5.2 and 5.9, respectively. The RRR values reported here are on the same order of magnitude as found in a Co_2MnSi single crystal (6.5), which the highest RRR reported for a Heusler compound [37, 71]. The RRR of the Czochralski grown crystal is lower by a factor of 2 and relatively poor performance compared to the zone melted and polycrystal, indicating a lower degree of order, in line with our NMR results. However, all RRR presented here indicate excellent crystallinity and homogeneity compared to previously reported RRR for Heusler compounds or alloys in general (see e.g. [38, 70, 72, 73]). The observation of a higher resistivity in the polycrystal compared to the floating zone single crystal might arise from mi-

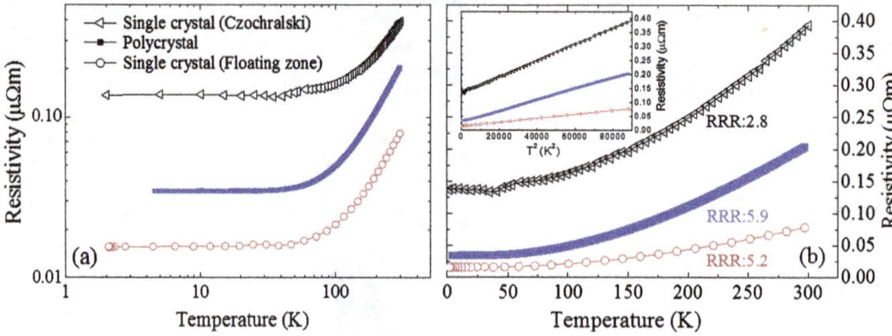

Fig. 2.9 (**a**) Resistivity as a function of temperature on a logarithmic scale to show the plateau at temperatures below 50 K. (**b**) Resistivity as a function of temperature for the Czochralski grown single crystal (*triangles*), the polycrystal (*squares*) and the zone melted single crystal. The *inset* shows the resistivity as a function of T^2 to demonstrate the linear dependence and, thus, the T^2 power law dependence of the resistivity as a function of temperature

croscopic cracks at the grain boundaries present in the polycrystal, leading to longer transport paths.

Figure 2.9(a) shows the resistivity as a function of temperature on a logarithmic scale to demonstrate the temperature independence of the resistivity below 50 K. Above 50 K, the resistivity curves of all crystals follow a T^N power law (fit not shown here), where $\mathrm{N} = 1.9$ for the zone molten single crystal, $\mathrm{N} = 2.1$ for the polycrystal and $\mathrm{N} = 1.6$ for the Czochralski grown single crystal, respectively. The resistivity for all three crystals is reasonably well described by an appropriate T^2 behavior in the high temperature regime, a transition regime and the temperature independent regime below 50 K, while the range of the transition regime is slightly different for the different crystals. The resistivity curves of the zone molten single crystal and the polycrystal approximately scale with each other by a factor of 2.55. The validity of the T^2 behavior in the high temperature regime is further confirmed by a plot of the resistivity as a function of T^2, demonstrating a linear dependence (see inset of Fig. 2.9). A T^2 behavior is expected for a conventional ferromagnet due to coherent one-magnon scattering processes [74]. Assuming one-magnon scattering of conduction electrons, there must be the possibility for spinflips, and thus, both the spin-up and the spin-down electrons are present at the Fermi-level [74]. Consequently, one expects the absence of one-magnon scattering in half-metallic ferromagnets, where only one spin channel contributes to the electrical transport. The absence of any T^2 contributions at temperatures below 50 K might suggest the absence of a one-magnon channel and thus half-metallic ferromagnetism in Co_2FeSi at temperatures below 50 K. The cross-over to more conventional ferromagnetic transport behavior above 50 K indicates the onset of spin scattering which needs to be taken into account (besides interface and impurity scattering) for understanding the temperature dependence of the tunneling magnetoresistance in Co_2FeSi devices [48, 50, 65, 68, 75, 76].

In summary, excellent crystals of the $L2_1$ ordered half-metallic Heusler compound Co_2FeSi confirmed by Laue diffraction, NMR (not shown here, see [77] for details) and the residual resistivity were obtained. The ratio between the residual resistivity at 300 K and 2 K is on the same order of magnitude as the best residual resistivity ratio previously reported for a Heusler compound. The resistivity as a function of temperature roughly follows a T^2 behavior in the high temperature regime, as expected for ferromagnets. Remarkably, the resistivity is temperature independent at low T, which might indicate half-metallic ferromagnetism of Co_2FeSi in the low temperature regime. The synthesis of such high quality Heusler single crystals opens a promising route to exploit the rich physical properties that are realized in various Heusler compounds.

2.6 Heusler Compounds with 4*d* Valence Electrons

Most of the work being reported above was concentrated on the Co_2-based Heusler compounds, whereas compounds based on 4*d* transition metals are rarely investigated. There are only a few results published on the two series $Ru_{2-x}Fe_xCrSi$ [78], and $Ru_{2-x}Fe_xCrGe$ [79]. These compounds exhibit the $L2_1$ structure but the reported saturation magnetization is lower than that predicted by theory. A reason for this could be that these compounds contain Cr, and the magnetic moment of Cr is reduced, as in the $Co_2Cr_{1-x}Fe_xAl$ series, because of complete, or at least partial, ferrimagnetic order (antiparallel orientation of the moments of different Cr atoms, or between X and Cr atoms). Therefore, several Heusler compounds based on 4*d* transition metals that contain other 3*d* transition metals instead of Cr on the Y position were synthesized.

2.6.1 *Investigation of* Rh_2MnGe

Two very well investigated Heusler alloys are Co_2MnZ (Z = Si and Ge) with a calculated band gap of 0.4 eV [35] to 0.8 eV [80] and a high Curie temperature above 900 K [36]. A strong hybridization of the Mn and Co 3*d* states in this system causes the Mn 3*d* states being itinerant to a certain degree [81]. In this section we report about investigation on the isoelectronic alloy Rh_2MnGe with 29 valence electrons per formula unit as in the case of Co_2MnGe. The main difference between Rh_2MnGe and Co_2MnGe is given by the smaller width of the Rh 4*d* states relative to the Co 3*d* states and the larger lattice constant of the Rh-based system. The Mn atom is effectively oversized because of the strongly widened lattice due to the large Rh atoms. The hybridization between the Rh and Mn atoms is smaller than between the Co and Mn atoms and the majority/minority spin splitting for Mn is strongly favored. Therefore, Rh_2MnZ compounds are discussed as systems with fully localized magnetic moments, in contrast to the Co_2MnZ-type compounds where the Co magnetic moment can obviously not be neglected. The gap in the minority-spin states of Co_2MnZ is also predicted for Rh_2MnZ but this gap apparently becomes broader and the Fermi level is no longer found in the gap [82] and the total magnetic moment is not expected to be an integer number.

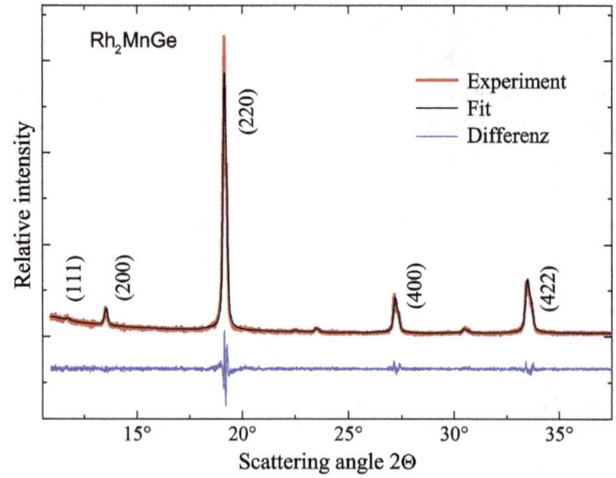

Fig. 2.10 X-ray powder diffraction pattern in Bragg–Brentano geometry for Rh_2MnGe, indicating an $L2_1$ structure. The simulated pattern and the difference between measured and simulated pattern are given for comparison

Structural characterization has been performed with XRD of powders. Due to the larger differences of the scattering factors between Rh, Mn and Ge, XRD provides a slightly better structural information compared to e.g. Co_2MnGe. The strong sharp (220)-peak confirms the presence of a single cubic phase. The comparatively small (10 % of the (220)-peak) (111) and (200) superstructure peaks (see Fig. 2.10) are decisive for the $L2_1$ Heusler structure. The simulated powder diffraction pattern of Rh_2MnGe shows the expected peaks for the defect-free structure. These superlattice peaks both vanish for the case of randomly occupied lattice sites ($A2$ structure). In the case of randomly occupied Mn sites with Mn and Ge ($B2$) only the (200) superlattice peak would be seen, while the (111) peak vanishes. A quantitative evaluation of the spectra limits the disorder to less than 10 % contribution of the $B2$ structure with a lattice parameter of 6.04 Å.

Additionally, XAS and XMCD spectra at the Rh $M_{3,2}$ edges and at the Mn $L_{3,2}$ edges have been measured of the ferromagnetic Heusler alloy Rh_2MnGe. The Rh spectrum shows no multiplet features and can be reproduced by LDA calculations. From this observation we conclude that the Mn moment is strongly localized. The Rh moment is an order of magnitude smaller than the Mn moment. The temperature dependence of the Mn and Rh moments is almost equal to each other. The contribution of the orbital magnetic moment is very small for both elements with orbital to spin moment ratios of 0.01 ± 0.02 for Mn and 0.20 ± 0.15 for Rh. For more details about the XAS and XMCD investigation of Rh_2MnGe see the chapter about XMCD and [83].

2.6.2 Investigation of Ru_2YZ

Half-metallic system with 24 valence electrons are of great interest because of their predicted semiconducting nature. In a similar vein, two new half-metallic Heusler

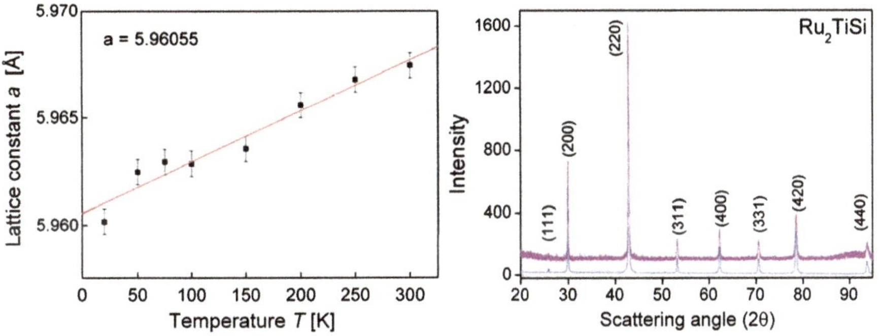

Fig. 2.11 XRD spectrum for Ru_2TiSi. The spectrum was excited by Cu-K_α radiation (*right*). Temperature dependency of the cubic cell volume V for Ru_2TiSi (*left*). The fit of the data to $a(T^3)$ dependence is shown as a *full line*

compounds namely Ru_2TiSi and Ru_2VAl with 24 valence electrons were synthesized.

The crystalline structure was determined by XRD using Cu-K_α radiation. Ru_2TiSi exhibits the $L2_1$ structure. The (111) superstructure reflection is not very strong and is very hard to see. Typical powder diffraction patterns are shown in Fig. 2.11(right). The lattice parameter found from the Rietveld-refinement is $a = 5.96$ Å.

Additionally, the structural parameter of Ru_2TiSi has been investigated in dependence of the temperature between 20 K and 300 K. Figure 2.11(left) shows the temperature dependency of the cell volume $V = a^3$ of the cubic cell. The volume decreases by about 0.5 % if the temperature is lowered from room temperature to 20 K. The change of the cell volume with temperature is clearly non-linear and evidence is given that the volume increases with an approximate T^3 law.

The Ru_2VAl compound exhibits a $B2$ structure, only the (200) superlattice peak of the $B2$ structure type was observed, while the (111) peak was vanished. This type of disorder shows up as a (200) superlattice peak with higher intensity than the (111) peak. The lattice parameter found from the Rietveld-refinement is $a = 5.99$ Å.

Both compounds are—as expected from the Slater–Pauling rule—diamagnetic, which was confirmed by various SQUID measurements.

2.7 Quaternary Heusler Compounds

In the following section, the properties of the $CoFe_{1+x}Ti_{1-x}Al$ and $CoMn_{1+x}V_{1-x}Al$ solid solutions are investigated. It is expected that the partial replacement of Ti by additional Fe or V by additional Mn will keep the $F\overline{4}3m$ and one has a transition from the Y to the X structure rather than to the $L2_1$ structure. The pure compounds $x = 0$ carry 24 valence electrons in the primitive cell and are expected to exhibit no net magnetic moment. Therefore, the onset of the magnetic phenomena with increasing valence electron concentration can be studied by the degree of substitution.

An advantage is that the particular way of quaternary substitution as used here does not change the crystal symmetry either global or local.

2.7.1 Structural Properties of $CoMn_{1+x}V_{1-x}Al$ and $CoFe_{1+x}Ti_{1-x}Si$

For the pure compounds CoFeTiAl and CoMnVAl (XRD not shown here) the *fcc* typical (111) and (200) reflections appear in the XRD pattern of both compounds. No impurities are detected by XRD. A refinement of the data reveals the lattice parameter $a_{FeTi} = 5.8509$ Å for CoFeTiAl and $a_{MnV} = 5.8045$ Å for CoMnVAl, both at 300 K. The best R-values were obtained for the Y structure (CoFeTiAl: $R_p = 6.6$; $R_{wp} = 8.7$ and CoMnVAl: $R_p = 7.2$; $R_{wp} = 9.9$). The difference between measured and refined data shows some deviation at the (200) reflection that might be caused by a small amount of disorder. However, a fit for the X structure with intermixing of Fe and Ti or Mn and V has worth results with R_p (R_{wp}) being larger by 0.3 (1.0) for both compounds. Other types of disorder—that are swapping of Co–Ti, Co–V, or any intermixing of Al with one of the transition metals—were leading to much higher R values and can be excluded as possible types of disorder. Only Co–Fe or Co–Mn interchange or a slight deficiency of Al results in R-values similar or slightly lower compared to the pure Y structure. However, it is not possible to distinguish either these two types of disorder or the correct amount of disorder from measurements at one single photon energy. The reason is that the number of parameters for the fit becomes larger than the number of distinguishable reflections.

Figure 2.12 compares the XRD data of the two series $CoFe_{1+x}Ti_{1-x}Al$ and $CoMn_{1+x}V_{1-x}Al$. The *fcc* typical (111) and (200) reflections are small as already observed for the pure compounds. However, they are clearly resolved in most of the samples. The (111) reflection is absent in $CoFe_{1.5}Ti_{0.5}Al$ what is explained by anti-site disorder. As for the pure compounds, the particular type of disorder cannot be determined from measurements at one single photon energy. The lattice parameter stay unchanged within the error of the measurement. This points out that the lattice parameter is dominated by the Co–Fe and Co–Mn sub-lattice interaction rather than by the interaction between transition metal and main group element.

2.7.2 Magnetic Properties of $CoMn_{1+x}V_{1-x}Al$ and $CoFe_{1+x}Ti_{1-x}Si$

The magnetic moment m of Heusler compounds, especially Co_2-based half-metallic ferromagnets, follows the Slater–Pauling rule where one has for localized moment systems an average magnetic moment (in multiples of the Bohr magneton μ_B) per atom of:

$$m_{av} = n_{av} - 6 - 2n_{sp}. \tag{2.2}$$

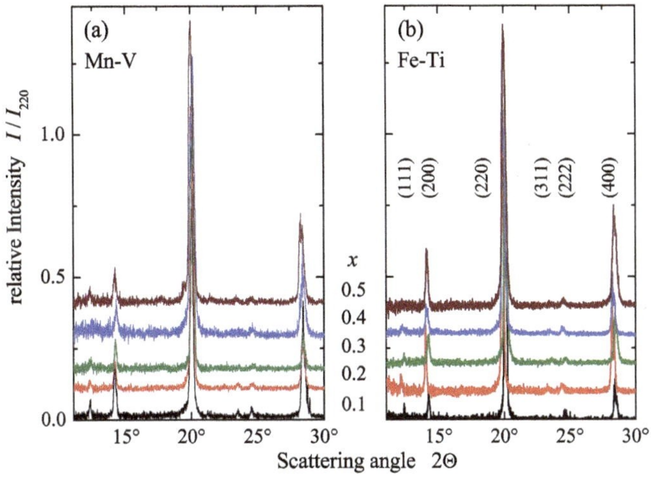

Fig. 2.12 XRD data of $CoMn_{1+x}V_{1-x}Al$ (**a**) and $CoFe_{1+x}Ti_{1-x}Al$ (**b**). The data are taken with Mo K_α radiation and normalized to the intensity of the (220) reflection. Data for $x > 0.1$ are plotted with an offset for better comparison

n_{av} is the mean number of valence electrons per atom in an alloy or compound and n_{sp} arises from the average number of unbalanced minority *sp* electrons. The number six arises from the fact that the *d* electrons are constrained such that the Fermi energy falls into a minimum (or gap) between occupied and unoccupied *d* states and therefore minimizes the total energy. In half-metallic ferromagnets with a gap in one of the spin densities or in quasi closed shell compounds all *sp* electrons are occupied and the n_{sp} term vanishes. In 2:1:1 or 1:1:1:1 Heusler compounds one has four atoms in the primitive cell and thus an overall magnetic moment of

$$m = (N_v - 24)\mu_B. \tag{2.3}$$

N_v is the accumulated number of valence electrons in the primitive cell (for details see [57, 81, 84, 85]). Here one has $2 + n(d)$ for each transition metal and $2 + n(p)$ for the main group element, where two arises in both cases from the *s* electrons and $n(d)$ and $n(p)$ are the numbers of available *d* and *p* valence electrons, respectively. At $N_v = 24$ the materials are not ferromagnetic according to the Slater–Pauling rule. The reason is that a quasiclosed shell character is reached at $N_v = 24$. It is caused by successive filling of the $a1$, $t1$, (two *s*, six *p* electrons) e, $t2$ (ten *d* electrons) bands, followed by subsequent complete filling of an additional $t2$ band (six *d* electrons). Deviations from (2.3) will appear when the compound is not in a half-metallic state and unbalanced *sp* or *d* electrons are present. For some of the $L2_1$ ordered compounds, 24 valence electrons lead to the phenomenon of half-metallic completely compensated ferrimagnetism [86], where the moments are ordered in a way that the total magnetic moment vanishes even though individual magnetic moments have-contrary to antiferromagnets—different magnitudes [87].

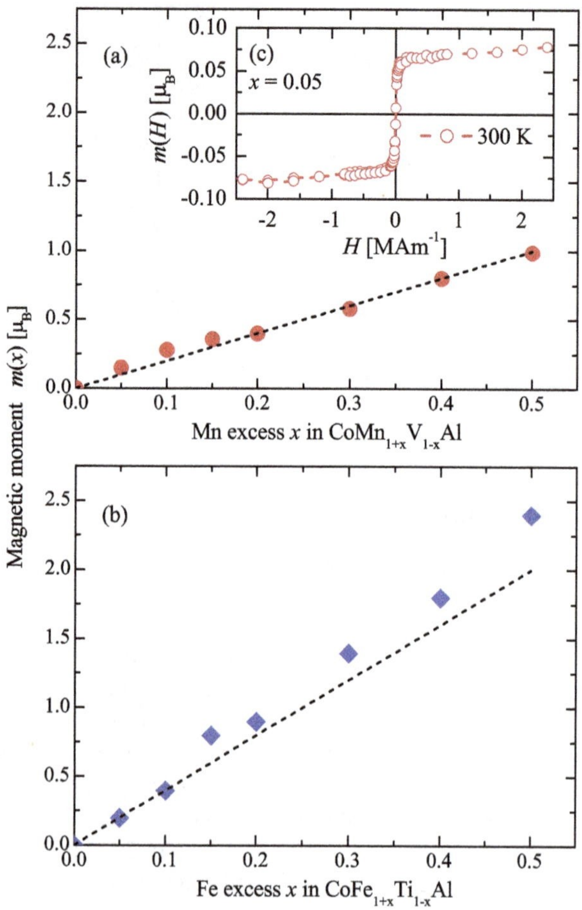

Fig. 2.13 Magnetization data of $CoT'_{1+x}T''_{1-x}Al$. Shown is the concentration dependence of the saturation magnetic moments of $CoMn_{1+x}V_{1-x}Al$ in (**a**) and of $CoFe_{1+x}Ti_{1-x}Al$ in (**b**). The measurements were performed at 5 K. The *lines* correspond to the expected values from the Slater–Pauling rule. The *inset* (**c**) shows the hysteresis of $CoMn_{1.05}V_{0.95}Al$ measured at 300 K

The magnetic data of the two series $CoFe_{1+x}Ti_{1-x}Al$ and $CoMn_{1+x}V_{1-x}Al$ are compared in Fig. 2.13.

The inset (c) in Fig. 2.13(a) shows the hysteresis of $CoMn_{1.05}V_{0.95}Al$ measured at 300 K. This demonstrates that the Curie temperature is above room temperature already at the lowest Mn substitution of 5 %. Similar to this composition, all samples of both series exhibited for $x > 0.05$ a ferromagnetic type hysteresis at 300 K. The shape of the hysteresis was in all cases soft-magnetic as for the shown sample. From the Slater–Pauling rule one expects that the saturation magnetization m increases with the concentration x and the given difference in the number of valence electrons by

$$m(x) = (\Delta N_v * x)\mu_B. \tag{2.4}$$

The differences ΔN_v are 4 e^- for the Fe–Ti and 2 e^- in the Mn–V compounds. This results in magnetic moments of 2 μ_B or 1 μ_B at $x = 0.5$. The linear increase with increasing number of valence electrons is clearly seen in Fig. 2.13 for both series of compounds. The observed deviations—here the higher values in the Ti–Fe compounds at large x—are caused by antisite disorder in the samples. Such disorder is very often observed in Al containing Heusler compounds. The low hybridization strength between Al and the surrounding transition metal atoms allows easily for a swapping between Al and transition metal atoms. The formation of an increase of nearest neighbor transition metals leads in particular for compounds containing Co and Fe to an enhancement of the magnetic moment. From the better agreement of the magnetization of $CoMn_{1+x}V_{1-x}Al$ with the Slater–Pauling values, it is expected that this series is a good candidate for half-metallic ferromagnetism.

In summary, the quaternary Heusler compounds without inversion symmetry $CoFe_{1+x}Ti_{1-x}Al$ and $CoMn_{1+x}V_{1-x}Al$ were synthesized and analyzed by means of X-ray diffraction and magnetometry. The pure compounds ($x = 0$) crystallize in the Y structure with $F\overline{4}3m$ symmetry, due to the 24 valence electrons in the primitive cell they do not carry magnetic moments either in the primitive cell or at the individual sites. This is remarkable—for CoFeTiAl—because CoFe is the transition metal alloy that carries the largest magnetic moment and exhibits the highest known Curie temperature—for CoMnVAl—because Mn carries a localized magnetic moment in all other Heusler compounds. Electronic structure calculations revealed also that the pure compounds do not show magnetic order, in agreement with the experiments. With increasing substitution of Ti by Fe or V by Mn a linear increase of the magnetic moment with the number of valence electrons is observed that agrees with the Slater–Pauling rule, in both experiment and calculations. This demonstrates that the Slater–Pauling behavior of Heusler compounds is not restricted to the $L2_1$ structure with $Fm\overline{3}m$ symmetry but is valid as well in $F\overline{4}3m$ symmetry.

2.8 Transport Properties of Co_2YZ Heusler Compounds

As already mentioned above, the Heusler alloys have attracted a lot of interest as suitable materials for spintronics applications [88] in the recent years. A huge amount of studies investigating the half-metallic properties theoretically and experimentally and enhancing the performance of the compounds and devices for different type of applications were done (see above). The recent observation of the spin Seebeck effect allows to pass a pure spin current over a long distance [89] and is directly applicable to the production of spin voltage generators which are crucial for driving spintronics devices [90–92].

For an effective generation of a spin current the spin Seebeck coefficient and the spin voltage namely the difference in the chemical potential of the spin up $\mu_\uparrow$ and spin down $\mu_\downarrow$ should be large [93]. This section reports on the investigation of Co_2-based Heusler compounds and their potential as spin voltage generators. The high application potential of the compounds is demonstrated. For a further optimization the electronic and magnetic properties of the compounds can be designed easily.

SQUID measurements were carried out to distinguish the saturation magnetization and temperature dependent measurements were done to distinguish the Curie

Table 2.2 Magnetic moments, Curie temperatures, and Seebeck coefficients of the investigated Co_2YZ compounds. The compounds are grouped by their number of valence electrons

Compound	T_c [K]	M_{sat}(5 K) [μ_B]	S(300 K) [μV K^{-1}]
Co_2TiAl	128	0.75	−55
Co_2TiSi	380	1.96	−27
Co_2TiGe	380	1.94	−22
Co_2TiSn	355	1.97	−34
Co_2MnAl	693	3.96	−4
Co_2MnSi	985	4.97	−7
Co_2MnGe	905	4.98	−15
Co_2MnSn	829	5.03	−33
Co_2FeSi	1100	5.97	−12

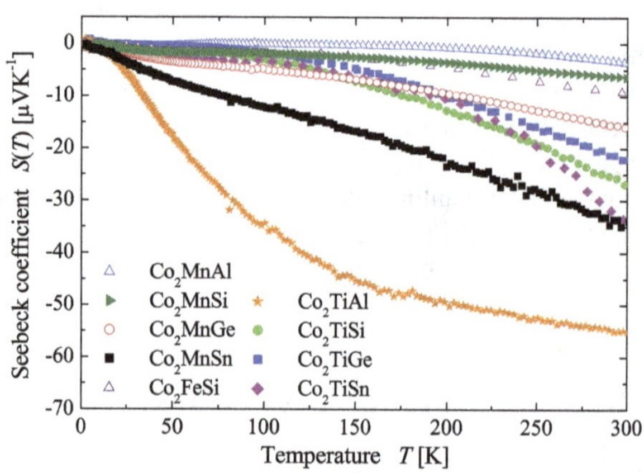

Fig. 2.14 Measured Seebeck coefficients (S) for selected Heusler compounds from 2 K to 300 K

temperatures, the values are summarized in Table 2.2. With the increase of the number of valence electrons and therefore as well the increase of the magnetic moment the Curie temperature increases. One can nicely see the increase of the Curie temperature for the samples with higher numbers of valence electrons. These behavior is well known for Heusler compounds and was reported already some years ago [84, 94]. This measurements show the easy tunability of the magnetic properties of the Heusler compounds. By controlling the number of valence electrons one can design the properties of the materials.

In Fig. 2.14 the measured Seebeck coefficients (S) are displayed for temperatures from 2 K to 300 K. The absolute values increase with increasing temperature. The values at 300 K of all the measurements are summarized in Table 2.2. For an increase of the valence electron count the absolute value of the Seebeck coefficient is decreased. This is explained by the increase of the electron concentration in the bands. By increasing the number of valence electrons additional electrons are added to the d-band at the Fermi energy [95]. This leads to an increase of the carrier concentration. The increase of the carrier concentration leads to a decrease of the Seebeck coefficient [96]. The interrelationship between carrier concentration and Seebeck coefficient can be seen from relatively simple models of electron transport. For simple metals or degenerate semiconductors with parabolic bands and energy-independent scattering the Seebeck coefficient S is given by

$$S = \frac{8\pi^2 k_B^2}{3eh^2} m^* T \left(\frac{\pi}{3n}\right)^{3/2}, \tag{2.5}$$

where n is the carrier concentration and m^* is the effective mass of the carrier. It can be clearly seen that S depends on the carrier concentration and on the effective mass m^*. The latter depend on the shape of the bands. Although the Seebeck coefficient is decreasing with increasing carrier concentration it is remarkable that Sn containing compounds yield large absolute values for the Seebeck coefficient. This effect is related to changes in the band structure and consequential changes in the effective mass m^*. The Seebeck coefficient of the investigated compounds is negative over the entire temperature range. The found absolute values are quite large compared to elemental metals ($S_{\mathrm{Co}} = -30.8$, $S_{\mathrm{Mn}} = -9.8$, $S_{\mathrm{Ti}} = +9.1$, $S_{\mathrm{Al}} = -1.66$, $S_{\mathrm{Sn}} = -1$, values at $T = 300$ K and in μV K^{-1}). Especially the Sn containing compounds show high absolute Seebeck coefficients with high Curie temperatures. This makes them attractive candidates for materials used in spin voltage generators.

2.9 Summary

In summary, we reported about the design of materials for spintronics applications. It has been shown that the combination of synthesis and electronic structure calculations is a powerful tool to pick out suitable materials. Starting with $Co_2Cr_{0.6}Fe_{0.4}Al$ and then Co_2FeSi we developed an understanding of the design of new materials and investigated the series $Co_2Mn_{1-x}Fe_xSi$ and $Co_2FeAl_{1-x}Si_x$. From the combination of experimental and theoretical findings (robust gap at $x \approx 0.5 \pm 0.25$) it is concluded that a compound with an intermediate Si concentration close to $x = 0.5 \ldots 0.7$ would be best suited for spintronics applications, especially for GMR and TMR applications. This findings were recently verified by Shan et al. [55] who demonstrated experimentally the half-metallicity of $Co_2FeAl_{0.5}Si_{0.5}$ at room temperature in thin film devices.

In order to separate intrinsic and extrinsic properties, high quality single crystals were grown, especially Co_2FeSi. All Co_2FeSi crystals show excellent ordering, resulting in outstanding electrical behavior with low residual-resistivity and high residual-resistivity-ratio. All Co_2FeSi crystals show constant resistivity below 50 K, which might point to half-metallic ferromagnetism. The cross-over from this unusual to more conventional transport (T^2 dependence) around 50 K indicates the onset of spin flip scattering and thus is indispensable for understanding the strong temperature dependence of Co_2FeSi tunneling magnetoresistance-devices.

Furthermore, it has been shown that synchrotron radiation techniques can solve the problem of crystallographic structure determination, even in the case where the scattering factors of elements in a compound are very similar in response to excitation by traditional lab sources. It is expected that the combination of EXAFS and anomalous XRD may lead to a better understanding of the structure of multilayer thin films in order to improve the quality of TMR-junctions.

As examples of quaternary Heusler compounds without inversion symmetry $CoFe_{1+x}Ti_{1-x}Al$ and $CoMn_{1+x}V_{1-x}Al$ were synthesized and analyzed by means of X-ray diffraction and magnetometry. The pure compounds ($x = 0$) crystallize in the Y structure with $F\overline{4}3m$ symmetry, due to the 24 valence electrons in the primitive cell they do not carry magnetic moments either in the primitive cell or at the individual sites. This is remarkable—for CoFeTiAl—because CoFe is the transition metal alloy that carries the largest magnetic moment and exhibits the highest known Curie temperature—for CoMnVAl—because Mn carries a localized magnetic moment in all other Heusler compounds. Electronic structure calculations revealed also that the pure compounds do not show magnetic order, in agreement with the experiments. With increasing substitution of Ti by Fe or V by Mn a linear increase of the magnetic moment with the number of valence electrons is observed that agrees with the Slater–Pauling rule, in both experiment and calculations. This demonstrates that the Slater–Pauling behavior of Heusler compounds is not restricted to the $L2_1$ structure with $Fm\overline{3}m$ symmetry but is valid as well in $F\overline{4}3m$ symmetry.

In the last section, Co_2-based Heusler compounds have been investigated on their potential use as spin voltage generators. The observed Seebeck coefficients were all negative in the whole temperature range. The absolute values are quite large compared to simple metals. Especially the Sn containing compounds show high Seebeck coefficients with high Curie temperatures. This makes them attractive candidates for materials used in spin voltage generators.

Acknowledgements This work was financially supported by the Deutsche Forschungsgemeinschaft (project P 1 in DfG research unit FOR 559). Additional financial support by the DAAD (D06/33952), CAPES PROBRAL (167/04), and Stiftung Innovation Rheinland-Pfalz is gratefully acknowledged. The authors thank Y. Hwu (Taipei, Taiwan), K. Kobayashi (Hyogo, Japan), H.-J. Lin (Hsinchu, Taiwan), J. Morais, M.C.M. Alves, F. Bernardi, (Porto Alegre, Brazil), R. Seshadri (Santa Barbara, USA) as well as V. Alijani, J. Barth, L. Basit, S. Berinskat, C.G.F. Blum, F. Caspar, T. Gasi, T. Graf, V. Jung, H.C. Kandpal, V. Ksenofontov, S. Ouardi, W. Tremel, J. Winterlik, and S. Wurmehl and all the student assistants during the last six years for their help with theory and experiments, and for fruitful discussions. Assistance by the staff of the synchrotron facilities BESSY (Berlin, Germany), LNLS (Campinas, Brazil), NSRRC (Hsinchu, Taiwan) and SPRING-8 (Hyogo, Japan) is gratefully acknowledged.

References

1. Prinz GA (1998) Science 282:1660
2. de Groot RA, Müller FM, Engen PGv, Buschow KHJ (1983) Phys Rev Lett 50:2024
3. Coey JMD, Venkatesan M, Bari MA (eds) (2002) Half-metallic ferromagnets. Lecture notes in physics, vol 595. Springer, Heidelberg

4. Youn SJ, Min BI (1995) Phys Rev B 51:10436
5. Ristoiu D, Nozieres JP, Borca CN, Borca B, Dowben PA (2000) Appl Phys Lett 76:2349
6. Ristoiu D, Nozieres JP, Borca CN, Komesu T, Jeong HK, Dowben PA (2000) Europhys Lett 49:624
7. Zhu W, Sinkovic B, Vescovo E, Tanaka C, Moodera JS (2001) Phys Rev B 64:R060403
8. Kübler J, Williams AR, Sommers CB (1983) Phys Rev B 28:1745
9. Webster PJ, Ziebeck KRA (1973) J Phys Chem Solids 34:1647
10. Heusler F (1903) Verh Dtsch Phys Ges 5:219
11. Block T, Felser C, Jakob G, Ensling J, Mühling B, Gütlich P, Beaumont V, Studer F, Cava RJ (2003) J Solid State Chem 176:646
12. Block T, Wurmehl S, Felser C, Windeln J (2006) Appl Phys Lett 88:202504
13. Inomata K, Okamura S, Goto R, Yezuka N (2003) Jpn J Appl Phys 42:L419
14. Kelekar R, Clemens BM (2004) J Appl Phys 96:540
15. Hirohata A, Kikuchi M, Tezuka N, Inomata K, Claydon JS, Xu YB (2005) J Appl Phys 97:10C308
16. Hirohata A, Kurebayashi H, Okamura S, Kikuchi M, Masaki T, Nozaki T, Tezuka N, Inomata K (2005) J Appl Phys 97:103714
17. Jakob G, Casper F, Beaumont V, Falka S, Auth N, Elmers HJ, Felser C, Adrian H (2005) J Magn Magn Mater 290–291:1104
18. Inomata K, Tezuka N, Okamura S, Kurebayashi H, Hirohata A (2004) J Appl Phys 95:7234
19. Marukame T, Kasahara T, Matsuda KI, Uemura T, Yamamoto M (2005) Jpn J Appl Phys 44:L521
20. Auth N, Jakob G, Block T, Felser C (2003) Phys Rev B 62:024403
21. Conca A, Falk S, Jakob G, Jourdan M, Adrian H (2005) J Magn Magn Mater 290–291:1127
22. Clifford E, Venkatesan M, Gunning R, Coey JMD (2004) Solid State Commun 131:61
23. Rodrìgues-Carval J (1993) Physica B 55:192
24. Ferreira FF, Granado E, Carvalho W Jr., Kycia SW, Bruno D, Droppa R Jr. (2006) J Synchrotron Radiat 13:46
25. Wurmehl S, Fecher GH, Kandpal HC, Ksenofontov V, Felser C, Lin HJ, Morais J (2005) Phys Rev B 72:184434
26. Ouardi S, Gloskovskii A, Balke B, Jenkins CA, Barth J, Fecher GH, Felser C, Gorgoi M, Mertin M, Schäfers F, Ikenaga E, Yang K, Kobayashi K, Kubota T, Oogane M, Ando Y (2009) J Phys D, Appl Phys 42:084011
27. Graf T, Fecher GH, Barth J, Winterlik J, Felser C (2009) J Phys D, Appl Phys 42:084003
28. Barth J, Fecher GH, Balke B, Ouardi S, Graf T, Felser C, Shkabko A, Weidenkaff A, Klaer P, Elermers HJ, Yoshikawa H, Ueda S, Kobayashi K (2010) Phys Rev B 81:064404
29. Wurmehl S, Alves MCM, Morais J, Ksenofontov V, Teixeira SR, Machado G, Fecher GH, Felser C (2007) J Phys D, Appl Phys 40:1524
30. Wurmehl S, Fecher GH, Kandpal HC, Ksenofontov V, Felser C, Lin HJ (2006) Appl Phys Lett 88:032503
31. Wurmehl S, Morais J, Alves MdCM, Teixeira SR, Fecher GH, Felser C (2006) J Alloys Compd 423:159
32. Wurmehl S, Fecher GH, Felser C (2006) Z Naturforsch 61b:749
33. Wurmehl S, Fecher GH, Kroth K, Kronast F, Dürr HA, Takeda Y, Saitoh Y, Kobayashi K, Lin HJ, Schönhense G, Felser C (2006) J Phys D, Appl Phys 39:803
34. Wurmehl S, Fecher GH, Ksenofontov V, Casper F, Stumm U, Felser C, Lin HJ, Hwu Y (2006) J Appl Phys 99:08J103
35. Fuji S, Sugimura S, Ishida S, Asano S (1990) J Phys Condens Matter 2:8583
36. Brown PJ, Neumann KU, Webster PJ, Ziebeck KRA (2000) J Phys Condens Matter 12:1827
37. Ravel B, Raphael MP, Harris VG, Huang Q (2002) Phys Rev B 65:184431
38. Geiersbach U, Bergmann A, Westerholt K (2002) J Magn Magn Mater 240:546
39. Kämmerer S, Heitmann S, Meyners D, Sudfeld D, Thomas A, Hütten A, Reiss G (2003) J Appl Phys 93:7945

40. Singh LJ, Barber ZH, Miyoshi Y, Bugoslavsky Y, Branford WR, Cohen LF (2004) Appl Phys Lett 84:2367
41. Wang WH, Przybylski M, Kuch W, Chelaru LI, Wang J, Lu F, Barthel J, Meyerheim HL, Kirschner J (2005) Phys Rev B 71:144416
42. Wang WH, Przybylskia M, Kuch W, Chelaru LI, Wang J, Lu YF, Barthel J, Kirschner J (2005) J Magn Magn Mater 286:336
43. Schmalhorst J, Kämmerer S, Sacher M, Reiss G, Hütten A, Scholl A (2004) Phys Rev B 70:024426
44. Schmalhorst J, Kammerer S, Reiss G, Hütten A (2005) Appl Phys Lett 86:052501
45. LeClair P, Swagten HJM, Kohlhepp JT, de Jonge WJM (2000) Appl Phys Lett 76:3783
46. Kandpal HC, Fecher GH, Felser C, Schönhense G (2006) Phys Rev B 73:094422
47. Balke B, Fecher GH, Kandpal HC, Felser C, Kobayashi K, Ikenaga E, Kim JJ, Ueda S (2006) Phys Rev B 74:104405
48. Inomata K, Okamura S, Miyazaki A, Kikuchi M, Tezuka N, Wojcik M, Jedryka E (2006) J Phys D, Appl Phys 39:816
49. Oogane M, Sakuraba Y, Nakata J, Kubota H, Ando Y, Sakuma A, Miyazaki T (2006) J Phys D, Appl Phys 39:834
50. Ebke D, Schmalhorst J, Liu NN, Thomas A, Reiss G, Hütten A (2006) Appl Phys Lett 89:162506
51. Tezuka N, Ikeda N, Miyazaki A, Sugimoto S, Kikuchi M, Inomata K (2006) Appl Phys Lett 89:112514
52. Tezuka N, Ikeda N, Sugimoto S, Inomata K (2006) Appl Phys Lett 89:252508
53. Fecher GH, Felser C (2007) J Phys D, Appl Phys 40:1582
54. Kobayashi K, Umetsu RY, Kainuma R, Ishida K, Oyamada T, Fujita A, Fukamichi K (2004) Appl Phys Lett 85:4684
55. Shan R, Sukegawa H, Wang WH, Kodzuka M, Furubayashi T, Ohkubo T, Mitani S, Inomata K, Hono K (2009) Phys Rev Lett 102:246601
56. Furubayashi T, Kodama K, Nakatani TM, Sukegawa H, Takahashi YK, Inomata K, Hono K (2010) J Appl Phys 107:113917
57. Kübler J, Fecher GH, Felser C (2007) Phys Rev B 76:024414
58. Galanakis I, Dederichs PH (eds) (2005) Half-metallic alloys. Lecture notes in physics, vol 676. Springer, Berlin
59. Ravel B, Cross JO, Raphael MP, Harris VG, Ramesh R, Saraf V (2002) Appl Phys Lett 81:2812
60. Tampo H, Fons P, Yamada A, Kim KK, Shibata H, Matsubara K, Niki S, Yoshikawa H, Kanie H (2005) Appl Phys Lett 87:141904
61. Zhu Q, Li L, Masteller MS, Corso GJD (1996) Appl Phys Lett 69:3917
62. Chow GM, Sun CJ, Soo EW, Wang JP, Lee HH, Noh DY, Cho TS, Je JH, Hwu YK (2002) Appl Phys Lett 80:1607
63. Cross JO, Newville M, Rehr JJ, Sorensen LB, Bouldin CE, Watson G, Gouder T, Lander GH, Bell MI (1998) Phys Rev B 58:11215
64. Graf T, Casper F, Winterlik J, Balke B, Fecher GH, Felser C, Anorg Z (2009) Z Anorg Allg Chem 635:976
65. Gercsi Z, Rajanikanth A, Takahashi YK, Hono K, Kikuchi M, Tezuka N, Inomata K (2006) Appl Phys Lett 89:082512
66. Ishikawa T, Marukame T, Kijima H, Matsuda Ki, Uemura T, Yamamoto M (2006) Appl Phys Lett 89:192505
67. Tsunegi S, Sakuraba Y, Oogane M, Takanashi K, Ando Y (2008) Appl Phys Lett 93:112506
68. Oogane M, Shinano M, Sakuraba Y, Ando Y (2009) J Appl Phys 105:07C903
69. Chioncel L, Sakuraba Y, Arrigoni E, Katsnelson MI, Oogane M, Ando Y, Miyazaki T, Burzo E, Lichtenstein AI (2008) Phys Rev Lett 100:086402
70. Ambrose T, Krebs JJ, Prinz GA (2000) Appl Phys Lett 76:3280
71. Raphael MP, Ravel B, Willard MA, Cheng SF, Das BN, Stroud RM, Bussmann KM, Claassen JH, Harris VG (2001) Appl Phys Lett 79:4396

72. Schneider H, Jakob G, Kallmayer M, Elmers HJ, Cinchetti M, Balke B, Wurmehl S, Felser C, Aeschlimann M, Adrian H (2006) Phys Rev B 74:174426
73. Paudel MR, Wolfe CS, Patton HMA, Simonson J, Dubenko I, Ali N, Stadler S (2009) J Appl Phys 105:07E902
74. Otto MJ, van Woerden RAM, van der Valk PJ, Wijngaard J, van Bruggen CF, Haas C (1989) J Phys Condens Matter 1:2351
75. Schmalhorst J, Ebke D, Weddemann A, Hütten A, Thomas A, Reiss G, Turchanin A, Gölzhäuser A, Balke B, Felser C (2008) J Appl Phys 104:043918
76. Inomata K, Ikeda N, Tezuka N, Goto R, Sugimoto S, Wojcik M, Jedryka E (2008) Sci Technol Adv Mater 9:014101
77. Blum CGF, Jenkins CA, Barth J, Felser C, Wurmehl S, Friemel G, Hess C, Behr G, Büchner B, Reller A, Riegg S, Ebbinghaus SG, Ellis T, Jacobs PJ, Kohlhepp JT, Swagten HJM (2009) Appl Phys Lett 95:161903
78. Matsuda K, Hiroi M, Kawakami M (2005) J Phys Condens Matter 17:5889
79. Kusakari Y, Kanomata T, Fukushima K, Nishihara H (2007) J Magn Magn Mater 310:e607
80. Picozzi S, Continenza A, Freeman AJ (2002) Phys Rev B 66:094421
81. Galanakis I, Dederichs PH, Papanikolaou N (2002) Phys Rev B 66:174429
82. Pugacheva M, Jezierski A (1995) J Magn Magn Mater 151:202
83. Klaer P, Kallmayer M, Elmers HJ, Basit L, Thöne J, Chadov S, Felser C (2009) J Phys D, Appl Phys 42:084001
84. Fecher GH, Kandpal HC, Wurmehl S, Felser C, Schönhense G (2006) J Appl Phys 99:08J106
85. Kübler J (2000) Theory of itinerant electron magnetism. Oxford University Press, Oxford
86. Wurmehl S, Kandpal HC, Fecher GH, Felser C (2006) J Phys Condens Matter 18:6171
87. Pickett WE (1998) Phys Rev B 57:10613
88. Felser C, Fecher GH, Balke B (2007) Angew Chem, Int Ed 46:668
89. Uchida K, Takahashi S, Harii K, Ieda J, Koshibae W, Ando K, Maekawa S, Saitoh E (2008) Nature 455:778
90. Wolf SA, Awschalom DD, Buhrman RA, Daughton JM, von Molnar S, Roukes ML, Chtchelkanova AY, Treger DM (2001) Science 294:1488
91. Zutic I, Fabian J, Sarma SD (2004) Rev Mod Phys 76:323
92. Chappert C, Fert A, Van Dau FN (2007) Nat Mater 6:813
93. Ong NP (2008) Nature 455:741
94. Kandpal HC, Felser C, Fecher GH (2007) J Magn Magn Mater 310:1626
95. Ouardi S, Balke B, Gloskovskii A, Fecher GH, Felser C, Schönhense G, Ishikawa T, Uemura T, Yamamoto M, Sukegawa H, Wang W, Inomata K, Yamashita Y, Yoshikawa H, Ueda S, Kobayashi K (2009) J Phys D, Appl Phys 42:084010
96. Snyder GJ, Toberer ES (2008) Nat Mater 7:107

Chapter 3
Crystal Structure of Heusler Compounds

Tanja Graf and Claudia Felser

Abstract Heusler compounds are promising materials in many fields of contemporary research. The spectrum of their possible applications ranges from magnetic and magneto-mechanical materials over semiconductors and thermoelectrics to superconductors. An important feature of the Heusler compounds is the possibility of controlling the valence electron concentration by partial substitution of elements. On the other hand, the properties also depend on the degree of ordering of the crystal structure. In general, Heusler compounds crystallize in the Cu_2MnAl-type structure but in many cases certain types of disorder are observed. In this chapter, a detailed description of the crystal structure as well as different types of atomic disorder are given. Furthermore, the relationship of the chemical ordering and the spin polarization is discussed and useful experimental methods for the structural analysis of Heusler compounds are presented.

3.1 Introduction

Heusler compounds first attracted interest among the scientific community in 1903, when F. Heusler found that the compound Cu_2MnAl becomes ferromagnetic, although non of its constituent elements is ferromagnetic by itself [1, 2]. However, it took three decades until the crystal structure was determined to be ordered with a face centered, cubic lattice [3, 4]. Unfortunately, Heusler compounds faded almost in oblivion in the following decades, and only few reports on the synthesis of new Heusler compounds were published in the 1970s [5, 6]. It was not until the prediction of half-metallic ferromagnetism in Co_2MnSn in 1983 by Kübler et al. [7] that scientific interest returned to Heusler compounds.

T. Graf · C. Felser
Institut für Anorganische Chemie und Analytische Chemie, Johannes Gutenberg – Universität, 55099 Mainz, Germany

C. Felser (✉)
Max-Planck-Institut für Chemische Physik fester Stoffe, 01187 Dresden, Germany
e-mail: felser@cpfs.mpg.de

C. Felser, G.H. Fecher (eds.), *Spintronics*, DOI 10.1007/978-90-481-3832-6_3,

This chapter is focusing on the structural properties of Heusler compounds. A detailed description of their crystal structure is given. Furthermore, the most prominent types of atomic disorder and their influence of the electronic properties are discussed. Finally, different experimental methods, which are essential for a thorough structural analysis of Heusler compounds, are illustrated.

3.2 Crystal Structure

Heusler compounds X_2YZ crystallize in the cubic space group $Fm\bar{3}m$ (225) with Cu_2MnAl ($L2_1$) as prototype. The X atoms occupy the Wyckoff position 8c (1/4, 1/4, 1/4), the Y and the Z atoms are located at 4a (0, 0, 0) and 4b (1/2, 1/2, 1/2), respectively. This structure consists of four interpenetrating *fcc* sublattices, two of which are equally occupied by X. A rock salt-type lattice is formed by the elements with largest difference in electronegativity (Y and Z). These elements are coordinated octahedrally, which is electrostatically favorable, due to the ionic character of their interaction. All tetrahedral holes are filled by X. This structure can also be understood as a zinc blende-type sublattice, build up by one X and Z, the second X occupies in the remaining tetrahedral holes, and Y is located in the octahedral holes. These relations are illustrated in Fig. 3.1. In the literature, Heusler compounds are often described by a CsCl-like superstructure. This can be understood, if the unit cell edges of the Heusler cell are shifted by (1/4, 1/4, 1/4) with respect to the $Fm\bar{3}m$ cell. The combination of both X-site *fcc* lattices leads to a simple cubic lattice. The Y and the Z atoms alternatingly occupy the centers of the simple cubic lattice, which results in the CsCl-like superstructure. The shifted Heusler cell, as well as the CsCl-structure, are displayed in Fig. 3.2.

In addition to the structure described above, an inverse Heusler structure is observed, if the atomic number of Y is higher than the one of X from the same period ($Z(Y) > Z(X)$), however, it may also appear in compounds with transition metals from different periods. In all cases, the element X is more electropositive than Y. Consequently, X and Z form a rock salt lattice to achieve an octahedral coordination for X. The remaining X atoms and Y atoms fill the tetrahedral holes with fourfold symmetry. This structure is still described by four interpenetrating *fcc* sublattices, the X atoms, however, do not form a simple cubic lattice. They are placed on the Wyckoff positions 4a (0, 0, 0) and 4d (3/4, 3/4, 3/4), while the Y and the Z atoms are located at 4b (1/2, 1/2, 1/2) and 4c (1/4, 1/4, 1/4), respectively. The prototype of this structure is $CuHg_2Ti$ with space group $F\bar{4}3m$ (116). It is possible to emphasize the difference to regular Heusler compounds by expressing the formula as $(XY)X'Z$. A well-studied example is the compound Mn_2CoSn or (MnCo)MnSn [8, 9]. In case of quaternary Heusler compounds, there are two different elements X and X'. They are located at the 4a and 4d position, respectively, Y is placed on 4b and Z on 4c. This structure has the prototype LiMgPdSn. An illustration of the inverse Heusler structure and the quaternary variant is given in Fig. 3.3.

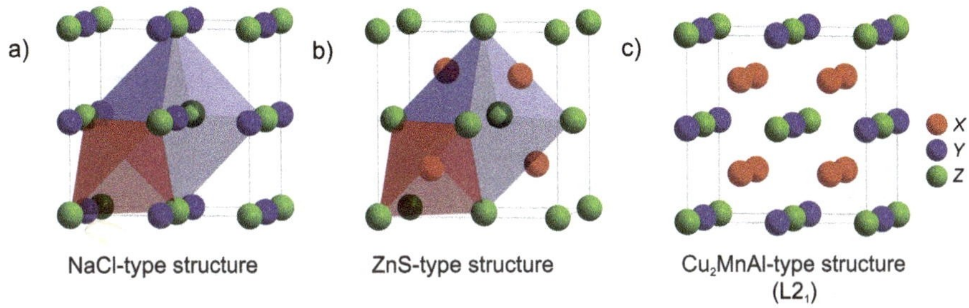

Fig. 3.1 (**a**) Rock salt-type structure, (**b**) zinc blende-type structure, and (**c**) Heusler structure

Fig. 3.2 (**a**) CsCl structure and (**b**) the Heusler structure which is shifted by (1/4, 1/4, 1/4) with respect to the standard cell to make the CsCl superstructure visible. Reprint with kind permission of Elsevier

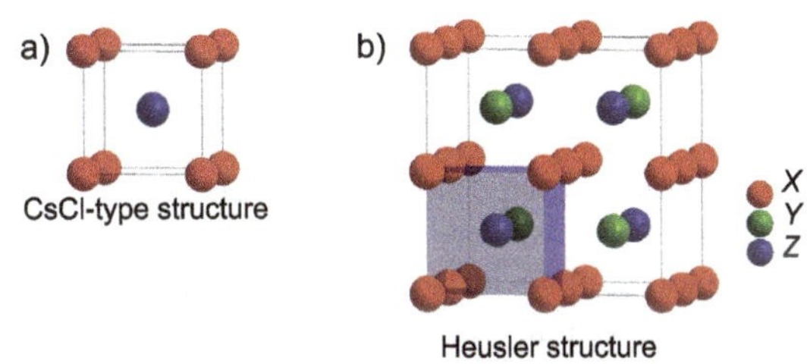

Fig. 3.3 (**a**) The inverse Heusler structure $CuHg_2Ti$ and (**b**) the quaternary version LiMgPdSn. Reprint with kind permission of Elsevier

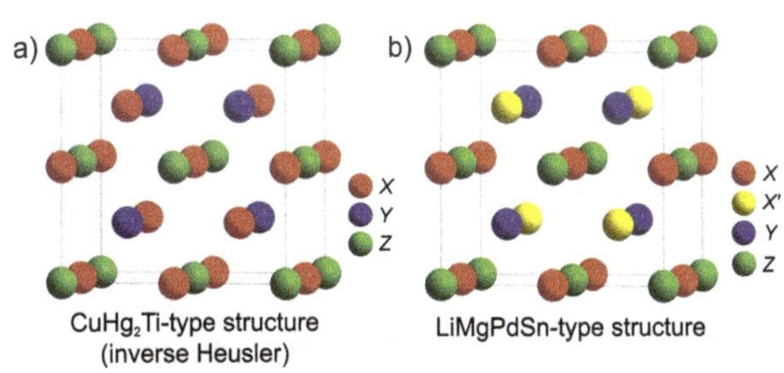

3.3 Tetragonal Distortion of Heusler Compounds

In addition to the well-known cubic structures of Heusler compounds, tetragonally distorted Heusler compounds have recently attracted great scientific interest in the field of spintronics, especially for spin-torque applications [10, 11]. A tetragonal distortion is observed for $Mn_{3-x}Ga$ compounds. The Mn atoms occupy two different lattice sites, one with tetragonal and one with octahedral coordination. Theoretical investigations by Kübler showed that the Mn atom on the octahedral site is formally in an oxidation state of +3 (Mn^{3+}, d^4) [7]. The electronic configuration for a single d^4 high spin ion in an octahedral environment, according to crystal field theory, is displayed in Fig. 3.4(a). The triple degenerated t_{2g} orbitals and one of the double degenerated e_g orbitals are occupied with single electrons. In fact, this electron configuration is energetically not favored, and energy can be gained by a distortion of the octahedron. Both, an elongation and a compression are possible, as shown in Figs. 3.4(b) and (c). These distortions lead to a lowering of the occu-

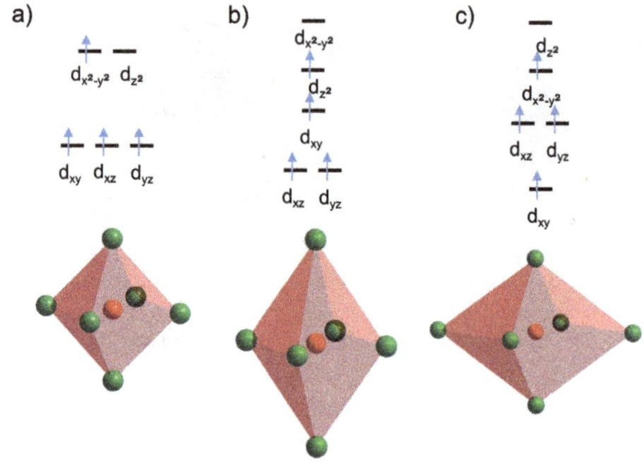

Fig. 3.4 Crystal field splitting for a d^4 ion in an octahedral coordination sphere: (**a**) non-distorted octahedron, (**b**) elongated octahedron, (**c**) compressed octahedron. The distortion in (**b**) and (**c**) is also known as Jahn–Teller distortion. Reprint with kind permission of Elsevier

pied orbitals, which results in an energy gain. The described model is well-known as Jahn–Teller distortion. In the case of $Mn_{3-x}Ga$ compounds, the cubic unit cell undergoes an elongation along the *c* axis, as shown in Fig. 3.5(a) [12, 13]. Consequently, the symmetry of the crystal changes form the cubic space group $F\bar{4}3m$ to the tetragonal space group $I4/mmm$. Figures 3.5(b) and (c) illustrate the relation between the tetragonal and the cubic unit cell. The tetragonal unit cell can be placed into the cubic cell, the cell edges, however, are rotated by 45°. The resulting tetragonal structure, with the prototype Al_3Ti, is displayed in Fig. 3.5(d). The Mn atoms are located at the Wyckoff position 2b (0, 0, 1/2), and at 4d (0, 1/2, 1/4). The Ga atom occupies the 2a position (0, 0, 0).

3.4 Order–Disorder Phenomena

The properties of Heusler compounds are strongly dependent on the atomic order. Band structure calculations show that already small amounts of disorder within the distribution of the atoms on the lattice sites cause distinct changes in their electronic structure, and thus also in their magnetic and transport properties [14–16]. Therefore, a careful analysis of their crystal structure is essential to understand the structure-to-property relation of Heusler compounds.

Figure 3.6 shows the transition from the ordered to the most prominent disordered Heusler structures, which will be explained in the following [17–21]: If the Y and the Z atoms are evenly distributed, the 4a and 4b positions become equivalent. This leads to a CsCl-like structure, also known as B2-type disorder. The symmetry is reduced and the resulting space group is $Pm\bar{3}m$. The random distribution of X and Y or X and Z leads to a BiF_3-type disorder (Space group no. 216: $Fm\bar{3}m$, DO_3). Different from these types of disorder, the NaTl-type structure is observed very rarely. In this structure type the X-atoms, which occupy one of the *fcc* sublattices, are mixed with the Y atoms, whereas the X atoms on the second sublattice are mixed with the Z atoms. This kind of disorder is also known as B32a disorder

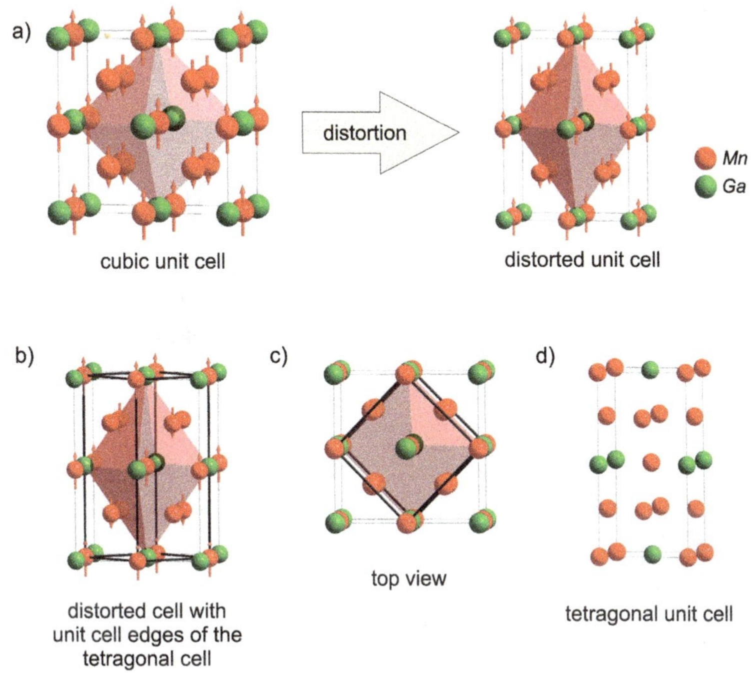

Fig. 3.5 Relation between the inverse cubic Heusler cell and a tetragonal distorted cell for Mn_3Ga. (**a**) Transition of the cubic Heusler cell to a tetragonal distorted cell with elongation along the c axis, (**b**) the unit cell edges of the tetragonal unit cell are marked on the cubic cell, (**c**) top view of the 45° rotation between the cubic and the tetragonal unit cell, (**d**) tetragonal unit cell with space group $I4/mmm$. Please, note that the Mn atoms are coupled ferrimagnetically as indicated by the *arrows*

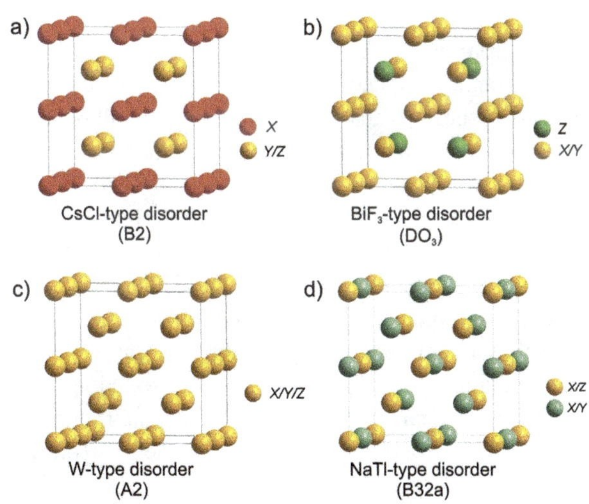

Fig. 3.6 Overview of the different types of disorder occurring in the Heusler structure: (**a**) CsCl-type disorder, (**b**) BiF_3-type disorder, (**c**) NaTl-type disorder, and (**d**) tungsten-type disorder. Reprint with kind permission of Elsevier

Table 3.1 Site occupancy and general formula for different atomic orders of Heusler compounds. The notations according to the Inorganic Crystal Structure Database (ICSD), the Strukturberichte (SB), the Pearson database, as well the space group are given

Site occupancy	General formula	Structure type ICSD	SB	Pearson	Space group
X, X', Y, Z	$XX'YZ$	LiMgPdSn	Y	cF16	$F\bar{4}3m$ (no. 216)
$X = X, Y, Z$	X_2YZ	Cu_2MnAl	$L2_1$	cF16	$Fm\bar{3}m$ (no. 225)
$X, X' = Y, Z$	XX'_2Z	$CuHg_2Ti$	X	cF16	$F\bar{4}3m$ (no. 216)
$X = X' = Y, Z$	X_3Z	BiF_3	DO_3	cF16	$Fm\bar{3}m$ (no. 225)
$X = X', Y = Z$	X_2Y_2	CsCl	B2	cP2	$Pm\bar{3}m$ (no. 221)
$X = Y, X' = Z$	$X_2X'_2$	NaTl	B32a	cF16	$Fd\bar{3}m$ (no. 227)
$X = X' = Y = Z$	X_4	W	A2	cI2	$Im\bar{3}m$ (no. 229)

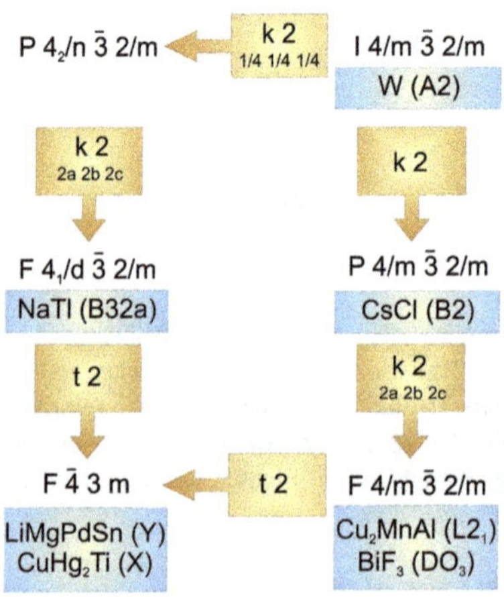

Fig. 3.7 Bärnighaus tree for Heusler compounds which illustrates the group-subgroup relations between different ordering variants. The indices of *klassengleiche* (k) and *translationsgleiche* (t) as well as the unit cell transformations and the origin shifts are given

(Space group no. 227, $Fd\bar{3}m$). The X atoms are placed at the Wyckoff position 8a (0, 0, 0), and the Y and Z are randomly distributed at position 8b (1/2, 1/2, 1/2). In contrast to these partial disorder phenomena all positions become equivalent in the tungsten-type structure with a *bcc* lattice and reduced symmetry ($Im\bar{3}m$ (A2)).

Figure 3.7 shows the group–subgroup relationship between all possible types of simple disorder in the Heusler structure. The numbers behind t and k specify the index of the reduction in symmetry. One should notice that there are two atoms per unit cell for the tungsten-type and the CsCl structure. For all other structure types, there are 16 atoms per unit cell due the doubling of all cell axes.

Table 3.1 summarizes the different ordering variants of Heusler compounds. The site occupancy is correlated with the corresponding general formula. Different notations of the crystal structures according to the Inorganic Crystal Structure Database (ICSD), the Strukturberichte (SB), and the Pearson database, as well as the space group are given.

3.5 Relationship of Atomic Disorder and Spin Polarization

To understand the properties of Heusler compounds and to correlate experimental and theoretical results, a detailed study of the crystal structure of the compounds is essential. Already small amounts of anti-site disorder cause changes in the electronic structure close to the Fermi energy, which lead to altered physical properties.

The tunneling magnetoresistance effect (TMR) was observed for the first time at room temperature in magnetic tunnel junctions with CsCl-type (B2) disordered $Co_2Cr_{0.6}Fe_{0.4}Al$ (CCFA) electrodes [22]. To clarify the relationship between atomic disorder and the spin polarization, Miura et al. investigated the electronic structure of disordered $Co_2Cr_{1-x}Fe_xAl$ theoretically, based on first-principle density functional calculations with the Korringa–Kohn–Rostoker coherent-potential approximation [14, 23]. In the parent phase Co_2CrAl, the spin polarization remains high (more than 90 %), even for a complete interchange of Cr and Al. A detailed analysis of the density of states (DOS) and the atom orbital projected local density of states (LDOS) of Co 3*d* revealed that the energy gap of the minority DOS near the Fermi energy level is mainly constructed of Co 3*d* states. Thus, the disorder between Cr and Al does not significantly affect the electronic structure near the Fermi level, and therefore, the semi-conducting character of the minority bands is still kept even in the disordered B2 structure. The spin polarization of $Co_2Cr_{1-x}Fe_xAl$ is reduced with increasing the Fe concentration for both, ordered $L2_1$ and disordered B2 structures. For CCFA the spin polarization is evaluated as 90 % and 77 % for the $L2_1$ and B2-type structures, respectively [23]. Contrary to the Cr–Al interchange, the Co–Cr interchange gives rise to a considerable reduction of the spin polarization, due to the appearance of anti-side Co 3*d* states in the minority-spin band gap. The total magnetic moment is reduced linearly from 3 μ_B per formula unit with increasing amount of Co–Cr disorder. Wurmehl et al. showed that this reduction of the total magnetic moment can be attributed to ferrimagnetic order with an anti-parallel alignment of the anti-side Cr moments to the other magnetic moments on the ordinary Cr and Co sides [24].

Investigations on Co_2CrGa by first-principles calculations, using the linear muffin-tin orbital (LMTO) method, combined with the atomic sphere approximation (ASA) and the coherent sphere approximation (CPA), yield similar results [25]. A spin polarization of 95 % and 84 % was determined for the $L2_1$ ordered and the B2 disordered structures, respectively. Differential scanning calorimetry measurements on bulk Co_2CrGa reveal that the order-disorder ($L2_1$–B2) phase transition temperature is 1050 K, which is much higher than the Curie temperature ($T_C = 495$ K). Metallurgically, it is important to note that the $L2_1$ ordered phase of Co_2CrGa is much more stable, compared with that of $Co_2Cr_{1-x}Fe_xAl$.

In-gap states arising from anti-site Co defects were also observed in Co_2MnSi and Co_2MnGe by Picozzi et al. on the basis of the full-potential linearized augmented plane-wave (FLAPW) method [16]. Due to their low formation energies, both, Mn and Co anti-sites are likely to be formed in a concentration as high as 8 %; on the other hand, atomic swaps (such as Co–Mn and Mn–Si) have lower defect densities. Half-metallicity, in fact, is preserved in all cases of atomic disor-

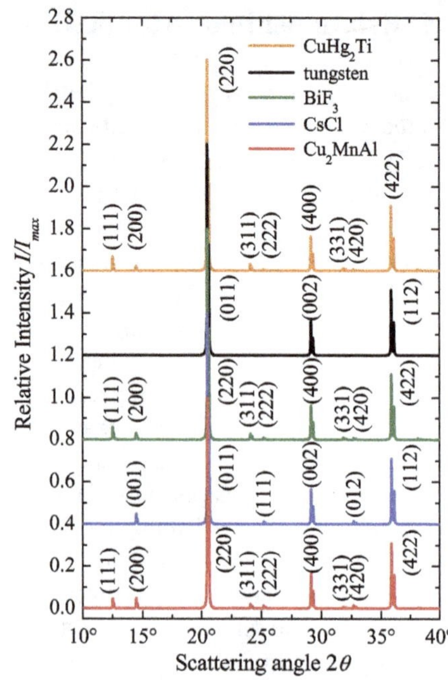

Fig. 3.8 Theoretical XRD patterns for Co_2MnSi under the assumption of different crystal structures

der, except for the Co antisites, where a defect-induced peak arises at the Fermi level.

3.6 Structure Determination

In the laboratory, powder X-ray diffraction (XRD) is the easiest experimental method to check both structure and purity of a sample. In case of Heusler compounds this is often not sufficient to exclude certain types of anti-side disorder: Theoretical X-ray diffraction patterns of Co_2MnSi, which were simulated under the assumption of different crystal structures, are displayed in Fig. 3.8 [21]. The ordered Cu_2MnAl-type structure is identified by the occurrence of the *fcc*-typical (111) and (200) reflections, and their relation to the (220) reflection. However, the intensity of these two *fcc*-typical reflections is very low in many of the investigated compounds. This is, in particular, the case, if all elements contained in the compound have a similar atomic number. The intensity of the (111) and the (200) reflections can then be below 1 % of the scattered intensity of the (220) reflection, which almost leads to the disappearance of the (111) and the (200) reflections. Table 3.2 provides a survey of the relative intensities of the reflexes for the XRD patterns displayed in Fig. 3.8. Comparing the Cu_2MnAl-type and the BiF_3-type structure, a difference in the intensities of the (111) and (200) reflections is evident. In contrast, the Cu_2MnAl and the $CuHg_2Ti$ structure are hardly distinguishable by X-ray

Table 3.2 Relative intensities of the reflections for Co_2MnSi in different structures

Structure	(111)	(200)	(220)	(311)	(222)	(400)
Cu_2MnAl	4.30	4.67	100	2.01	1.30	16.46
CsCl	–	4.67	100	–	1.30	16.46
BiF_3	0.79	0.41	100	0.33	0.09	16.58
Tungsten	–	–	100	–	–	16.58
$CuHg_2Ti$	6.53	2.27	100	3.26	0.57	16.46
NaTl	4.38	–	100	2.26	–	16.46

diffraction. Much care has to be taken in the structural analysis, as both have general *fcc*-like symmetry. In the CsCl-type structure only the (111) reflex vanishes, whereas in the NaTl-type structure the (200) reflection disappears. In the tungsten-type structure, however, both, the (111) and (200) reflections disappear. Of course, most samples are not completely ordered or disordered, small amounts of disorder in an ordered compound or a low degree of order in a disordered structure may also occur. Unfortunately, this is not always detectable by standard XRD methods. In such cases, anomalous XRD investigations with synchrotron radiation can be performed. This allows for the direct observation of anti-site disorder, e.g. Co occupation on the Mn sublattice could be quantified directly in the case of Co_2MnGe thin films [26].

The quaternary alloy $Co_2FeAl_{1-x}Si_x$ is one example, in which increasing order is observed with the substitution of Si for Al [27]. Co_2FeAl is CsCl-type disordered, i.e. Fe and Al atoms are randomly distributed. With increasing Si concentration, the amount of Cu_2MnAl-type order increases, as indicated by the increasing intensity of the (111) reflection (compare Fig. 3.9(a)). Differential scanning calorimetry measurements revealed that the phase transition temperature decreases with increasing Si concentration as shown in Fig. 3.9(b). Together with band structure calculations, these results lead to the conclusion that a composition of 50 % Al and 50% Si yields a stable and well ordered half-metallic ferromagnet, and therefore, this material is now used in magnetic tunneling junctions [28].

A very useful method to further investigate the atomic disorder is spin echo nuclear magnetic resonance (NMR). These measurements are able to probe the direct local environments of the active atoms. NMR investigations provide a tool to obtain the local environment by measuring the resonance frequencies, and consequently, probing the local hyperfine magnetic fields. Thus, the first neighboring shells of the active atoms are revealed [29–31]. The quaternary substitution series $Co_2Mn_{1-x}Fe_xSi$, which was theoretically predicted to show half-metallic ferromagnetism, was studied using the NMR technique by Wurmehl et al. [32, 33]. The local environment of ^{55}Mn nuclei was investigated in detail, and the obtained ^{55}NMR spectrum of $Co_2Mn_{0.5}Fe_{0.5}Si$ is displayed in Fig. 3.10. In agreement with the expected random distribution of Mn and Fe on the 4b position, several resonance lines are found. Each line can be correlated with the one particular Fe configuration in the third Mn coordination shell. This important precondition for quaternary alloys

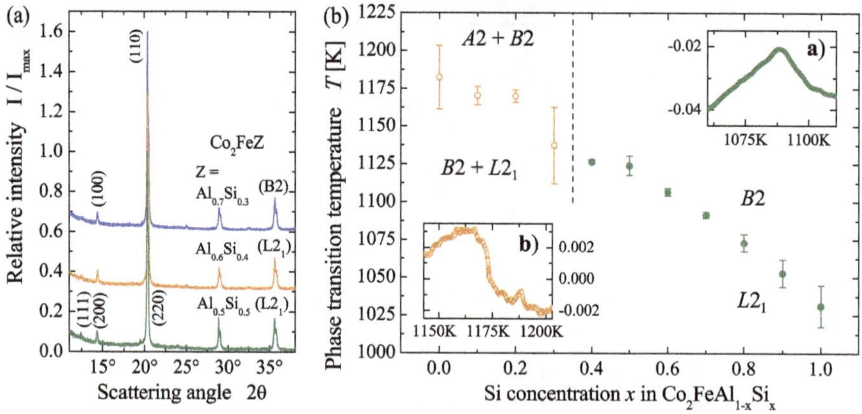

Fig. 3.9 (**a**) Powder diffraction of $Co_2FeAl_{1-x}Si_x$. Shown are the powder patterns measured with Mo Kα at room temperature for selected compositions with $x = 0.3, 0.4$, and 0.5. (**b**) Phase transitions in $Co_2FeAl_{1-x}Si_x$. Shown is the composition dependence of the phase transition temperature. The length of the vertical bars corresponds to the experimental hysteresis. The insets (**a**) and (**b**) display typical DSC curves in low (0.1) and high (0.7) Si content compounds, respectively. Data taken from Ref. [27]. Reprint with kind permission of Elsevier

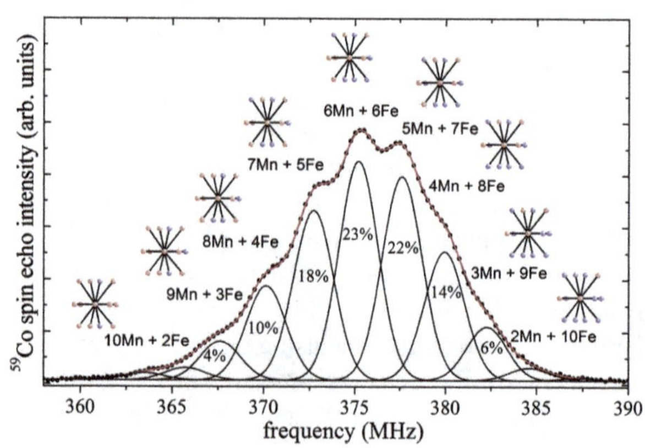

Fig. 3.10 Spin echo intensity as a function of frequency for ^{55}Mn in $Co_2Mn_{0.5}Fe_{0.5}Si$ (*black dots*) and the fit (*red line*) resulting from Gaussians (*black lines*). The distribution of Mn and Fe atoms in the third coordination shell of the ^{55}Mn, as well as the relative areas obtained from Gaussians are given for each line. Data taken from Ref. [32]. Reprint with kind permission of Elsevier

was proved by the NMR method, and thus, these compounds may show stable half-metallicity, including a very high spin polarization. In the case of Co_2FeSi thin films, which were determined by XRD to display the ordered Cu_2MnAl-type structure, the ^{59}Co NMR experiment has shown Fe anti-sites on the Si sublattice, which corresponds to CsCl-type disorder [34].

Additionally, Mößbauer spectroscopy is a useful characterization technique, which measures the hyperfine field at the core of Mößbauer active atoms (commonly ^{57}Fe). The nature of the obtained the spectra give an indication on the local environment of the probed atomic species. As an example, the $Co_{(2-x)}Fe_{(1+x)}Si$ series exhibits almost identical XRD patterns (see Fig. 3.11(a)), which impedes a clear differentiation between $CuHg_2Ti$ and the Cu_2MnAl-type structure. A definite

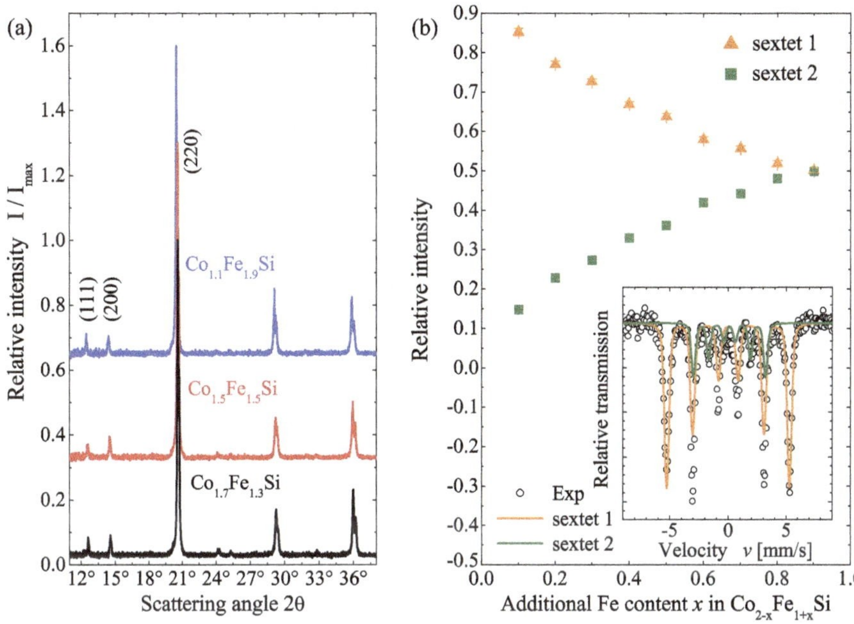

Fig. 3.11 (**a**) Powder XRD of $Co_{2-x}Fe_{1+x}Si$. Shown are the powder pattern measured with Mo Kα at room temperature for selected compositions with $x = 0.2$, 0.5, and 0.9. (**b**) ^{57}Fe Mößbauer spectra of $Co_{2-x}Fe_{1+x}Si$. Shown are the relative intensities of the two hyperfine magnetic fields related to the sextets I and II. The *inset* displays the ^{57}Fe Mößbauer spectrum of $Co_{1.8}Fe_{1.2}Si$. Data taken from Ref. [35]. Reprint with kind permission of Elsevier

assignment, however, is possible using Mößbauer spectroscopy [35]. All Mößbauer spectra exhibit two sextets, which means that the Fe atoms are present in two different magnetic environments. Figure 3.11(b) displays the spectrum of $Co_{1.8}Fe_{1.2}Si$ as well as the relative intensities of both sextets. It can be seen that the intensity of the second sextet increases with increasing Fe content, because the additional Fe atoms occupy the 4d lattice position. This leads to the conclusion that Co_2FeSi and $CoFe_2Si$ crystallize in the Cu_2MnAl-type ($L2_1$) and the $CuHg_2Ti$-type structure, respectively.

For the series $Co_2Mn_{1-x}Fe_xAl$, the existence of a short range order was detected, although XRD studies indicated a complete disorder on the Y and Z positions [36]. To get an insight into the structural and magnetic properties of thin films, ^{57}Fe conversion electron Mößbauer spectroscopy (CEMS) is a very powerful method. CEMS spectra of as-deposited epitaxial $Co_2Cr_{0.6}Fe_{0.4}Al$ (CCFA) thin films with a Fe buffer layer show distinguishable subspectra originating from the CCFA and the Fe buffer layer (Fig. 3.12(a)) [37]. Figures 3.12(b)–(d) present the spectra of thin films without Fe buffer after annealing at 450, 550, and 600 °C, respectively. Spectra of samples with a buffer layer have broader lines and look similar, except for the presence of a Fe subspectrum, which is reduced by increasing annealing temperature, and finally disappears after annealing above 600 °C. This result indi-

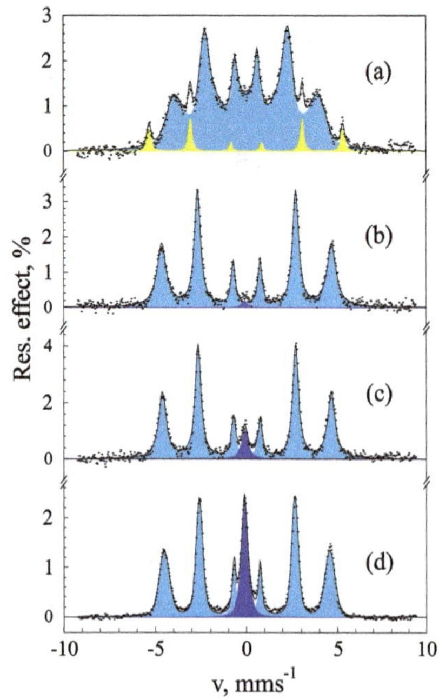

Fig. 3.12 ^{57}Fe CEMS spectra of 100 nm CCFA thin films (*light blue*) deposited on MgO substrates with a 10 nm Fe buffer layer (*yellow*) without annealing (**a**); spectra of 100 nm CCFA thin films (*light blue*) deposited on MgO substrates without a buffer layer annealed at (**b**) 450 °C, (**c**) 550 °C, and (**d**) 600 °C. The central peak *dark blue* corresponds to Fe atoms occupying Co sites. Data taken from Ref. [37]. Reprint with kind permission of Elsevier

cates a diffusion of Fe from the buffer layer into the CCFA thin film, and apparently the related diffusion of Cr atoms from the CCFA thin film into the Fe buffer layer occurs simultaneously. The central peak, which is observed in the spectra (b)–(d), results from Fe in a nonmagnetic surrounding, i.e. from Fe atoms substituting Co atoms in the B2-type structure. Summarizing, CEMS studies demonstrate that annealing of CCFA thin films (i) monotonously improves their crystallographic order in the temperature range of 450–700 °C, (ii) causes diffusion of Fe atoms from the Fe buffer layer into the CCFA, changing its composition, and (iii) favors anti-site Co_2Fe disorder.

Furthermore, extended X-ray-absorption fine structure (EXAFS) is a sensitive method to determine short range chemical order around atoms, which has also been successfully applied to investigate Heusler compounds. In Co_2MnSi, a distinct amount of disorder between the Co and the Mn sites has been revealed, which is an explanation for the rather low measured spin polarization of 55 %, although band structure calculations predict half-metallic ferromagnetism for the ordered compound [38]. Another example that demonstrates the enormous potential of this characterization technique for Heusler compounds is displayed by the Co_2FeZ (Z = Al, Si, Ga, Ge) system. For Z = Al and Si the complete structure could be revealed by XRD. However, for Z = Ga and Ge XRD studies showed only a cubic structure without the degree of ordering (compare Fig. 3.13(a)) [39]. EXAFS measurements were performed at the Co K edges and the Fe K edges. The spectra could be fitted with a CsCl-like disordered structure for Co_2FeAl, but for the remaining samples the fitting procedure yielded well ordered structures as shown in Figs. 3.13(b) and (c).

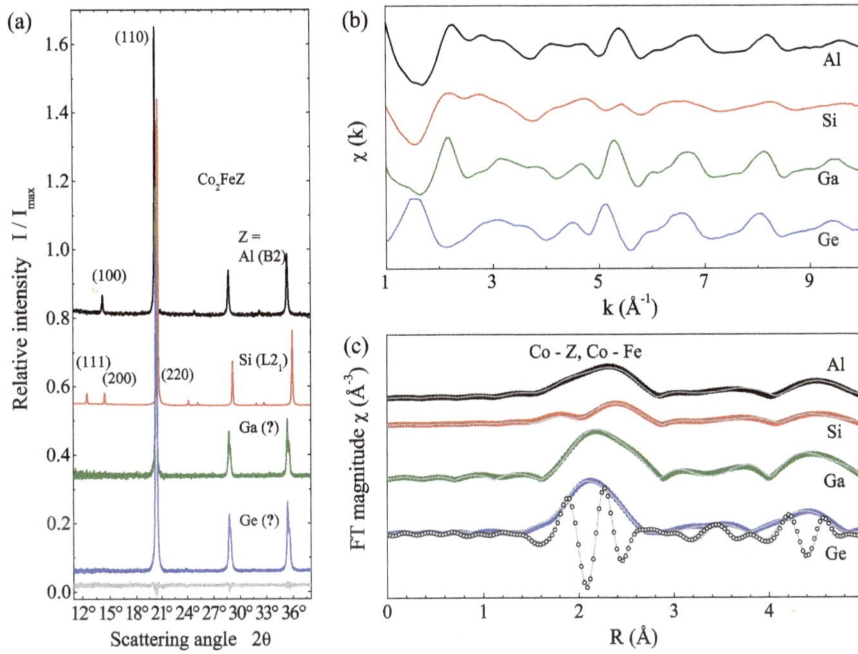

Fig. 3.13 (**a**) XRD of Co_2FeZ with Z = Al, Si, Ga, Ge. The XRD pattern have been excited by Mo Kα radiation. Note the different indexing of the reflections in the *sc* (Z = Al) and *fcc* (Z = Si) crystal systems. The line below Z = Ge is the difference between the measured data and the $L2_1$ Rietveld refinement for Co_2FeGe. (**b**) EXAFS oscillations extracted from the X-ray absorption measurements at the Co K edge. (**c**) Corresponding Fourier transforms (symbols) and best fitting results (*gray line*). The imaginary part of the Fourier transform is displayed for the Co_2FeGe compound (*open circles*). Data taken from Ref. [39]. Reprint with kind permission of Elsevier

3.7 Conclusion

The crystal structure of Heusler compounds is reviewed from various viewpoints. It is emphasized that the electronic properties depend strongly on the atomic order of the material. Since disordered structures are frequently observed, a careful structure analysis is inevitable to understand the structure to property relations. In many cases, standard XRD methods are not sufficient to determine the atomic arrangement in Heusler phases. Therefore, additional methods have to be applied, as described above. It should be noted that, in some cases, disorder effects do not sustainably perturb the electronic structure. To clarify the influence of the crystal structure on the electronic states, detailed theoretical investigations for each specific type of material need to be performed. Well-controlled synthetic methods can circumvent problems associated with atomic disorder and open the door to innumerable applications.

Acknowledgements The authors are grateful to B. Balke, C.G.F. Blum, F. Casper, G.H. Fecher, V. Jung, V. Ksenofontov, J. Winterlik, and S. Wurmehl for providing data and for fruitful discussions. Financial support by the Deutsche Forschungsgemeinschaft (research unit 559) and by the Graduate School of Excellence "Material Science in Mainz" is gratefully acknowledged.

References

1. Heusler F (1903) Verh Dtsch Phys Ges 5:219
2. Heusler F, Starck W, Haupt E (1903) Verh Dtsch Phys Ges 5:220
3. Heusler O (1934) Ann Phys 19:155
4. Bradley AJ, Rodgers JW (1934) Proc R Soc A 144:340
5. Webster PJ (1971) J Phys Chem Solids 32:1221
6. Brooks JS, Williams JM (1975) Phys Status Solidi A 32:413
7. Kübler J, Williams AR, Sommers CB (1983) Phys Rev B 28:1745
8. Surikov VV, Zhordochkin VN, Astakhova TY (1990) Hyperfine Interact 59:469
9. Lakshmi N, Pandey A, Venugopalan K (2002) Bull Mater Sci 25:309
10. Wu F, Mizukami S, Watanabe D, Naganuma H, Oogane M, Ando Y, Miyazaki T (2009) Appl Phys Lett 95:122503
11. Wu F, Mizukami S, Watanabe D, Sajitha EP, Naganuma H, Oogane M, Ando Y, Miyazaki T (2010) IEEE Trans Magn 46:1863
12. Balke B, Fecher GH, Winterlik J, Felser C (2007) Appl Phys Lett 90:152504
13. Winterlik J, Balke B, Fecher GH, Felser C (2008) Phys Rev B 77:054406
14. Miura Y, Nagao K, Shirai M (2004) Phys Rev B 69:144413
15. Kandpal HC, Ksenofontov V, Wojcik M, Seshadri R, Felser C (2007) J Phys D, Appl Phys 40:1587
16. Picozzi S, Continenza A, Freeman AJ (2004) Phys Rev B 69:094423
17. Webster PJ (1969) Contemp Phys 10:559
18. Bacon GE, Plant JS (1971) J Phys F, Met Phys 1:524
19. Webster PJ, Ziebeck KRA (1988) Landolt-Börnstein new series group III, vol 19c, p 75
20. Ziebeck KRA, Neumann KU (2001) Landolt-Börnstein new series group III, vol 32c, p 64
21. Graf T, Casper F, Winterlik J, Balke B, Fecher GH, Felser C, Anorg Z (2009) Z Anorg Allg Chem 635:976
22. Inomata K, Okamura S, Goto R, Yezuka N (2003) Jpn J Appl Phys 42:L419
23. Miura Y, Shirai M, Nagao K (2004) J Appl Phys 95:7225
24. Wurmehl S, Fecher GH, Kroth K, Kronast F, Dürr HA, Takeda Y, Saitoh Y, Kobayashi K, Lin HJ, Schönhense G, Felser C (2006) J Phys D, Appl Phys 39:803
25. Umetsu RY, Kobayashi K, Kainuma A, Fujita R, Fukamichi K, Ishida K, Sakuma A (2004) Appl Phys Lett 85:2011
26. Ravel B, Cross JO, Raphael MP, Harris VG, Ramesh R, Saraf V (2002) Appl Phys Lett 81:2812
27. Balke B, Fecher GH, Felser C (2007) Appl Phys Lett 90:242503
28. Tezuka N, Ikeda N, Miyazaki A, Sugimoto S, Kikuchi M, Inomata K (2006) Appl Phys Lett 89:112514
29. Schaf J, Campbell IA, Le Dang K, Veillet P (1983) J Magn Magn Mater 36:310
30. van Roy W, Wojcik M, Jedryka E, Nadolski S, Jalabert D, Brijs B, Borghs G, De Boeck J (2003) Appl Phys Lett 83:4214
31. Wieldraaijer H, de Jonge WJM, Kohlhepp JT (2005) Phys Rev B 72:155409
32. Wurmehl S, Kohlhepp JT, Swagten HJM, Koopmans B, Wojcik M, Balke B, Blum CGF, Ksenofontov V, Fecher GH, Felser C (2007) Appl Phys Lett 91:052506
33. Wurmehl S, Kohlhepp JT, Swagten HJM, Koopmans B, Wojcik M, Balke B, Blum CGF, Ksenofontov V, Fecher GH, Felser C (2008) J Appl Phys 103:07D706
34. Inomata K, Wojcik M, Jedryka E, Ikeda N, Tezuka N (2008) Phys Rev B 77:214425
35. Jung V, Balke B, Fecher GH, Felser C (2008) Appl Phys Lett 93:042507

36. Jung V, Fecher GH, Balke B, Ksenofontov V, Felser C (2009) J Phys D, Appl Phys 42:084007
37. Ksenofontov V, Herbort C, Jourdan M, Felser C (2008) Appl Phys Lett 92:262501
38. Ravel B, Raphael MP, Harris VG, Huang Q (2002) Phys Rev B 65:184431
39. Balke B, Wurmehl S, Fecher GH, Felser C, Alves MCM, Bernardi F, Morais J (2007) Appl Phys Lett 90:172501

Chapter 4
Substitution Effects in Double Perovskites: How the Crystal Structure Influences the Electronic Properties

Haitao Gao, Alexandra Jung, Irene Bonn, Vadim Ksenofontov, Sergey Reiman, Claudia Felser, Martin Panthöfer, and Wolfgang Tremel

Abstract We systematically studied substituted Sr_2FeReO_6 with respect to experimental characterization and theoretical band structure calculations. In the framework of the tight-binding approach, hole- or electron-doping of $Sr_2MM'O_6$ were performed at the M or M' positions either by transition or main group metals. Hole-doping, rather than electron-doping, has a favorable effect to improve the half-metallicity (Curie temperature and saturation magnetization) of the parent compound. When M is substituted by another metal, the original M' metal will serve as a redox buffer (and vice versa). Substituting M by another metal with a size similar to that of the metal at M' position causes disorder, which has high impact on the properties of the starting compound. Main group metals block the super-exchange pathways that underlie the half-metallic properties in Sr_2FeReO_6. Thus a Mott-insulating and spin-frustrated state is produced in an ordered phase due to the geometrical arrangement, e.g. in Sr_2InReO_6. However, M/M' disorder is significant in the main group elements containing double perovskites, this triggers electronic conductivity arising from electron hopping from Re to adjacent Re ions as observed in Sr_2GaReO_6.

4.1 Introduction

After the discovery of the powder magnetoresistance effect in Sr_2FeReO_6 at room temperature and low magnetic fields, Sr_2FeReO_6 has indicated the potential of compounds from the so-called double perovskite group for spintronic applications [1, 2]. The extreme flexibility of the double perovskite structure type, in terms of structure/symmetry and the elements involved, allows the substitution of Sr, Fe, and Re by a wide variety of metal cations without the risk of a phase separation. Therefore, by substituting the metal components of the parent Sr_2FeReO_6 with elements that

H. Gao · A. Jung · I. Bonn · V. Ksenofontov · S. Reiman · C. Felser · M. Panthöfer · W. Tremel (✉)
Institut für Anorganische Chemie und Analytische Chemie, Johannes Gutenberg – Universität, 55099 Mainz, Germany
e-mail: tremel@uni-mainz.de

C. Felser, G.H. Fecher (eds.), *Spintronics*, DOI 10.1007/978-90-481-3832-6_4,

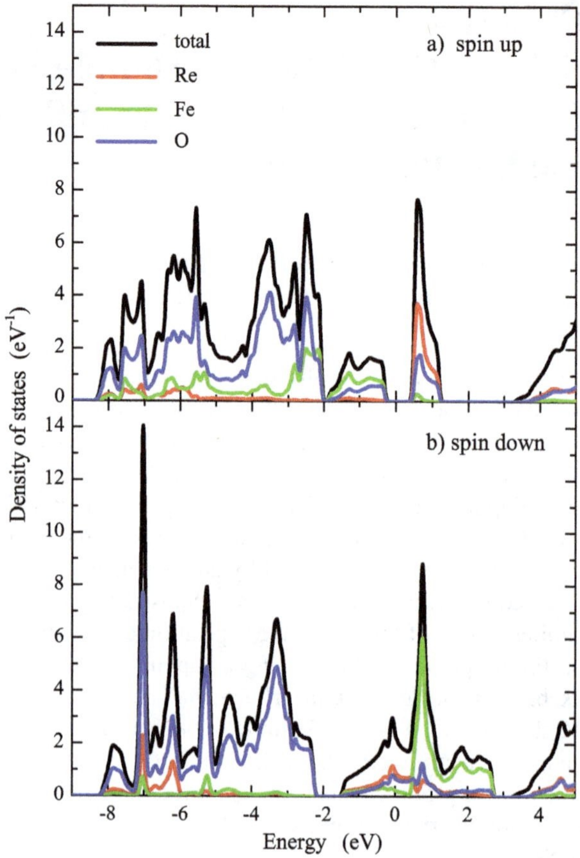

Fig. 4.1 Density of states (DOS) of Sr_2FeReO_6

provide a different number of valence electrons, the magnetization, Curie temperature (T_C) and PMR effect can be varied in a systematic manner.

Sr_2FeReO_6 crystallizes in the space group *I4/m*. The tolerance factor $t = 0.905$ for Sr_2FeReO_6 would indicate an ideal cubic or a tetragonal distorted double perovskite structure [3]. The size difference of the Fe and Re cations ($r(Fe^{3+}) = 65$ pm, $r(Re^{5+}) = 58$ pm) prevents antisite disorder in Sr_2FeReO_6 [4].

Sr_2FeReO_6 was reported to be a half-metallic ferrimagnet with 100 % spin polarized charge carriers [2, 5–8]. Due to the boundary condition of charge neutrality, i.e. $n + m = 8$ for the valence states Fe^{n+} and Re^{m+}, seven electrons are distributed within the Fe and Re *d* states. In the spin-up direction, the Fe 3*d* states are filled with five electrons, and the corresponding Re-5*d* states are empty (Fig. 4.1). In contrast to the band gap in the spin-up direction, Sr_2FeReO_6 is metallic in the spin-down direction. The two remaining electrons are delocalized within the Fe/Re t_{2g} conduction band. The electrons localized on the Fe atoms and the itinerant electrons distributed over the Fe and Re sites are coupled antiferromagnetically. This arrangement is mediated by means of a Zener double exchange along the -Fe-O-Re- O-Fe- units, which leads to an overall ferrimagnetic behavior [9–11]. As a result, the valence states of Fe and Re are difficult to determine. NMR- and Mößbauer spectroscopic studies report unanimously that the valence state of Fe is between 2.5+ and 3+ [2, 5–8, 12–14]. The experimental saturation magnetization ($M_S = 2.7\ \mu_B$/f.u.) is in good agreement with the magnetization that is expected from the antiparallel alignment of the itinerant and localized electrons ($M_{S,calc} = 3\ \mu_B$/f.u.). The ferromagnetic to paramagnetic transition occurs at $T_C = 401$ K.

The low-field magnetoresistance effect of Sr_2FeReO_6 is based on a powder magnetoresistance effect, i.e. the properties of the grain boundaries determine the properties of the sample. Maximum MR values of 21 % at 4.2 K and 70 kOe were reported. Depending on the synthesis, Sr_2FeReO_6 was reported to exhibit metallic or semiconducting transport behavior [2, 5–8, 12, 13].

We have studied the substitution effect at the M or M' positions ($Sr_2MM'O_6$) on the electronic and magnetic properties in Sr_2FeReO_6. The results are presented in terms of substitution positions by main group or transition metals, and finally summarized using a scheme.

4.2 Substitution with Transition Metals

4.2.1 M Position

In doped double perovskites $Sr_2Fe_{1-x}M_xReO_6$ ($0 \leq x \leq 1$, M = Zn, Cr), strong correlations have been found between the electronic, structural and magnetic properties. The interplay between the van Hove singularity and the Fermi level plays a crucial role for the magnetic properties (Fig. 4.2). To reach the coincidence of the van Hove singularity and Fermi energy, the conduction band must be either filled or depleted. Thus, this electronic tuning may lead to a larger saturation magnetization due to the larger number of electronic states close to E_F, thereby enhancing the Curie temperature and the MR effect. As a clear example, electron-doping has been successful for Sr_2FeMoO_6, e.g. $Sr_{2-x}La_xFeMoO_6$ was reported to exhibit higher Curie temperatures [15]. In case of Sr_2FeReO_6, hole-doping is supposed to be favorable.

4.2.1.1 Cr-Series (Hole-Doping)

Hole-doping of the parent compound Sr_2FeReO_6 with Cr leads to a non-monotonic increase of the saturation magnetization for doping levels up to 10 %. The Curie temperatures monotonically increase from 401 to 616 K (Fig. 4.2(c)). Superimposed on the electronic effects is the structural influence which can be explained by size effects modeled by the tolerance factor t. For Cr substitution levels higher than 40 %, we observed a tetragonal–cubic transformation [16].

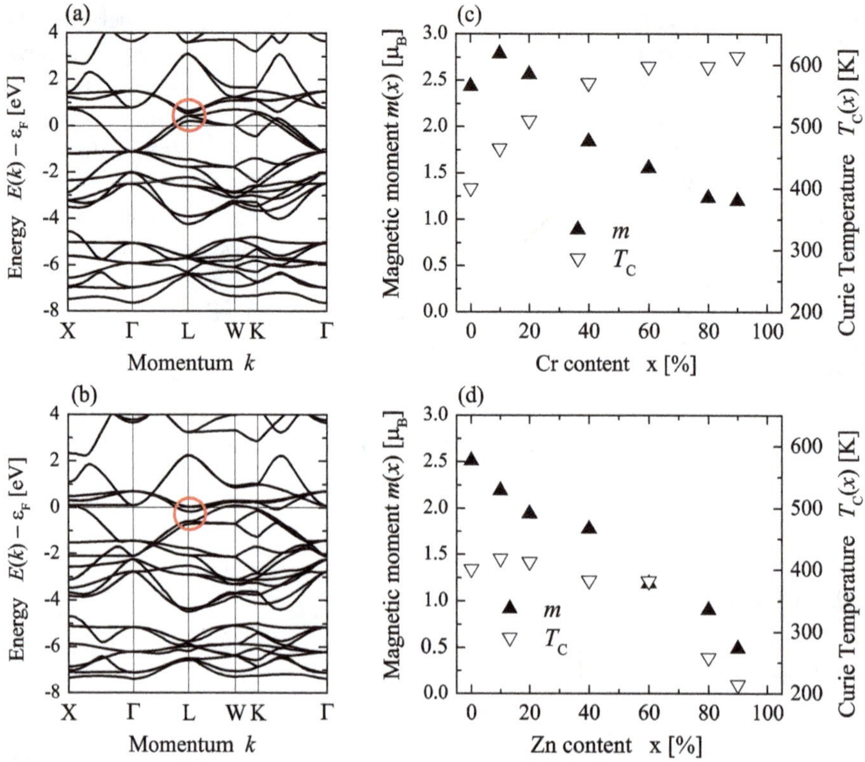

Fig. 4.2 Band structures of the metallic spin direction of (**a**) Sr_2FeReO_6, (**b**) Sr_2FeMoO_6; saturation magnetization of measured at 5 K (*black triangles*) and T_C measured at magnetic field of 0.1 T (∇) versus metal content: (**c**) $Sr_2Fe_{1-x}Cr_xReO_6$, (**d**) $Sr_2Fe_{1-x}Zn_xReO_6$

4.2.1.2 Zn-Series (Electron-Doping)

In contrast to Cr substitution, Zn (electron rich) substitution leads to a continuous decrease of the saturation magnetization and the Curie temperatures (Fig. 4.2(d)). For Zn-series samples the tetragonal distortion linearly increases with increasing Zn content. The valence state of iron is determined by Mößbauer spectroscopy to be +2.7. Upon doping, the Fe valence remains constant, suggesting that Re serves as a redox buffer [16].

4.2.2 M' Position

In Fe substituted (electron-doping) solid solutions of $Sr_2FeRe_{1-x}Fe_xO_6$ ($0 \le x \le 0.5$), the series undergoes a structural transition from the ordered parent phase ($x = 0$) to various highly disordered phases upon increasing degree of substitution. This situation is similar to that observed in $Sr_2FeRe_{1-x}Ga_xO_6$ (Fig. 4.3(a)).

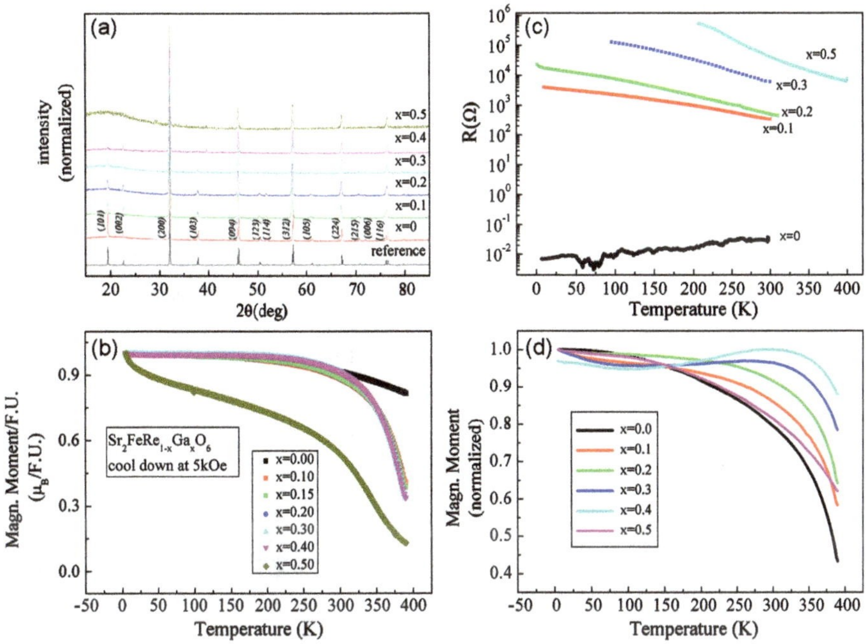

Fig. 4.3 (**a**) XRD powder patterns of $Sr_2FeRe_{1-x}Fe_xO_6$, (**b**) magnetic susceptibility of $Sr_2FeRe_{1-x}Ga_xO_6$, (**c**) conductivity measurement of $Sr_2FeRe_{1-x}Fe_xO_6$, (**d**) magnetic susceptibility of $Sr_2FeRe_{1-x}Fe_xO_6$

The (101) and (002) reflections in the XRD powder pattern of the tetragonal parent phase are indicative of M/M' ordering and vanish upon substitution. Fe-substituted solid solutions undergo magnetic transitions from ferro/ferri-magnetic to antiferromagnetic ordering, suggesting an increasing amount of -Fe-O-Fe- units (Fig. 4.3(d)). As shown in Fig. 4.3(c), upon substituting the metal M' in Sr_2FeReO_6, the samples become semiconducting (Fig. 4.3(c)). Mößbauer spectroscopy shows that the oxidation state of Fe increases gradually to 3+, and thereafter remains constant while Re serves as a valence buffer [17].

4.3 Substitution with Main Group Metals

4.3.1 M Position

4.3.1.1 Ga-Doping

In the case of $Sr_2Fe_{1-x}Ga_xReO_6$ $(0 \leq x \leq 0.7)$, an unexpected phase separation makes magnetic effects more complicated compared with a simple magnetic dilution without such disorder (Fig. 4.4(a)). An unexpected finding was the observation that Ga not only occupies the Fe but also the Re lattice site. This is in con-

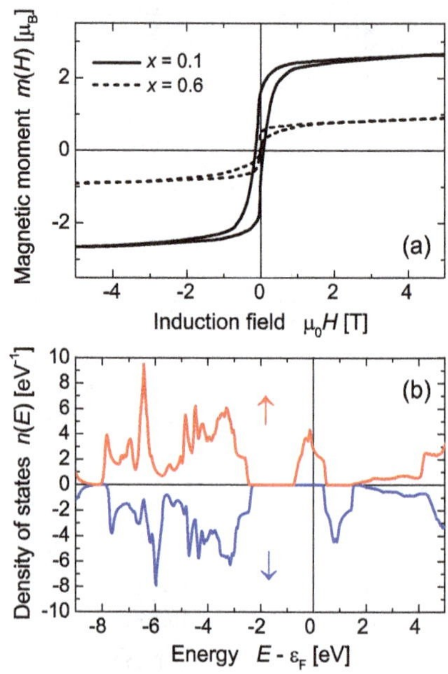

Fig. 4.4 (**a**) Magnetic hysteresis loops of $Sr_2Fe_{1-x}Ga_xReO_6$ measured at 5 K, (**b**) DOS of Sr_2GaReO_6

trast to the observation of virtually no Fe/Re (M/M') disorder in the parent compound Sr_2FeReO_6. On the other hand, Fe^{3+} and Re^{5+} can easily be substituted by Ga^{3+} due to the similarity of the ionic radii. For this reason Ga^{3+} is used for the diamagnetic dilution of ferric oxides or sulfides. The experimental data indicate the formation of two phases, a ferrimagnetic tetragonal and a paramagnetic cubic phase. Below 20 % Ga content, Ga statistically enters the -Fe-O-Re-O-Fe-double exchange pathways. Phase separation occurs for a Ga substitution greater than 20 %. For a Ga content between 20 % and 40 %, the paramagnetic Ga-based phase appears to contain no Fe (Fig. 4.5). The Fe containing cubic and paramagnetic phases detected by Mößbauer spectroscopy appear first for a composition $Sr_2Fe_{0.6}Ga_{0.4}ReO_6$. Mößbauer spectroscopy also shows that the valence state of Fe remains unchanged with the substitution level, indicating that Re acts here as a redox buffer. Interestingly, the end member of the substitution series $Sr_2Fe_{1-x}Ga_xReO_6$, Sr_2GaReO_6, was predicted by Jung et al. to be a half-metallic ferromagnet although it contains no magnetic elements (Fig. 4.4(b)) [18].

4.3.1.2 Sr_2GaReO_6

Considering the disorder effect in Ga-substituted Sr_2FeReO_6 shown above, a high degree of disorder was expected in Sr_2GaReO_6. The structural analysis in terms of both X-ray diffraction and neutron diffraction indicated that the disorder amounts to 40 % in Sr_2GaReO_6. This large disorder triggers the conductivity in Sr_2GaReO_6 (Fig. 4.6(c)). Compared with the iso-electronic and insulating compound, Sr_2InReO_6 (Figs. 4.6(a), (b)), the magnetic moment in Sr_2GaReO_6 is greatly

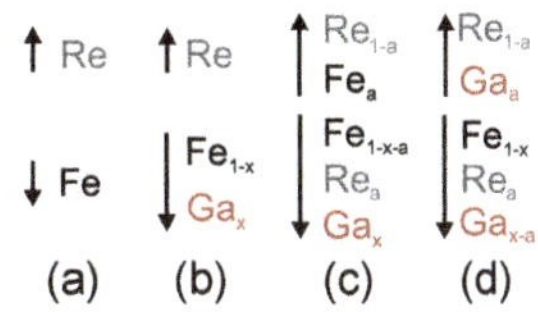

Fig. 4.5 Distribution of the elements Fe, Ga and Re over the two sublattices as obtained by Rietveld refinement for (**a**) the unsubstituted Sr_2FeReO_6, (**b**) the fully ordered $Sr_2Fe_{1-x}Ga_xReO_6$, (**c**) the Fe/Re disordered $Sr_2Fe_{1-x}Ga_xReO_6$ and (**d**) the Ga/Re disordered $Sr_2Fe_{1-x}Ga_xReO_6$. x indicates the Ga content, a represents the amount of disorder. The *up* (↑) and *down* (↓) *arrows* indicate the spin direction of the respective magnetic sublattices

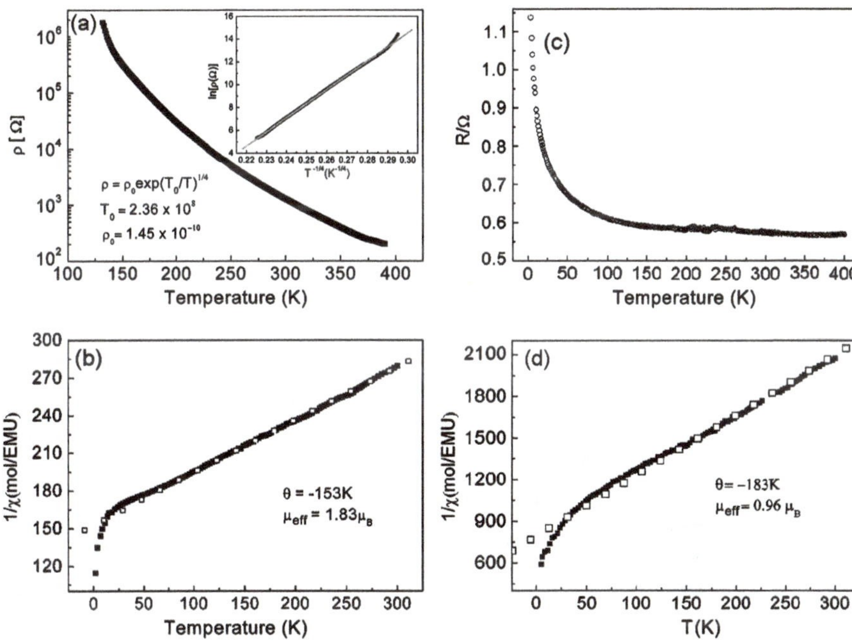

Fig. 4.6 (**a**) Conductivity and (**b**) magnetic susceptibility of Sr_2InReO_6, (**c**) conductivity measurement and (**d**) magnetic susceptibility of Sr_2GaReO_6. In (**a**), variable range hopping conductivity is observed

reduced (Fig. 4.6(d)), indicating a hopping behavior of the Re-5*d* electrons. The conductivity and magnetic measurement suggest that ordered Sr_2GaReO_6 is an inherent semiconductor rather than a half metal as predicted previously [19].

4.3.1.3 Sr_2InReO_6

Due to larger difference in the ionic radii between Re and In, Sr_2InReO_6 has an ordered structure. It crystallizes in the monoclinic cryolite type (space group, $P2_1n$).

Sr_2InReO_6 is a Mott-insulator, because In blocks the electronic hopping in -In-O-Re-In-O-Re-O- (Fig. 4.6(a)). Magnetometry indicates two unpaired and localized electrons on Re^{5+}, giving rise to an antiferromagnetic coupling. However, due to a nearly fcc arrangement of the octahedral 21 units (Re^{5+}, $5d^2$), spin-frustration is observed and a long-range antiferromagnetic ordering is hindered (Fig. 4.6(b)) [20].

4.3.2 M' Position

4.3.2.1 Ga-Series

$Sr_2FeRe_{1-x}Ga_xO_6$ ($0 \leq x \leq 0.5$) undergoes a transition from double perovskite to disordered simple cubic perovskite, similar to $Sr_2FeRe_{1-x}Fe_xO_6$. For $x = 0.5$, an equal distribution of Ga at both, M and M', positions was found. Already at a substitution level of $x = 0.1$ the conductivity drops below an experimentally significant value. In contrast to $Sr_2FeRe_{1-x}Fe_xO_6$, $Sr_2FeRe_{1-x}Ga_xO_6$ does not show any magnetic transition (Fig. 4.3(b)). This suggests that no adjacent Fe-O-Fe motifs are present [21].

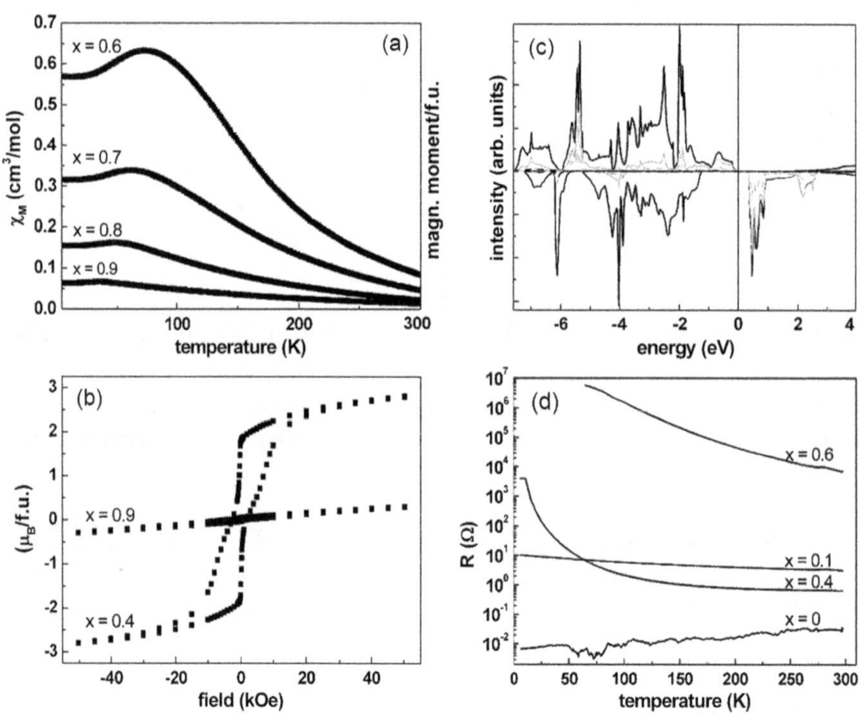

Fig. 4.7 Molar susceptibility measured at 5 kOe in dc mode of and (**b**) field dependence of the magnetization at 5 K of $Sr_2FeRe_{1-x}Sb_xO_6$, (**c**) DOS of Sr_2FeSbO_6, (**c**) conductivity of $Sr_2FeRe_{1-x}Sb_xO_6$

4.3.2.2 Sb-Series

Compounds of the solid solution series $Sr_2FeRe_{1-x}Sb_xO_6$ crystallize in the tetragonally distorted double perovskite structure type. Antisite disorder is as low as 2 %. For large Sb contents ($x \geq 0.6$) a macroscopic antiferromagnetic behavior is observed (Figs. 4.7(a), (b)). The absence of antisite disorder precludes this antiferromagnetism to originate from a superexchange via the Fe-O-Fe units. Mößbauer spectroscopy indicates the simultaneous presence of ferri- and antiferromagnetic clusters at the microscopic level in all samples, except for Sr_2FeReO_6 and $Sr_2FeRe_{0.1}Sb_{0.9}O_6$, which are purely ferri, and antiferromagnetic, respectively. Sb rich samples were found to be semiconductors or insulators by band structure calculations and conductivity measurements (Figs. 4.7(c), (d)). Thus, replacing Re by Sb leads to a transition from an itinerant ferrimagnet for low substitution levels to a localized antiferromagnet for high substitution levels. As a consequence of the bleaching of the double exchange pathways by a substitution of Re by Sb, the magnetic behavior of $Sr_2FeRe_{1-x}Sb_xO_6$ is determined by its Fe^{3+} component. Another finding from Fe Mößbauer spectroscopy is that the valence state of Fe was increased with substitution, indicating of a role of Fe as a redox buffer [22].

4.4 Summary

Summary of substitution effects in Sr_2FeReO_6 is shown in Fig. 4.8.

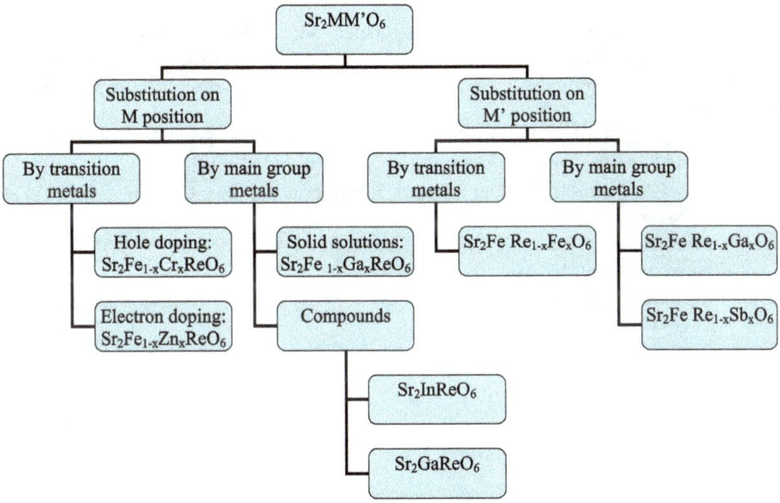

Fig. 4.8 Summary of substitution effects in Sr_2FeReO_6

Acknowledgements This work was supported by the Deutsche Forschungsgemeinschaft (Forschergruppe 559) and the Materials Science Center (MWFZ) at the Johannes-Gutenberg Universität Mainz. Dr. Dominik Maschke is specially acknowledged for his constructive discussions concerning the structure arrangement of the paper.

References

1. Kobayashi KI, Kimura T, Sawada H, Terakura K, Tokura Y (1998) Nature 395(6703):677
2. Kobayashi KI, Kimura T, Tomioka Y, Sawada H, Terakura K, Tokura Y (1999) Phys Rev B 59(17):11159
3. West A (1999) Basic solid state chemistry, 2nd edn. Wiley, New York
4. Shannon R (1976) Acta Crystallogr, Ser A 32(5):751
5. Abe M, Nakagawa T, Nomura S (1973) J Phys Soc Jpn 35:1360
6. Blanco JJ, Hernandez T, Rodriguez-Martinez LM, Insausti M, Barandiaran JM, Greneche JM, Rojo T (2001) J Mater Chem 11(2):253
7. Gopalakrishnan J, Chattopadhyay A, Ogale SB, Venkatesan T, Greene RL, Millis AJ, Ramesha K, Hannoyer B, Marest G (2000) Phys Rev B 62(14):9538
8. Nakamura S, Tanaka M, Kato H, Tokura Y (2003) J Phys Soc Jpn 72(2):424
9. Zener C (1951) Phys Rev 82(3):403
10. Zener C (1951) Phys Rev 83(2):299
11. Zener C (1951) Phys Rev 81(3):440
12. Auth N, Jakob G, Westerburg W, Ritter C, Bonn I, Felser C, Tremel W (2004) J Magn Magn Mater 272–276(Supplement 1):E607
13. Bramnik KG, Ehrenberg H, Dehn JK, Fuess H (2003) Solid State Sci 5(1):235
14. Westerburg W, Lang O, Ritter C, Felser C, Tremel W, Jakob G (2002) Solid State Commun 122(3–4):201
15. Navarro J, Frontera C, Balcells L, Martínez B, Fontcuberta J (2001) Phys Rev B 64(9):092411
16. Jung A, Bonn I, Ksenofontov V, Melnyk G, Ensling J, Felser C, Tremel W (2005) J Mater Chem 15(17):1760
17. Gao H, Ksenofontov V, Barth J, Panthoefer M, Adler P, Felser C, Tremel W $Sr_2FeRe_{1-x}Fe_xO_6$: transitions of structure, transport and magnetism (to be submitted)
18. Jung A, Bonn I, Ksenofontov V, Panthofer M, Reiman S, Felser C, Tremel W (2007) Phys Rev B 75(18):184409
19. Gao H, Llobet A, Barth J, Felser C, Panthöfer M, Tremel W Sr_2GaReO_6: when disorder triggers electric conductivity in an intrinsic insulator (to be submitted)
20. Gao H, Llobet A, Panthöfer M, Dai X, Barth J, Winterlik J, Felser C, Tremel W (2011) Phys Rev B 83:134406
21. Gao H, Ksenofontov V, Felser C, Adler P, Panthoefer M, Tremel W $Sr_2FeRe_{1-x}Ga_xO_6$: how Ga distributed on M and M' positions (to be submitted)
22. Jung A, Ksenofontov V, Reiman S, Therese H, Kolb U, Felser C, Tremel W (2006) Phys Rev B 73(14):144414

Chapter 5
Half-Metallic Ferromagnets

Jürgen Kübler, Gerhard H. Fecher, and Claudia Felser

Abstract Half-metallic ferromagnetism and the related ferrimagnets are reviewed from a theoretical point of view. Concentrating on Heusler compounds as well as the double perovskites we discuss the electronic structure, the nature of the magnetic moments, and the physics of the energy-gap in one of the spin-channels. The role of spin–orbit coupling is examined and a brief account is given of a tetragonal Heusler compound. Exchange interactions are determined and used for calculations of the Curie temperatures and spin-wave spectra. In view of the large Curie temperatures of the Heusler compounds and the double perovskites we discuss double-exchange and super-exchange models.

5.1 Introduction

Half-metallic ferromagnets (HMF) are metals with 100 % spin-polarized electrons at the Fermi edge, i.e. the electron states of one spin direction are gapped while for the other spin direction they are metallic. The term HMF was coined in 1983 by de Groot et al. [1] to describe the electronic structure of NiMnSb and PtMnSb. These are Heusler compounds that have the $C1_b$ crystal structure, which can be best described by starting with the zinc blende structure sketched on the left hand side of Fig. 5.1. When another set of atoms is added on an *fcc* lattice (right portion of the figure), thus completing the rock-salt structure for the dark shaded atoms, the

J. Kübler
Institut für Festkörperphysik, Technische Universität Darmstadt, Darmstadt, Germany
e-mail: juergen.kuebler@gmail.com

G.H. Fecher · C. Felser
Institut für Anorganische Chemie und Analytische Chemie, Johannes Gutenberg – Universität, 55099 Mainz, Germany

G.H. Fecher · C. Felser (✉)
Max-Planck-Institut für Chemische Physik fester Stoffe, 01187 Dresden, Germany
e-mail: felser@cpfs.mpg.de

G.H. Fecher
e-mail: fecher@cpfs.mpg.de

C. Felser, G.H. Fecher (eds.), *Spintronics*, DOI 10.1007/978-90-481-3832-6_5,

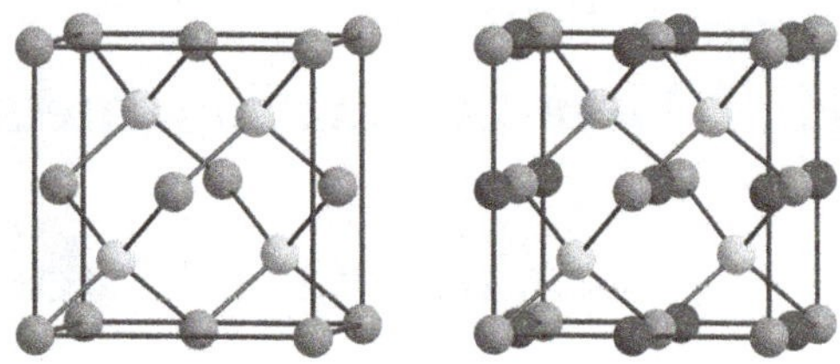

Fig. 5.1 Zinc blende crystal structure on the *left* and $C1_b$ Heusler structure on the *right*

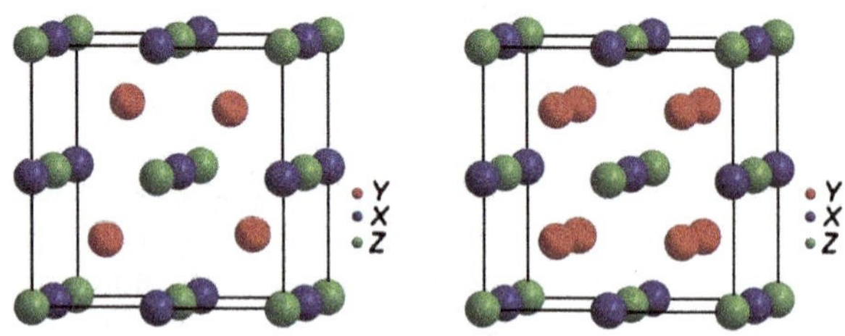

Fig. 5.2 $C1_b$ crystal structure on the *left* and $L2_1$ structure on the *right*

$C1_b$ Heusler structure is obtained. In this early work the emphasis was on magneto-optical applications [2, 3] for which the large Kerr angle found in PtMnSb appeared very promising. Although this is highly interesting physics we will not deal with this topic here any further.

Also in 1983 half-metallic ferromagnetism was theoretically predicted for other ternary systems: Co_2MnAl and Co_2MnSn [4]. These are also Heusler compounds, but they have the $L2_1$ crystal structure, which can again be visualized by means of Fig. 5.1 by filling the remaining tetrahedral site. This is sketched in Fig. 5.2, where the connection of the $C1_b$ structure with the $L2_1$ structure is shown. The half-metallic ferromagnetism of the Co-Heusler compounds received only little attention in this early work except for a remark that one might expect unusual transport properties. The emphasis was more on simple attempts to estimate the Curie (Néel) temperatures from the electronic structure calculations and on a description of the nature of the magnetic moment of manganese in these compounds, a topic we will take up again in the next section.

Recently, Heusler compounds had an astonishing comeback. Because of their high spin polarization and their high Curie temperatures they received exceptional attention for possible applications in the field of spin-electronics and magneto-resistive devices (spintronics). Recent review articles cover the basic physics of HMF's [5] and aspects of applications [6, 7]. Although the number of chemically different HMF's is quite substantive [5], we shall focus our attention in this chapter only on the Heusler compounds and the related double perovskites [8] apologizing already here for not being able to cite all relevant work on the topics chosen. In the description of the electronic ground-state structure we shall besides trends look at the role of spin–orbit coupling and at structural variants of the cubic Heusler compounds. Excited-states properties like spin-waves and the Curie-temperatures are obtained by means of frozen spiral spin configurations; they allow us to study trends again and discuss exchange mechanisms that might be helpful in understanding the size of the Curie temperatures.

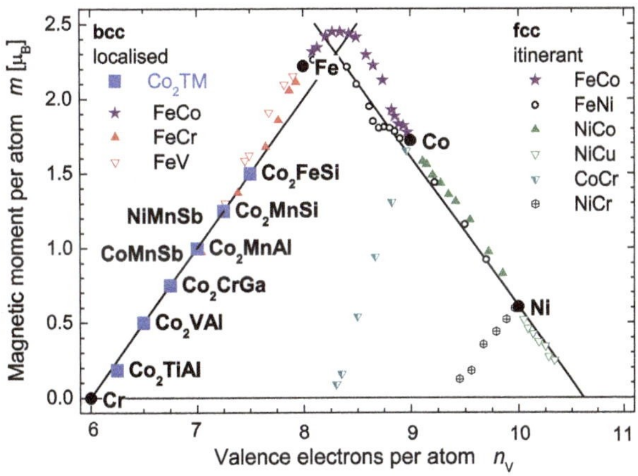

Fig. 5.3 Slater–Pauling curve for the $3d$ transition metals and their alloys. Experimental values for a subset of the Co_2-based and two $C1_b$ Heusler compounds are included. For more information on CoMnSb, see Sect. 5.2.4

5.2 Electronic Structure of Heusler Compounds

5.2.1 Slater–Pauling Behavior

We begin this section by orienting ourselves employing the *Slater–Pauling* curve, where the average saturation magnetization of ferromagnetic alloys is plotted as a function of the electron concentration [9]. The regularities and trends in the Slater–Pauling curve shown in Fig. 5.3 are perhaps the most prominent properties of itinerant-electron ferromagnets; they have from the very beginning been ascribed to broad band-structure features. Originally conceived for binary alloys only the Slater–Pauling curve has been generalized here to include a subset of Heusler compounds.

The history of the Slater–Pauling curve is somewhat involved as outlined in Ref. [10]. However, the basic physical facts are simple to state: The magnetic moment M is given in terms of the number of up- and down-spin electrons, $N_\uparrow$ and $N_\downarrow$ by $M = N_\uparrow - N_\downarrow$, whereas the number of valence electrons is $N_V = N_\uparrow + N_\downarrow$. So if the up-spin count is constant then the two equations are rewritten as $M = 2N_\uparrow - N_V$, which obviously is the case for the right hand side of the Slater–Pauling curve labeled *fcc "itinerant"*. It is rewarding to read the ground-breaking paper of Friedel [11] which gives the physical reasons for the constancy of the up-spin count in a lucent way.

Here we are concerned with the left hand side labeled *bcc "localized"*. (The acronyms *fcc* and *bcc* refer to the dominant coordination of the atoms). It is the down-spin count that is constant in this case, so that the magnetization is given by $M = N_V - 2N_\downarrow$. As we will see, this constancy of the down-spin count in the Heusler compounds collected in Fig. 5.3 comes about through the band gap in the down-spin electrons, which is the defining feature of HMF.

The magnetism of the Heusler compounds, especially of Cu_2MnAl and Cu_2MnSn, was traditionally alluded to as being localized Refs. [12, 13] and an explanation for localized behavior in the itinerant-electron picture is certainly of

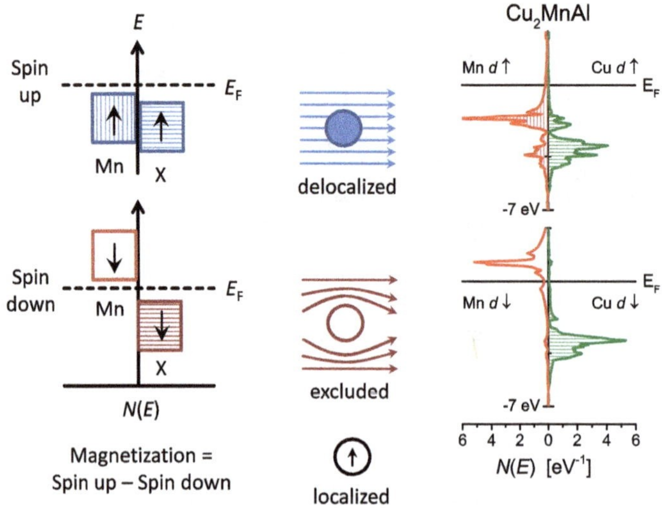

Fig. 5.4 Localized magnetic moment from delocalized electrons. Schematic diagram of up-spin and down-spin electrons in Heusler compounds X_2MnY derived from calculated density-of-states curves like that shown on the *right*. Reproduced with kind permission from Kübler et al. [4]. Copyright (1986) by the American Physical Society

interest. In Fig. 5.4, taken from [4], one sees that the Mn spin-up *d* states are almost completely occupied and the bandwidths indicate that they are just as delocalized as the *d* electrons of X (Cu). The Mn down states, however, are nearly empty. So the spin-up *d* electrons of the Mn atoms join those of the X (Cu) atoms in forming a common *d* band, whereas the spin-down *d* electrons are almost completely excluded from the Mn site. The result of this localized exclusion is an equally localized region of magnetization. We therefore have completely localized moments composed of completely itinerant electrons [4]. We emphasize that the Mn atom occupies the rock-salt structure.

5.2.2 Band Structure of the Co_2-Heusler Compounds

We begin by drawing attention to the magnetic moments per unit cell of our selection of Heusler compounds graphed as a function of the number of valence electrons per unit cell, N_V. This is shown in Fig. 5.5 taken from Ref. [14] which, clearly shows a portion of the left hand-side of the Slater–Pauling curve discussed above. The coordinates used here can easily be transformed to those used in Fig. 5.3. Figure 5.5 also reveals that the calculations are in good agreement with measured values of the magnetic moments. We furthermore can see from this figure that most of these compounds possess integer or near-integer valued magnetic moments. Indeed, these are half-metallic ferromagnets (HMF), as is revealed by Figs. 5.6, 5.7, and 5.8 where the densities of states (DOS) are shown for increasing number of valence electrons.

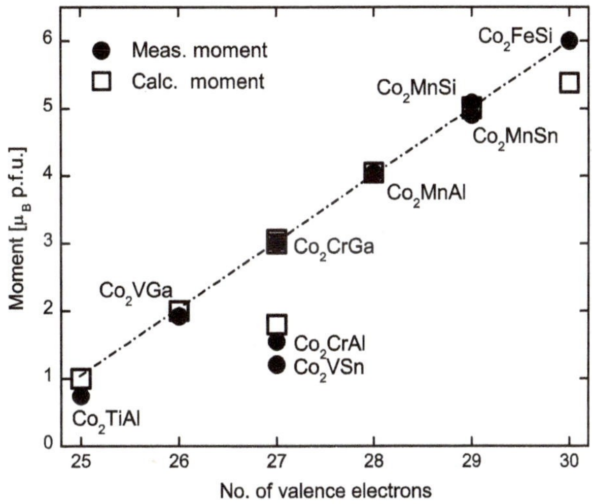

Fig. 5.5 Section of the Slater–Pauling curve describing a set of Co_2-Heusler compounds; Reproduced with permission from Kübler, J., Fecher, G.H., and Felser, C.: *Phys. Rev. B* 76, 024414 [14]. Copyright (2007) by the American Physical Society

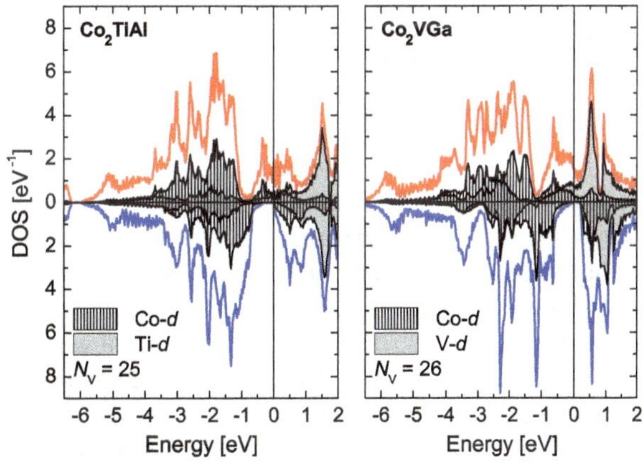

Fig. 5.6 Spin and site resolved density of states of Co_2TiAl and Co_2VGa, *upper panel* is spin up, *lower panel* is spin down. Reproduced with permission from Kübler, J., Fecher, G.H., and Felser, C.: *Phys. Rev. B* 76, 024414 [14]. Copyright (2007) by the American Physical Society

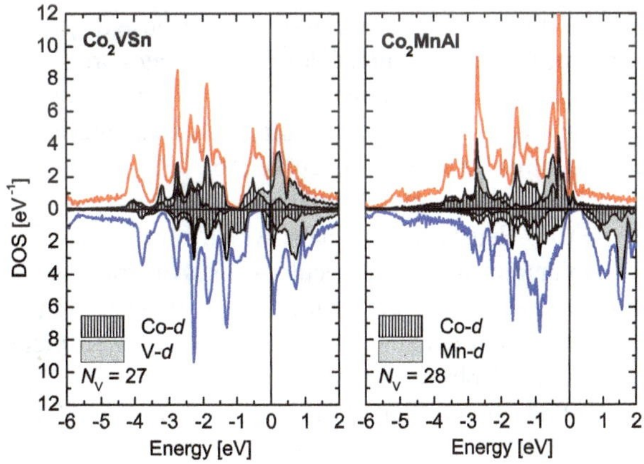

Fig. 5.7 Spin and site resolved density of states of Co_2VSn and Co_2MnAl, *upper panel* is spin up, *lower panel* is spin down. Reproduced with permission from Kübler, J., Fecher, G.H., and Felser, C.: *Phys. Rev. B* 76, 024414 [14]. Copyright (2007) by the American Physical Society

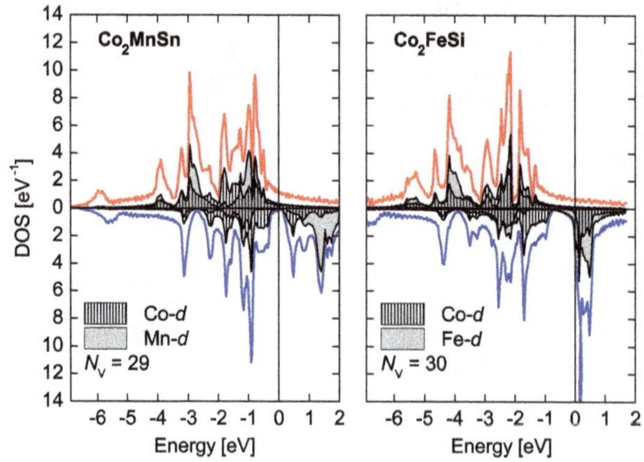

Fig. 5.8 Spin and site resolved density of states of Co_2MnSn and Co_2FeSi, *upper panel* is spin up, *lower panel* is spin down. Reproduced with permission from Kübler, J., Fecher, G.H., and Felser, C.: *Phys. Rev. B* 76, 024414 [14]. Copyright (2007) by the American Physical Society

The deviation from the straight line in Fig. 5.5 for $N_V = 27$ is seen to arise from the occupations of down-spin states above the gap (Fig. 5.7 for Co_2VSn). Another deviation is seen to occur for $N_V = 30$, Co_2FeSi, where the *calculated* value of the magnetic moment is too low. Again, in our GGA calculations the down-spin states above the gap are occupied. According to Ref. [15] the experimental value of the magnetic moment is 6 μ_B and not as calculated 5.4 μ_B, and the compound is half-metallic. Both the standard LSDA and the GGA calculations [10], which we used here for most of the calculated results shown, do not succeed in describing this state correctly. Wurmehl et al. [15] showed that by employing so-called LDA+U calculations agreement with the measured values is obtained, in particular the half-metallic state. The LDA+U increases the strength of the electron–electron repulsion (U) beyond the amount that is accounted for by the local density functional approximation (LDA). In Fig. 5.9 we show the results of these calculations, for more details see Refs. [7, 15]. For certain applications in e.g. tunnel junctions it is of importance to tune the Fermi energy to fall into a desired position of the gap. For this a high-moment case is most desirable. Therefore two of these authors [16] determined the band structure and the magnetic moment of $Co_2FeAl_{1-x}Si_x$ for representative values of x, for which super-cells can be constructed with a moderate effort. This results in electron doping because Si possesses one more valence electron than Al. The calculated densities of states are shown in Fig. 5.10, for which the GGA+U method (a variant of LDA+U) was employed. A rather small value of U is sufficient to obtain these results. The magnetic moments for these alloys follow precisely the Slater–Pauling curve [16].

We now attempt to explain the reason for the gap in the down-spin electron bands of the Co_2-Heusler compounds, which supply nearly all HMF's in the $L2_1$ family. An explanation using a molecular-orbital diagram was given by Galanakis et al. [17] some time ago. However, a direct check of their assumptions by examining the wave functions in our calculations did not support their picture. So here we give another qualitative but physical explanation.

We first observe that the gap seen in Figs. 5.6–5.8 (except for Co_2FeSi) is bordered entirely by Co-states, which we therefore concentrate on in our explanation.

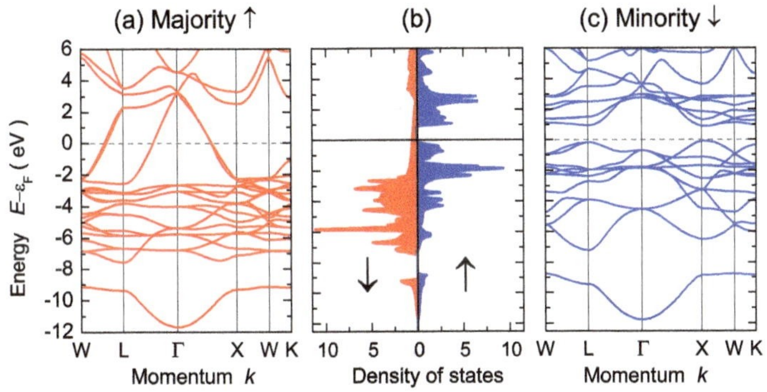

Fig. 5.9 LDA+U band structure and densities of states (DOS) of Co_2FeSi calculated by means of Wien2k. The effective Coulomb-exchange parameters were set to $U_{\mathrm{eff}} = 4.8$ eV for Co and $U_{\mathrm{eff}} = 4.5$ eV for Fe. Reproduced with permission from Wurmehl, S., Fecher, G.H., Kandpal, H.C., Ksenofontov, V., Felser, C., Lin, H.-J., and Morais, J.: *Phys. Rev. B* 72, 184434 [15]. Copyright (2005) by the American Physical Society

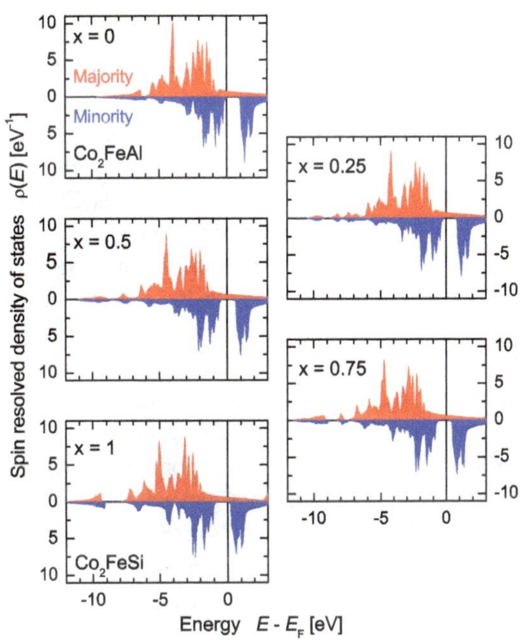

Fig. 5.10 Density of states for $Co_2FeAl_{1-x}Si_x$ obtained by using the GGA+U method employing a moderate value of U of about 2 eV. Details of the calculation are reported in [16]

Next we recall a simple argument by Söderlind et al. [18] for an explanation of the observed sequence of crystal structures for Mn, Fe, Co and Ni. They assumed that the structural energy differences are predominantly determined by the down-spin electrons because the up-spin bands are nearly filled and therefore mostly inert for bonding. The reader interested in the details of the different crystal structures ob-

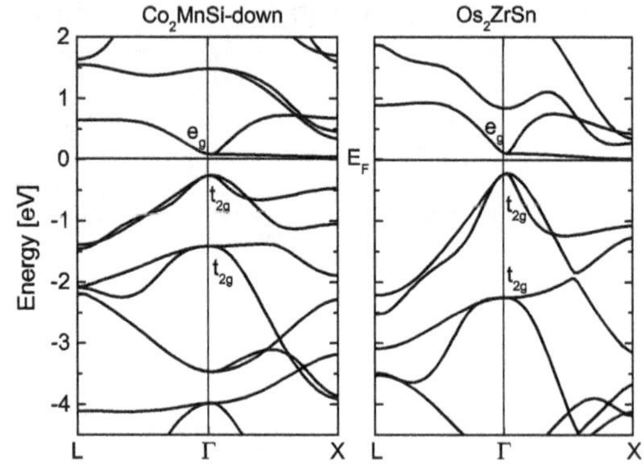

Fig. 5.11 Comparison of the down-spin band structure of Co_2MnSi with that of non-magnetic Os_2ZrSn also listing the representation labels of the d-electrons

served for the transition metals can find a lucid explanation in the paper by Söderlind et al. [18]. Here we make another use of their observations.

Examining the number of d-electrons in Co_2MnSi (which we take as an example for our discussion here) we find that the up-spin count is about $n_{d\uparrow} = 4.4$ whereas the down-spin count is $n_{d\downarrow} = 3.3$. The latter is to be multiplied with 2, which gives a d count of about 6.6 near that of Ru or Os. Ignoring furthermore the localized Mn-magnetism we therefore look for an Os–Heusler compound, which we find in Os_2ZrSn having a valence electron count of 24. Its electronic structure is easily obtained and found to be insulating; its d-electron count is 6.7 near enough to the doubled $n_{d\downarrow}$-count of Co_2MnSi so that Os_2ZrSn can be taken as a model-system for the down-spin Co_2-Heusler compounds.

In Fig. 5.11 we compare representative sections of the band-structure of down-spin Co_2MnSi with that of Os_2ZrSn obtained using the LDA. The gross features are indeed very similar, including the representation of the d-electrons bordering the gap. Since our argument depends on the specific d-electron count of magnetic Co in the Heusler compound, it is unique for the Co_2-system. This is, of course, not a stringent theory, but it singles out the Co compounds, as desired. We add that we might have chosen Fe_2TiSn for comparison just as well; it is non-magnetic but the e_g- and t_{2g}-states are not gapped, the band structure in all other respects being nearly identical with that of Os_2ZrSn. We hasten to draw attention to the early work of de Groot et al. [1] who explained the band-gap of the spin-down electrons of the $C1_b$-Heusler compound NiMnSb and pointed out the similarity of its band structure with that of GaAs.

The next question concerns the experimental proof of half-metallicity. There is no dedicated experiment except the spin-resolved positron annihilation measurements by Hanssen et al. [19] for NiMnSb. This is proven by these authors to be a HMF to the precision of the experiment which was better than one-hundredth of an electron. Since this is a tedious and expensive experiment, NiMnSb will probably remain the only proven HMF for some time. But there is much indirect experimental evidence that half-metallic magnetism is physical and not merely numerical reality.

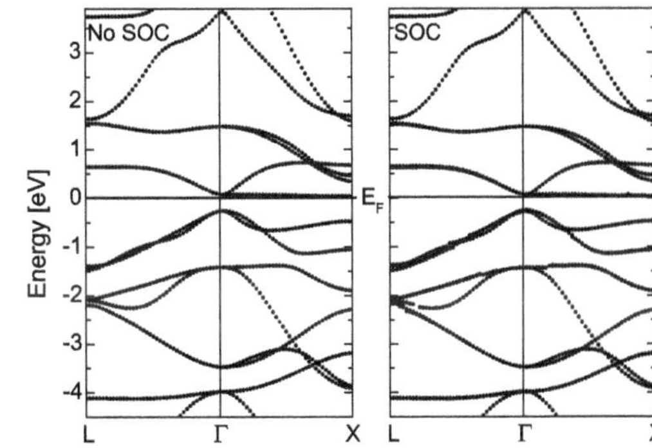

Fig. 5.12 The role of spin–orbit coupling (SOC) in the band-structure of Co_2MnSi. The *left-hand side* shows the down-spin states obtained without SOC. The *right-hand side* was calculated including SOC showing the minority-spin electron states

5.2.3 Spin–Orbit Coupling

It has been known since the work of Brooks [20] that the main effect of spin–orbit coupling (SOC) is the magneto-crystalline anisotropy. This is the energy involved in rotating the magnetization from a direction of low energy—the so-called easy axis—to a direction of high energy—the so-called hard axis. The SOC breaks the rotational invariance with respect to the spin quantization axis. At this stage, however, we are not attempting to determine the magneto-crystalline anisotropy of the Heusler compounds. It rather is the effect of SOC on the band structure that we concentrate on here.

In the computations described now, the spin–orbit Hamiltonian is included in the local spin-density functional calculations as a second variation, which is brought to self-consistency. The procedure is explained in detail in Chaps. 2.4 and 3.4.7.3 of Ref. [10].

As is well known, in the spin–orbit coupled states spin is no longer a good quantum number. So distinguishing spin-up and spin-down states is not a priori possible. Therefore, to interpret the results of our SOC-calculations a spin-filter is employed. This is constructed by computing the norm at each **k**-point projecting out the up- and down-spin components. The norm at each **k**-point for a given band-index i is thus written as $n_i(\mathbf{k}) = n_i^{\mathrm{up}}(\mathbf{k}) + n_i^{\mathrm{down}}(\mathbf{k}) = 1$. An energy state $\varepsilon_i(\mathbf{k})$ having $n_i^{up}(\mathbf{k}) > n_i^{\mathrm{down}}(\mathbf{k})$ we call a *majority state* whereas $\varepsilon_i(\mathbf{k})$ having $n_i^{\mathrm{up}}(\mathbf{k}) < n_i^{\mathrm{down}}(\mathbf{k})$ is called *minority state*.

When the densities of states computed this way are plotted, the spin–orbit coupled states look basically the same as those obtained without SOC. This is different when the band-structure is examined. In Fig. 5.12 we compare the down-spin band-structure with the minority band-structure of Co_2MnSi. The SOC band-splitting is very small and the band gap is unchanged, but the minority band-structure looks somewhat "spiky". The reason for the discontinuities that give rise to the "spikiness" can be seen by examining Fig. 5.13, where both majority and minority states are graphed in part A. The discontinuities originate from band crossings of minority- and majority states. At these points, which we call *hot spots*, the projection of the spins changes drastically. This is demonstrated in part C of Fig. 5.13 where the norm

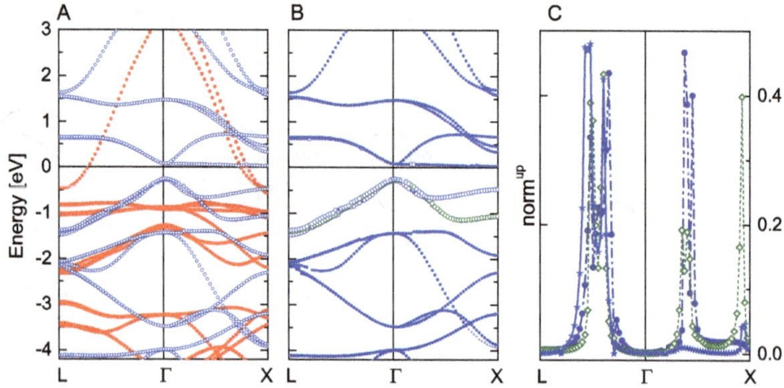

Fig. 5.13 The role of spin–orbit coupling (SOC) in the band-structure of Co_2MnSi. Part **A** gives all states of the band structure, but filtered such that the majority (*solid dots*) and minority states (*open circles*) are recognizable. Part **B** is identical with the minority states given in Fig. 5.12; the bands 9, 10, and 11 are marked with *open circles*. Part **C** is the up-spin-norm in the minority states for bands 9, 10 and 11

of the up-spins in the minority states is shown for the three t_{2g}-bands numbered 9, 10, and 11 to be found just below the gap. It is to be remembered that these norms ($n_i^{\mathrm{up}}(\mathbf{k})$) vanish in the absence of SOC. It is seen clearly that near the hot spots these norms rise rapidly indicating the spin-reversal that eventually leads to the hot-spot state to appear in the majority bands. It appears quite plausible that experimentally these states should be observable: at these crystal momenta **k** spin reversal costs no energy.

5.2.4 Half Metallic Compensated Ferrimagnets

For spin torque applications, among other things, it is advantageous to design magnetic materials with ferrimagnetic arranged compensated moments. de Groot [21] first suggested to follow the trend predicted by the Slater–Pauling curve for the $C1_b$-compounds, see Fig. 5.3, to reach a suitable electron concentration that results in a vanishing total moment. He therefore calculated the electronic structure of CrMnSb and found the desired results: the large Mn-moment cancels exactly the moment of Cr. Unfortunately, CrMnSb cannot be realized experimentally. More recent experimental results demonstrated that this line of arguing is misleading even in the first step, namely the case of CoMnSb. Although placed correctly on the Slater–Pauling curve the experiments of Ksenofontov et al. [22] showed that it does not have the $C1_b$-crystal structure; it actually is represented as an alternation of Co_2MnSb and MnSb structure units. Furthermore, band structure calculations showed that CoMnSb is not a half-metallic ferromagnet [22].

Both conceptually and experimentally more successful was an approach starting with the $L2_1$-Heusler compounds. Using the Slater–Pauling curve and appreciating

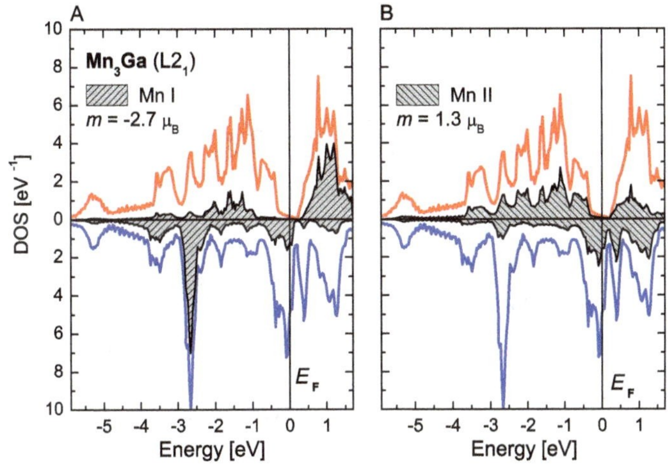

Fig. 5.14 Densities of states (DOS) of Mn_2MnGa in the $L2_1$-crystal structure. **A**: the states of Mn I on the rock-salt sites (*gray*), **B**: the states of Mn II (*gray*)

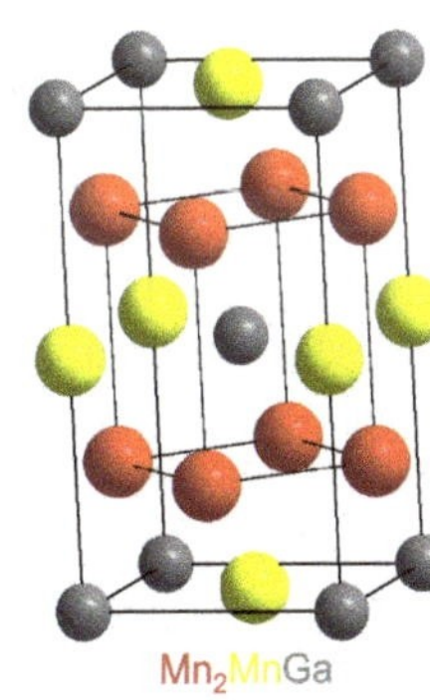

Fig. 5.15 The DO_{22}-crystal structure for Mn_2MnGa

the local moment of Mn occupying the rock-salt sites, Wurmehl et al. [23] first designed the Heusler compound Mn_2MnGa (written simply as Mn_3Ga) computationally; later on Balke et al. [24] synthesized and investigated the compound. A similar approach was taken by Galanakis et al. [25] for Cr_2MnZ (Z = P, As, Sb, and Bi).

The computational design yielded indeed a totally compensated ferrimagnetic half-metallic $L2_1$-Heusler compound with Mn^{3+}, called Mn I, at the octahedral (rock-salt) sites having $3 \times 7 + 3 = 24$ valence electrons. The two magnetic moments of the tetrahedral Mn II cancel the moment of Mn I, as is show in Fig. 5.14.

When the compound was synthesized [24], however, it turned out to be tetragonal having the DO_{22}-structure, the compensation no longer being perfect and the spin-polarization being about 80 %. A sketch of the DO_{22}-structure is graphed in Fig. 5.15. The density of state of the tetragonal version of Mn_3Ga is shown in Fig. 5.16A. The magnetic moment of Mn I here is $M = -2.83\ \mu_B$ and that of Mn

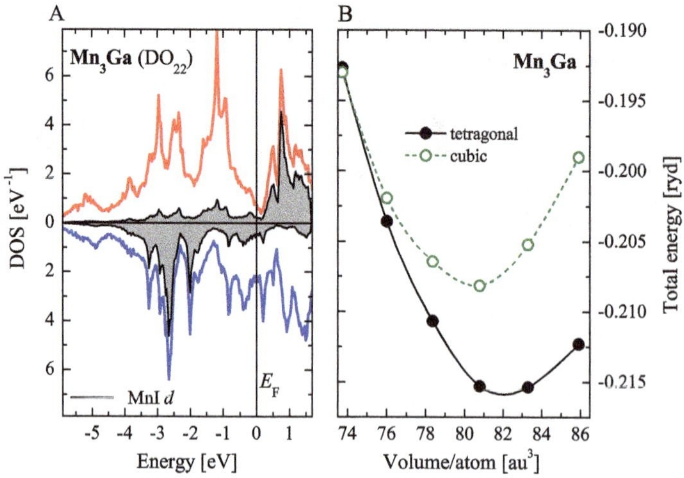

Fig. 5.16 **A**: Densities of states (DOS) of Mn_2MnGa in the DO_{22}-crystal structure; the states of Mn I are shaded gray. **B**: A comparison of the total energies (counted from an arbitrary origin) of cubic and tetragonal Mn_3Ga

II is $M = 1.14\ \mu_B$ so that a residual moment of $0.55\ \mu_B$ remains. (Only the relative signs are relevant here.)

Also shown in Fig. 5.16B is a comparison of the total energies of tetragonal and cubic Mn_3Ga. It is seen that under pressure the tetragonal form will go over to the cubic $L2_1$-structure. Similar to the properties calculated for ambient pressure the magnetic moments will be entirely compensated and the spin polarization will be near 100 %.

The tetragonal Heusler alloys, of which Mn_3Ga is but one example, are under intense investigation presently because they possess high perpendicular anisotropy and low magnetic damping, at the same time having high Curie temperatures (see below, also Ref. [26]) and containing, among other things, shape memory compounds.

5.3 Curie Temperatures and Spin Waves

The success of the local spin-density functional (LSDA) approximation for the determination of the electronic structure of solids need not be emphasized here. Even though some important many-body effects require special treatment and corrections to the LSDA are necessary, like the enhancement of the electron–electron repulsion in the LDA+U method, it is clear that the LSDA together with suitable computer codes provides a powerful tool to understand ground-state properties like those covered in Sect. 5.2.

For the determination of excited-states properties like Curie temperatures and spin waves the use of the LSDA is somewhat problematic. Still, one may invoke

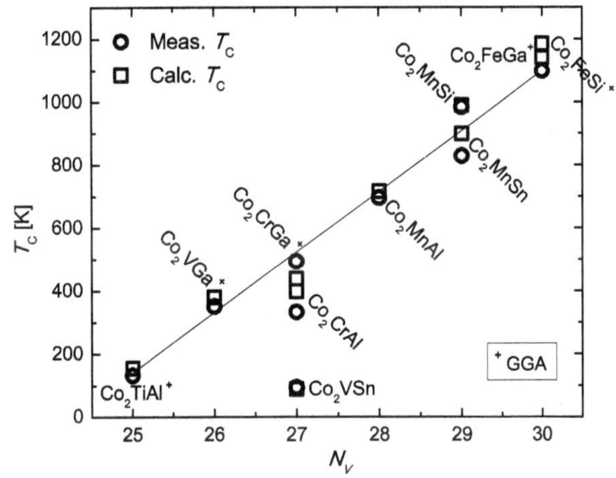

Fig. 5.17 Measured and calculated Curie temperatures T_C of a representative set of Co_2-Heusler compounds as a function of the number of valence electrons, N_V. For the calculations to be explained the LSDA or the GGA was used; the latter are marked with a +. Revised data from Ref. [14] are displayed

the adiabatic approximation and freeze non-collinear spin arrangements to obtain energy changes that are assumed to reflect excitation energies. It is the success of such an approach that warrants its application and justifies its use.

5.3.1 Curie Temperatures

The challenge is to explain the large Curie temperatures, T_C, of the Heusler HMF's and the nearly linear trend of T_C with the number of valence electrons, N_V, shown in Fig. 5.17. This trend is quite remarkable; it even features exceptions similar to the Slater Pauling behavior graphed in Fig. 5.5.

At the center of the theory for the Curie temperatures is the determination of what are generally called exchange constants. A more suitable terminology should replace *constant* by *function* because they vary and decay in real or reciprocal space, whichever is the chosen method. The real-space approach is connected with Liechtenstein et al. [27] and has been used for Heusler compounds by Kurtulus et al. [28] and Thoene et al. [29]. The reciprocal-space approach has been used for the Heusler compounds in Refs. [14, 26, 30]. In this method total-energy changes are obtained from spiral spin configurations, for which the theoretical basis can be found for instance in the very readable review article by Sandratskii [31]. One must keep in mind, however, that the symmetry of the spin-spiral is broken by spin–orbit coupling, which is therefore neglected for the purpose at hand. This might be acceptable for estimates of the Curie temperatures because their energy equivalents are much larger than the spin–orbit coupling energies.

Once the exchange functions have been determined, different methods can be employed to deal with the spin-Hamiltonian and find approximations for the Curie temperature. A well-known and standard approach is the mean-field approximation which has been used for Heusler compounds by, for instance, Thoene et al. [29]. It also is used in Ref. [30] where it is compared with the more precise RPA. The

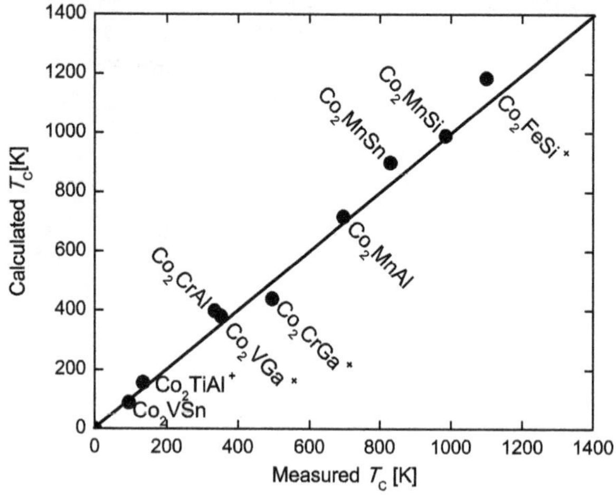

Fig. 5.18 Calculated Curie temperatures T_C as a function of the measured T_C, for which the spherical approximation [10] was used. The data used are a revised version of those given in Ref. [14]

latter is related to the spherical approximation, favored in Refs. [14, 26]. We do not intent do go into much detail for which the interested reader can turn to the original literature. The main purpose here is to find an explanation for the linear trend, Fig. 5.17, and to discuss the physics of the exchange interaction.

Figure 5.18 demonstrates the quality of the calculations of the Curie temperatures by means of the LSDA (GGA) for the exchange functions and the spherical approximation for an estimate of the T_C's [14]. A similar figure of comparable quality is shown in the paper by Thoene et al. [29]. A possible interpretation of the obvious trend in Figs. 5.17 and 5.18 proceeds as follows.

First, for Heusler compounds of the form Co_2XY, X being the other magnetic atom, a spin-spiral characterized by the wave vector $\mathbf{k}$ defines four different exchange functions. These are $j_{11}(\mathbf{k})$ for the exchange interaction between the X atoms, $j_{22}(\mathbf{k})$ for Co on the same sublattice, $j_{12}(\mathbf{k})$ for the exchange interaction between Co and X atoms, and $j_{23}(\mathbf{k})$ for the exchange interaction between the Co atoms on different sublattices. For each $\mathbf{k}$-point in the Brillouin zone these are elements of a three by three matrix that is diagonalized to obtain the three eigenvalues $j_n(\mathbf{k})$, $n = 1, 2, 3$.

Second, in the spherical approximation [10, 14, 26] the Curie temperature is given by

$$k_B T_C = \frac{2}{3} \sum_\tau \mathscr{L}_\tau^2 \left[\frac{1}{N} \sum_{\mathbf{k}n} \frac{1}{j_n(\mathbf{k})} \right]^{-1} . \tag{5.1}$$

The quantity $\mathscr{L}_\tau$ describes the local moment of the atom at site τ. In principle it must be determined self-consistently, see (12) in [26], but an acceptable approximation is $\mathscr{L}_\tau = M_\tau$, where M_τ is the moment of the atom at site τ.

Returning to the trend in Fig. 5.14 we use (5.1) and look separately at the factors $\sum_\tau \mathscr{L}_\tau^2$ and $[\frac{1}{N} \sum_{\mathbf{k}n} 1/j_n(\mathbf{k})]^{-1}$. These are plotted as 'prefactor' and 'exchange average', respectively, in the insets of Fig. 5.19. Multiplying the numbers of one

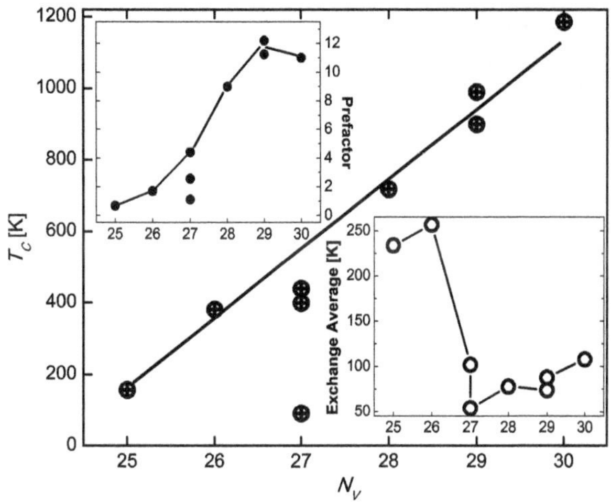

Fig. 5.19 Composition of the calculated Curie temperatures (*encircled crosses*) in terms of the two parts (*insets*) of (5.1), see text. Reproduced with permission from Kübler, J., Fecher, G.H., and Felser, C.: *Phys. Rev. B* 76, 024414 [14]. Copyright (2007) by the American Physical Society

inset with those of the other we obtain the calculated Curie temperatures plotted as encircled crosses.

Starting in Fig. 5.19 at the left side with $N_V = 25$ one sees the expected trend in the prefactor, where the lowest exceptions at $N_V = 27$ for Co_2VSn is clearly due to the fact that this is not a HMF (see Fig. 5.7). Quite remarkably, the exchange average is largest for $N_V = 25$ and 26 decreasing sharply and thus adding to the low Curie temperature of Co_2VSn. The compound Co_2CrAl is calculated to be a HMF. Its calculated T_C is 399 K in satisfactory agreement with the measured value of 334 K. It is below the linear trend, however, because the exchange average is low. Only in moderate agreement with experiment is Co_2CrAl because the exchange average is obtained too low as well. From $N_V = 27$ (Co_2CrGa) up to $N_V = 30$ the exchange average increases, most likely a little too much for Co_2FeSi, for which we underestimate the prefactor: since the electron–electron correlation is not considered here we do not see a HMF, in contrast to the LDA+U calculations (see Fig. 5.9). So, to the right of the break in the trend at $N_V = 27$ the linear behavior is restored and in both regions, low and high N_V, we conclude that it is dominated by the increase in the magnetic moment which for $N_V = 28, 29, 30$ is to some extend tuned down by the slow increase of the exchange average.

Next we comment on models for exchange mechanisms that are frequently discussed in the literature. It should be emphasized that the various mechanisms proposed appear in the calculations in mixed form and are not easily untangled. In particular, the reason for the large values of the Curie temperature does not simply drop out of a calculation, even though the large values themselves do.

In contrast to this is the model-Hamiltonian approach where different perturbation treatments give different exchange interactions, depending on the relative magnitude of the relevant exchange parameters. Still, it perhaps helps our understanding if an analysis is attempted on the basis of calculated trends. These might reveal RKKY-type interactions, which we assume are well-known to the reader, or concepts like *antiferromagnetic super-exchange* [32, 33] and *ferromagnetic double exchange* through charge carriers, concepts that are connected with the names Zener, Anderson, and de Gennes [34, 35].

It is most likely Zener's theory that provides the key concept and not RKKY. In the wording of de Gennes [35] it is summarized by three points:

(1) Intra-atomic exchange is strong so that the only important configurations are those where the spin of each carrier is parallel to the local ionic spin.
(2) The carriers do not change their spin orientation when moving; accordingly they can hop from one ion to the next only if the two ionic spins are not antiparallel.
(3) When hopping is allowed the ground-state energy is lowered (because the carriers are then able to participate in the bonding). This results in a lower energy for ferromagnetic configurations.

This exchange mechanism through carriers is called double exchange, sometimes also kinetic exchange. It obviously applies to metals and, in particular, to half-metallic ferromagnets.

In contrast to this is the concept of Anderson's super-exchange [32, 33] that in its pure form applies to insulators and explains antiferromagnetism. It operates through an intermediary non-magnetic ion, often, but not necessarily, oxygen. In the oxygen ion the spins are antiparallel so that the electrons of one of the magnetic ion can move over to the other magnetic ion via the oxygen bridge only if their spins are antiparallel. If the spins were parallel the Pauli principle would partly forbid such a transition. This mechanism does not apply here but might be of importance for the double-perovskites.

Turning thus to Fig. 5.19 (and ignoring the break at $N_V = 27$) we see that the Curie temperatures increase with increasing valence electron concentration, N_V, for two reasons, first, because the magnetic moments increase, second, because the exchange averages increase, most likely so because of the double exchange mechanism (DE). We conclude that *the high Curie temperatures are due to the DE mechanism that is enhanced by the large magnetic moments.* Both effects are brought about by the electron structure characteristic for HMF.

5.3.2 Spin Waves

The theory of spin waves in HMF is at present rather approximative. On the one hand one assumes the validity of the Heisenberg Hamiltonian [36] and on the other hand takes again excitation energies from spiral spin-configurations, which force us to neglect spin–orbit coupling. This is perhaps more of a problem here than in the case of estimating Curie temperatures, because the acoustic spin-waves have small energies. Nevertheless, we proceed by stating that the spin-wave spectrum as a function of the wave vector $\mathbf{k}$, $\omega(\mathbf{k})$, is obtained by diagonalizing a matrix $\mathscr{A}$

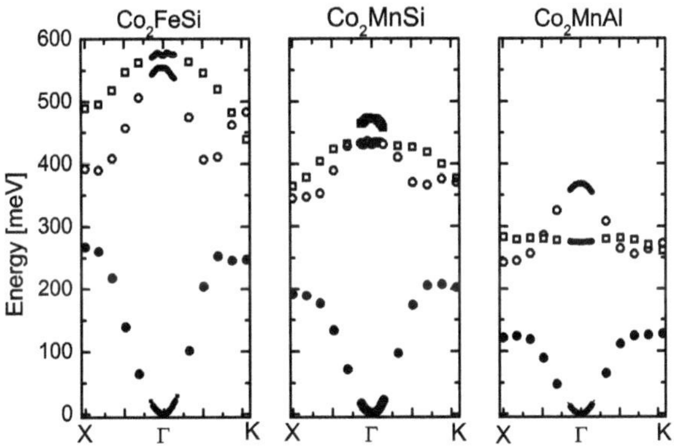

Fig. 5.20 Calculated spin-wave spectra of Co_2FeSi, Co_2MnSi, and Co_2MnAl, along two symmetry lines in the Brillouin zone. A fit around the zone center gives the spin-wave stiffness constants: 705 meV Å^2 for Co_2FeSi, 770 meV Å^2, and 570 meV Å^2 for Co_2MnAl

whose elements are given by

$$(A)_{\nu\mu} = 4\sqrt{M_\nu}\tilde{j}_{\nu\mu}(\mathbf{k})\sqrt{M_\mu}, \tag{5.2}$$

where the quantity $\tilde{j}_{\nu\mu}(\mathbf{k})$ is related to the exchange functions $j_{\nu\mu}(\mathbf{k})$ by

$$\tilde{j}_{\nu\mu}(\mathbf{k}) = j_{\nu\mu}(\mathbf{k})\cos\big[\mathbf{k}\cdot(\boldsymbol{\tau}_\mu - \boldsymbol{\tau}_\nu)\big] - \delta_{\mu\nu}\sum_\lambda j_{\mu\lambda}(0)M_\lambda/M_\mu. \tag{5.3}$$

The result is derived in Refs. [10, 36].

Taking the exchange functions used in the previous section and doing the required diagonalization for the three by three matrix $\mathscr{A}$ we obtain the results shown in Fig. 5.20. Unfortunately, to our knowledge no experimental data are available to verify the calculated spectra, except for measurements of the spin-wave stiffness constant [37], which basically is the curvature of the acoustic spectrum at the Γ-point. The calculated spin-wave stiffness constant of Co_2FeSi of 705 meV Å^2 is in (perhaps fortuitous) agreement with the measured value of 715 meV Å^2, for Co_2MnSi we compare the value of 770 meV Å^2 with the measured 580 meV Å^2, and for Co_2MnAl we compare 570 meV Å^2 with measured 140 meV Å^2. Although the decreasing Curie temperatures and hence decreasing exchange functions are reflected in the over-all scale of the parts of Fig. 5.20, the large measured decrease of the stiffness constants cannot be reproduced by our calculations. The reasons for this disagreement are presently unclear.

We close this subsection by a rough estimate of the decrease of the spin polarization by exciting spin waves, i.e. by warming the crystal. Taking for example Co_2MnSi we inspect the acoustic part of the spin-wave spectrum in Fig. 5.20 and read off roughly the value of $\omega(\mathbf{k})$ at about 25 meV for k $\simeq 0.2$ (in units of $2\pi/a$), corresponding approximately to room temperature. Therefore we calculate the density of states in the presence of a spin-spiral of wave vector $k = 0.2$ and for clarity also for $k = 0.3$, thus estimating the changes for about 500 K, too. The result is shown in Fig. 5.21, where one can approximately read off the corresponding polarization decrease to be about 20 % and about 40 % for the higher temperature. That is, there still remains a polarization near 80 % at room temperature, a value that is near the observed polarization.

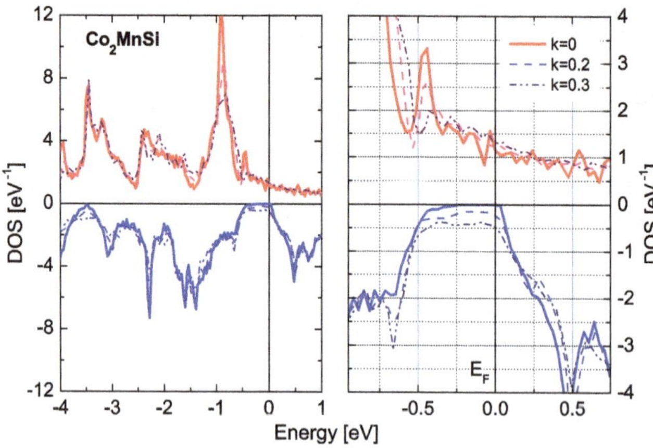

Fig. 5.21 Demonstrating polarization changes in the density of states (DOS) of Co_2MnSi through application of spin-spirals of wave vector 0.2 and 0.3 in units of $2\pi/a$. On the *right* is the zoomed DOS

Fig. 5.22 Connection of the $L2_1$-Heusler crystal structure on the *left* with the perovskite structure on the *right*. For better visibility the oxygen octahedra are only drawn around three atoms

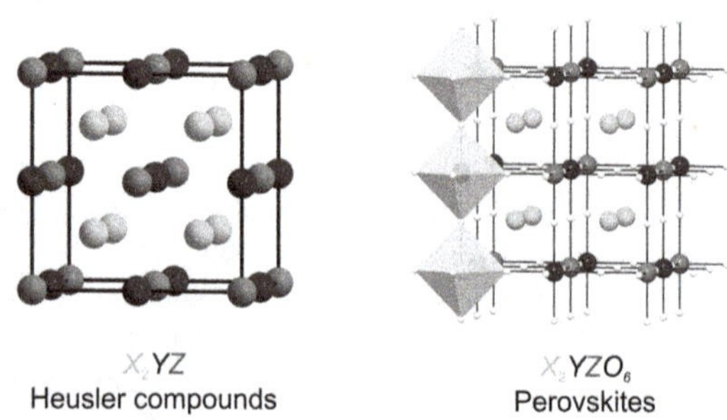

5.4 Electronic Structure of the Double Perovskites

The crystal structure of the double perovskites is closely related with the Heusler compounds and can easily be derived from the $L2_1$ structure as is shown in Fig. 5.22. If we denote the double perovskite by the formula X_2YZO_6, where X = Sr, Ba, Ca, Y = Cr, Mn, Fe, and Z = Mo, W, Re, Os, then the X-atoms occupy the corners of the inscribed cube (8 light-gray spheres), Y and Z occupy the fcc-sites that all together form a rock-salt structure. In addition, corner-sharing oxygen octahedra are introduced such that their centers coincide with the Y- and Z-atoms, thus placing an oxygen atom between these two atoms. Often the cubic structure is tetragonally distorted, sometimes even forming lower symmetric structures [8].

The interest in these compounds was excited by the report of room-temperature colossal magneto-resistance in the ordered double perovskite (DP) Sr_2FeMoO_6 [38]. The high degree of spin polarization in this and other DP's is believed to be due to the half-metallic nature of the electronic structure [38, 39]. In Fig. 5.23 we give an overview of a set of double perovskites arranged according to the Curie temperatures and the number of valence electrons, N_V, which are supplied by the transition-metal

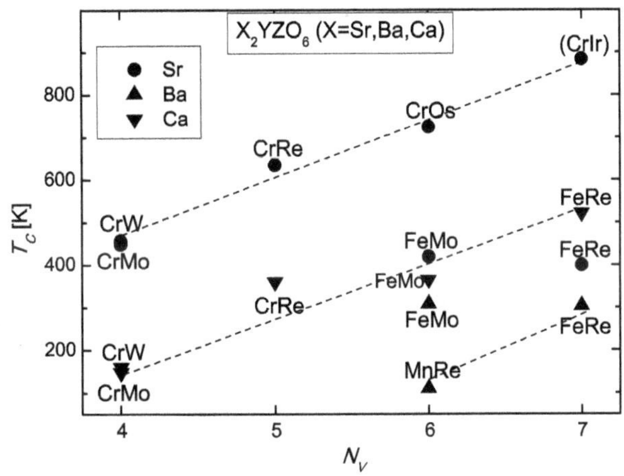

Fig. 5.23 Measured Curie temperatures (T_C) versus the number of valence electrons N_V. The names of the elements Y and Z in the double perovskites X_2YZO_6 appear near the data points. The *value in parentheses* is a projected one. Reproduced with permission from Mandal, T.K., Felser, C., Greenblatt, M., and Kübler, J.: *Phys. Rev. B* 78, 134431 [41]. Copyright (2008) by the American Physical Society

elements Y and Z. The remaining number of electrons from the X and O atoms are the same in each series and are therefore suppressed in the notation. This figure is not intended to be complete; there are a great number of other double perovskites described in the literature, see for instance Refs. [8, 40].

5.4.1 Band Structure of the Double Perovskites

Largely following Ref. [41] we now discuss the band structure of a selection of double perovskites, arranged in a way that brings out the relevant trends.

Beginning with Fig. 5.24 we show the spin- and site-resolved densities of states, DOS, of the Sr-series with Cr and the $5d$-transition elements W, Re, and Os. Different shadings distinguish the DOS due to the different atoms. Going from left to right in this figure the magnetic moment per formula unit decreases from 2 μ_B to 0 μ_B. Clearly, Sr_2CrWO_6 and Sr_2CrReO_6 are half-metallic magnets and Sr_2CrOsO_6 is an insulator. The magnetic moment of Cr in these compounds is about 2.5 μ_B to 2.4 μ_B; the moments of the $5d$-metals (and a smaller contribution from oxygen) are oppositely oriented, thus causing the decrease of the total moment to zero in Sr_2CrOsO_6. Hence, these double perovskites are half-metallic ferrimagnets. The reason for the opposite orientation of the moments is clearly the oxygen-ion in between Cr and the $5d$-elements. The occupation of the various states can easily be read off from Fig. 5.24: the three up-spin t_{2g}-states of Cr are all occupied, in W there is one electron in the down-spin t_{2g}-states, in Re there are two, and in Os the down-spin t_{2g}-states are filled completely with three electrons.

The compound Sr_2CrOsO_6 was synthesized by Krockenberger et al. [42] who also determined the electronic structure verifying the insulating property both experimentally and theoretically, albeit with a smaller gap in the up-spin spectrum. Sr_2CrOsO_6 and Sr_2CrRuO_6, the latter being similar to the former, were also discussed by Lee and Pickett [43], who obtained a gap for both systems, although a smaller one in Sr_2CrRuO_6, in agreement with the calculations of Mandal et al. [41].

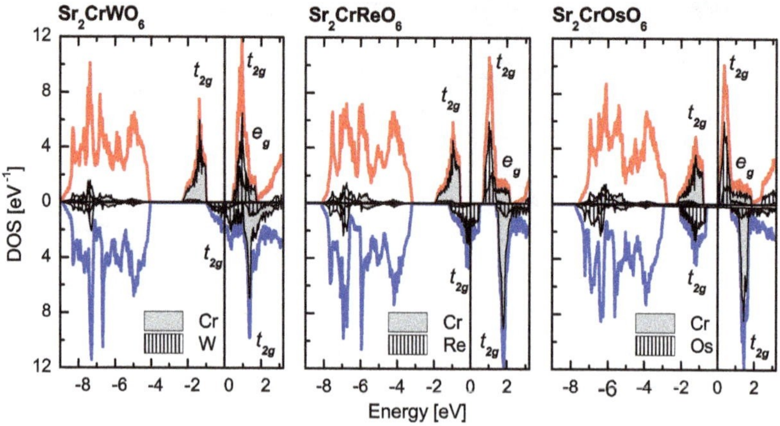

Fig. 5.24 The density of states (DOS) of Sr_2CrWO_6, Sr_2CrReO_6 and Sr_2CrOsO_6 (Reproduced with permission from Mandal, T.K., Felser, C., Greenblatt, M., and Kübler, J.: *Phys. Rev. B* 78, 134431 [41]. Copyright (2008) by the American Physical Society). *Upper panel* is spin up, *lower panel* spin down. The partial DOS for Cr, W, Re, and Os is graphed with different shadings. From left to right the number if valence electrons is 4, 5, and 6, while the magnetic moment is 2 μ_B, 1 μ_B, and 0 μ_B per formula unit. The labels t_{2g} and e_g give the symmetries of the dominant d-states. The calculations here employ the GGA

It must be emphasized that the vanishing magnetic moment is a result of complete cancellation. If spin–orbit coupling is employed the gap remains but the cancellation is no longer complete. A small ferromagnetic component appears of about 0.2 μ_B. Thus we here see a *weak ferromagnetism* in an insulator, which is a rather rare occurrence.

We now turn to the Sr-double perovskites that contain Fe. A series of three densities of states is shown in Fig. 5.25, where we selected Sr_2FeMoO_6, Sr_2FeReO_6, and Sr_2FeRuO_6. Here the number of valence electrons increases from 6 to 8 going through the figure panels from left to right; the magnetic moments are 4 μ_B for Sr_2FeMoO_6 and 3 μ_B for Sr_2FeReO_6. But then the expected trend is broken by Sr_2FeRuO_6, for which a ferromagnetic solution is found to be unstable. This instability is observed through a spin-wave branch that becomes negative. A search for another ground state leads via an LDA+U calculation to an antiferromagnet Mott–Hubbard insulator with a small gap. The ordering wave-vector is found to be $\mathbf{K} = (0.5, 0.5, 0.5)$ (AFII) with the Ru-moment orthogonal to the Fe-moment. Similar calculations were undertaken by Fang et al. [44] for Sr_2FeZO_6 (Z = Mo,W,Re), who found an AFII order for Sr_2FeWO_6.

The density of states for Sr_2FeMoO_6 shown in Fig. 5.25 is in good agreement with previous calculations [38, 44]. Here and in Sr_2FeReO_6 one sees that the filled up-spin t_{2g} states of Fe have merged with the oxygen states and the up-spin e_g-states are filled also. Oxygen enforces the moments of Mo and Re to be antiparallel to the Fe moments, which in these compounds have a value of about 3.7 μ_B; therefore, the down-spin t_{2g}-states of Mo are occupied with one and those of Re with two electrons. In the Ru case the t_{2g}-states are filled completely, but spread over the

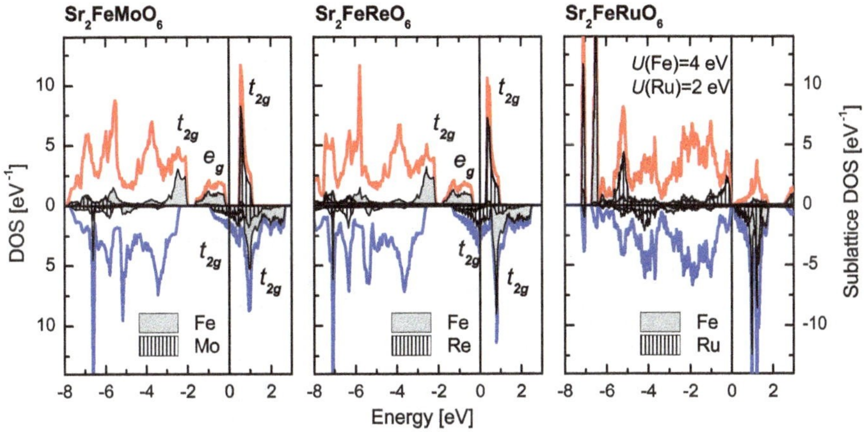

Fig. 5.25 The density of states (DOS) of Sr_2FeMoO_6, Sr_2FeReO_6 and Sr_2FeRuO_6 (Reproduced with permission from Mandal, T.K., Felser, C., Greenblatt, M., and Kübler, J.: *Phys. Rev. B* 78, 134431 [41]. Copyright (2008) by the American Physical Society). *Upper panel* is spin up, *lower panel* spin down. The partial DOS for Fe, Mo, Re, and Ru is graphed with different shadings. From left to right the number if valence electrons is 6, 7, and 8, while the magnetic moment is 4 μ_B, 3 μ_B, and 0 μ_B per formula unit. The labels t_{2g} and e_g give the symmetries of the dominant d-states. The calculations employ the LSDA, except for Sr_2FeRuO_6, where the LDA+U method was used. The latter case is an antiferromagnetic Mott–Hubbard insulator

entire range of the oxygen density of states. Note that in this case the sublattice DOS is graphed in Fig. 5.25.

We close this section with a discussion of Ba_2FeMoO_6 and Ba_2MnReO_6, for which the density of states are shown in Fig. 5.26. The number of valence electrons is in both cases 6 and the magnetic moment is 4 μ_B. In the case of Ba_2FeMoO_6 the GGA results in the half-metallic ferrimagnet shown, the DOS being very similar to the case of Sr_2FeMoO_6. This is in contrast to Ba_2MnReO_6, where both LSDA and GGA calculations return a ferrimagnetic metal with spin-up Mn-t_{2g}-states and spin-down Re-e_g-states at the Fermi energy. This is in stark contrast to measurements by Popov et al. [45] who found this compound to be of high resistivity. We therefore applied the LDA+U method and obtained a half-metallic ferrimagnet. The DOS is shown in Fig. 5.26. Comparing with the case of Ba_2FeMoO_6 we see that all spin-up Mn states are well below E_F and the states at E_F are almost purely spin-down Re states.

5.4.2 Curie Temperatures of the Double Perovskites

In Fig. 5.27 measurements of the Curie temperatures of a set of double perovskites are collected and compared with ab initio calculations [41]. These were carried out by determining up to six exchange functions by means of spin-spiral total energies, as was described for Heusler compounds in Sect. 5.2. For the Curie temperatures

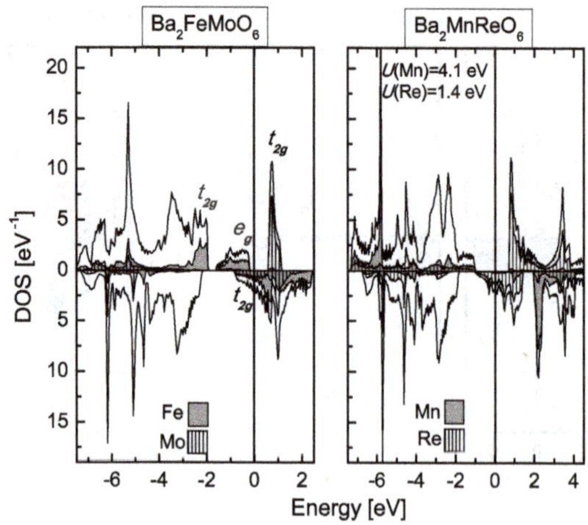

Fig. 5.26 The density of states (DOS) of Ba_2FeMoO_6 and Ba_2MnReO_6 (Reproduced with permission from Mandal, T.K., Felser, C., Greenblatt, M., and Kübler, J.: *Phys. Rev. B* 78, 134431 [41]. Copyright (2008) by the American Physical Society). *Upper panel* is spin up, *lower panel* spin down. The partial DOS for Fe, Mo, Re, and Mn is graphed with different shadings. In both cases the number of valence electrons is 6 while the magnetic moment is 4 μ_B per formula unit. The labels t_{2g} and e_g give the symmetries of the dominant d-states. The calculations employ the LSDA for Ba_2FeMoO_6 and the LDA+U method for Ba_2MnReO_6

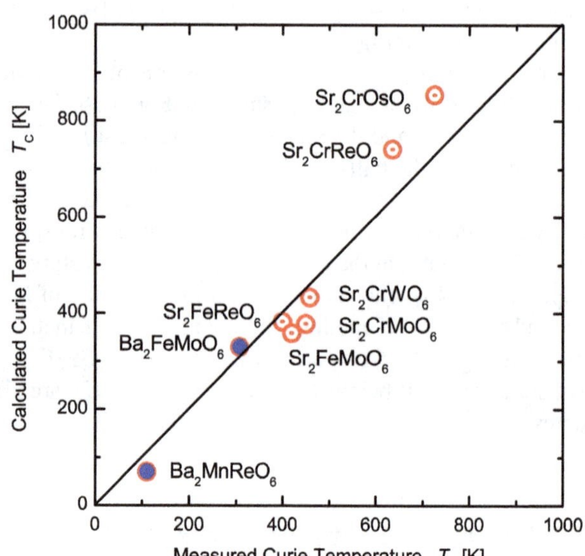

Fig. 5.27 Calculated versus measured Curie temperatures of a set of double perovskites. Reproduced with permission from Mandal, T.K., Felser, C., Greenblatt, M., and Kübler, J.: *Phys. Rev. B* 78, 134431 [41]. Copyright (2008) by the American Physical Society

the spherical approximation was used as given in (5.1). Some notable details are described in the following.

When the induced magnetic moments of the oxygen atoms is smaller than 0.1 μ_B it was assumed that three exchange functions are sufficient to obtain a fair estimate of the Curie temperatures; these are (i) the exchange functions between the Y-atoms, (ii) between the Z-atoms and (iii) between the Y- and Z-atoms, denoting the double perovskite by X_2YZO_6. This results in two eigenvalues to be used in (5.1). In the case of Ba_2FeMoO_6 the induced oxygen moments were found to be slightly larger than 0.1 μ_B. For this case exchange interactions between the O-atoms, between the O- and the Y-atoms, and between the O- and the Z-atoms were also determined from spiral spin configurations, which led to an additional eigenvalue in (5.1).

In Fig. 5.27 it is seen that the Curie temperature of Ba_2MnReO_6 is slightly underestimated (71 K calculated versus 110 K measured); this calculation was done using the LDA+U method as described above. For Ba_2FeMoO_6 the calculations with six exchange constants overestimates the Curie temperature slightly (331 K calculated versus 308 K measured).

For the Mott–Hubbard insulator Sr_2FeRuO_6, one of the three cases for which the densities of states were shown in Fig. 5.25, the determination of the Néel temperature failed because one of the spin-wave branches went negative. This perhaps indicates that the calculated ground state is incorrect. The Curie temperatures of the other two half-metallic ferrimagnets, Sr_2FeMoO_6 and Sr_2FeReO_6 are in moderate to fair agreement with the measured values.

Of the four Sr_2CrZO_6 Curie temperatures shown in Fig. 5.27 two appear considerably overestimated, including the interesting magnetic insulator Sr_2CrOsO_6, where the calculations give $T_C = 881$ K which is to be compared with the measured value of $T_C = 725$ K. Still, the calculations of the Curie temperatures reproduced the over-all trend shown in Fig. 5.23. A glance at Fig. 5.27, however, reveals deviations notably in the calculations where the GGA had to be used. It is emphasized that the neglected structural distortions and anti-site disorder can influence the measured magnetic properties so that a comparison with the idealized calculations may only result in qualitative understanding, especially for the case of the magnetic insulator Sr_2CrOsO_6.

Finally, the large Curie temperatures, especially for Sr_2CrReO_6 and Sr_2CrOsO_6, require remarkably large ferromagnetic exchange interactions between the Cr cations. This issue was raised by Sarma et al. [46] and in large detail by Fang et al. [44] who invoked a mechanism that is comparable to double exchange (DE) discussed in some detail in Sect. 5.3.1. This mechanism involves charge carriers and is thus operative for the half metallic ferrimagnets. For the magnetic insulators Sr_2CrOsO_6 and Sr_2CrRuO_6 and the Mott–Hubbard insulator Sr_2FeRuO_6 we resort to the antiferromagnetic super-exchange (SE), also discussed in Sect. 5.3.1 which through the $4d$- and $5d$-cations results in ferromagnetic exchange between the $3d$-cations. The interaction path that is typical for both DE and SE possesses a distinctly visible fingerprint in Figs. 5.24, 5.25, and 5.26 where the oxygen-transition-metal hybridization is clearly visible at an energy of about −7 eV to −6 eV.

Acknowledgements The financial support by the Deutsche Forschungsgemeinschaft in research unit FOR 559 is gratefully acknowledged. The authors thank Tanja Graf and Benjamin Balke for their kind help with the figures.

References

1. de Groot RA, Mueller FM, van Engen PG, Buschow KHJ (1983) Phys Rev Lett 50:2024
2. de Groot RA, Mueller FM, van Engen PG, Buschow KHJ (1984) J Appl Phys 55:2151
3. Oppeneer PM, Antonov VN, Kraft T, Eschrig H, Yaresko AN, Perlov AY (1995) Solid State Commun 94:255
4. Kübler J, Williams AR, Sommers CB (1983) Phys Rev B 28:1745
5. Katsnelson MI, Irkhin VY, Chioncel L, de Groot RA (2008) Rev Mod Phys 80:315
6. Felser C, Fecher GH, Balke B (2007) Angew Chem Int Ed 46:668
7. Balke B, Wurmehl S, Fecher GH, Felser C, Kübler J (2008) Sci Technol Adv Mater 9:014102
8. Serrate D, De Teresa JM, Ibarra MR (2007) J Phys Condens Matter 19:023201
9. Bozorth RM (1951) Ferromagnetism. Van Nostrand Company, New York
10. Kübler J (2009) Theory of itinerant electron magnetism. Oxford University Press, Oxford
11. Friedel J (1958) Nuovo Cimento 10(Suppl. 2):287
12. Ishikawa Y (1977) Physica 91B:130
13. Hamcić A, Azomoza R, Campbell IA (1981) J Phys F 11:1441
14. Kübler J, Fecher GH, Felser C (2007) Phys Rev B 76:024414
15. Wurmehl S, Fecher GH, Kandpal HC, Ksenofontov V, Felser C, Lin H-J, Morais J (2005) Phys Rev B 72:184434
16. Fecher GH, Felser C (2007) J Phys D, Appl Phys 40:1582
17. Galanakis I, Dederichs PH, Papanikolaou N (2002) Phys Rev B 66:174429
18. Söderlind P, Ahuja R, Eriksson O, Wills JM, Johansson B (1994) Phys Rev B 50:5918
19. Hanssen KEHM, Mijnarends PE, Rabou LPLM, Buschow KHJ (1990) Phys Rev B 42:1533
20. Brooks H (1940) Phys Rev 58:909
21. de Groot RA (1991) Physica B 172:45
22. Ksenofontov V, Melnyk G, Wojcik M, Wurmehl S, Kroth K, Reiman S, Blaha P, Felser C (2006) Phys Rev B 74:134426
23. Wurmehl S, Kandpal HC, Fecher GH, Felser C (2006) J Phys Condens Matter 18:6171
24. Balke B, Fecher GH, Winterlik J, Felser C (2007) Appl Phys Lett 90:152504
25. Galanakis I, Özdogan K, Sasioglu E, Aktas B (2007) Phys Rev B 75:172405
26. Kübler J (2006) J Phys Condens Matter 18:9795
27. Liechtenstein AI, Katsnelson MI, Antropov VP, Gubanov VA (1987) J Magn Magn Mater 67:65
28. Kurtulus Y, Dronskowski R, Samolyuk GD, Antropov P (2005) Phys Rev B 71:014425
29. Thoene J, Chadov S, Fecher GH, Felser C, Kübler J (2009) J Phys D, Appl Phys 42:084013
30. Sasioglu E, Sandratskii LM, Bruno P, Galanakis I (2005) Phys Rev B 72:184415
31. Sandratskii LM (1998) Adv Phys 47:91
32. Anderson PW (1950) Phys Rev 79:350
33. Anderson PW (1963) In: Magnetism I. Rado G, Suhl H (eds) Academic Press, New York. Chap 2
34. Zener C (1951) Phys Rev 82:403
35. de Gennes P-G (1960) Phys Rev 118:141
36. Halilov SV, Eschrig H, Perlov AY, Oppeneer PM (1998) Phys Rev B 58:293
37. Trudel S, Gaier O, Hamrle J, Hillebrands B (2010) J Phys D, Appl Phys 43:193001
38. Kobayashi K-I, Kimura T, Sawada H, Terakura K, Tokura Y (1988) Nature (London) 395:677
39. Park JH, Vescovo E, Kim HJ, Kwon C, Ramesh R, Ventkatesan T (1998) Nature (London) 392:794
40. Pardo V, Pickett WE (2009) Phys Rev B 80:054415
41. Mandal TK, Felser C, Greenblatt M, Kübler J (2008) Phys Rev B 78:134431
42. Krockenberger Y, Mogare K, Reehuis M, Tovar M, Jansen M, Vaitheeswaran G, Kanchana V, Bultmark F, Delin A, Wilhelm F, Rogalev A, Winkler A, Alff L (2007) Phys Rev B 75:020404(R)
43. Lee K-W, Pickett WE (2008) Phys Rev B 77:115101
44. Fang Z, Terakura K, Kanamori J (2001) Phys Rev B 63:180407(R)

45. Popov G, Greenblatt M, Croft M (2003) Phys Rev B 67:024406
46. Sarma DD, Mahadevan P, Saha-Dasgupta T, Ray S, Kumar A (2000) Phys Rev Lett 85:2549

Chapter 6
Correlation and Chemical Disorder in Heusler Compounds: A Spectroscopical Study

Jürgen Braun, Hubert Ebert, and Ján Minár

Abstract The first part of this study deals with the effects of local electronic correlations and alloying on the properties of the Heusler compound $Co_2Mn_{1-x}Fe_xSi$. The analysis has been performed by means of first-principles band-structure calculations based on the local approximation to spin-density functional theory (LSDA) as well as photoemission calculations within the one-step model of photoemission. Correlation effects are treated using the Dynamical Mean-Field Theory (DMFT) and the LSDA+U approach. The formalism is implemented within the Korringa–Kohn–Rostoker (KKR) Green's function method. In satisfactory agreement with available experimental data the magnetic and spectroscopic properties of $Co_2Mn_{1-x}Fe_xSi$ are explained in terms of strong electronic correlations. In addition the correlation effects have been analyzed separately with respect to their static or dynamical origin. To achieve a quantitative description of the electronic structure of $Co_2Mn_{1-x}Fe_xSi$ both static and dynamic correlations must be treated on equal footing. Furthermore, we report on our investigation of the spin-dependent electronic structure of ordered NiMnSb as well as of the disordered $Ni_xMn_{1-x}Sb$ alloy system. As a first point we studied the magneto-optical Kerr effect in ordered NiMnSb to extract information on the bulk-related electronic structure of this compound. In addition the influence of chemical disorder on the unoccupied electronic density of states was investigated by use of the ab-initio Coherent Potential Approximation method. These results are used for a detailed discussion of spin-resolved Appearance Potential Spectroscopy measurements. Our theoretical approach describes the spectra as the fully relativistic self-convolution of the matrix-element weighted, orbitally resolved density of states. The analysis is completed by one-step photoemission calculations focusing on the surface electronic structure of ordered NiMnSb(001).

J. Braun · H. Ebert · J. Minár (✉)
Department Chemie, Ludwig-Maximilians-Universität München, Butenandtstr. 11, 81377 Munich, Germany
e-mail: jan.minar@cup.uni-muenchen.de

J. Braun
e-mail: juergen.braun@cup.uni-muenchen.de

H. Ebert
e-mail: hubert.ebert@cup.uni-muenchen.de

C. Felser, G.H. Fecher (eds.), *Spintronics*, DOI 10.1007/978-90-481-3832-6_6,

6.1 Introduction

In 1903, the first Heusler alloys have been singled out as a special group of magnetic materials that exhibit ferromagnetism in compounds containing only non-magnetic elements [1]. Currently, the Mn-based half-Heusler alloys with the generic formula $X\text{Mn}Z$ (with X being a $3d$ metal and Z belonging to the III or IV group) are well known magnetic systems that crystallize in the $C1_b$ structure, which is closely related to the zinc blende structure. Especially, the half-metallic ferromagnet NiMnSb [2] has been triggered a variety of experimental and theoretical investigations due to the search for materials to be used as spin-injectors. These materials are defined as magnetic materials with a band gap at the Fermi level for electrons of one spin direction. Band structure calculations for bulk NiMnSb show a gap of about 0.5 eV at T = 0 K for the minority electrons, which means 100 % spin polarization at the Fermi level E_F. In fully relativistic calculations a reduced but not vanishing gap and 99 % spin polarization at E_F are predicted [3]. As in the case of the group III–V semiconductors, the crystal structure is the reason for the occurrence of the band gap. It is, therefore, called a covalent band gap [4]. As a consequence, the crystal structure and the site occupation within the given structure are important for the appearance of the gap. Atomic disorder, especially Ni–Mn interchange, results in a strong reduction of the spin polarization at E_F [5]. However, such an interchange is comparable in energy with the evaporation energy of the metallic constituents [4]. In consequence, half-metallic behavior can be seen as an idealization hindered due to the influence of structural inhomogeneities [6–8] and through the decrease of the spin polarization of spin carriers with increasing temperature. Also, a vanishing of the semiconducting band gap for minority spin electrons takes part in real crystals and interfaces caused by finite temperature effects.

One of the most interesting class of materials for magneto-electronic applications is found in the family of Co_2YZ half-metallic Heusler ferromagnets [9, 10]. Special attention has been paid to Co_2MnSi compound because of a predicted large minority band gap (0.4 eV) and a very high Curie temperature (985 K) among the variety of Heusler compounds [11–14]. However, an even more interesting material, Co_2FeSi, was found in recent investigations [15–17] exhibiting a Curie temperature of about 1100 K and a very high magnetic moment of 6 μ_B in the primitive cell. Later on it was shown that the complete substitutional series $Co_2Mn_{1-x}Fe_xSi$ crystallizes in the Heusler type $L2_1$ structure with a high degree of structural order [18]. Also confirmed was a Slater–Pauling behavior [13, 19, 20] of the magnetic moment ranging linearly from 5 μ_B to 6 μ_B with increasing Fe concentration x.

The detailed understanding of the electronic and magnetic structure of half-metallic ferromagnets, their surfaces and corresponding nanostructures is directly connected with an appropriate theoretical description based on first-principle methods. From previous studies [21, 22] it follows that quantitative calculations of the electronic structure of $Co_2Mn_{1-x}Fe_xSi$ compounds require a correct incorporation of local correlation effects. For example, a former study [21] applying the LSDA+U method [23, 24] was able to reproduce the experimental band gap as well as the correct values of the magnetic moments when using reasonable values for the parameters U and J and including appropriate double-counting corrections. In the present analysis electronic correlation effects have been treated in the framework of the Dynamical Mean-Field Theory [25], which quantitatively takes into account dynamical correlation effects, in particular spin-flip processes induced by fluctuations. In combination with the LSDA (LSDA+DMFT) this formalism provides a very reliable approach to deal with a wide range of ground state and spectroscopic properties of 3*d* transition metals. These are, for example, total energies, magnetic moments as well as magneto-optical and photoemission spectra [26–31].

The most serious complication when combining LSDA with DMFT arises from the double-counting problem. As has been found by numerous DMFT studies, the static many-body effects which are typically over counted in LSDA+DMFT calculations are relatively small in 3*d* transition metals. Therefore, the established procedure concerning the description of spectroscopic properties is to neglect the dynamical part of the self-energy, setting this term to zero at the Fermi level [26, 27, 32]. However, it was recently shown [30] that this approximation is not sufficient for an accurate description of angle-resolved photoemission spectra of Ni. In this case an additional static polarization term, which is spin- and orbital-dependent had to be included in the calculations. Later it was demonstrated [31] that using LSDA+U as a static limit for DMFT leads to an improved description of the orbital and spin magnetic moments in a wide range of the 3*d* transition metal compounds.

A recent investigation on Co_2MnSi [33], based on the assumption that static correlations are already accounted by the LSDA, thus considering only dynamical contributions in the self-energy, supports the existence of the so-called non-quasiparticle states [34]. However, the insufficient treatment of static correlations resulted in a wrong position of the Fermi level, which in the case of Co_2MnSi [21, 22] must be located closer to the lower edge of the band gap. Similar findings were obtained in former investigations [35] based on plain LSDA calculations. Therefore, the position of the Fermi level with respect to the band gap is controlled more by the static rather than by the dynamic correlations. In the present work, both static and dynamic correlations are taken into account simultaneously by combining them as done in [31].

As a second example, we present spin-integrated and spin-resolved APS spectra from NiMnSb(001) and discuss by comparison with experimental data the degree of spin polarization as a function of chemical disorder. APS is a surface sensitive tool to study the unoccupied density of states (DOS) with elemental resolution [36]. For spin-resolved APS, the sample is bombarded with a spin-polarized electron beam of variable energy while the total yield of emitted X-rays or electrons is monitored [37, 38]. The APS signal depends on the spin of the exciting electron in the case of ferromagnets because of the spin-dependent unoccupied DOS. The elemental resolution comes from the fact that core levels are involved whose energies are characteristic for the various elements.

Finally we present fully relativistic one-step photoemission calculations of the ordered alloy NiMnSb(001) focusing on the surface electronic structure and in consequence on the degree of spin-polarization that can be measured from Heusler compounds like NiMnSb.

6.2 Theoretical and Computational Details

The electronic structure calculations were performed using the relativistic full-potential Korringa–Kohn–Rostoker Green function method (SPR-KKR) [39]. The DMFT-solver consists in the relativistic version of the so-called Spin Polarized T-Matrix Plus Fluctuation Exchange (SPTF) approximation [32, 40] using the real-energy axis formalism [41]. According to this scheme the local Green function is obtained by the corresponding site projection of the full KKR Green's function. The local Green function is needed to obtain the bath Green function for the Anderson impurity model via the saddle-point equation. The bath Green function is used as input for the SPTF DMFT-scheme to calculate the local self-energy. The latter appears as an additional energy-dependent potential in the radial Dirac equation which is solved in order to calculate the new full KKR Green function. This procedure is repeated until self-consistency in both the self-energy and the charge density is achieved. This scheme has been already successfully applied for the description of magneto-optical [42] and magnetic [31] properties of $3d$ transition metals. Also photoemission intensities [43], including the corresponding matrix element effects, were obtained in quantitative agreement with the experiment.

The static double-counting is treated in the form of the so-called "around atomic limit" (AAL) [24] which is appropriate in the case of integer orbital occupation numbers and therefore well applicable to half-metals. Although different mechanisms exist that lead to half-metallicity, in the case of locally correlated half-metals the AAL tends to increase the spin magnetic moment, which is often underestimated in plain LSDA calculations [15]. The self-energy within the DMFT is parameterized by the averaged screened Coulomb interaction U and the Hund exchange interaction J. For the latter the screening is usually not crucial; the value of J can be calculated directly from LSDA and is approximately equal to 0.9 eV for all $3d$ elements. This value has been adopted for all calculations presented here. Different methods in calculating the screened Coulomb interaction U for the metallic $3d$ transition metal systems lead to U-values in the range between 2–3 eV. In our analysis we keep the U parameter fixed at 3 eV which is often used for $3d$ transition metals.

The random substitutional order of the Mn-Fe compounds is treated within the so-called Coherent Potential Approximation (CPA). Since the CPA approach can be formulated in the multiple scattering formalism [44, 45] it has a rather straightforward implementation within the KKR method [46]. As it follows from the experiment the lattice constant does not change substantially its value within the whole range of concentrations x. The lattice parameter was fixed to 5.658 Å in the present study.

Concerning the calculation of the ground state properties few additional details have to be mentioned at this stage. First of all we stress the importance of the full potential treatment, which often leads to some improvements in the description of the band gap already for plain LSDA calculations. In addition, one has to increase the l-value for the expansion in spherical harmonics. This typically improves the calculated value for the band gap and for the magnetic moments. In our case we used $l_{\max} = 3$ (f electrons).

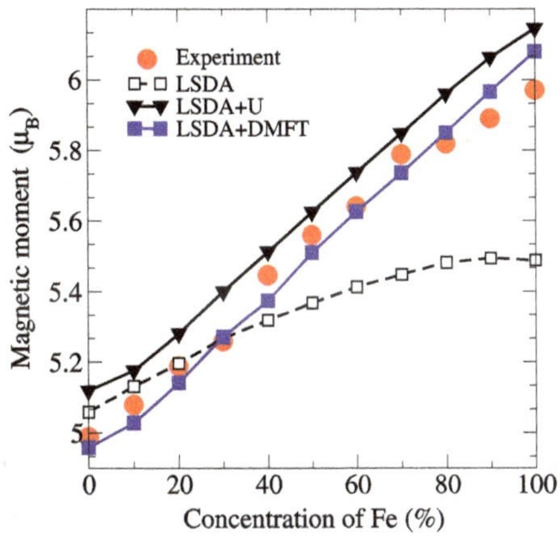

Fig. 6.1 Comparison of the total magnetic moments for $Co_2Mn_xFe_{1-x}Si$ compounds calculated within LSDA (*black dashed line, opened squares*), LSDA+U (*black triangles*) and LSDA+DMFT (*blue squares*) with the results of SQUID magnetic measurements (*red circles*) [52]. Reproduced with permission from S. Chadov, G.H. Fecher, C. Felser, J. Minar, J. Braun, and H. Ebert, *J. Phys. D: Appl. Phys.* 42, 084002. Copyright (2009), IOP Publishing

The 8 keV valence-band photoemission spectra (VB-XPS) were calculated using the formalism derived by Ebert and Schwitalla [47], which is based on the so-called one-step model of photoemission. An important aspect which one has to consider in calculations of hard X-ray photoemission is the correct treatment of the X-ray cross-section as a function of the photon energy. It follows from the calculations [22] that the cross-sections for p and d electrons are nearly the same for kinetic energies of 3–4 keV. However, at 8 keV their ratios are completely different, in particular the d cross-section becomes much smaller. In the present work these effects are taken into account implicitly within the one-step formalism. On the other hand the uncertainty in the calculation of the final state increases with increasing photon energy, because of an increasing number of oscillations in the final state radial wave functions. Therefore, one needs to account for a much larger number of points on the radial mesh in order to reproduce these variations correctly.

As the calculational procedure concerning APS is found in detail in [48, 49] we restrict ourselves to a short description of some computational details. The fully relativistic DOS calculations have been performed within the SPRKKR computer program [50] running in the CPA-mode [51]. The chemical disorder introduced in the spectroscopical analysis consists of an interchange between Mn atoms on Ni sites and vice versa with concentration values x of 0 %, 5 %, 10 % and 15 %. This means we have calculated $(Ni_{1-x}Mn_x)(Mn_{1-x}Ni_x)Sb$ in the $L2_1$ crystal structure using the experimental lattice constant 5.91 Å. In Figs. 6.2 and 6.3 we use the compact notation $Ni_{1-x}Mn_xSb$. Figure 6.1 shows the spin-resolved total DOS for NiMnSb. Clearly visible is the closing of the spin-dependent gap at the Fermi energy caused by chemical disorder. Even at 5 % interchange of Mn and Ni atoms the gap completely vanishes because spectral weight is shifted towards the Fermi level. A significant enhancement of spectral weight around the Fermi level is also

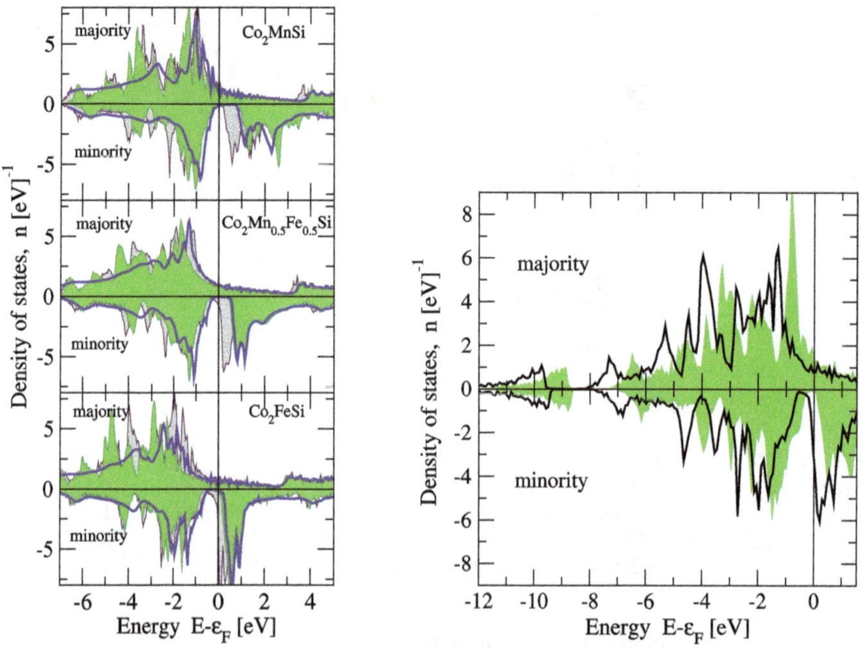

Fig. 6.2 *Left panel*: The total spin-resolved DOS curves for Co_2MnSi, $Co_2Mn_{0.5}Fe_{0.5}Si$ and Co_2FeSi calculated within LSDA (*light/gray filled area*), LSDA+U (*dark/green filled area*) and LSDA+DMFT (*blue line*). *Right panel*: Comparison of the total spin-resolved (LSDA) DOS curves for Co_2MnSi in case of the hole created in $1s$ core level of Mn (*black line*) and without (*green filled area*). Reproduced with permission from S. Chadov, G.H. Fecher, C. Felser, J. Minar, J. Braun, and H. Ebert, *J. Phys. D: Appl. Phys.* 42, 084002. Copyright (2009), IOP Publishing

observable in the majority DOS and therefore affect the amount of spin polarization that is measured in an APS experiment. The corresponding partial densities of states together with the spin-dependent potentials serve as input quantities for the spectroscopical calculations. In our case we probed the local DOS (LDOS) for the Mn atom in NiMnSb. The surface sensitivity is a result of the short mean free path of the electrons in the solid for energies that correspond to the Mn_{2p} excitation (≈640–650 eV). The elemental resolution results from the fact that a core level with an element-specific binding energy is involved. Due to the excitation with spin-polarized electrons the APS method becomes a magnetically sensitive technique. The calculated spin-dependent APS-intensities result from the weighted self-convolution of the corresponding LDOS. The weighting factor itself is given by the spin-dependent effective cross section consisting of the variety of relativistic Coulomb matrix elements. For a quantitative comparison between APS measurements and theoretical spectra one has to take into account various lifetime effects and the experimental broadening. Electron lifetime effects have been included in our analysis in a phenomenological way introducing a parametrized complex inner potential with an imaginary part of $V_{0i} = 30$ eV. The core hole lifetime is set to zero. Lifetime effects in the valence-band states are accounted via a Lorentzian with

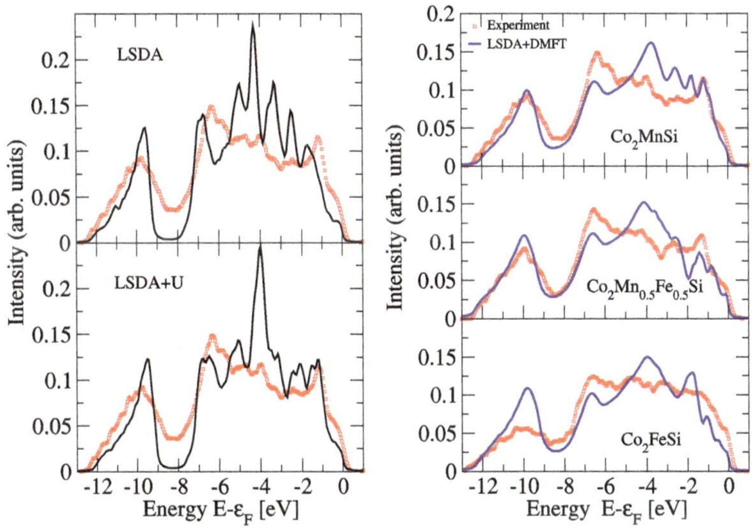

Fig. 6.3 *Left panel*: Comparison of the theoretical (*solid lines*) 8 keV VB-XPS spectra for Co_2MnSi calculated within plain LSDA and LSDA+U to the experiment (*red squares*) [22]. The area under each curve is normalized to unity. *Right panel*: The 8 keV VB-XPS spectra for Co_2MnSi, $Co_2Mn_{0.5}Fe_{0.5}Si$ and Co_2FeSi calculated within LSDA+DMFT (*solid line*) compared to experiment (*squares*) [22]. Reproduced with permission from S. Chadov, G.H. Fecher, C. Felser, J. Minar, J. Braun, and H. Ebert, *J. Phys. D: Appl. Phys.* 42, 084002. Copyright (2009), IOP Publishing

an energy-dependent width. For the explicit parametrization of the Lorentzian the reader is referred again to [48]. Strong correlations manifest themselves in form of the simple scaling factor $\lambda = 0.7(E - E_F)$. Furthermore, due to experimental conditions the first derivative of the APS signal has to be calculated and the apparatus broadening is considered by convoluting the calculated spectra by a Gaussian of FWHM = 1.4 eV.

6.3 Correlation Effects in Co_2MnSi

6.3.1 Magnetic Properties of Co_2MnSi

As was mentioned above, a quantitative treatment of static correlations is important for a reliable description of the minority band gap as well as of the magnetic moments in correlated half-metals. More details can be found in preceding publications [15]. Our calculations show a noticeable increase of the magnetic moments in analogy to the Slater–Pauling curve within the whole range of concentrations (see Fig. 6.1). By construction, the dynamical part of the local self-energy Σ in the vicinity of the Fermi level describes Fermi liquid behavior, i.e. $\mathrm{Re}\Sigma(\varepsilon) \sim -(\varepsilon - \varepsilon_F)$,

$\mathrm{Im}\Sigma(\varepsilon) \sim -(\varepsilon - \varepsilon_F)^2$. Therefore, we do not expect a substantial change for the integral quantities by accounting only for dynamical correlations. Indeed, as follows from Fig. 6.1, the magnetic moments calculated with the LSDA+DMFT scheme are only slightly reduced comparing to those obtained from LSDA+U. This small decrease may be attributed to the spin-flip events induced by the dynamical fluctuations leading to the so-called non-quasiparticle states absent in the LSDA treatment [33]. At the same time the effect of static correlations on the magnetic moment is more pronounced for the Fe-rich limit of the concentration x. The dynamical correlations appear to be more significant for the spectroscopic properties. As it follows from Fig. 6.2, the LSDA+U shifts the d states away from the Fermi level creating a band gap in the minority-spin channel. At the same time adding dynamical correlations tend to narrow the bandwidth by shifting the states back to the Fermi level, however, without influencing the band gap itself. One also can observe a substantial broadening of the DOS within the range between -2 to -8 eV, which is due to the imaginary part of the energy-dependent self-energy.

It follows from the DOS curves that the enhancement of the magnetic moment with the increase of the Fe concentration (see Fig. 6.2) is connected to the corresponding shift of the Fermi energy across the band-gap. In both limiting cases one observes a non-vanishing minority-spin DOS at the Fermi level. This indicates that both Co_2MnSi and Co_2FeSi may not represent fully spin polarized ferromagnets in contrast to the compounds with intermediate Fe concentrations. The influence of correlation effects in VB-XPS is illustrated in Fig. 6.3. It follows that the consideration of static correlations (LSDA+U) substantially improves the plain LSDA result in the range above -5 eV, reproducing the -1 eV and -4 eV peaks, which are shifted compared to plain LSDA by about 0.5 eV to higher binding energies. Furthermore, tiny features such as the local minima at -1.5 eV, -2.5 eV and the peak splittings are well reproduced. Not only the peak positions but also the intensities above -5 eV were improved in comparison to experiment, while only the -4 eV peak appears to be much too intense. Various attempts to treat the final states more accurately did not change this situation substantially. Therefore, it is concluded that the origin of this discrepancy is found in a band-structure effect caused by the Si p states and by the d states of Mn and Co. Indeed, the corresponding total DOS spectra (see Fig. 6.2) show the most intensive peak at about -3 eV binding energy, which is formed by the corresponding p and d states. As discussed above one has to scale the DOS energy axis by a factor of 1.1 in order to compare with the VB-XPS. At high photon energies the cross-section of the p states is larger than for the d states, however, the latter ones have larger spectral weight. Therefore, the contributions of p and d states are comparable.

The result of the complete correlation scheme is illustrated in the top panel of Fig. 6.3. In conclusion, it follows that inclusion of DMFT makes some important improvements for the VB-XPS spectra by correcting the amplitudes and the positions of the -1 eV and -10 eV peaks. Also the intermediate regime between -10 and -6 eV binding energy shows a very close agreement with experiment. The position of the -6 eV peak formed by p states of Si is still not perfectly reproduced. Nevertheless it is moved in the proper direction compared to the LSDA+U result.

The pronounced spectral feature observable at −4 eV is substantially reduced in intensity, but it is still overestimated by theory. The position is slightly shifted towards the Fermi level induced by the real part of the energy-dependent self-energy. The corresponding shift of the DOS is seen in Fig. 6.2. The energy regime ranging from −8.5 to −4.5 eV binding energy corresponds to the maximal amplitude of the imaginary component of the self-energy. From Fig. 6.3, it follows that this causes some Lorentzian over-broadening of the features at −4.5 eV for all considered concentrations. One the other hand it perfectly describes the so-called Heusler gap at about −8.5 eV. Thus, it is concluded that a non-zero intensity in the Heusler gap is mostly determined by the electron ground states existing in this region, rather than by background effects.

6.4 Electronic Structure of the Prototype Half Heusler Alloy NiMnSb

6.4.1 Magneto Optical Kerr Effect of NiMnSb

Complementary to photoemission, much information on the bulk electronic structure of magnetic solids is gained by optical and magneto-optical spectroscopy. This type of measurements are only weakly surface sensitive and for that reason primarily reflect the bulk electronic structure. However, measured optical and magneto-optical spectra can hardly be analyzed without a theoretical interpretation. For the last two decades ab-initio calculations of optical and magneto-optical properties of solids based on the LSDA often lead to a quantitative agreement between theoretical and experimental spectra [54]. The situation is very different when we consider more strongly correlated materials because in all calculations the LSDA eigenenergies are implicitly interpreted as one-particle excitation energies. Therefore, it is desirable to supplement LSDA calculations by many-body approaches to arrive at a realistic description of the one-particle excitations in correlated systems. Furthermore, corresponding theoretical works often use the so-called joined density of states for comparison with experimental data ignoring the influence of matrix elements. Only in a few cases LSDA+DMFT electronic structure calculations were complemented by an appropriate evaluation of the corresponding matrix elements which are necessary to obtain a realistic optical conductivity and in consequence the spin–orbit driven magneto optical Kerr effect (MOKE) [28, 42, 55]. In the following we examine the calculated MOKE spectra of the weakly correlated NiMnSb Heusler alloy [56].

Since the discovery of the giant magneto-optical Kerr effect in PtMnSb [57] magneto-optical properties became an important issue for the Mn-based family of Heusler alloys [58–66]. However, despite the similar structure, the group of the isoelectronic alloys PtMnSb, NiMnSb and PdMnSb show quite different maximum amplitudes in their MOKE spectra [57, 58]. A theoretical description of these differences is qualitatively possible within ab-initio band-structure calculations [64, 66–68], but one should notice that several systematic discrepancies between experiment

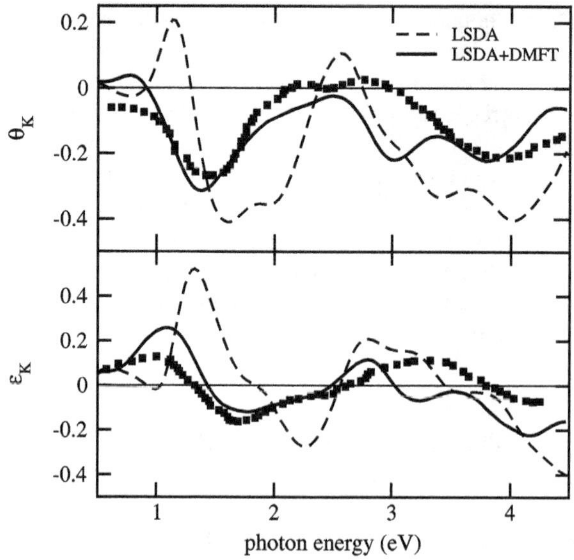

Fig. 6.4 Magneto-optical spectra for NiMnSb. *Upper panel*: Kerr rotation angle; *lower panel*: Kerr ellipticity. *Broken line*: LSDA calculations; *solid line*: LSDA+DMFT calculations. The square points represent the experimental results given in Refs. [57, 58]. Reproduced with permission from S. Chadov, J. Minar, H. Ebert, A. Perlov, L. Chioncel, M.I. Katsnelson, and A.I. Lichtenstein, *Phys. Rev. B* 74, R140411. Copryright (2006) by the American Physical Society

and theory exist generally ascribed to the use of the LSDA. In Fig. 6.4 we compare LSDA and LSDA+DMFT calculations of Kerr rotation and ellipticity with the corresponding experiments. It is clearly observable that the low-energy peak of the Kerr rotation spectrum located at about 1.4 eV is shifted significantly to lower energies around 1.6–2 eV. Also, a noticeable deviation of the peak intensity is observable at an energy of about 4 eV as well as in the intermediate energy regime. The discrepancies encountered in LSDA-based spectra are expected to appear due to an insufficient treatment of electronic correlations. In the case of NiMnSb the main contribution to the optical transitions came from the d-shell of Mn which supplies the unoccupied part of the density of states. At the same time d-electrons of Mn should be treated as locally correlated [56]. Consequently, we applied the LSDA+DMFT method on the Mn d-states with an effective Coulomb interaction parametrized by $U = 3$ eV and $J = 0.9$ eV (solid line in Fig. 6.4). By use of the LSDA+DMFT method the agreement with the experimental data is much improved. In consequence one may state that the consideration of local correlations is essential for a correct description of the energetic positions and intensities of the low- and high-energy Kerr rotation peaks (situated at 1.4 eV and 4 eV). This, of course, is a direct consequence of the complex self-energy that modifies the interband contributions to the MOKE spectra.

6.4.2 Chemical Disorder in NiMnSb

Figure 6.5 (left panel) presents calculated spin-integrated AP spectra for NiMnSb(001) in comparison with APS data. The spectra are normalized to equal

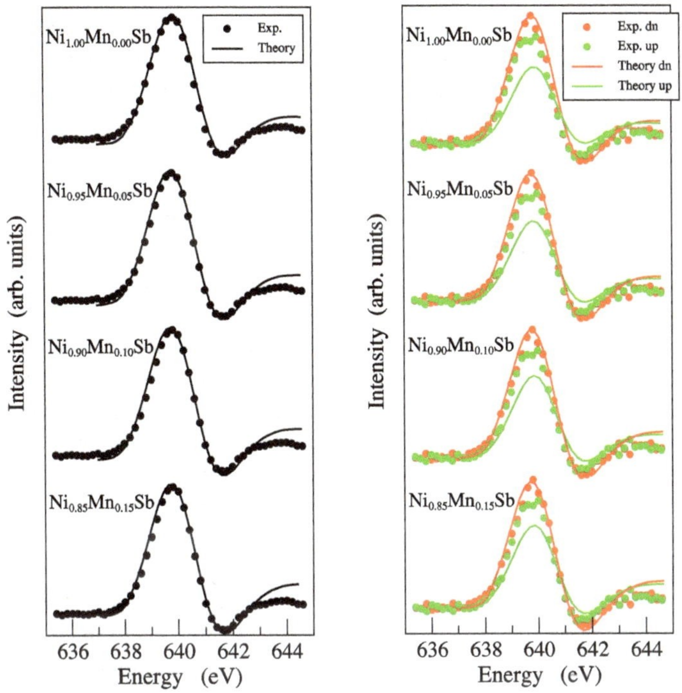

Fig. 6.5 *Left panel*: Spin-integrated AP spectra for NiMnSb(001) at the Mn $2p_{3/2}$ threshold for a varying degree of chemical disorder. Experimental data are given by *open circles*. *Right panel*: Spin-resolved AP spectra for NiMnSb(001) at the Mn $2p_{3/2}$ threshold as a function of chemical disorder. Experimental data are given by *filled circles*. Reproduced with permission from J. Minar, J. Braun, S. Bornemann, and H. Ebert, *J. Phys. D: Appl. Phys.* 42, 084009. Copyright (2009) IOP Publishing

maximum intensity. The theoretical spectra have been calculated with respect to the Fermi level. To be comparable with the experiment the calculated spectra were shifted along the energy axis until the peak maxima of the measured and calculated spectra coincidence. As it can be seen good agreement between experiment and theory is achieved. The almost quantitative agreement suggests that the DOS as calculated resembles the Mn local DOS in NiMnSb quite well in the spin-integrated case. The spectra show a pronounced structure at 639.7 eV from the Mn $2p_{3/2}$ threshold. These features originate from the self-convoluted local density of unoccupied $3d$ states. The spectral feature at the high-energy side of the main line corresponds to the maximum of the sp-like DOS, whose appearance and energetic position is sensitive to the short-range crystallographic order around the atom where the local excitation occurs. As expected, the effect of chemical disorder is nearly negligible concerning the spin-integrated DOS. Only slight deviations in the APS line shapes between the spectra calculated for the ordered and for the disordered structures are observable.

By adding spin resolution (see spectra in Fig. 6.5, right panel), we are able to obtain information about the spin-dependent DOS. The experimental data for majority (↑) and minority (↓) spins are presented by black and gray open circles, respectively. First of all, the APS lines show a clear spin asymmetry, A, between the spin-dependent intensities $I_\uparrow$ and $I_\downarrow$ $[A = (I_\uparrow - I_\downarrow)/(I_\uparrow + I_\downarrow)]$. The experimental spin-resolved spectra of Fig. 6.5 represent an average from many different surface regions. By taking into account the reduced remanent magnetization [48], we end up with an estimated spin asymmetry of $A = -0.115 \pm 0.012$ for a NiMnSb sample with saturated magnetization. The negative asymmetry reflects the high density of unoccupied minority states in NiMnSb, i.e. the unoccupied minority Mn $3d$ states. The theoretical spectra calculated for the ordered structure, however, predict a higher negative spin asymmetry in the main APS peak of $A = -0.250$. This discrepancy between experiment and theory by a factor of two and more is in line with the findings of earlier spin-resolved electron spectroscopic experiments. Introducing chemical disorder the calculated value for the spin asymmetry decreases. This has to be expected because of the pronounced increase of spectral weight in the majority spin DOS around the Fermi energy that overcompensates the shift of spectral weight in the minority spin DOS. The concentration dependence of the spin asymmetry is demonstrated by the series of spin-dependent APS data shown in Fig. 6.5. The spin asymmetry in the main peak is calculated to $A = -0.230$ for 5 % interchange of Ni and Mn atoms and decreases from $A = -0.215$ for 10 % disorder to $A = -0.200$ for 15 % interchange of Ni and Mn atoms. Therefore, our analysis supports the result found by Orgassa et al. that chemical disorder significantly lowers the spin polarization at E_F. On the other hand, introducing the effect of chemical disorder to our APS calculations, we end up with an improved agreement between experiment and theory but the predicted spin asymmetry is still to high by nearly a factor of two. Our analysis indicates that other mechanisms besides chemical disorder must be taken into account for a quantitative explanation of the strongly reduced spin polarization in NiMnSb(001).

6.4.3 Surface Emission from NiMnSb

To learn more about the influence of surface states on the spin-polarization in Heusler alloys we calculated spin-resolved E($\mathbf{k}_\parallel$) contour plots for photoemission intensities along the main symmetry directions of the NiMnSb(100) surface. The spectral intensities are based on LSDA electronic structure calculations obtained self-consistently within the relativistic full-potential Korringa–Kohn–Rostoker (SPR-KKR) method [39]. Figures 6.6 and 6.7 show the spin- and angle-resolved photoemission (ARUPS) data from both $\overline{G}$–$\overline{X}$, i.e. (Fig. 6.6), and $\overline{G}$–$\overline{M}$, i.e. (Fig. 6.7) directions of $\mathbf{k}_\parallel$. We calculated the data over the entire Brillouin zone in both directions for a photon energy of $h\nu = 21.2$ eV. Using the one-step model for photoemission calculations it is easy to separate the surface related emission from the bulk-like intensities. This is because of the theoretical formulation of the

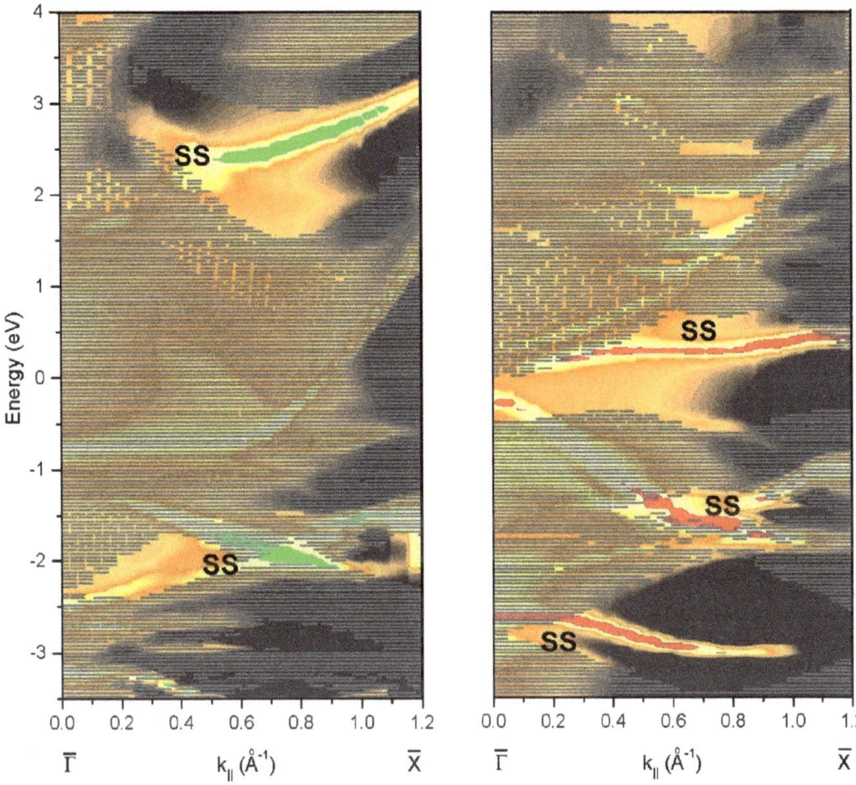

Fig. 6.6 Spin-resolved ARUPS data from NiMnSb(100) in $\overline{G}$–$\overline{X}$ direction. *Green (red) color* indicates surface related majority (minority) spin features. The *shaded areas* indicate the projected bulk bands in order to define the gaps in the bulk related electronic structure

photocurrent first introduced by Pendry and coworkers [70, 71]. In Figs. 6.6 and 6.7 we show the surface related intensity distributions in green (red) color for the majority (minority) spin channel. For both directions it is clearly observable that surface related features exist in the corresponding surface Brillouin zone in both spin channels [72]. Inspecting the most interesting region around the Fermi level one can see that in the minority spin channel dispersing surface states cross the Fermi level and therefore close the spin-dependent gap of the half Heusler compound NiMnSb. In consequence the amount of spin-polarization is significantly reduced. Similar effects could be expected, for example, from interface states existing at the boundary between a thin Heusler film and a certain substrate [73, 74].

Our photoemission calculations have revealed an additional mechanism which may be responsible for a reduced spin-polarization at the Fermi level in the half Heusler alloy NiMnSb. In combination with effects that result from chemical disorder and from a distorted magnetic structure in the surface region, which we deduced from our APS-investigation, we expect an effective spin-polarization occurring in NiMnSb that is far below 100 %.

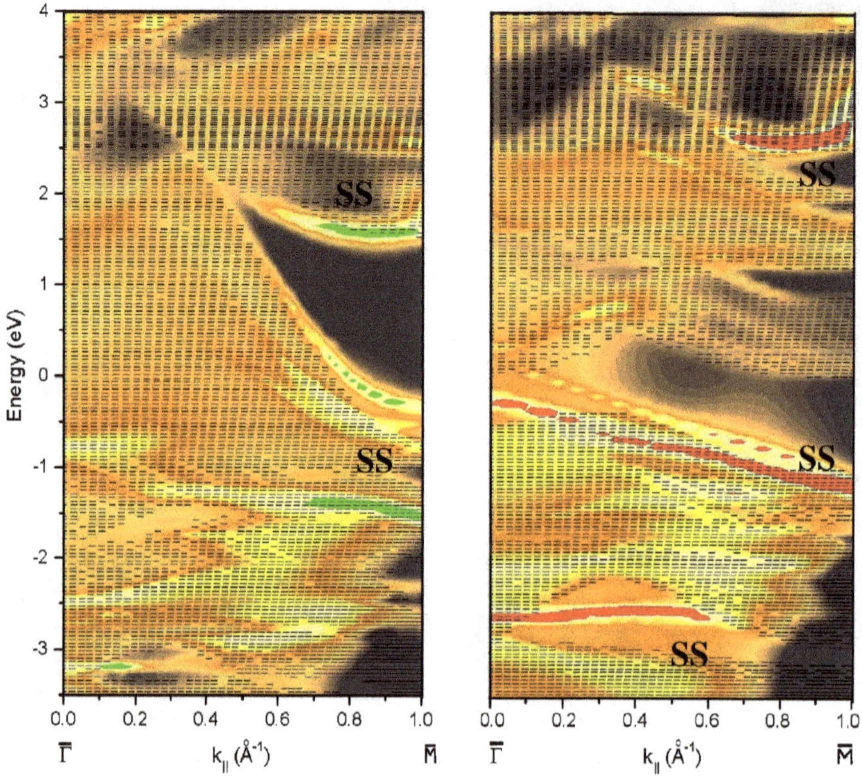

Fig. 6.7 Spin-resolved ARUPS data from NiMnSb(100) in $\overline{G}$–$\overline{M}$ direction. *Green (red) color* indicates majority (minority) spin features. The *shaded areas* indicate the projected bulk bands in order to define the gaps in the bulk related electronic structure

6.5 Summary

From our analysis it follows that $Co_2Mn_{1-x}Fe_xSi$ is a very pronounced locally correlated material for which plain LSDA calculations are insufficient to describe its electronic and magnetic properties correctly. The values for the magnetic moments and the details of the minority band gap are mostly determined by static correlations. In particular, this influence becomes more significant for the magnetic moments with increasing Fe concentration. A satisfactory description of this effect can be given on the basis of the LSDA+U approach within the AAL type of double-counting correction.

It turned out that it is not sufficient to explain all spectral properties accounting for static correlations only. Peak positions and intensities are significantly improved by the incorporation of dynamical correlations. These correlations can be described quantitatively by the LSDA+DMFT approach.

Obviously, the improvement is not completely perfect and agreement with experiment gets worse with increasing Fe concentration. One reason may be found in the perturbative nature of the used DMFT-solver that affects the ground-state description. However, this is difficult to verify, since more accurate solvers are much more time-consuming and restricted to only very small systems. In principle, the DMFT itself becomes exact only in the limit of infinite coordination numbers [25]. This implies that non-local correlations may also play a role [75]. The uncertainty of the double-counting and values of U and J has to be considered as well. Another source of errors is the insufficient description of the excitation processes, in particular the uncertainty in dealing with the final states. On the other hand, a very important excitation effect, namely the influence of the core holes was recognized, quantified and taken into account in the present work.

Calculated MOKE spectra of Mn-based Heusler alloys like NiMnSb show a very satisfying agreement with the experimental data when using of the LSDA+DMFT method. This can be directly explained in terms of local correlations which are essential for the energetic behavior and for the intensity distributions of the low- and high-energy Kerr rotation peaks.

Furthermore, we have presented spin-integrated and spin-resolved APS data for the unoccupied electronic states of $Ni_xMn_{1-x}Sb$ as a function of chemical disorder. The calculated series of spin-resolved and concentration dependent APS spectra demonstrates that chemical disorder influences the spin asymmetry of the unoccupied states of NiMnSb. Concerning the spin polarization at the Fermi level, one must note that APS is not particularly sensitive in that region. The nature of the APS signal does not allow one to resolve small energy gaps of the order of 0.5 eV giving rise to a positive spin asymmetry at E_F. The APS signal is rather dominated by the minority d states above E_F leading to a negative spin asymmetry. One the other hand the significantly reduced spin polarization observed in the experiment clearly indicates an incomplete spin polarization with a value well below 100 %. Therefore, the expected 100 % spin polarization at the Fermi level is not verified, as in a number of spin-polarized electron spectroscopic experiments before. Furthermore, one has to take into account other mechanisms than chemical disorder to be able to explain quantitatively the unexpected low value for the measured spin asymmetry. Therefore, we conclude that the following issues must be examined in future studies on well-defined samples. First the magnetization within the surface layers compared with the bulk should be investigated in more detail. Also surface phases with respect to composition and/or crystallographic order different from the bulk could be important and the position and width of the band gap should be analyzed in more detail in the calculations. A first study of NiMnSb(110) showed the formation of microstructures upon preparation, which has consequences for the magnetic properties of the surface [76]. Even with a stoichiometric surface, the number of nearest neighbors is changed at the surface and therefore surface/interface states may influence the gap [73, 74]. In the case that the free surface will not provide 100 % spin polarization, there is still hope that specific interfaces may open the way for 100 % spin-polarized charge-injection [77] in spintronic devices. Our theoretical ARUPS investigation supports these findings and in consequence suggests a spin-polarization far below 100 %. Only under very special conditions we would expect a spin-polarization value near 100 % in NiMnSb. Other materials with half-metallic behavior must be considered as well in order to design functional materials for future applications [78].

Finally, it is emphasized that the LSDA+DMFT scheme used in this work has significantly improved the description of the magnetism and VB-XPS in the presented series of Heusler compounds with $L2_1$ structure. This result is in accordance with former studies on $3d$ transition metals.

Acknowledgements The authors gratefully acknowledge financial support by the Deutsche Forschungsgemeinschaft DFG (FOR1346) and the BMBF (05K10WMA).

References

1. Heusler F (1903) Verh Dtsch Phys Ges 5:219
2. de Groot RA, Mueller FM, van Engen PG, Buschow KHJ (1983) Phys Rev Lett 50:2024
3. Mavropoulos Ph, Sato K, Zeller R, Dederichs PH, Popescu V, Ebert H (2004) Phys Rev B 69:054424
4. Fang CM, de Wijs GA, de Groot RA (2002) J Appl Phys 91:8340
5. Orgassa D, Fujiwara D, Schulthess TC, Butler WH (1999) Phys Rev B 60:13237
6. Soulen RJ, Byers JM Jr., Osofsky MS, Nadgorny B, Ambrose T, Cheng SF, Broussard PR, Tanaka CT, Nowak J, Moodera JS, Barry A, Coey JMD (1998) Science 282:85
7. Raphael MP, Ravel B, Willard MA, Cheng SF, Das BN, Stroud RM, Bussmann KM, Claassen JH, Harris VG (2001) Appl Phys Lett 79:4396
8. Brown PJ, Neumann KU, Webster PJ, Ziebeck KRA (2000) J Phys: Condens Matter 12:1827
9. Kübler J, William AR, Sommers CB (1983) Phys Rev B 28:1745
10. Felser C, Fecher GH, Balke B (2007) Angew Chem Int Ed 46:668
11. Fuji S, Sugimura S, Ishida S, Asano S (1990) J Phys Cond Matt 2:8583
12. Brown PJ, Normann KU, Webster PJ, Ziebeck KRA (2000) J Phys Cond Matt 12:1827
13. Fecher GH, Kandpal HC, Wurmehl S, Felser C, Schönhense G (2006) J Appl Phys 99:08J106
14. Kübler J, Fecher GH, Felser C (2007) Phys Rev B 76:024414
15. Wurmehl S, Fecher GH, Kandpal HC, Ksenofontov V, Felser C, Lin HJ, Morais J (2005) Phys Rev B 72:184434
16. Wurmehl S, Fecher GH, Kandpal HC, Ksenofontov V, Felser C, Lin HJ (2006) Appl Phys Lett 88:032503
17. Kandpal HC, Fecher GH, Felser C (2006) Phys Rev B 73:094422
18. Balke B, Fecher GH, Kandpal HC, Felser C, Kobayashi K, Ikenaga E, Kim JJ, Ueda S (2006) Phys Rev B 74:104405
19. Kübler J (2000) Theory of itinerant electron magnetism. Clarendon Press, Oxford
20. Garibay-Alonso R, Dorantes-Davila J, Pastor GM (2002) J Appl Physics 91:8254
21. Kandpal HC, Fecher GH, Felser C, Schönhense G (2006) Phys Rev B 73:094422
22. Fecher GH, Balke B, Ouardi S, Felser C, Schönhense G, Ikenaga E, Kim J, Ueda S, Kobayashi K (2007) J Phys D: Appl Phys 40:1576
23. Anisimov VI, Zaanen J, Andersen OK (1991) Phys Rev B 44:943
24. Czyżyk MT, Sawatzky GA (1994) Phys Rev B 49:14211
25. Kotliar G, Savrasov SY, Haule K, Oudovenko VS, Parcollet O, Marianetti CA (2006) Rev Mod Phys 78:865
26. Lichtenstein AI, Katsnelson MI, Kotliar G (2001) Phys Rev Lett 87:067205
27. Grechnev A, Di Marco I, Katsnelson MI, Lichtenstein AI, Wills J, Eriksson O (2006) cond-mat, page 0610621
28. Perlov A, Chadov S, Ebert H (2003) Phys Rev B 68:245112

29. Minár J, Ebert H, De C, ı N, Brookes NB, Venturini F, Ghiringhelli G, Chioncel L, Katsnelson MI, Lichtenstein AI (2005) Phys Rev Lett 95:166401
30. Braun J, Minár J, Ebert H, Katsnelson MI, Lichtenstein AI (2006) Phys Rev Lett 97:227601
31. Chadov S, Minár J, Katsnelson MI, Ebert H, Ködderitzsch D, Lichtenstein AI (2008) Europhys Lett 82:37001
32. Katsnelson MI, Lichtenstein AI (2002) Eur Phys J B 30:9
33. Chioncel L, Sakuraba Y, Arrigoni E, Katsnelson MI, Oogane M, Ando Y, Miyazaki T, Burzo E, Lichtenstein AI (2008) Phys Rev Lett 100:086402
34. Irkin VYu, Katsnelson MI, Lichtenstein AI (2007) J Phys Cond Matt 19:315201
35. Ishida S, Fuji S, Kashiwagi S, Asanor S (1995) J Phys Soc Jap 64:2152
36. Park R, Houston JE (1972) Phys Rev 6:1073
37. Ertl K, Vonbank M, Dose V, Noffke J (1993) Solid State Commun 88:557
38. Reinmuth J, Passek F, Petrov VN, Donath M, Popescu V, Ebert H (1997) Phys Rev B 56:12893
39. Minár J, Chioncel L, Perlov A, Ebert H, Katsnelson MI, Lichtenstein AI (2005) Phys Rev B 72:45125
40. Pourovskii LV, Katsnelson MI, Lichtenstein AI (2005) Phys Rev B 72:115106
41. Drchal V, Janiš V, Kudrnovsky J (2002) Physica B 312–313:519
42. Chadov S, Minár J, Ebert H, Perlov A, Chioncel L, Katsnelson MI, Lichtenstein AI (2006) Phys Rev B 74:R140411
43. Minár J, Chadov S, Ebert H, Chioncel L, Lichtenstein A, de Nadaï C, Brookes NB (2005) Nucl Inst Meth Phys Res A 547:151
44. Gyorffy BL (1972) Phys Rev B 5:2382
45. Butler WH (1985) Phys Rev B 31:3260
46. Ebert H, Battocletti M (1996) Solid State Commun 98:785
47. Ebert H, Schwitalla J (1997) Phys Rev B 55:3100
48. Kolev H, Rangelov G, Braun J, Donath M (2005) Phys Rev B 72:104415
49. Donath M, Rangelov G, Braun J, Grentz W (2005) Magnetization, spin polarization and electronic structure of NiMnSb surfaces. In: Donath M, Nolting W (eds) Local-moment ferromagnets: unique properties for modern applications. Lecture notes in physics, vol 678. Springer, Berlin, p 261
50. Ebert H et al The Munich SPR-KKR package, version 3.6. http://olymp.cup.uni-muenchen.de/ak/ebert/SPRKKR
51. Ebert H (2000) Fully relativistic band structure calculations for magnetic solids – formalism and application. In: Dreyssé H (ed) Electronic structure and physical properties of solids. Lecture notes in physics, vol 535. Springer, Berlin, p 191
52. Balke B, Fecher GH, Kandpal HC, Felser C (2006) Phys Rev B 74:104405
53. Chadov S, Fecher GH, Felser C, Minar J, Braun J, Ebert H (2009) J Phys D: Appl Phys 42:084002
54. Ebert H (1996) Rep Prog Phys 59:1665
55. Tomczak JM, Biermann S (2009) Phys Rev B 80:085117
56. Katsnelson MI, Irkhin VYu, Chioncel L, Lichtenstein AI, de Groot RA (2008) Rev Mod Phys 80:315
57. van Engen PG, Buschow KHJ, Jongebreur R, Erman M (1983) Appl Phys Lett 42:202
58. van Engen PG (1983) Ph.D. thesis, Tech. Uni. Delft
59. Ohyama R, Koyanagi T, Matsubara K (1987) J Appl Phys 61:2347
60. van Engelen PPJ, de Mooij DB, Wijngaard JH, Buschow KHJ (1994) J Magn Magn Mat 130:247
61. van Ek J, Maclaren JM (1997) Phys Rev B 56:R2924
62. Kautzky MC, Clemens BM (1995) Appl Phys Lett 66:1279
63. Oppeneer PM, Antonov VN, Kraft T, Eschrig H, Yaresko AN, Perlov AY (1995) Solid State Comm 94:255
64. Antonov VN, Oppeneer PM, Yaresko AN, Perlov AYa, Kraft T (1997) Phys Rev B 56:13012

65. Gao X, Woolam JA, Kirby RD, Sellmyer DJ, Tanaka CT, Nowak J, Moodera JS (1999) Phys Rev B 59:9965
66. Picozzi S, Continenza A, Freeman AJ (2006) J Phys D: Appl Phys 39:851
67. Kulatov ET, Uspenskii YA, Halilov SV (1994) Phys Lett A 125:267
68. Uspenskii YA, Kulatov ET, Khalilov SV (1995) JETP 80:952
69. Minar J, Braun J, Bornemann S, Ebert H (2009) J Phys D: Appl Phys 42:084009
70. Hopkinson JFL, Pendry JB, Titterington DJ (1980) Comp Phys Commun 19:69
71. Braun J (1996) Rep Prog Phys 59:1267
72. Correa JS, Eibl Ch, Rangelov G, Braun J, Donath M (2006) Phys Rev B 73:125316
73. Jenkins SJ, King DA (2001) Surf Sci 494:L793
74. Jenkins SJ, King DA (2002) Surf Sci 501:L185
75. Schäfer J, Hoinkis M, Rotenberg E, Blaha P, Claessen R (2005) Phys Rev B 72:155115
76. Eickhoff Ch, Kolev H, Donath M, Rangelov G, Chi LF (2007) Phys Rev B 76:205440
77. de Wijs GA, de Groot RA (2001) Phys Rev B 64:020402
78. Fonin M, Dedkov YuS, Rüdiger U, Güntherodt G (2005) Growth and room temperature spin polarization of half-metallic epitaxial CrO_2 and Fe_3O_4 thin films. In: Donath M, Nolting W (eds) Local-moment ferromagnets: unique properties for modern applications. Lecture notes in physics, vol 678. Springer, Berlin, p 289

Chapter 7
Theory of the Half-Metallic Heusler Compounds

Gerhard H. Fecher, Stanislav Chadov, and Claudia Felser

Abstract Multicomponent systems play a key role in materials design providing the multifunctionality and tunability of materials. At present, the ternary Heusler materials form probably the most thoroughly studied class of multicomponent systems. In present work, results of ab-initio band structure calculations for A_2BC Heusler compounds that consist of A and B sites occupied by transition metals and C by a main group element are presented. These systems cover an extremely wide range of properties from half-metallic ferromagnets to non-magnetic semiconductors. The calculations have been performed in order to understand the properties of the band gap in the minority spin channel, the peculiar transport properties and magnetic behaviour found in these materials. Among the interesting aspects of the electronic structure of these materials are the contributions from both A and B atoms to the total magnetic moment. In several classes of these compounds the magnitudes of the total magnetic moment show a trend consistent with the Slater–Pauling behaviour. The total magnetic moment depends as well on the kind of C atoms although they do not directly contribute to it. In Co_2 compounds, a change of the C element changes the contribution of the t_{2g} states to the moment at the Co sites. The localised moment in these magnetic compounds resides at the B site. Other than in the classical Cu_2-based Heusler compounds, the A atoms in Co_2-, Fe_2- and Mn_2-based compounds may contribute to the total magnetic moment significantly. It is shown that the inclusion of electron–electron correlation in the form of LDA+U/DMFT calculations helps to understand the magnetic properties of those compounds that exhibit already a minority gap in calculations where it is neglected. Beside the large group

G.H. Fecher · S. Chadov · C. Felser
Institut für Anorganische Chemie und Analytische Chemie, Johannes Gutenberg – Universität, D-55099 Mainz, Germany

S. Chadov
e-mail: chadov@uni-mainz.de

G.H. Fecher · S. Chadov · C. Felser (✉)
Max-Planck-Institut für Chemische Physik fester Stoffe, D-01187 Dresden, Germany
e-mail: felser@cpfs.mpg.de

G.H. Fecher
e-mail: fecher@cpfs.mpg.de

C. Felser, G.H. Fecher (eds.), *Spintronics*, DOI 10.1007/978-90-481-3832-6_7,

of Co_2 compounds, half-metallic ferromagnetism was here found only in such compounds that contain Mn. In parallel we consider the different types of chemical disorder often accompanying multicomponent systems and characterise its influence on spin-polarisation and magnetic properties.

7.1 Introduction

The Heusler compounds [1] are ternary intermetallics with a 2:1:1 stoichiometry and the chemical formula A_2BC. They usually consist of two transition metals (A, B) and a main group element (C). They first attracted the interest of the magnetism community when F. Heusler et al. had shown that the compound Cu_xMn_yAl becomes ferromagnetic in the 2:1:1 form ($x = 2$ and $y = 1$), even none of its constituents is ferromagnetic by itself [2]. However, it took three decades before their structure was explained to be that of an ordered compound with a face centred cubic structure [3, 4]. The general crystal structure of the Heusler compounds is displayed in Fig. 7.1.

The main interest during the first decades after their discovery was concentrated on Cu and Mn containing compounds. Co_2 based compounds were synthesised and investigated in the 1970s [5]. Kübler et al. [6] recognised that the minority spin densities at the Fermi energy (ε_F) nearly vanish for Co_2MnAl and Co_2MnSn. The authors concluded that this should lead to peculiar transport properties in these Heusler compounds because only the majority density contributes to the states at ε_F. At the same time, de Groot et al. [7] proposed the concept of the so-called half-metallic ferromagnets (HMF) that are materials predicted to exhibit 100 % spin polarisation at ε_F. This exceptional property would make the HMF ideal candidates for spin injection devices to be used in spin electronics [8].

The calculation of the electronic structure plays an important role in determining the magnetic properties of Heusler compounds and, in particular, for predicting half-

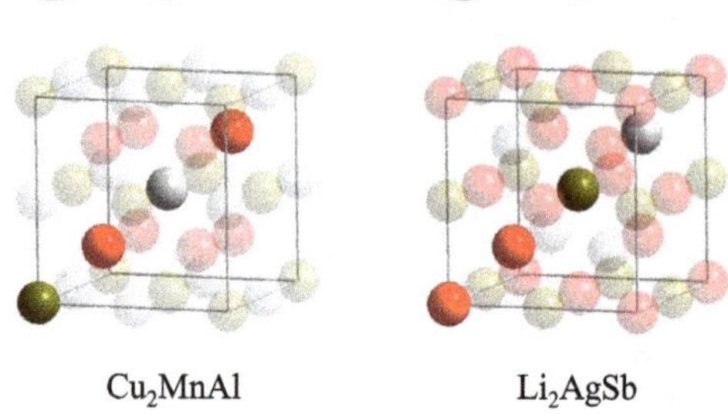

Fig. 7.1 Crystal structure of the A_2BC Heusler compounds. Shown are the primitive (*upper row*) and cubic (*lower row*) cells of the regular $L2_1$ and inverted X type structures of the Cu_2MnAl and Li_2AgSb prototypes

metallic ferromagnetism. Therefore, the band structure calculations must be performed very carefully. The first attempt to calculate the band structure of some Co_2 based compounds (Co_2MnSn, Co_2TiSi and Co_2TiAl) did not indicate half-metallic ferromagnetism [9]. These calculations displayed a minimum of the minority density of states (DOS) at ε_F but not a gap. At that time, the calculations were based on spherical potentials, and the exchange-correlation potential of the local spin density approximation (LSDA) was used in a rather simple form [10–13]. The first clear indication of half-metallic ferromagnetism in Co_2 based Heusler compounds was reported by Ishida et al. [14, 15] for Co_2MnC and Ru_2MnC (C = Al, Si, Sn, and Sb). Using full symmetry potentials, Mohn et al. [16] found the magnetic ground state of Co_2TiC (C = Al and Sn), but not a half-metallic state. Galanakis et al. [17] reported half-metallic behaviour in various A_2BC compounds, but not for the Co_2 compounds with Ti or Fe. The results were compatible with those found for the Mn compounds as calculated by Picozzi et al. [18] using the generalised gradient approximation instead of the pure LSDA. The generalised gradient approximation (GGA), as introduced by Perdew et al. [19–22], accounts for gradients of the density that are absent in the pure LSDA parameterisation of the exchange-correlation functional [10–13]. Using spherical potentials and the GGA, a half-metallic state could not be verified for Co_2FeAl [23, 24]. A half-metallic ferromagnetic ground state was also found for the complete series $Co_2Cr_{1-x}Fe_xAl$, when the full symmetry potentials were used along with the GGA in the calculations [25]. This clearly indicates that one principally needs both the full symmetry potentials and the generalised gradient approximation to find the correct electronic structure and ground state for the Heusler compounds.

With this information, the properties of the reported transition metal based Heusler compounds were calculated in the present work using both the GGA and the full symmetry potentials. Even at that state of sophistication, however, for some compounds the magnetic properties could not be explained properly. The reason was found in an underestimation of the electron–electron on-site correlation in the LSDA or GGA calculations [26, 27]. Therefore, the LDA+U method [28, 29] was used for selected compounds to study the effect of on-site correlation at the transition metal sites on the magnetic and half-metallic properties of the Heusler compounds.

A systematic examination of the electronic and the magnetic structure of the Heusler compounds was carried out in this work. The results from the electronic structure calculations are compared to study the effect of the different kind of atoms and valence electron concentration on the magnetic properties and in particular on the band gap in the minority states.

7.1.1 Spin Polarisation and Magnetic Moments of Half-Metallic Ferromagnets

For both scientific and technological reasons it is useful to define the electron spin polarisation at the Fermi energy of a material, although it is difficult to measure [30].

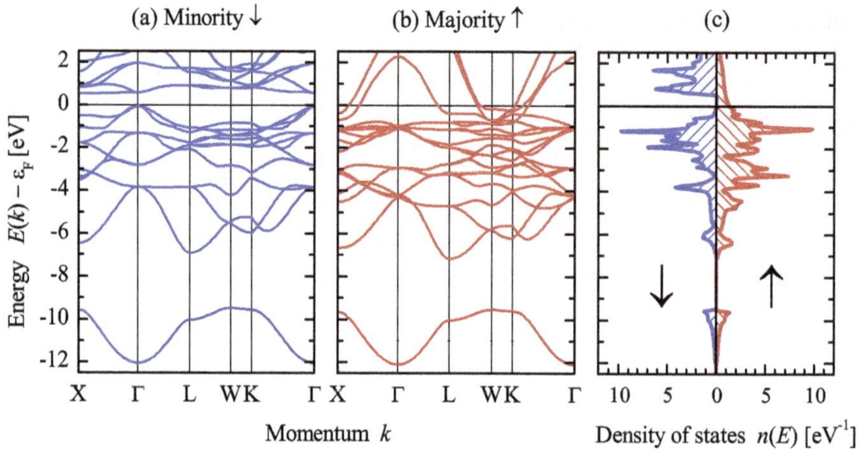

Fig. 7.2 Typical electronic structure of a half-metallic (Type Ia) compound (Co_2MnGe). Panels (**a**) and (**b**) represent the minority- and majority-spin band structures, respectively. Panel (**c**) shows the corresponding spin resolved DOS. ↑ corresponds to minority-, ↓ to majority-spin channels

The energy dependence of the electron spin polarisation $P_e(E)$ is given by

$$P_e(E) = \frac{n_\uparrow(E) - n_\downarrow(E)}{n_\uparrow(E) + n_\downarrow(E)}, \tag{7.1}$$

where $n_\uparrow(E)$ and $n_\downarrow(E)$ are the spin dependent density of states. The arrows ↑ and ↓ assign states of opposite spin that are majority and minority states, respectively (see Fig. 7.2 for an example). P vanishes for paramagnetic or in antiferromagnetic materials even below the magnetic transition temperature. However, it has a finite value in ferromagnetic or ferrimagnetic materials below the Curie temperature. The electrons at the Fermi energy ($E = \varepsilon_F$) are fully spin polarised ($P = 100$ %) when either $n_\uparrow(\varepsilon_F)$ or $n_\downarrow(\varepsilon_F)$ equals zero.

Concerning experiments, the term *spin polarisation* is not unique, the result depends on the way it was measured. In more general, the spin polarisation at the Fermi energy may be defined by:

$$P(\varepsilon_F) = \frac{v_\uparrow^x n_\uparrow(\varepsilon_F) - v_\downarrow^x n_\downarrow(\varepsilon_F)}{v_\uparrow^x n_\uparrow(\varepsilon_F) + v_\downarrow^x n_\downarrow(\varepsilon_F)}, \tag{7.2}$$

where $v_\uparrow$ and $v_\downarrow$ are the Fermi velocities for majority and minority electrons, respectively. x is the power with which the Fermi velocity contributes to the property being measured.

The photoexcitation process does not depend on the Fermi velocity of the electrons and one has $x = 0$ ($v^0 \equiv 1$) such that the spin polarisation in photoemission is purely given by the majority ($n_\uparrow$) and minority ($n_\downarrow$) densities. The situation is, however, different for transport quantities that may depend on the electron momentum $(m_e \cdot v)$ such that $x = 1$. Examples are the electrical resistivity (ρ) or magnetoresistance (MR), wherefore the values from different measurements cannot be compared directly but only their trends. It is worthwhile to note that in half-metallic ferromagnets—defined so far—one has $P(\varepsilon_F) = 1$ independent of the method.

The definition of the tunneling magnetoresistance (TMR) ratio by $r_{\mathrm{TMR}} = (R_{\uparrow\downarrow} - R_{\uparrow\uparrow})/R_{\uparrow\uparrow}$ (where R is the resistance for parallel ($\uparrow\uparrow$) and antiparallel ($\uparrow\downarrow$) configuration of the magnetic layers) causes an additional problem as:

$$P_{\mathrm{TMR}} = \frac{r_{\mathrm{TMR}}}{(r_{\mathrm{TMR}} + 2)} \tag{7.3}$$

needs that the TMR ratio becomes infinite $r_{\mathrm{TMR}} = \infty$ to result in $P_{\mathrm{TMR}} = 1$, that is, for half-metallic ferromagnets. Further, the resistance R and thus the TMR ratio does not only depend on the spin dependence of the transport resistivity ρ of the ferromagnetic material but also on the kind of tunneling barrier. In case of coherent tunneling through single crystalline MgO, only particular electrons that match the Δ_1 band contribute and lead to an enhancement of the TMR and thus the spin polarisation (for examples see Refs. [31, 32]). On the other hand, incoherent tunneling may reduce the effective spin polarisation by spin scattering while passing the tunneling barrier (for example amorphous AlO_x). Therefore, measured TMR ratios may be bad indicators for the electron spin polarisation of particular materials in particular when (7.3) is used.

In the present work, the classification scheme for half-metals as proposed by Coey et al. is used [8, 33]. The first class is Type I, where only one type of spin polarised electrons (either $\uparrow$ or $\downarrow$) does contribute to the conductivity. This is the situation described by de Groot et al. [7], where one has—for one spin direction—a gap in the density of states that includes the Fermi energy. The half-metallic gap appears in the minority density of states in Type Ia half-metals such that only itinerant majority electrons contribute to the conductivity (see Fig. 7.2). In Type Ib, the majority density of states vanishes at ε_F. The electrons at ε_F are itinerant in Type I half-metals, whereas they are localised in Type II. In Type III half-metals no gap appears at the Fermi energy but one has at ε_F itinerant electrons for one spin direction and localised electrons for the other. In Type IV half-metals, the conductivity is semi-metallic rather than metallic like in Type III. Type IIIa (IVa) and Type IIIb (IVb) distinguish half-metals with the localised (or itinerant) states in opposite spin channels. In Type III or IV half-metals, a gap may appear in one of the spin channels that is located below or above ε_F. In Types I and II half-metals the spin polarisation P is always 100 %, whereas lower values appear in Types III and IV.

It is worthwhile to note that half-metallic conductivity, as explained above, is different from semi-metallic conductivity. Semi-metals, of which graphite is a textbook examples, have small and equal numbers of electrons and holes due to a small overlap between valence and conduction bands and they exhibit a pronounced directional anisotropy in the conductivity. Most often, semi-metals exhibit high conductivity (metallic) in one direction and low conductivity (semiconductor) in another.

The Slater–Pauling curve [34, 35] is a simple way to study for ferromagnetic alloys the interrelation between the valence electron concentration and the magnetic

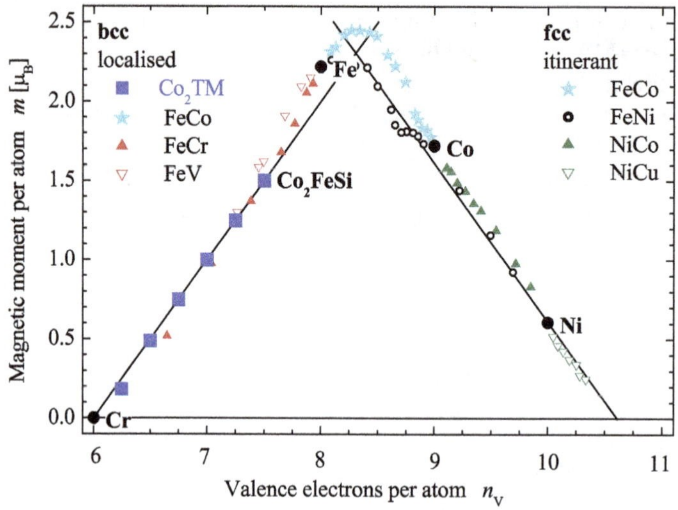

Fig. 7.3 Slater–Pauling curve for 3d transition metals and their alloys. Experimental values for *selected* Co_2 based Heusler compounds are given for comparison. (Note: The $A_{1-x}B_x$ alloys are given as AB in the legend, for short.)

moments (see Fig. 7.3). It is well known that Heusler compounds based on Co_2 follow the Slater–Pauling rule for predicting their total spin magnetic moment [17, 36, 37] that scales linearly with the number of valence electrons. The Co_2 based compounds are found on the so-called *localised* part of the Slater–Pauling curve [36, 37] where the magnetic moment increases with an increasing number of valence electrons. In this part of the curve one has in majority materials with bcc or bcc-derived structures. Like for the example of iron, the electronic structure of these alloys exhibits a minimum in the minority density of states and the Fermi energy is pinned in this minimum. The minimum in the minority spin density constrains the number of occupied electrons in the minority bands to be approximately three such that the number of majority electrons increases proportional to the total number of electrons and so does the magnetic moment as a direct consequence (for more details see for example [36, 37]).

Half-metallic ferromagnets, like the Co_2 based Heusler compounds, are supposed to exhibit not only a minimum but a real gap in the minority density of states and the Fermi energy is pinned inside of that gap. From this point of view, the Slater–Pauling rule is strictly fulfilled with

$$m_{\mathrm{HMF}} = n_V - 6 \tag{7.4}$$

for the mean magnetic moment per atom (m_{HMF}). n_V is the mean number of valence electrons per atom found by averaging over all atoms and 6 is two times the mean number of occupied minority states. The advantage of this equation is that it neither depends on the number of atoms in the compound nor relies on integer site occupancies.

For ordered compounds with different kinds of atom it might be more convenient to work with all atoms of the primitive cell. In the case of four atoms per primitive cell, as in Heusler compounds, one has to subtract 24 (6 multiplied by the number of atoms) from the accumulated number of valence electrons in the primitive cell N_V (s, d electrons for the transition metals and s, p electrons for the main group element) to find the magnetic moment per primitive cell (m):

$$m = N_V - 24 \tag{7.5}$$

with N_V denoting the accumulated number of valence electrons in the primitive cell. In the case of Heusler compounds, the number 24 arises from the number of completely occupied minority bands that has to be 12 in the half-metallic state. In particular these are one s (a_{1g}), three p (t_{1u}) and eight d bands [36, 38]. The last consist of two triply degenerate bands with t_{2g} symmetry and one with e_g symmetry (note that the given assignments of the irreducible representations are only valid at the Γ-point and neglecting the spin of the electrons).

This kind of rule was first noted by Kübler et al. [38] for $C1_b$ compounds with three atoms per primitive cell ($m_{C1_b} = N_V - 18$). In both cases the magnetic moment per primitive cell becomes strictly integer (in multiples of Bohr magnetons μ_B) for Type I or II half-metals, what may be seen as an advantage of the *valence electron rule* (7.5) compared to the original *Slater–Pauling* approach (7.4) even so it suggests the existence of different laws. It is worthwhile to note that (7.5) leads only for ternary 2:1:1 compounds to integer magnetic moments but not for quarternary derivatives like $A_2B_{1-x}B'_xC$ or $A_2BC_{1-x}C'_x$ as reported in [39] or [40], respectively. In those cases, one observes from (7.5) non-integer values of the magnetic moment even in the half-metallic case due to the non-integer site occupancy.

The Slater–Pauling curve is shown in Fig. 7.3. Experimental values of the magnetic moments in selected Co_2 based Heusler compounds are compared to $3d$ transition metals and their alloys. This comparison is only possible if the Slater–Pauling rule in the formulation of (7.4) is used. The itinerant part of the curve is included for clarity about the behaviour of the Heusler compounds in comparison to other ferromagnetic alloys. For a detailed discussion of that part see Refs. [33, 36, 37].

7.2 Crystal Structure and Calculational Details

The A_2BC Heusler compounds crystallise in the cubic $L2_1$ structure (space group no. 225: $F\,m\bar{3}m$), the prototype is Cu_2MnAl. The structure was first explained by O. Heusler [3] as well as Bradley and Rodgers [4]. In general, the A and B atoms are transition metals and C is a main group element. In some cases, B is replaced by a rare earth element. The A atoms are placed on the Wyckoff position 8c ($\frac{1}{4}, \frac{1}{4}, \frac{1}{4}$). The B and C atoms are located on 4a (0, 0, 0) and 4b ($\frac{1}{2}, \frac{1}{2}, \frac{1}{2}$) positions, respectively. Thus, the cubic $L2_1$ structure consists of an *fcc* lattice with a basis of four atoms. The two A-site *fcc* sublattices combine to form a simple cubic sublattice.

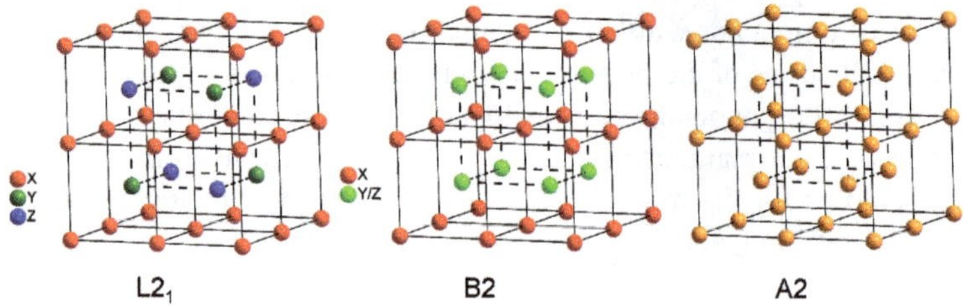

Fig. 7.4 Frequent structures for Heusler compounds. Shown are the ordered $L2_1$ as well as the disordered $B2$ and $A2$ structures. (Note: Here A_2BC is assigned by the atoms X, Y, Z, or X, Y/Z. The atoms of the $L2_1$ structure are shifted by (1/4, 1/4, 1/4) compared to Fig. 7.1 to make the CsCl superstructure better visible.)

The B and C atoms occupy alternatingly the centre of the simple cubic sublattice resulting in a CsCl-type super structure. The crystal structure of Heusler compounds (compare also Fig. 7.1) is illustrated in Fig. 7.4 with the positions shifted by $(\frac{1}{4}, \frac{1}{4}, \frac{1}{4})$ with respect to the standard $F\,m\bar{3}m$ cell to make the CsCl superstructure better visible. The Γ point of the paramagnetic $L2_1$ structure has the symmetry O_h. The symmetry of 4a and 4b positions is O_h, whereas the 8c position has T_d symmetry. However, the wave functions at the Γ point have to be described at least by the symmetry reduced to C_{4h} in the ferromagnetic state to account for the correct transformation of the electron spin.

The cubic A_2BC compounds are not only found with the Cu_2MnAl ($L2_1$) type structure but also with the X-type structure [41] (prototype: Li_2AgSb).[1] The Li_2AgSb type structure exhibits T_d symmetry (space group no. 216: $F\,\bar{4}3m$). In that structure the A_2 atoms occupy the non equivalent 4a, 4c Wyckoff positions at $(0, 0, 0)$ and $(\frac{1}{4}, \frac{1}{4}, \frac{1}{4})$. The B and C are located on 4b $(\frac{1}{2}, \frac{1}{2}, \frac{1}{2})$ and 4d $(\frac{3}{4}, \frac{3}{4}, \frac{3}{4})$ positions, respectively. All four positions adopt T_d symmetry and there is no position with O_h symmetry. This structure is similar to the ABC compounds with $C1_b$ structure, but with the vacancy filled by an additional A atom. This structure is frequently observed if the nuclear charge of the B element is larger than the one of the A element from the same period, that is, $Z(B) > Z(A)$ for two $3d$ transition metals. The structure may also appear in compounds with transition metals from different periods. However, the two structures may be hardly distinguishable by X-ray diffraction and much care has to be taken in the structural analysis, as both have the general fcc-like symmetry. It is interesting to note that both types of structure result in the $A1$ structure (bcc-like) if the atoms are randomly distributed over the sites of the primitive cell, that is, for complete alloying. In that case the magnetic moment might still follow the localised part of the Slater–Pauling curve displayed in

[1]Note that the classification of the cubic A_2BC compounds with Cu_2MnAl ($L2_1$) or Li_2AgSb (X) type structures is sometimes not uniquely given in *Pearson's Handbook* [42]. The compounds with Li_2AgSb (X) type structure [41] are assigned in the printed version of [42] to the prototype $CuHg_2Ti$. It is not clear from the original work [43], however, whether or not the $CuHg_2Ti$ compound de facto exhibits the X type crystalline structure.

Fig. 7.3 even so the half-metallic character is lost and the gap in the minority bands is closed and only a minimum appears in the minority density of states. In this work, only calculations for compounds of the Cu_2MnAl type are respected to stay within the O_h symmetry.

In the present work, the electronic structure calculations based on density functional theory were performed using the full-potential linearised augmented plane wave (FLAPW) code WIEN2K [44]. The exchange-correlation functional was evaluated within the generalised gradient approximation, using the Perdew–Burke–Ernzerhof [21] parameterisation. The energy threshold between the core and the valence states was set to −81.6 eV. The muffin-tin-radii (R_{MT}) were chosen to ensure nearly touching spheres and minimising the interstitial space. $R_{MT} \times k_{max} = 7$ was used for the number of plane waves and the expansion of the wave functions was set to $l = 10$ inside of the muffin tin spheres. The self-consistent calculations employed a grid of 455 k points in the irreducible Brillouin zone taken from a 25×25×25 mesh. It turned out that this number of irreducible k points is a good compromise to balance the quality of the integration and the speed of the calculation. The energy convergence criterion was set to 10^{-5} Ry and the charge convergence was monitored simultaneously. (Note: The calculations were restarted if the charge convergence was after energy convergence still above 0.01 electrons.)

As multicomponent systems the Heusler materials are subject to chemical disorder. There are basically three simple structures realised in Heusler compounds that are $L2_1$, $B2$ and $A2$. Figure 7.4 compares the differences between the fully ordered ($L2_1$), the partially disordered ($B2$) and the fully alloyed ($A2$) structures.

$L2_1$ is the complete ordered structure of the Cu_2MnAl prototype [3, 4], where Cu occupies the 8c Wyckoff position, Mn the 4b position and Al the 4a position of the cubic fcc cell with $F\,m\bar{3}m$ symmetry. For the $A2$ type a complete disorder is observed and all atoms are randomly distributed over all positions resulting in a bcc like lattice with $I\,m\bar{3}m$ symmetry. This is the alloy state of the Heusler compounds. The $B2$ type may be realised in different ways [45]. In A_2BC Heusler compounds, most often complete random exchange of the B and C atoms in an unchanged A_2 sublattice is observed resulting in a CsCl like lattice with $I\,m\bar{3}m$ symmetry. Indeed, other mixtures like DO_3 (not shown here), X, or combinations like $B2 + L2_1$ may appear, too (see [46, 47]). DO_3 means a complete random exchange of A_2 and B or A_2 and C atoms resulting for example in $(A_{2/3}B_{1/3})_3C$ without changing the symmetry, while in the X structure only one distinguished A atom is completely exchanged with the B (or C) atom resulting in a lattice with $F\,\bar{4}3m$ symmetry. In nearly all cases of disorder the half-metallic ferromagnetic behaviour is destroyed. That is why the study of the local structure in Heusler compounds is of fundamental interest.

The band structure in its usual definition has no meaning for alloys with random disorder, due to absence of periodicity. Therefore, it make sense to discuss only the integrated quantities as magnetic moments, total energy, etc., which depend more directly on the DOS rather than the particular form of the dispersion of the electronic bands.

The practical way to account for the random chemical disorder is the so-called coherent potential approximation (CPA) theory [48–50]. The idea of the CPA is to

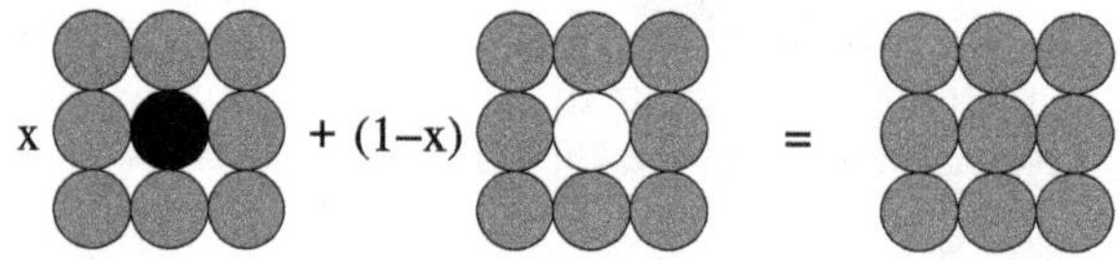

Fig. 7.5 The CPA model: The Green function of the effective medium (*gray atoms*) is obtained as an average of the partial impurities Green functions (*black and white atoms*)

replace the random array of real on-site potentials by an ordered array of effective potentials. The scattering properties of the effective potential are than determined self-consistently in terms of the local mean-field theory with the condition that the total Green function of the effective system does not change upon replacement of the single effective potential with the real one. This idea is sketched in Fig. 7.5.

The computational cost of CPA calculations is significantly lower than supercell approaches where the disorder is approximated by a randomly generated configurations within a large number of large supercells. The important advantage is that the CPA does not affect the translational (as well as the point group) symmetry of the primitive cell whereas the supercells must be chosen sufficiently large to avoid effects from the assumed order or periodic images, which are caused by an artificial translational symmetry. The CPA allows investigation of the electronic structure as a continuous function of the substitution level, which is very important describing phase transitions as well as in studying the evolution of the electron structure with concentration. The CPA also provides type-resolved contributions of the different local quantities. The disadvantage is that as any mean-field theory, the standard CPA does not include the local environment effects such as preferential ordering [51], the Invar effect [52, 53] and lattice relaxations around the impurity site. The self-consistent band structure calculations within CPA approach are carried out by the spin polarised, full relativistic Korringa–Kohn–Rostocker method (SPRKKR) provided by Ebert et al. [54, 55]. The exchange correlation functional was taken within the parametrisation of Vosko Wilk and Nussair [13]. All Brillouin zone integrations were performed on base of a $22 \times 22 \times 22$ mesh of k-points. An imaginary part of 0.001 Ry was added to the energy when calculating the density of states.

It was recently reported that the LSDA and GGA schemes are not sufficient to describe the electronic structure correctly for Co_2FeSi [27]. Therefore, the LDA+U method [29, 56, 57] was used also here to account for on-site correlation at the transition metal sites. The LDA+U method accounts for an orbital dependence of the Coulomb and exchange interaction that is absent in the pure LDA or GGA. In particular, the effective Coulomb-exchange interaction ($U_{\mathrm{eff}} = U - J$) was used for the present calculations. The LDA+U self-interaction correction (SIC) scheme is used in WIEN2K to account for double-counting corrections. The use of $U - J$ neglects, however, multipole terms in the expansion of the Coulomb interaction. It should be mentioned that the $+U$ was used on top of GGA rather than LSDA parameterisation of the exchange correlation functional. No significant differences were observed using one or the other of the parameterisations.

The U_{eff} values being used for the calculations are summarised in Table 7.1 for the different $3d$ elements. The values for Co, Mn and Fe are able to explain the mag-

Table 7.1 U_{eff} values used for the different elements in the LDA+U calculations of Heusler compounds

Element	U_{eff} [eV]
Ti	1.36
V	1.34
Cr	1.59
Mn	1.69
Fe	1.80
Co	1.92

netic moment in $Co_2Mn_{1-x}Fe_xSi$ over the whole range of Fe concentration, as was found in previous calculations [39]. All values correspond to about 7.5 % of those calculated for 3d ions using Cowan's program [58] and are close to the values for the Coulomb interaction U_{dd} for d electrons in the elemental 3d transition metals as reported in [59]. Comparing the semi-empirical values of Table 7.1 to the values found in constrained LDA calculations (for details see [39]), it was found that the latter are too high to explain the magnetic moments. Additional calculations for the elemental 3d transition metals revealed that all values for U_{eff} found in constrained LDA calculations are considerably too high to explain those metallic systems correctly. This is despite the fact that such calculations may result in reliable values for Mott insulators [29]. The constrained LDA calculations revealed that the U_{eff} values at the Co site may vary by about ± 10 % depending on the atoms on B or C sites. As U_{eff} was here used as a parameter for LDA+U, fixed values have been preferred for better comparison.

Whilst the static (or energy-integrated) properties as e. g. magnetic moment can be substantially improved by taking into account the static correlations via LDA+U, the recent analysis [60] of the photoemission spectra in $Co_2Mn_{1-x}Fe_xSi$ reveals the importance of the dynamical corrections as well. As follows from [60] this can be done within the so-called dynamical mean-field theory (DMFT) [61], which quantitatively takes into account dynamical correlation effects in particular spin-flip processes induced by fluctuations. In combination with the LDA (LDA+DMFT) this formalism provides a very reliable approach to deal with a wide range of static and spectroscopic properties of 3d transition metals. [62–67].

It was found theoretically by Mavropoulos et al. [68] that the spin–orbit interaction has only a weak influence on the half-metallic ferromagnetism in Heusler compounds. Later Picozzi et al. [18] and Galanakis [69] reported the same for Co_2 based Heusler compounds, therefore spin–orbit interaction was typically neglected in the calculations discussed here. To verify this assumption, test-calculations including spin–orbit interaction as well as fully relativistic calculations were performed for several Co_2 based Heusler compounds. These calculations revealed a ratio of the orbital to spin moment (m_l/m_s) in the order of only 10^{-2}. These values are in agreement to those earlier reported by Ishida et al. [70]. Therefore, the present calculations include only the mass velocity and Darwin terms to correct for relativistic effects.

7.3 Ordered Co_2TM Compounds

In the half-metallic Heusler compounds discussed here, the gap stays with few exceptions in the minority spin channel, whereas ε_F cuts through bands in the majority spin channel. For the majority spin channel, the position of ε_F is in the region of the d derived bands. These states are shifted to lower energies with respect to the corresponding minority spin states by the exchange splitting. Half-metallic behaviour has been already predicted for a large number of Heusler compounds which are members of the series of ferromagnetic compounds. In general the B (here $= T$) atoms and in some cases also the A (here = Co) atoms carry the magnetic moments in these compounds. It is well known that the magnetic properties of these compounds depend sensitively on the valence electron concentration [17, 23, 36]. The magnetic properties depend on whether the C (here $= M$) component is a member of the $3A$ or $4A$ main group, with the latter group resulting in higher magnetic moments and Curie temperatures. Spin polarised electronic structure calculations indicate that the moments are predominantly of $3d$ origin. The density of state of the minority states is almost zero at the Fermi energy whereas the majority d density can have a peak or a valley close to ε_F. It is filling and emptying of the majority states that is assumed to produce the change in the size of the magnetic moment and the Curie temperature. In order to verify this, and to obtain further evidence for half-metallic behaviour, a series of calculations for $A_2BC = Co_2TM$ compounds was performed and is presented in the following.

7.3.1 Calculated Magnetic and Electronic Structure

In the following sections the electronic structure and magnetic properties of the Heusler compounds will be discussed in more detail. Starting point are the Co_2 based Heusler compounds.

Starting with the Co_2 based compounds, all the information about the calculated total magnetic moments, atom-resolved moments and hyperfine fields, as well as the experimental lattice parameters and magnetic moments are summarised in Table 7.2. For each compound, calculations were carried out using the experimental lattice parameters, and most of them exhibit at least nearly a HMF type character (gap in the minority or majority states). This is clear from Figs. 7.6 and 7.7 where the spin projected density of states of nine selected compounds are shown. (Note that details will be discussed later).

In most of the cases given in Table 7.2, the calculated total spin magnetic moment is exactly integer as expected for a half-metallic system. The experimental magnetic moments m_{exp} are also given in Table 7.2 for comparison with the calculations. In most cases, the calculated magnetic moments m_{cal} are in good agreement with the experimental results. Deviations between experimental and calculated values appear particularly for compounds carrying a high magnetic moment like Co_2FeSi or Co_2FeGe where the calculations result in too low values.

Table 7.2 Magnetic properties of Co_2BC Heusler compounds. The calculated total magnetic moments m_{calc} are compared to experimental values m_{exp}. All magnetic moments are given in μ_B. N_V is the number of valence electrons per formula unit (within parentheses). Total magnetic moments are given per primitive cell. m_{Co} and m_B are the site resolved magnetic moments on Co and B sites, respectively. Most of the experimental values of the lattice parameter a_{exp} and magnetic moment m_{exp} are taken from Refs. [42, 46, 47, 71–74], additional experimental data are found in Refs. [26, 75–105]

Compound	N_V	Exp.		GGA		
		a_{exp}	m_{exp}	m_{calc}	m_{Co}	m_B
Co_2TiAl	25	5.847	0.74	1.00	0.67	−0.18
Co_2TiGa	25	5.850	0.75	1.00	0.63	−0.15
Co_2TiSi	26	5.743	1.65	2.00	1.03	−0.01
Co_2TiGe	26	5.807	1.59	1.97	1.05	−0.06
Co_2TiSn	26	6.077	1.96	1.96	1.08	−0.07
Co_2VAl	26	5.722	1.95	2.00	0.94	0.22
Co_2VGa	26	5.779	1.92	2.01	0.97	0.16
Co_2VSi	27	5.657	–	3.00	1.09	0.79
Co_2VSn	27	5.960	1.21	3.03	1.10	0.86
Co_2CrAl	27	5.727	1.55	3.00	0.83	1.47
Co_2CrGa	27	5.805	3.01	3.05	0.76	1.63
Co_2CrIn	27	6.060	1.10	3.20	0.67	1.98
Co_2MnAl	28	5.749	4.04	4.04	0.76	2.75
Co_2MnGa	28	5.767	4.05	4.12	0.75	2.78
Co_2MnSi	29	5.645	4.90	5.00	1.00	3.00
Co_2MnGe	29	5.749	4.93	5.00	1.02	3.06
Co_2MnSn	29	5.984	5.08	5.03	0.97	3.23
Co_2FeAl	29	5.730	4.96	4.98	1.23	2.80
Co_2FeGa	29	5.737	5.04	5.02	1.20	2.81
Co_2FeSi	30	5.640	6.00	5.59	1.40	2.87
Co_2FeGe	30	5.738	5.90	5.70	1.42	2.92

An inspection of the site resolved magnetic moments (m_{Co} and m_B) reveals that the Co and B atoms may both possess high spin magnetic moments in this series of compounds. In the case of small magnetic moment compounds, the Co atoms contribute mostly to the moment, if comparing to the compounds with large magnetic moments. While going from the low to the high magnetic moment side, the B atoms contribute an increasing moment.

It should be noted that m_{cal} is the calculated total spin magnetic moment of the compound found by integration over the entire cell. Therefore, it is not just the combination of the moments at the A (2 times), B and C sites but respects also the moment of the interstitial between the sites. The interstitial is due to the calculational scheme using non overlapping spheres to define the sites. That is the reason why the site resolved values m_A and m_B alone are not summing up to result in m_{cal}. The missing or excess of the total moment is found at the C sites and to some extend also in the interstitial between the sites where the wave functions are expanded as plane waves. It should also be pointed out that there might be a discrepancy to measurements of the X-ray magnetic circular dichroism (XMCD) in photoabsorption as this method integrates strictly over the extend of the initial state wave functions that are the $2p$ core level in absorption at the L edges.

In all reported compounds, the C atoms carry a negligible magnetic moment that does not contribute much to the overall moment. In most of the compounds it is aligned antiparallel to the A and B moments. It emerges from hybridisation with the transition metals and is caused by the overlap of the electron wave functions. Here, the small moments found at the C sites are mainly due to a polarisation of these atoms by the surrounding, *magnetically active* atoms on the A and B sites. This polarisation results also in non-vanishing hyperfine fields at the nucleus of the atoms on C sites.

In summary, the magnetic moment of most of the compounds shown in Table 7.2 exhibits a linear behaviour with the number of valence electrons and thus follows the Slater–Pauling rule (7.5). However, there are some deviations from this trend: for example, Co_2FeAl, Co_2VGa, Co_2TiGa and Co_2ScSi. That means, there are some cases where the pure LSDA or GGA approximation does not give the magnetic moments expected for half-metallic ferromagnetism, and ε_F may not fall into the minority gap.

In all comparison between the experimental and the calculated quantities, lattice parameter or magnetic moments, one should carefully check whether the compound under investigation has had the required $L2_1$ structure in the experiments.

The following will be concentrated on selected compounds, these nine compounds are Co_2BC Heusler compounds with B = Ti, V, Cr, Mn, Fe and C = Al, Ga and Si. They will be used as representatives of the compounds as listed in Table 7.2. By making such a choice, one covers both a range of electronically different kinds of B and C atoms with Co atoms at the A site. It cannot be excluded that some of the experimental lattice parameters a_{exp} given in Table 7.2 were determined for disordered compounds and not for the required $L2_1$ structure. The use of wrong lattice parameters in the calculations may lead to the prediction of a wrong ground state. Therefore, the lattice parameters of the two series of compounds Co_2BAl and Co_2BSi (B = Sc, Ti, V, Cr, Mn, Fe) were optimised. The lattice parameter a_{opt} minimising the total energy defines the optimal a and is tabulated in Table 7.3. These values were taken as input for further calculations.

The optimised lattice parameters were found to be on average within <0.5 % of the experimentally obtained lattice parameters, as is seen from Table 7.3. This implies that the electronic structure is in most cases independent of the use of the experimental or optimised lattice parameters. A large deviation is found for Co_2FeAl that is known to exhibit in experiments a disordered $B2$ structure. The deviation is even larger for Co_2MnAl what may also point on some type of disorder. Differences in the element specific magnetic moments are mainly due to the different Muffin-tin radii that were fixed to the lowest needed value during the optimisation procedure while varying only a. In the two cases, Co_2MnSi and Co_2FeSi, the optimised lattice parameter was found to match the experimental one within the uncertainty of experiment and calculation. However, the magnetic moment of the Fe containing compound is still too low.

In Table 7.3, the total and element specific magnetic moments for Co and the atoms on the B sites are displayed together with the optimised lattice parameters. In

Table 7.3 Magnetic data for the series Co_2BC (C = Al, Si) with optimised lattice parameter. Listed are the optimised lattice parameters (a in Å), total magnetic moments (m_{tot}) and atom resolved magnetic moments (m_{Co} and m_B). $\Delta(a)$ is the change of the lattice parameter in % with respect to the experimental value. All magnetic moments are given in μ_B

Compound	a_{opt}	$\Delta(a)$	m_{tot}	m_{Co}	m_B
Co_2ScAl	5.960	–	0	0.00	0.00
Co_2TiAl	5.828	−0.3	1	0.62	−0.13
Co_2VAl	5.754	+0.6	2	0.94	0.23
Co_2CrAl	5.708	−0.3	3	0.80	1.52
Co_2MnAl	5.695	−0.9	4	0.77	2.67
Co_2FeAl	5.692	−0.7	5	1.22	2.79
Co_2ScSi	5.865	–	1	0.60	−0.10
Co_2TiSi	5.760	+0.3	2	1.03	−0.02
Co_2VSi	5.688	+0.5	3	1.10	0.80
Co_2CrSi	5.647	–	4	1.00	2.03
Co_2MnSi	5.643	0	5	1.06	2.99
Co_2FeSi	5.634	0	5.55	1.39	2.85

the Co_2BAl series, the magnetic moment at the Co site first increases and then stays at about 1 μ_B, while the magnetic moment at the B site starts from small negative values and then increases linearly with the valence electron concentration. In all compounds, the induced magnetic moment at the Al and Si sites is very small and antiparallel to Co. This is a result of polarisation caused by the overlap of the wave functions of the main group element with those of the surrounding transition metal atoms. It seems that the moment of Co is fixed at about 1 μ_B and compels the B moment to result in the overall magnetic moment according to the valence electron concentration.

In detail, the magnetic moment at the Co site stays at about 1 μ_B from Ti to Mn and diverges from this value only for Sc and Fe containing compounds. The magnetic moment at the B site increases linearly from Ti to Mn compounds such that the total magnetic moment follows the Slater–Pauling rule, but deviates from this trend for Fe. Co_2ScAl has 24 valence electrons and the magnetic moment vanishes at all sites. The result is a vanishing total magnetic moment as expected from (7.4). A ferrimagnetic behaviour with compensating moments, as reported for Mn_2 based compounds [106], was not observed. The reason is that the single d electron at the Sc site is not able to form a localised moment by filling a majority e_g band completely.

In the Co_2BSi series, the trends are the same as in the just discussed Al series. The only difference is the detail of the magnetic moments at Co and B atoms. In particular in Co_2ScSi, the Co_2 sublattice starts to order ferromagnetically whereas the Sc atoms do not contribute to the magnetic moment. The formation of the magnetic Co_2 sub lattice is completely different from the classical Heusler compounds like Cu_2MnAl where the magnetic order is mediated only by the strong local moment at the Mn atom. Finally, one observes again the break of the trend of increasing moments at the B site for Fe. Here, the moment at the iron site even decreases compared to Mn. Seemingly this decrease of the moment at Fe cannot be compensated by the Co atoms resulting in an overall moment being to small to be compatible to the Slater–Pauling rule.

In summary the Co atoms contribute in the Co_2BC (C = Al, Si) compounds about 1 μ_B to the magnetic moment and *drive* the properties of the B atoms to have a magnetic moment according to the number of valence electrons of the compounds. The properties of the Co_2BC compounds are dominated by the nature of the Co atoms which force the *extent of localisation* of the electrons and the resulting magnetic moment at the B site. The result is that the total magnetic moment follows the Slater–Pauling rule (compare (7.4) and (7.5)). The behaviour of the compounds with other elements on the C site is similar as can be seen from Table 7.2. The increase of the magnetic moments at the Co and B sites is attributed to the rearrangement of the electrons and will be discussed briefly in the next section. Obviously, the calculated magnetic moments of the compounds Co_2FeSi and Co_2FeGe are too small compared to the experimental value. This fact will be discussed later together with on-site correlation.

7.3.2 Electronic Structure and Density of States

As seen in the previous section, the Co_2BC compounds can be distinguished in two classes, one with small magnetic moments (less than 4 μ_B) and the remainder with high magnetic moments. The class of compounds with small magnetic moments will be called "*low-m*" and the one with high magnetic moments "*high-m*". In the following sections, the density of states and the distribution of the electrons over the various states and atoms will be discussed.

The DOS for the low-m compounds Co_2TiAl, Co_2TiSi, Co_2VAl and Co_2VSi are displayed in Fig. 7.6. The upper part of each panel displays the majority spin densities and the lower one the minority spin densities.

The increase of the splitting of the minority states at the Fermi energy is clearly visible. In Co_2TiAl is close to the minority valence states, whereas it is positioned in the upper half of the gap close to the minority conduction band for the remaining compounds. The size of the gap increases if replacing Al by Si. At the same time the size of the Heusler-typical gap at below -6 eV increases considerably from some hundred meV to about 2 eV. This points on a much stronger hybridisation in the Si containing compounds compared to the Al containing compounds. As typical for low-m compounds, the majority states exhibit a rather high density in the vicinity of ε_F that is caused by weakly dispersing, flat bands. The width of the d bands is almost the same in all four compounds. Indeed, in all four compounds the states at and close to ε_F are strongly spin polarised and all four systems exhibit a gap in the minority states. The spin polarisation is found to be 100 % which characterises the systems to be half-metallic.

In the next step, the six compounds Co_2BC (B = Cr, Mn, Fe and C = Al, Ga) with intermediate and high m values were chosen. The DOS of these compounds is displayed in Figs. 7.7(a)–(f). The situation of the high-m compounds with Mn and

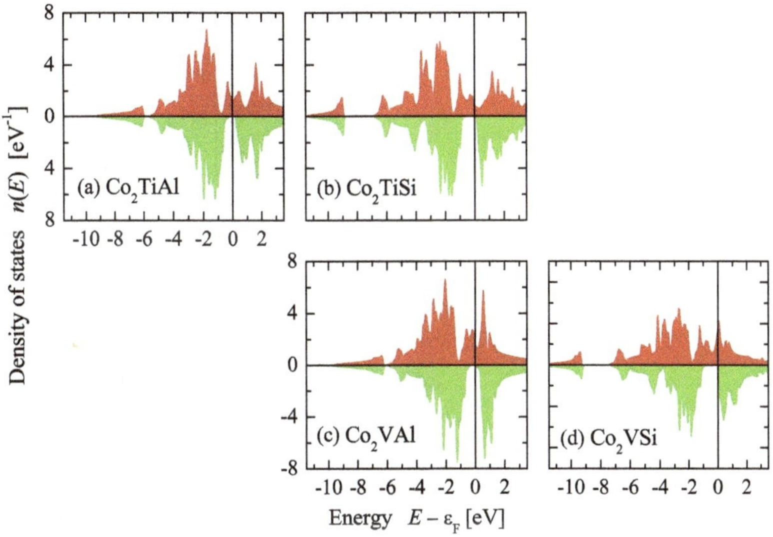

Fig. 7.6 Density of states of selected low m Co_2BC compounds. Shown are the spin resolved DOS for compounds with B = Ti, V and C = Al, Si. (Note: *Upper* and *lower parts* of the panels show for each compound the majority $n(E)_\uparrow$ and minority $n(E)_\downarrow$ DOS, respectively.)

Fe is somewhat different from the low-m compounds. All listed compounds exhibit almost the same band width with a gap in the minority states. However, there are distinct differences in the shapes of the densities.

Some pronounced changes are observed in the DOS while going from the low-m compounds to the high-m compounds. A decrease in the contribution from majority states just at ε_F is observed. In the Cr compounds, with about 3 μ_B, one observes still a high majority density of states close to ε_F. In the Mn compounds, with about 4 μ_B, this high majority density is shifted just below ε_F and finally one observes only a rather low density in the Fe compounds with 5 μ_B that emerges from strongly dispersing bands (compare to the band structure of Co_2CrSi discussed above). This low density of the majority states close to ε_F is nevertheless arising from d-electrons. The particular shape of the majority DOS of the high-m compounds is finally the reason for the observed deviation of the magnetic moments in Co2FeSi or Co2FeGe. The filling of one more electron by replacing Al by Si (or Ga by Ge) will need a much larger energy range to add up 1 μ_B. It also will shift the states with high density much further away from ε_F what is obviously not possible if attributing the shift only to an exchange splitting. That might be one reason why LSDA fails to find the correct moments of 6 μ_B but needs some additional interaction between the electrons. This interaction can be interpreted in the underestimation of on-site correlation [27]. However, if it plays a role in the high-m compounds it should do so in the low-m compounds, too. This fact will be discussed later.

At that point, it should be noted that the magnetic moment does not depend on the density of states at the Fermi energy. Neither does it depend much on the particular shape of the DOS or band structure as it is an integrated quantity, namely the

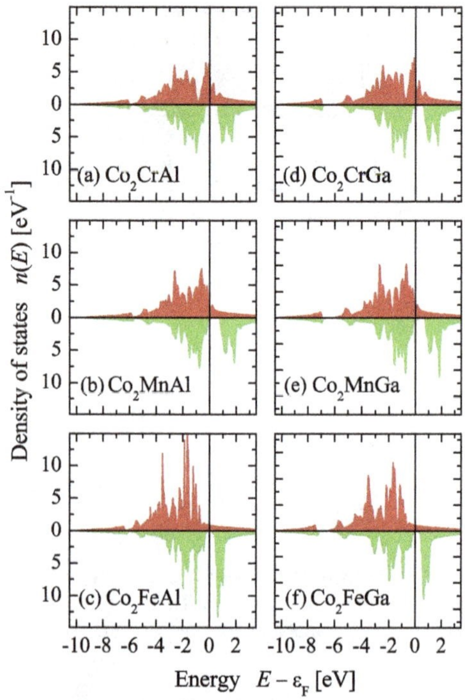

Fig. 7.7 Density of states for low to high m Co_2BC compounds. Shown are the spin resolved DOS for compounds with $B =$ Cr, Mn, Fe and $C =$ Al, Ga

difference between the number of occupied majority and minority electrons. From that point of view the magnetic moment may help to find half metallic ferromagnets, however, it is not suited to prove the existence of band gaps in the minority states.

In the following, some remarks will be made on the distribution of the electrons and influence of the main group element C. After all, the Heusler compounds are metals with delocalised electrons, therefore it should be noted that many of the electrons are located in the interstitial space between the muffin-tin spheres describing the atoms of the primitive cell and thus cannot be specifically attributed either to one or another of the contributing elements or a specific orbital momentum. At this point—and before continuing with the magnetic moments—recall that, from the objective point of view of the solid, all electrons are indistinguishable whereas only a subjective observer divides the solid arbitrarily into spheres or other objects of a given but virtual size. Strictly speaking, the dispersion $E(k)$, total density of states and total magnetic moments are the only well-defined ground-state quantities but they are not the site-resolved ones.

On careful examination of the magnetic moments on the Co sites in both series (Co_2BC with $C =$ Al and Si), one finds that the Co site is carrying an average of 0.8 μ_B (Al series) and 1.0 μ_B (Si series). At this point it is not quite clear why the magnetic moment of Co is changing with the exchange of Al by Si. Thus the question arises how C affects the overall magnetic moment without directly contributing to it? To check this, it is necessary to inspect the electron distribution of the system.

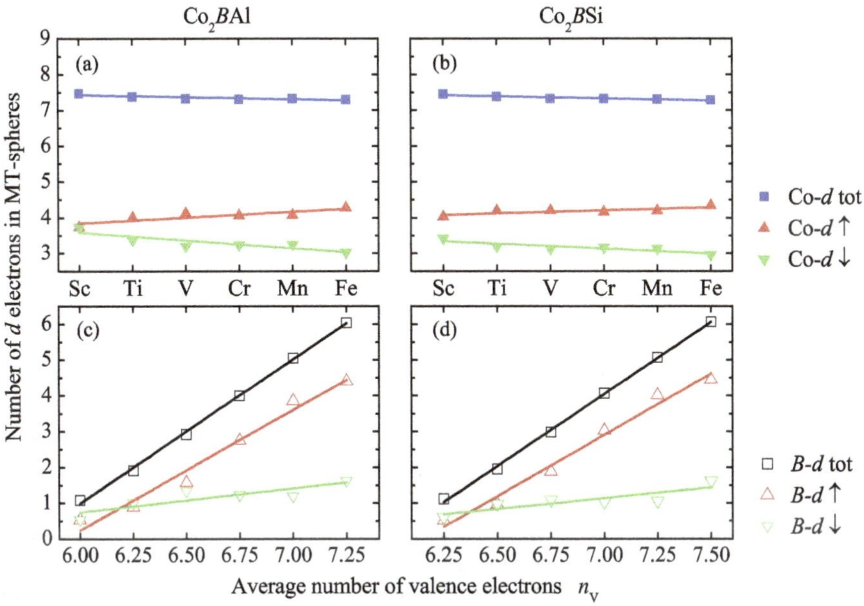

Fig. 7.8 Valence d electrons in Co_2BC compounds. Shown are the total (tot) and spin resolved ($\uparrow$ = majority, $\downarrow$ = minority) numbers of d electrons in Co_2BAl and Co_2BSi with B = Sc, Ti, V, Cr, Mn and Fe. (**a**), (**b**) are for Co and (**c**), (**d**) for atoms at the B site. The lines are results of a linear fit

The total magnetic moment in the series is defined by the number of occupied d electrons in the majority and minority states because the number of occupied s and p states is rather independent of the spin (note that the total number of valence electrons N_V appears in (7.5) only because $N_V = N_\uparrow + N_\downarrow$ such that $m = N_\uparrow - N_\downarrow$ can be rewritten to be $m = N_V - 2N_\downarrow$). The total number of d electrons at the Co site is about 7.5 including majority and minority electrons (see Figs. 7.8(a) and (b)). The Co d states for majority and minority spin are almost constant and contain about four and three electrons, respectively. Their values are nearly constant and do not change much if going from Ti to Mn. This is one reason why the magnetic moment of Co is fixed at about 1 μ_B.

The distribution of the d electrons in majority and minority states for the B element is shown in Figs. 7.8(c), (d). It is clear that filling of both majority and minority d electrons increases in parallel up to V and thereafter the electrons start to fill more majority d states as compared to minority d states. Al and Si contribute very less to the d states and in-fact they do not contribute directly to the overall magnetic moment. In general, it is found that the trends are the same in the Co_2BSi and Co_2BAl compounds.

In order to have a deeper insight, the distribution of electrons in states with different symmetry has been studied in the series of Co_2BAl and Co_2BSi. As determined earlier by Kübler [38], to maximise the number of unpaired electrons, the minority states should be filled by eight d electrons and the rest of the electrons fills the majority states. In the Co_2BAl series, overall (average) 7.85 d electrons are in the minority states. There should be eight d electrons to fill the d states completely and to have a gap. The missing part of the electrons is found in the interstitial. That is, they are completely de-localised and cannot be attributed to a particular atom. When inspecting the Co_2BSi series, the number of total minority d electrons is in average 7.5. the remaining are found in the interstitial and overlap with the Si s and p states. To proof this situation in detail, LMTO-ASA [107] calculations were performed to obtain the crystal orbital Hamiltonian population (COHP) of Co-Al and Co-Si. It was found that the bonding interaction between Co and Si is much stronger than between Co and Al. This is expected from the higher electro-negativity of Si compared to Al. Due to the stronger bonding interaction between Co and Si in Co_2BSi, some more electrons are in between the atoms and missing from the expected total count of eight. At the same time, the electrons in the majority states are redistributed with the result of a higher magnetic moment.

Replacement of Al by Si or exchange between other members of the $3A$ and $4A$ groups, plays an important role for the distribution of electrons in the various symmetry distinguished states (t_{2g} and e_g) at Co as well as at B sites. In fact, the overall number of d electrons remains the same. Addition of an extra electron by replacing Al by Si affects mostly all symmetry distinguished states except the e_g states at the Co site. The latter e_g states form the bonds between the Co sites in the (100)-like planes. The number of symmetry resolved e_g and t_{2g} states at B increases linearly. In particular, the e_g states at B are not affected by other states. The reason is that there is no possible direct overlap to other states of the same symmetry at nearest neighbours, whereas the t_{2g} states form the bonds with the atoms at the Co sites. The e_g states at the B site are mainly responsible for the localised magnetic moment at this site. In The Al compounds a moment attributed to the Co t_{2g} states is only for $B = \text{Fe}$ worth mentioning. The interesting point is, however, that the t_{2g} states at the Co site contribute much more to the magnetic moment in the Si compounds as compared to the Al compounds. This observation, finally, links the indirect connection between the site specific magnetic moment at Co and the hybridisation arising from the interaction between the electrons at the C sites with the neighbouring electrons in Co t_{2g} states.

Galanakis and co-workers [17] explained the half-metallicity of Co_2MnGe by a particular molecular orbital coupling scheme. To develop the model, they had to assume, however, that the Co atoms are in an environment with O_h symmetry. This assumption is incorrect even though the total symmetry of the cubic cell of $L2_1$ Heusler compounds is O_h. The sites occupied by Co have no inversion symmetry and carry T_d symmetry, as is evident from the primitive cell shown in Fig. 7.1. In contrast, the sites occupied by Mn and Ge exhibit O_h symmetry. Figure 7.9 shows schematically, without assumption of any particular molecular orbital coupling scheme, directly the order of the states at the Γ point of Co_2MnGe. The half-metallic character determined by the band filling of the minority states is obvious. The successive complete filling of the minority bands (a_{1g}, t_{1u}, e_g, $2 \times t_{2g}$) by 12 electrons leads to a quasi-closed-shell character. The energy scheme for the majority electrons is considerably different from that of molecular orbital models. It is

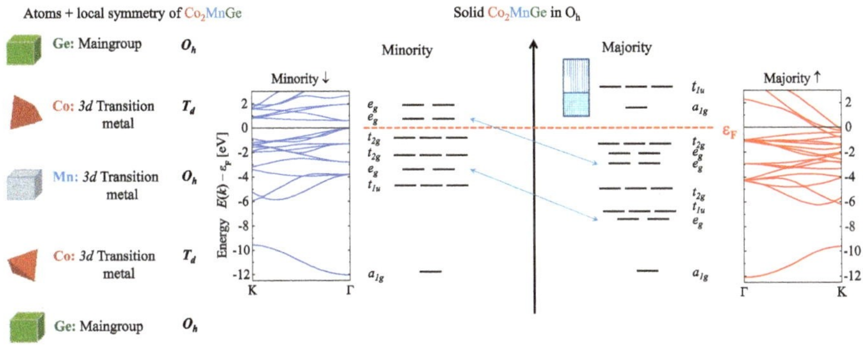

Fig. 7.9 Schematic illustration of the orbitals and states in Co_2MnGe. The local symmetries of the atoms in the primitive cell with O_h symmetry (see Fig. 7.1) are shown at the *left side*. The *right part* shows the Σ direction of the band structure (see Fig. 7.2) and a schematic of the states at the Γ point of the solid with O_h symmetry. The *box* in indicates that the unoccupied majority states with a_{1g} and t_{1u} character are delocalised and smeared over a wide band of energies in k space

seen that the additionally occupied doubly degenerate states are pushed at Γ below the triply degenerate states. Finally, the strongly dispersing high-lying (a_{1g} and t_{1u} character at Γ) majority bands start to cross ε_F and bring the majority electronic structure close to a simple metal (compare also Fig. 7.2). Both effects together show the limits for the application of molecular orbital schemes. The latter has another important consequence: When the two additional e_g states are occupied one reaches compounds with 28 valence electrons overall, which exhibit a total spin moment of 4 μ_B. At that valence electron concentration the "*localised*" weakly dispersing bands are used up and a further increase will need to fill "*delocalised*" states with strong band dispersion. The filling of these bands needs a rather large exchange interaction to split the occupied and unoccupied localised states. It seems that there is a limit at which the exchange interaction cannot be increased further. This limit is reached when the compound has 30 valence electrons, that is, when about two "*delocalised*" majority bands are filled. This explains why Heusler compounds with magnetic moments of considerably more than 6 μ_B are not known, in contrast to fully localised, nondispersing molecular orbital schemes that suggest 7 μ_B as a limit [17].

7.3.3 The Properties of the Minority Band Gap

The half-metallic ferromagnetism manifests itself in a band gap in one of the spin densities. However, the Fermi energy may fall inside or outside of that gap. The band structure of all calculated compounds was carefully examined to find the top of the valence band (E_{max}) and the bottom of the conduction band (E_{in}) enclosing the gap. Tables 7.4 and 7.5 summarise the results calculated for the minority band gap of the Co_2BC compounds. All of the listed compounds show a gap along Γ–X

Table 7.4 Co_2BC compounds that are Type I half-metals. Given are the data for compounds exhibiting in GGA calculations a Type I half-metallic ferromagnetism, that is, the minority band gap includes the Fermi energy. E_{min} and E_{max} are the values of the minimum energy of the conduction band and the maximum energy of the valence band. The band gap ΔE is the difference between these extremal energies. All energies are given in eV. $n_\uparrow(\varepsilon_F)$ is the density of states at ε_F for majority electrons in eV^{-1}. For lattice parameters see Table 7.2. Note that the last four compounds are very close to the Type III half metals

Compound	$E_{max}(\Gamma)$	E_{min} (X)	ΔE	$n_\uparrow(\varepsilon_F)$
Co_2TiAl	−0.24	0.215	0.456	1.37
Co_2TiSi	−0.606	0.179	0.785	0.99
Co_2TiGe	−0.401	0.194	0.595	0.93
Co_2TiSn	−0.222	0.282	0.504	1.05
Co_2VGa	−0.142	0.047	0.189	1.52
Co_2VSn	−0.397	0.151	0.548	5.34
Co_2CrAl	−0.118	0.63	0.748	4.84
Co_2MnSi	−0.292	0.506	0.798	1.27
Co_2MnGe	−0.048	0.533	0.581	1.29
Co_2VSi	−0.886	−0.067	0.072	3.50
Co_2FeAl	−0.138	−0.027	0.111	0.87
Co_2TiGa	0.06	0.216	0.210	1.44
Co_2FeGa	0.086	0.107	0.021	0.88

Table 7.5 Co_2BC compounds that are Type III half-metals. Shown are the data of Co_2BC compounds exhibiting band gaps in the minority states but with ε_F outside of the gap. In this class of compounds the minority band gap does not include ε_F. The compounds are grouped for the cases where ε_F from the GGA calculations is above (*upper part*) or below (*lower part*) of the gap. ($n_\downarrow(\varepsilon_F)$) is the density of states at ε_F for minority electrons. For lattice parameters see Table 7.2, for other quantities see Table 7.4)

Compound	E_{max}	E_{min}	ΔE	$n_\uparrow(\varepsilon_F)$	$n_\downarrow(\varepsilon_F)$
Co_2CrGa	0.203	0.628	0.425	2.8	0.2
Co_2CrIn	0.450	0.612	0.162	2.35	0.5
Co_2MnAl	0.172	0.716	0.664	1.05	0.2
Co_2MnGa	0.364	0.663	0.301	1.7	0.34
Co_2MnSn	0.183	0.594	0.411	1.22	0.16
Co_2VAl	−0.357	−0.119	0.238	1.6	0.03
Co_2FeSi	−0.735	−0.589	0.146	2.7	0.71
Co_2FeGe	−0.517	−0.43	0.087	2.3	0.74

that is in the Δ-direction of the paramagnetic state. The Δ-direction is perpendicular to the Co_2 (100)-planes. As was shown earlier [25], just the Δ-direction plays the important role for understanding of the HMF character and magnetic properties of Heusler compounds. This fact was also pointed out by Öĝüt and Rabe [108].

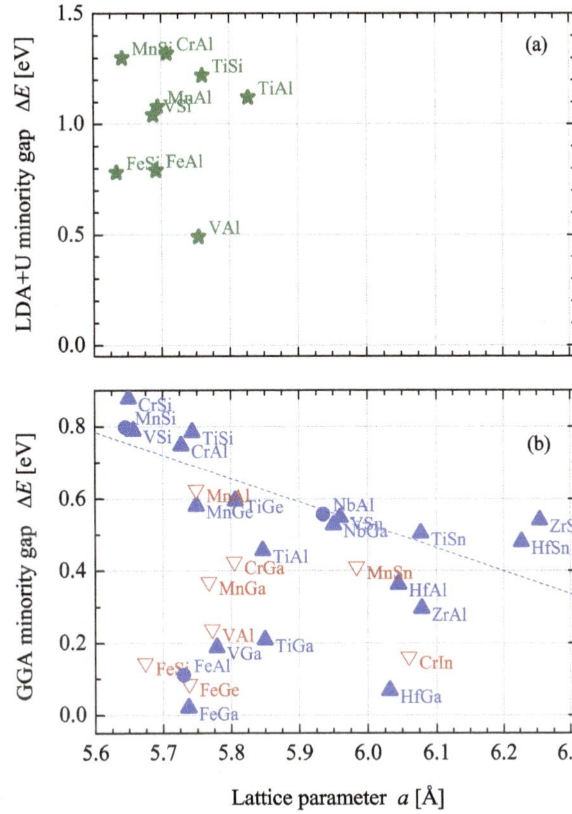

Fig. 7.10 Minority band gap in Co_2-based half-metallic ferromagnets. Shown is the size of the gap (ΔE) as function of the lattice parameter (a). The width of the gap was determined from the band structure calculated with GGA and is shown in (**b**). Type I half-metals are displayed as filled and other materials by open symbols. The *line* is drawn to guide the eye. (**a**) shows the minority band gap of the Type I half-metals containing Al and Si as revealed from LDA+U calculations

Table 7.4 is devoted to the properties of Type I half-metals where ε_F is located inside of the gap of the minority band structure. As a direct consequence, the minority density $n_\downarrow(\varepsilon_F)$ at the Fermi energy vanishes and the spin polarisation (7.1) is 100 %. The listed compounds (except the four shown at the bottom of the table) exhibit a clear gap at ε_F and behave like half-metallic ferromagnets, at least within GGA.

The minority gaps of the clear Type I half-metals have, with exception of Co_2VGa, a width of $\Delta E \approx 0.5 \ldots 0.8$ eV. The gap is considerably smaller in those compounds being close to Type III like behaviour (see also Fig. 7.10).

Only few compounds containing Cr or V have a remarkably high majority DOS at ε_F whereas the remainder exhibit lower values of about 1 eV^{-1}. As one approaches towards high magnetic moment compounds containing Fe, there is a reduction in the density of states of the majority spin states at ε_F. Instead of a high majority DOS they exhibit only a small DOS. Already the GGA calculations reveal this small DOS at ε_F. The important conclusion one can draw at this point is that the LSDA-GGA does not estimate the gap and magnetic moment (compare Table 7.2) of these compounds correctly and results in a Type III half-metal.

The data for the minority band gaps of the remaining Co_2BC compounds, which are Type III in the GGA calculations, are shown in Table 7.5. The minority DOS $n_\downarrow(\varepsilon_F)$ at the Fermi energy ε_F is listed in addition. In all cases one has from the GGA calculations $n_\uparrow(\varepsilon_F) > n_\downarrow(\varepsilon_F)$ resulting in a positive spin polarisation. The interesting point to note is—in some cases of Type III Heusler half-metals—that the gap remains at the Γ-point but is closed at the X-point. This means they are metals for one spin direction (majority) and semi-metals for the other (minority).

Summarising the discussion of the minority gap, Fig. 7.10(b) displays the size of the minority gap as function of the lattice parameter. From the GGA calculations in Fig. 7.10(b), there is a tendency seen that larger lattice parameter of the compounds lead to smaller band gaps for Type I half-metals. It is obvious that those materials deviate strongly from this trend where the Fermi energy is found to be located outside of the gap. This trend cannot be used quantitatively, but is useful as a starting guess to search for good candidates for HMF materials with a reasonable size of the gap. It is interesting to note that the Ti and V compounds exhibit a gap even in the case where they do not form a localised magnetic moment. This gives a clear advice that a localised magnetic moment at the B sites is not a necessary condition for the formation of half-metallic ferromagnetism in the Co_2-based Heusler compounds.

7.3.4 Correlation and the Electronic Structure Within LDA+U

In this subsection, the influence of correlation on the electronic structure of various compounds is discussed. LDA+U calculations were performed for Co_2BC compounds with B = Ti, V, Cr, Mn and Fe and C = Al and Si. The semi-empirical values corresponding to 7.5 % of the atomic values of the Coulomb-exchange parameter U_{eff} have been used for all compounds (see Table 7.1).

The results of the LDA+U calculations are summarised in Table 7.6 and visualised in Fig. 7.10(a). The total magnetic moment is in agreement with the Slater–Pauling rule. A strong deviation is found for Co_2MnAl, where the Fermi energy is

Table 7.6 Properties of Co_2BC Heusler compounds calculated by LDA+U. Magnetic moments (m) are given in μ_B and energies (E) are in eV. For lattice parameters see Table 7.2

Compound	N_V	m_{calc}	m_{Co}	m_B	E_{max}	E_{min}	ΔE
Co_2TiAl	25	1.00	0.84	−0.39	−0.206	0.914	1.12
Co_2TiSi	26	2.00	1.16	−0.18	−0.583	0.632	1.22
Co_2VAl	26	2.00	1.15	−0.09	−0.414	0.072	0.49
Co_2VSi	27	3.00	1.20	0.69	−0.905	0.138	1.04
Co_2CrAl	27	3.01	0.79	1.70	0.071	1.39	1.32
Co_2MnAl	28	4.28	0.73	3.17	0.424	1.505	1.08
Co_2MnSi	29	5.00	0.98	3.29	0.007	1.307	1.30
Co_2FeAl	29	5.00	1.22	2.97	0.022	0.811	0.79
Co_2FeSi	30	6.00	1.50	3.14	−0.81	−0.028	0.78

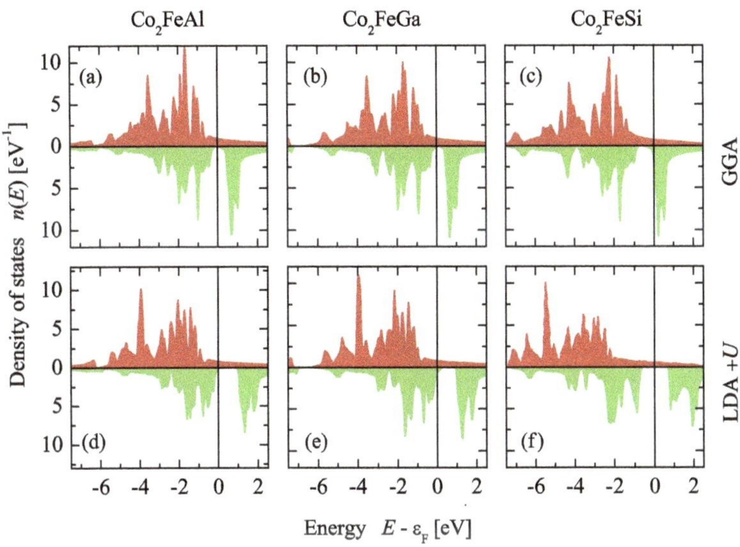

Fig. 7.11 Spin resolved density of states for Co_2FeAl, Co_2FeGa and Co_2FeSi. Shown are the spin resolved density of states for Co_2FeAl (**a**), (**d**), Co_2FeGa (**b**), (**e**), and Co_2FeSi (**c**), (**f**) calculated by GGA in the *upper part* (**a**), (**b**), (**c**) and LDA+U in the *lower part* (**d**), (**e**), (**f**). The *upper* and *lower parts* of each plot display the majority and minority densities, respectively. (See text for the particular values of U at the different sites.)

found far below the gap in the minority states. This deviation appears also if the optimised lattice parameter is used instead of the experimental one. In general, the width of the band gap increases with inclusion of U in the LDA scheme. However, the gap was already existing in all compounds when using GGA for the calculations. That means, U is not creating the half-metallic ferromagnetism but confirms the half-metallic property of those compounds, and helps to explain the magnetic properties correctly. The second impact of the on-site correlation on the electronic states is an increase of the splitting between bands of different symmetry. This causes that the Fermi energy is shifted with respect to the gap in the minority states, or vice versa depending on the point of view. The effect is that the half-metallic character is changed in some cases from Type III to Type I, indeed, the opposite may appear too as seen from a comparison of the data given in Tables 7.4, 7.5 and 7.6.

Major differences between experimental and calculated moments were found for the high-m compounds containing Fe. Out of these, the electronic structure of the two compounds Co_2FeAl and Co_2FeGa will be discussed in more detail to explain the effect of the on-site correlation. The spin resolved density of states for Co_2FeAl and Co_2FeGa that are calculated using the GGA and LDA+U approximations are shown in Fig. 7.11 together with Co_2FeGSi for comparison. There is no major difference between the LSDA and the GGA calculations (compare Fig. 7.6(c) and (f)). Both compounds exhibit a very small minority band gap at ε_F in pure LSDA calculations. This becomes considerably larger when LDA+U is used in the calculations. In both cases of the 29 valence electron compounds (Al, Ga), the Fermi energy stays close to the top of the minority valence band. The low lying majority e_g states, being responsible for the localised moment at the Fe site, are shifted from about −3.5 eV by about 0.5 eV to lower binding energies. At the same time the unoccupied minority d states are shifted from about 0.6 eV above the Fermi energy to 1.3 eV. The splitting of this band complex is also enlarged from about 0.3 eV to 0.45 eV. In average, the splitting between occupied majority bands and unoccupied minority bands is increased by about 1.2 eV.

Finally, it should be noted that just the high-m compounds are a sensitive test for the use of electron–electron correlation in the form of LDA+U. In some cases, it is required to explain their electronic and magnetic properties correctly. More details of the change in the electronic structure are discussed elsewhere for Co_2MnSi [39] and Co_2FeSi [26, 27, 39].

7.3.5 Further Co_2-Based Half-Metallic Ferromagnets

Half-metallic ferromagnetism may not only appear in those Co_2BC compounds where B is a $3d$ element but also if B is a $4d$ (here Zr or Nb) or $5d$ (here Hf) transition metal. Indeed, various Co_2-based Heusler compounds of this type are reported where B is not a $3d$ transition metal but one from a higher period. It was found in the present calculations that most of them can be expected to exhibit half-metallic ferromagnetism.

The magnetic properties of several, reported Co_2-based Heusler compounds with B being not a $3d$ transition metal are summarised in Table 7.7. There is obviously a large discrepancy between observed and calculated magnetic moments. However, it is not clear whether the experimentally reported samples exhibited the $L2_1$ structure (assumed in the calculations), the Li_2AgSb structure, or any other type of disorder. For Co_2NbSn, neither the measured nor the calculated magnetic moments come close to the value expected for a half-metallic state. It should be noted that this compound is reported to crystallise in the ordered $Fm\bar{3}m$ structure at room temperature and undergoes a structural transition at 235 K from cubic to orthorhombic

Table 7.7 Properties of Co_2-based Heusler compounds containing $4d$ or $5d$ elements on the B site. (For explanation of the quantities see Tables 7.2 and 7.4.) Experimental data are taken from Refs. [90, 95, 109–116]

Compound	N_V	a_{exp}	m_{exp}	m_{calc}	m_{Co}	m_B	ΔE
Co_2ZrAl	25	6.078	0.79	1.00	0.62	−0.10	0.296
Co_2HfAl	25	6.045	0.82	1.00	0.61	−0.09	0.363
Co_2HfGa	25	6.032	0.60	1.00	0.60	−0.09	0.068
Co_2NbAl	26	5.935	1.35	2.00	1.04	0.01	0.557
Co_2NbGa	26	5.95	1.39	2.00	1.04	−0.01	0.527
Co_2ZrSn	26	6.254	1.81	2.00	1.09	−0.09	0.540
Co_2HfSn	26	6.227	1.57	2.00	1.07	−0.07	0.480
Co_2NbSn	27	6.152	0.69	1.94	0.95	0.07	0.430

with $P\,mma$ symmetry [117, 118]. Spin–orbit interaction, which was not accounted in the present calculations, may also play already a role for the Hf containing compounds.

The most interesting point is, however, that in all those compounds only the Co atoms are responsible for the magnetic moment. That means, in none of these compounds a localised magnetic moment at the B atoms is present. This points on the important fact that the existence of a localised moment at the B atoms is not a necessary condition for the occurrence of half-metallic ferromagnetism in Co_2-based Heusler compounds, as also mentioned in the previous section.

7.4 $Co_2T_{1-x}T'_xM$, $Co_2TM_{1-x}M'_x$ Random Compounds and Disordered Alloys

7.4.1 $Co_2Cr_{1-x}Fe_xAl$

The properties of the $L2_1$ compounds and random alloys where calculated in $F\,m\overline{3}m$ symmetry using the experimental lattice parameter ($a = 5.727$ Å) as determined by X-ray powder diffraction. Co and Al where placed on 8c (1/4, 1/4, 1/4) and 4a (0, 0, 0) sites, respectively. Disorder between chromium and iron was accounted by setting the 4b (1/2, 1/2, 1/2) site occupations to $1-x$ for Cr and x for Fe. The SPRKKR method with CPA was used for the calculations reported in the following.

Several types of disorder have been considered. It turned out that the mostly occurring type of disorder is $B2$ like. This type is a result of disorder in the YZ (Y = Cr, Fe, Z = Al) planes. It is found by setting the $F\,m\overline{3}m$ site occupations of the 4a and 4b sites equally to 1/2 for Y and Z (assigned as $B2a$ in [45]). The resulting structure with reduced $P\,m\overline{3}m$ symmetry was used for the calculations. The lattice parameter is $a_{B2} = a_{L2_1}/2$. X was placed at the 1a site (origin of the cube) and the site occupations for Y and Z at 1b (centre of the cube) were set equally to 1/2. For the mixed Fe–Cr systems, the site occupation factors for Y and Y' have to be weighted by x and $(1-x)$, respectively. Complete disorder between all sites results in the $A2$ structure with reduced symmetry $I\,m\overline{3}m$.

The electronic structure of the pure and doped alloys will be discussed in the following. First, the electronic structure of the ordered alloys are presented, followed by the results for disordered alloys.

7.4.1.1 Electronic and Magnetic Structure of $L2_1$ Ordered Alloys

First, the electronic structure for the $L2_1$ ordered ternary compounds was calculated. The resulting density of states (DOS) of Co_2CrAl and Co_2FeAl is shown in Fig. 7.12 (a, c). These are the end members of the quaternary series of alloys. In the next step the electronic structure for the $L2_1$ ordered alloys $Co_2Cr_{1-x}Fe_xAl$ was

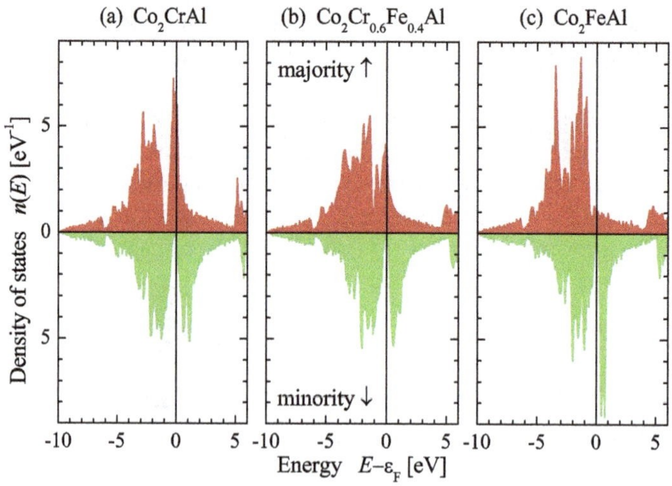

Fig. 7.12 Spin resolved density of states of $Co_2Cr_{1-x}Fe_xAl$ ($x = 0, 0.4, 1$)

calculated and the result for $x = 0.4$ is compared exemplary in Fig. 7.12 (b) to the pure compounds.

The low lying sp-bands are visible at about 6 eV to 10 eV below ε_F. The d-bands start at about 5.5 eV below ε_F. The gap between s- and d-bands is less pronounced compared to non relativistic calculations (see [119]). The majority DOS at the Fermi energy decreases with increasing iron concentration x. The density of majority electrons at ε_F is a crucial point for spectroscopic methods investigating the spin polarisation, like spin resolved photo emission. The small, non vanishing density in the half-metallic gap of the minority states at ε_F emerges mainly from the imaginary part added to the energy when calculating the DOS by means of the Greens function. However, a coupling between majority and minority states is always expected in the full relativistic calculations thus that a pure spin up state is not possible in general. See in Mavropoulos et al. [68, 120] for a discussion of the relativistic effects and their influence on the half-metallic gap.

Figure 7.12 reveals that the Fermi energy (ε_F) is pinned in the minimum of the density of the minority states. Further, it is seen that the majority density is shifted with increasing Fe concentration and thus with increasing number of valence electrons. Both together have finally the result that the magnetic moments increase with the number of valence electrons (see below). This behaviour, in particular the pinning of ε_F in a minimum of the minority density, is typical for Slater–Pauling like behaviour in the case of materials with localised magnetic moments [25, 35, 121, 122]. From the Slater–Pauling rule for ferromagnetic alloys with localised moments, the magnetic moment per atom is approximately given by $m = n_\uparrow - n_\downarrow = n_V - 2n_\downarrow \approx n_V - 6$. Here, $n_V = n_\uparrow + n_\downarrow$ is the mean number of valence electrons per atom and given by the sum of the accompanied majority ($n_\uparrow$) and minority ($n_\downarrow$) electrons.

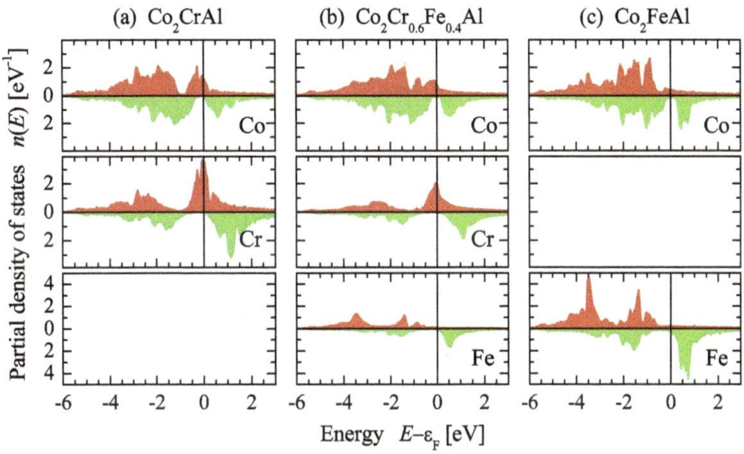

Fig. 7.13 Spin resolved partial density of states of $Co_2Cr_{1-x}Fe_xAl$ $x = 0$, 0.4, 1. Shown is the PDOS of the contributing 3*d* elements. (Note: The different scales of Fe compared to Co and Cr at the partial DOS.)

The spin resolved partial density of states (PDOS) is displayed in Fig. 7.13. Shown are the partial (site specific) densities for the majority and minority states of $Co_2Cr_{1-x}Fe_xAl$ for $x = 0, 0.4, 1$. In particular the gap in the minority DOS is better resolved. From the behaviour of the different PDOS depending on the Fe concentration x, it is clear that the electronic structure does not follow a rigid band like model. In particular it is seen that the Cr PDOS decreases with increasing iron content keeping its shape rather unchanged. In the same way the Fe PDOS increases with x. However, the maxima of the density at the Co sites are clearly shifted away from ε_F. The Al PDOS stays nearly unaffected and is rather independent of the Fe content.

From Fig. 7.13, it is seen that the high majority DOS at ε_F emerges from Cr. Both, Co and Fe exhibit only a small majority PDOS at and above ε_F. Overall, the change of the majority DOS of $Co_2Cr_{1-x}Fe_xAl$ around ε_F can be clearly attributed to the increasing amount of iron with respect to chromium. The minimum in the minority DOS around ε_F is mainly restricted by the shape of the Co PDOS. This indicates that the HMF like behaviour is mainly characterised by Co. The steep increase of the minority PDOS of Cr and Fe is mainly located in the unoccupied part above ε_F.

Doping with Fe changes not only the total DOS but also the PDOS of Co and much less pronounced the one of Cr. In particular, a very small shift of the Cr PDOS causes an additional decrease of majority states at ε_F. This shift increases with increasing Fe concentration (as was found from the PDOS with variation of x, not shown here). The energy shift of the PDOS results in a change of the local magnetic moments in particular at the Co sites.

Figure 7.14 compares the measured and the calculated magnetic moments. The calculated total magnetic moment $m_{tot} = m_s + m_l$ depends linearly on the com-

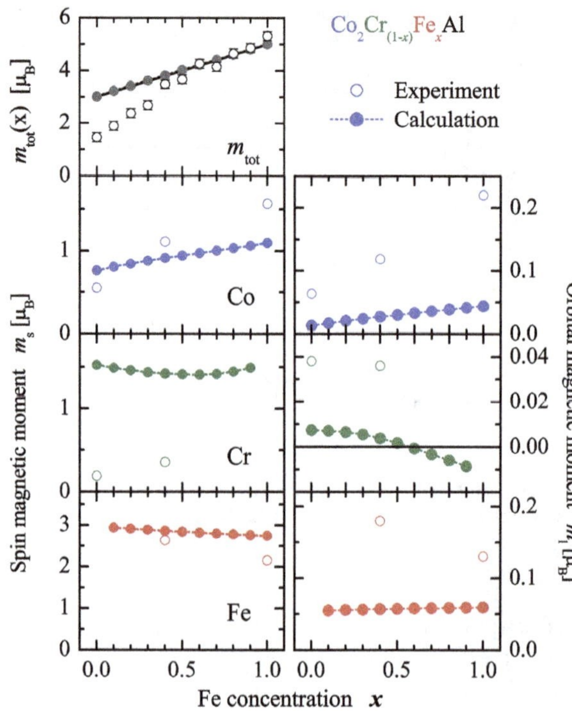

Fig. 7.14 Total and partial magnetic moments of the transition metals in $Co_2Cr_{1-x}Fe_xAl$

position x and follows the Slater–Pauling rule [25, 38, 122]. The measured total moments as determined by SQUID magnetometry at 5 K are smaller compared to the calculated values in particular for low Fe (or equivalently high Cr) content. This behaviour will be discussed in the next section about disorder.

Site specific spin (m_s) and orbital (m_l) magnetic moments were determined for $x = 0, 0.4, 1$ from XMCD measurements at the $L_{3,2}$ edges of Cr, Fe and Co. Details of the spectroscopy experiments are reported below. The values were found from a sum rule analysis neglecting the magnetic dipole term [123]. The calculated spin magnetic moments at the Co sites increase with increasing Fe concentration x as result of the shift of the majority PDOS of Co. The calculated spin moments at the Fe sites decrease slightly with increasing x. The principal dependence of the measured Co and Fe spin moments on x is the same as in the calculation. However, the changes in the measured moments are stronger. The calculated Al spin moment is negative and very small. It is induced by the surrounding polarised Co atoms. Most obvious, the measured spin magnetic moments at the Cr sites are smaller compared to the calculated ones. This causes finally the too small total moment observed by SQUID magnetometry. The calculated orbital magnetic moments increase with x at the Co sites by a factor of 4 and are nearly constant at Fe sites. The calculated Cr orbital moments are very small and change sign at $x = 1/2$. In contrast, the measured orbital moments are throughout larger. Both experiment and calculation exhibit, however, the same trend for m_l.

Most evidently the calculated ratios between orbital m_l and spin m_s magnetic moments are much smaller than the experimental values. For Cr, a vanishing of the orbital momentum is expected, as observed in the calculation.

The findings of the full relativistic calculations are throughout compatible to those made in non-relativistic calculations using ordered compounds [25]. Comparing the random alloy $Co_2Cr_{0.6}Fe_{0.4}Al$ and the nearly iso-electronic ordered compound $Co_4CrFeAl_2$ [25], it is found that there are no major differences between CPA and virtual crystal approximation calculations for the materials investigated here. In particular, the rather integrated quantities like magnetic moments are the same. The non rigid band like character of the electronic structure upon Fe doping is revealed in both methods. The band gaps in the minority densities are here smaller due to the additional splitting of bands caused by the spin–orbit interaction rather than by properties of the CPA scheme, as is seen from the pure compounds.

7.4.1.2 Random Alloys with Disorder

Table 7.8 summarises the results of the calculations for the magnetic moment in the well ordered ($L2_1$), partially disordered ($B2$) and completely disordered ($A2$) alloy $Co_2Cr_{0.6}Fe_{0.4}Al$ and compares them to the experimental values.

The magnetic moment of correctly ordered $Co_2Cr_{0.6}Fe_{0.4}Al$ is close to the value of 3.8 μ_B expected for a HMF ground state. It also concerns Co_2CrAl and Co_2FeAl alloys which exhibit in $L2_1$ a total moment of 3.0 μ_B and 5.0 μ_B, respectively.

From the calculations it is expected that the total magnetic moment of the $B2$ disordered compounds is lower compared to the one in the $L2_1$ structure. However, the calculations for complete $B2$ or $A2$ disorder are not able to explain the too low value of the magnetic moment found experimentally for Co_2CrAl. A reduction of the overall moment caused by an antiferromagnetic order of the Cr atoms could not be verified by the calculations for these structure types. Therefore, calculations were also performed for other type of disorder. In the DO_3-type disordered alloy the Co and Cr atoms in 8c and 4b positions of the $L2_1$ structure are mixed. The calculations revealed an overall moment of 2.0 μ_B. The moment of the Co atoms at the two different sites are 0.86 μ_B and 1.51 μ_B. The moments of the Cr atoms are aligned antiparallel with respect to each other and amount to -0.41 μ_B and $+0.44$ μ_B at 8c

Table 7.8 Magnetic moments in $Co_2Cr_{0.6}Fe_{0.4}Al$. Given are the site specific moments per atom and the overall magnetic moment per primitive cell, both in multiples of the Bohr magneton (μ_B). The overall magnetic moments for the disordered structures are converted to fit those of the $L2_1$ cell in number of atoms

Structure	Co	Cr	Fe	tot
$L2_1$	0.94	1.42	2.92	3.82
$B2$	0.57	−1.12	2.95	1.61
$A2$	1.44	0.19	2.24	3.83
Exp.	1.11	0.36	2.64	3.49

Table 7.9 Magnetic moments in disordered $Co_2Cr_{0.6}Fe_{0.4}Al$. All calculated values are in μ_B per atom. The mean value is found using the disorder site occupation factors

		8c	4a	4b	Mean
20 %$B2$	Co	0.99	–	–	0.99
	Cr	–	1.65	1.43	1.61
	Fe	–	2.88	2.93	2.89
20 %$A2$	Co	1.08	1.86	1.98	1.25
	Cr	−0.9	1.33	1.2	0.87
	Fe	0.96	2.79	2.87	2.43

and 4b sites, respectively. Exchange of only one of the Co atoms with the Cr atom leads to the X structure [45]. In this case the calculations revealed an antiparallel alignment of the Cr moments ($-0.93\ \mu_B$) with respect to the Co moments ($0.8\ \mu_B$ and $1.11\ \mu_B$) with a resulting overall moment of only $0.95\ \mu_B$ in the cell. From this approach, a tendency to ferrimagnetic order is concluded if Cr and Co atoms change sites resulting in a lower magnetic moment. It is worthwhile to note that disorder of this type closes the gap in the minority DOS and the half-metallic character is lost.

The measured overall magnetic moment of $Co_2Cr_{0.6}Fe_{0.4}Al$ is smaller compared to the calculated one for the ordered compound. Inspecting the site resolved moments one finds that the Co moment is slightly enhanced whereas the Cr and Fe moments are lowered compared to the $L2_1$ calculation. Such type of behaviour is also observed in the calculation for $A2$ disorder, but less pronounced. Thus it is obvious to assume a partial disorder for this type of samples. It is interesting to note that the calculations revealed a ferrimagnetic ground state for $B2$ disordered $Co_2Cr_{0.6}Fe_{0.4}Al$, indeed, in contrast to the experiment where the Cr moment was small but aligned parallel to the Co moment.

The results from a calculation for 20 % disordered systems are compared in Tab. 7.9. The total moment for (80 $\%L2_1 + 20\ \%B2$) is $4.04\ \mu_B$ and thus clearly above the experimental value. The Co moment is slightly lower compared to the ordered structure. The Cr moments are at 20 % $B2$ disorder still aligned parallel to the Co moments. For (80 $\%L2_1 + 20\ \%A2$), the Co and Fe moments are larger in the 4a and 4b positions compared to the 8c position. Most evidently the Cr moment exhibits an antiparallel alignment in the 8c position. The total magnetic moment in the cell is $3.96\ \mu_B$ and thus larger than the experimentally observed value. Comparing to the $L2_1$ ordered material, the enhancement of the Co moments with a simultaneous decrease of the Fe moments gives clear advice on $A2$ disorder in the experimentally examined samples.

In the experiment at Co_2FeAl, a larger total magnetic moment was observed in comparison to the expectation from the Slater–Pauling rule. At the same time the Co moment was enhanced and the Fe moment lowered compared to the calculations for $L2_1$. Comparing the values calculated for different type of disorder, one finds easily that the experimental values coincide with those expected for $A2$ type disorder. This means that the sample investigated by XMCD was to a large amount completely disordered.

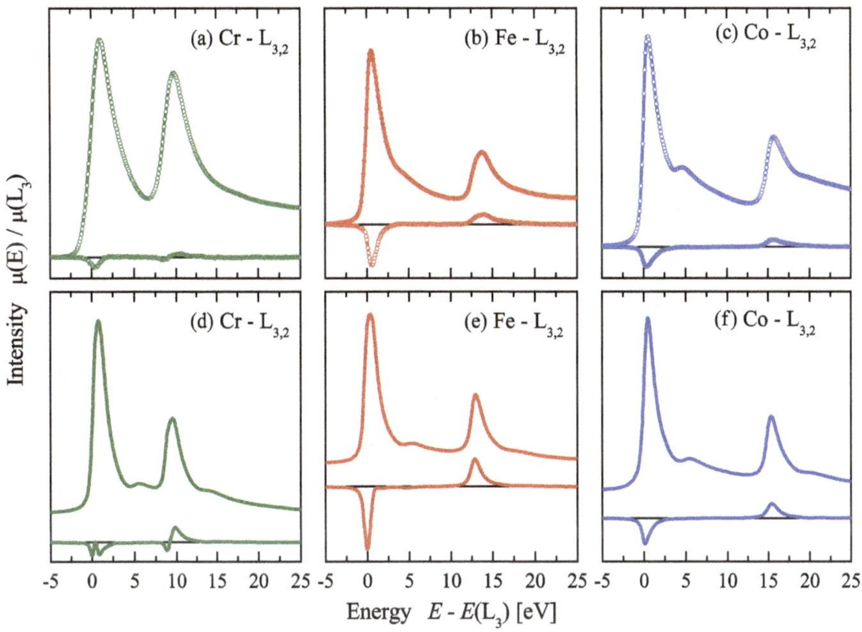

Fig. 7.15 Experimental (**a**)–(**c**) and calculated (**d**)–(**f**) X-ray absorption spectra and XMCD at the $L_{3,2}$ edges of Cr, Fe and Co in $Co_2Cr_{0.6}Fe_{0.4}Al$. The energy scales of the spectra are shifted with respect to the steepest point of the spectra (maximum of the derivative). The intensity scales are normalised to the maxima at the L_3 edges and shifted for sake of clarity

7.4.1.3 Spectroscopic Properties of $Co_2(Cr_{0.6}Fe_{0.4})Al$: Photo Absorption

X-ray absorption spectroscopy probes the unoccupied density of states above the Fermi energy. For the resonant excitation of the $2p$-states of the $3d$ transition elements the transition probability is proportional to the density of final states resulting from the $3d$ and $4s$ holes. The spectra, consisting mainly of the L_3 $(2p_{3/2})$ and L_2 $(2p_{1/2})$ white lines (see Fig. 7.15), reflect the high density of unoccupied states resulting from the $3d$ electrons.

Figure 7.15 compares measured and calculated absorption spectra at the $L_{3,2}$ edges of Cr, Fe and Co in $Co_2Cr_{0.6}Fe_{0.4}Al$. The monochromator resolution was set to 50 meV for the spectra in (a)–(c). The spectra (d)–(f) were calculated for $L2_1$ structures using a lifetime broadening of 136 meV. The measured white line at the Cr edge is considerably wider compared to the calculated spectra whereas the width of the Fe and Co white lines is comparable. This points on a shorter lifetime of the holes at the Cr sites. Similarly, the measured L_2 lines of Co and Fe hint on a slightly larger lifetime broadening compared to the L_3 edges. Overall it is found that the spectra are governed by the lifetime broadening rather than by the experimental resolution.

The experimental spectrum exhibits a prominent feature at the Co L_3 line for $Co_2Cr_{0.6}Fe_{0.4}Al$, in particular, which is shifted by about 4 eV with respect to the white line. This feature appears only as weak shoulder in the Fe spectrum. It is not revealed in the measured Cr spectra what may be partially attributed to the higher lifetime broadening. The same structure occurs in the simulated absorption spectra and can be related to the structural properties. A similar feature is observed for example in fcc Ni but not in bcc Fe or hcp Co. It reflects the onset of the high lying sd-bands (see also Figs. 7.12, 7.13). Here, its occurrence is characteristic for the highly ordered Heusler compounds. It vanishes for annealed and presumably disordered samples [123]. However, no particular features giving advice on half-metallic ferromagnetism are found in the absorption spectra what may partially be due to the relatively large width of the lines.

Figure 7.15 shows also the measured and calculated XMCD spectra of at the $3d$ L-edges for $Co_2Cr_{0.6}Fe_{0.4}Al$. The experimental XMCD spectra were taken at 300 K by switching a constant induction field of $B_0 = \pm 0.73$T (for field dependent measurements see [97]). The angle of photon incidence was 70° with a degree of circular polarisation of 85 %. The spectra (d)–(f) are calculated for parallel and antiparallel orientation between magnetisation and photon spin. Size and shape of the XMCD agree well at the Fe and Co edges comparing experiment (b), (c) and calculation (e), (f). It is clearly seen from the spectra at the Co edges (c), (f) that the structural feature does not contribute to the XMCD signal. The calculated Cr XMCD (d) exhibits a splitting at the L_3 edge that is not resolved in the experimental spectra. However, the change of sign close to the L_2 edge is clearly visible in experimental as well as theoretical spectra. Both, splitting and change of sign, are due to the high density of majority states directly at the Fermi energy and the energetically shifted high density of the minority d-states above the half-metallic gap (compare Fig. 7.13).

In summary, results from spectroscopy experiments (XMCD) at quaternary Heusler alloys $Co_2Cr_{1-x}Fe_xAl$ were analysed in the light of ab-initio calculations. Special attention was focussed on $Co_2Cr_{0.6}Fe_{0.4}Al$. The measured overall and site specific magnetic moments are in accordance with the calculations, if disorder is assumed. In particular, it was shown that Co-Cr disorder leads to the low magnetic moment observed in Co_2CrAl. It points on ferrimagnetic order with an antiparallel alignment of part of the Cr moments in disordered alloys.

7.4.2 $Co_2(Mn_{1-x}Fe_x)Al$

The substitutional series $Co_2Mn_{1-x}Fe_xAl$ with $0 < x < 1$ are isoelectronic to $Co_2Cr_{1-x}Fe_xAl$ with $x > 0.5$ while comparing the total number of valence electrons. The total magnetic moment expected from the Slater–Pauling rule [17, 36, 37] is $(4\ldots5)\ \mu_B$ for the $Mn_{1-x}Fe_x$ compounds compared to $(3\ldots5)\ \mu_B$ for the $Cr_{1-x}Fe_x$ compounds if respecting the full range of iron concentration. This point makes the two series attractive for comparative investigations. Studies of the $Co_2Mn_{1-x}Fe_xAl$ series are less often reported (see [124, 125]) compared to the $Co_2Cr_{1-x}Fe_xAl$ (see Sect. 7.4.1). In the $Co_2Mn_{1-x}Fe_xAl$ series investigated here

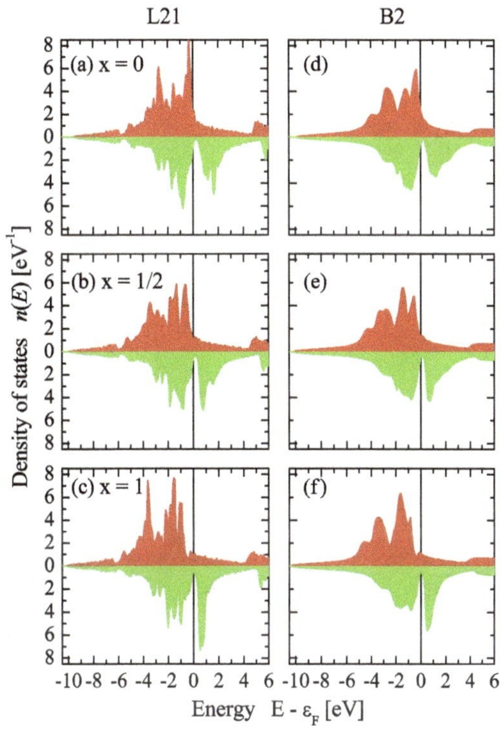

Fig. 7.16 Spin resolved density of states $Co_2(Mn_{1-x}Fe_x)Al$. (**a**)–(**c**) show the DOS for the $L2_1$ structure, (**d**)–(**f**) show the DOS for the $B2$ structure

a $B2$ structure is observed. This is very often the case in the Al containing Heusler compounds [40].

The SPRKKR method with CPA was used for the calculations reported in the following. The calculations were performed using the lattice parameters set for the $L2_1$ structure to [126]: $a_{L2_1}(x) = (5.754 - 0.051x)$ Å according to Vegard's law as found from a linear fit of the experimental data. The value for the $B2$ structure is $a/2$ due to the reduced CsCl type cell. An optimisation of the lattice parameter for the disordered structure would be very time and resource consumptive without bringing more insight in the problem. Due to thermal expansion (measurement at room temperature) the experimental lattice parameter will not correspond to the one in the ground state, anyway. Therefore, the use of the experimental lattice parameters was preferred.

Figure 7.16 displays the calculated spin resolved total DOS of $Co_2(Mn_{1-x}Fe_x)Al$ with $x = 0, 0.5$ and 1. Comparing the densities of the $L2_1$ structure with that of the $B2$ structure the smearing of the structures is obvious. Such smearing is typical for systems with a random distribution of atoms. The minority density exhibits in all cases a deep valley at the Fermi energy (ε_F). This distinct minimum is typical for compounds following the *localised* part of the Slater–Pauling curve [36, 37]. The Fermi energy is pinned in that minimum and the increasing number of valence electrons (while going from Mn to Fe) is filling majority states. The electronic

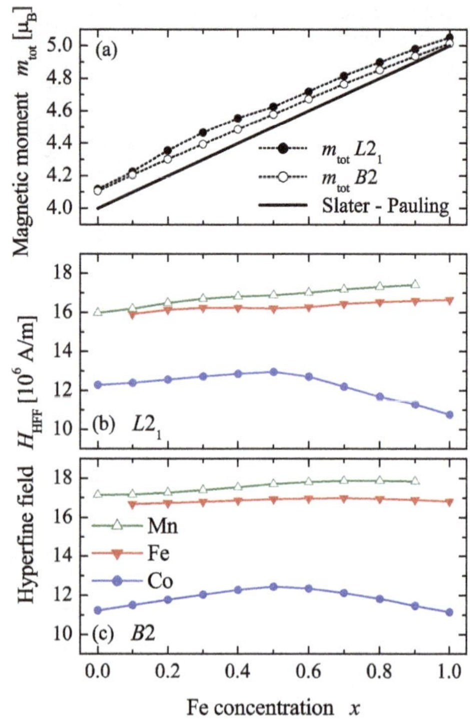

Fig. 7.17 Calculated total magnetic moments (**a**) and site resolved hyperfine fields (**b**), (**c**) of ordered ($L2_1$) and partially disordered ($B2$) $Co_2(Mn_{1-x}Fe_x)Al$

structure of those compounds is already well discussed above and in the literature [17, 37, 39, 127, 128], therefore a further discussion is omitted here.

Figure 7.17(a) displays the calculated dependence of the total magnetic moments of $Co_2(Mn_{1-x}Fe_x)Al$ on the Fe concentration x. The results for the $B2$ and $L2_1$ structures are compared to the Slater–Pauling behaviour. The magnetic moments increase in both cases linearly with the Fe content. The magnetic moments are slightly higher for the $L2_1$ structure compared to $B2$. For both cases, the total magnetic moment is higher compared to the one calculated from the Slater–Pauling rule ($m = 4 + x$), if assuming half-metallic ferromagnetism. The overall behaviour still follows a general Slater–Pauling law for systems with localised moments, that is, an increase of the magnetic moment with an increasing number of valence electrons. The deviation is, indeed, expected from the electronic structure as reported above. It should be noted that the total magnetic moments are considerably higher than the spin moments due to the additional orbital magnetic moments ($m_{tot} = m_s + m_l$). An inspection of the site resolved magnetic moments reveals that the orbital magnetic moment resides mainly at the Fe atoms ($m_l \approx 0.058\ \mu_B$). Overall, the magnetic moments behave very similar independent of the structure, $L2_1$ or $B2$.

Figures 7.17(b), (c) display the calculated hyperfine fields at the transition metal sites for the $B2$ and $L2_1$ structures. For the $L2_1$ structure, both hyperfine fields at the 4a site (Mn, Fe) increase with increasing iron content. On the other hand, the hyperfine field at Fe stays rather constant in the $B2$ structure. Both constituents exhibit higher hyperfine fields in the $B2$ structure compared to $L2_1$. More interesting is the behaviour of the hyperfine field at the Co sites. It exhibits a clear maximum with the change of the iron content. For both structures, the maximum appears at an iron concentration of about 50 %. The Co atoms exhibit for low Fe concentration ($x \leq 0.7$) lower hyperfine fields in the $B2$ structure compared to $L2_1$, differing from Mn and Fe. The similar behaviour of the two structure types makes it obvious that there is no direct relation between the hyperfine fields and the half-metallic character of the compounds. A detailed comparison with experiments [126], however, reveals that the CPA approach does not provide a complete description of the hyperfine fields, since it calculates them only for a mean surrounding medium but cannot distinguish different neighbours. The supercell approaches impose another problem, namely it results in one or more hyperfine fields for a very particular set-up of the environment and, moreover, it implies some artificial periodicity and symmetry. On the other hand, two extensions that might help to overcome the problems at least partially are proposed [126].

7.4.3 $Co_2(Mn_{1-x}Fe_x)Si$

This series spans the region between two very attractive Heusler systems. The first one, Co_2MnSi, exhibits a large half-metallic band gap (about 0.4 eV) and a very high Curie temperature (985 K) [74, 129–131]. The second, an even more interesting system, Co_2FeSi, exhibits a Curie temperature of about 1100 K and a very high magnetic moment of 6 μ_B in the primitive cell [26, 27, 104]. It was shown that the complete substitutional series $Co_2Mn_{1-x}Fe_xSi$ crystallise in the Heusler type $L2_1$ structure with a high degree of structural order [39]. Also confirmed was a Slater–Pauling behaviour (see [17, 36, 130]) of the magnetic moment ranging linearly from 5 μ_B to 6 μ_B with increasing Fe concentration x. However, both ordered edge systems are unstable as half-metallic ferromagnets due to proximity of the Fermi energy to the band gap edges [132]. This situation can be stabilised by selecting the intermediate alloys with the Fermi energy set in the middle of the band gap.

As was mentioned in Sect. 7.2 the adequate description of electronic structure for $Co_2(Mn_{1-x}Fe_x)Si$ series requires correct incorporation of local correlation effects which was carried out here by the so-called LDA+DMFT approach [60]. This is discussed in detail in the Chapter "*Correlation and chemical disorder in Heusler alloys: a spectroscopical study*" by J. Minár et al. Figure 7.18 shows the calculated density of states using LDA+DMFT [60]. A clear shift of the Fermi energy across the band gap is observed that leads to the increase of the magnetic moment with increasing Fe concentration as shown in Fig. 7.19.

Figure 7.19 compares calculated and measured magnetic moments. The values calculated in the LDA+U approach [44, 132] follow strictly the Slater–Pauling values. From Fig. 7.19 a noticeable increase of the magnetic moments in analogy to the Slater–Pauling curve within the whole range of concentrations is observed.

The behaviour of the minority spin band gap was investigated in detail in [132] by LDA+U calculations within FLAPW formalism [44]. It is seen from Fig. 7.18

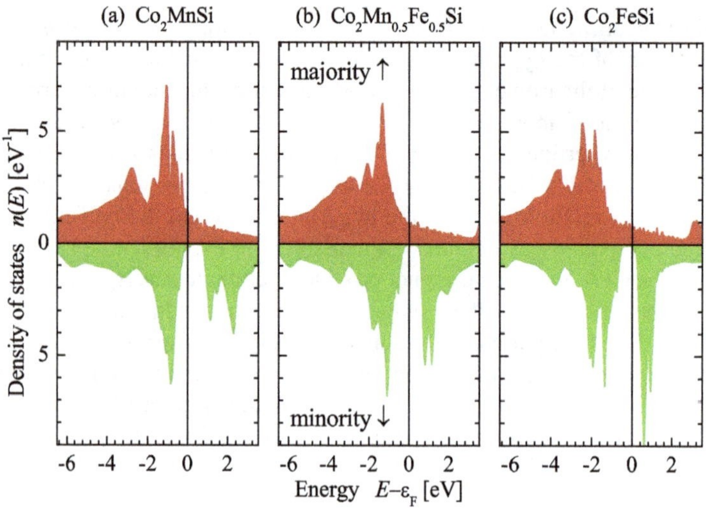

Fig. 7.18 Spin resolved density of states of Co_2MnSi, $Co_2Mn_{0.5}Fe_{0.5}Si$ and Co_2FeSi calculated within LDA+DMFT

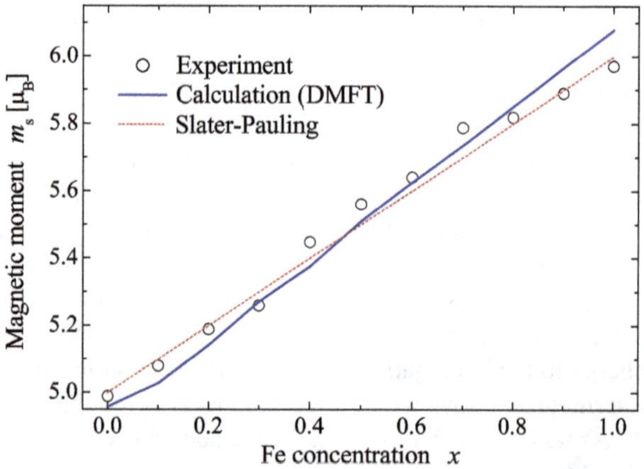

Fig. 7.19 Comparison of the total magnetic moments for $Co_2Mn_{1-x}Fe_xSi$ compounds as calculated by LDA+DMFT compared to results of SQUID measurements and Slater–Pauling behaviour

that the minority density of states is shifted with respect to the Fermi energy such that ε_F moves from the top of the minority valence bands at low x to the bottom of the minority conduction bands at high x. In general, from both studies [60, 132] it can be concluded that the additional electrons affect both majority and minority states. The largest gap in the minority states is found for Co_2MnSi. The size of the gap decreases with increasing Fe content x. At the same time, the position of the Fermi energy is moved from the top of the valence band to the bottom of the conduction band. It is also seen that the compounds with $x = 0$ and 1 are on the borderline to half-metallic ferromagnetism, as the Fermi energy just touches the top of the valence band or the bottom of the conduction band. The thermally most stable half-metallicity is expected at about $x = 0.5$.

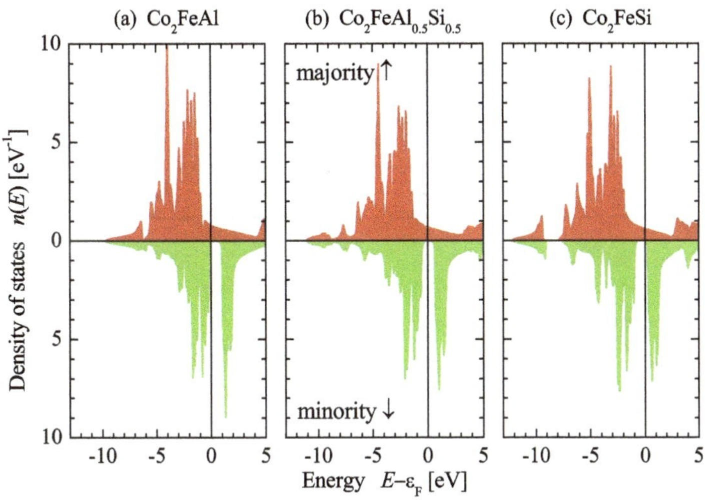

Fig. 7.20 Spin resolved density of states of $Co_2FeAl_{1-x}Si_x$ for $x = 0, 0.5$, and 1 calculated using LDA+U

7.4.4 $Co_2Fe(Al_{1-x}Si_x)$

The electronic structure of the substitutional series of the quaternary Heusler compounds $Co_2FeAl_{1-x}Si_x$ was investigated by means of FLAPW band structure calculations using the LDA and LDA+U approximations [40]. It was found that the $Co_2FeAl_{1-x}Si_x$ series of compounds exhibits half-metallic ferromagnetism if using the LDA+U scheme. Moderate Coulomb-interaction parameters of less than 2 eV were used. For the two end-members, Co_2FeAl and Co_2FeSi, the Fermi energy is close to the band edges of the minority states. The high densities at those band edges make the half-metallic character of both compounds rather unstable at finite temperatures above 0 K. This might be one reason explaining the low tunneling magnetoresistance ratio found in those compounds at room temperature.

Figure 7.20 shows the spin resolved density of states of $Co_2FeAl_{1-x}Si_x$ for $x = 0, 0.5$, and 1. For $x \approx 0.5$, the calculations predict that the Fermi energy is located in the middle of the gap of the minority states. This behaviour will make $Co_2FeAl_{0.5}Si_{0.5}$ stable against temperature variations.

7.5 Other Heusler Compounds Exhibiting Half-Metallic Ferromagnetism

So far, only the properties of Heusler compounds based on Co_2 were considered. However, there exist also half-metallic ferromagnets in the remaining large group of Heusler compounds with 2:1:1 stoichiometry. For example, Galanakis et al. [17] have proposed half-metallic ferromagnetism in Rh_2-based compounds. However,

Table 7.10 Properties of Heusler compounds with $A_2 = \mathrm{Mn}_2, \mathrm{Fe}_2, \mathrm{Ir}_2, \mathrm{Ru}_2$. Shown are compounds that exhibit half-metallic ferromagnetism or at least a band gap in one of the spin densities. Negative gap values are used if there is an indirect overlap of bands. SM is used for a semi-metal like band gap, CF for completely compensated half-metallic ferrimagnet, and HF for Type I half-metal. (For explanation of the other quantities see Tables 7.2 and 7.4.) Experimental values are taken from Refs. [91, 135–146]

Compound	N_V	a_{exp}	m_{exp}	m_{calc}	m_A	m_B	ΔE	Type
Mn_2VAl	22	5.897	1.82	1.99	1.52	−0.95	0.32	HF
Mn_3Al	24	5.804		0.00	−1.42	2.84	0.55	CF
Mn_3Ga	24	5.823		0.01	1.54	−3.03	0.17	CF
Mn_3Si	25	5.722		1.00	−0.88	2.69	0.62	HF
Fe_2TiAl	23	5.879		0.95	0.67	−0.28		
Fe_2VAl	24	5.761		0			−0.04	SM
Fe_2VGa	24	5.782		0			−0.20	SM
Fe_2VSi	25	5.675		0.81	0.55	−0.17		
Fe_2MnAl	26	5.816		2.00	−0.31	2.62	0.39	HF
Fe_2MnSi	27	5.671	2.33	3.00	0.20	2.63	0.63	HF
Ir_2MnAl	28	6.025		4.00	0.28	3.42	0.37	HF
Ru_2MnGe	27	5.985	3.8	3.03	−0.01	3.00	0.13	
Ru_2MnSi	27	5.887		3.00	0.02	2.92	0.10	
Ru_2MnSn	27	6.217		3.08	−0.06	3.21	0.14	
Ru_2MnSb	28	6.200	4.4	4.02	0.22	3.55	0.28	

only the compounds $\mathrm{Rh_2Mn}C$ with C = Ge, Sn, or Pb (with $N_V = 29$) crystallise in the required $L2_1$ structure [133, 134]. In the calculations, these compounds turned out to exhibit a magnetic moment of about 4.7 μ_B in good agreement with the experiment but incompatible to half-metallic ferromagnetism and a band gap did not appear in either of the spin densities.

From the remaining group of known Heusler compounds, half-metallic ferromagnetism was only found in Mn containing compounds (see Table 7.10). Mn_2VAl one of the few Heusler compound exhibiting a gap in the majority density of states, unlike the other Heusler compounds. The reason is that Mn_2VAl has only 22 valence electrons, that is, less than 24, therefore, the completely filled bands appear in the majority states. Indeed, fixing the number of occupied states in the spin channel exhibiting the gap has the result that the other channel has to have less electrons occupied. This is expressed in the Slater–Pauling rule where m becomes virtually negative if $N_V < 24$. It should be noted that Mn_3Si exhibits in experiments a complicated magnetic behaviour close to anti-ferromagnetism [135], this might destroy the band gap.

The DO_3 type structure is the binary derivative of the $L2_1$ Heusler structure. Mn_3Al and Mn_3Ga are, besides of Mn_3Si and many others, two of the binary com-

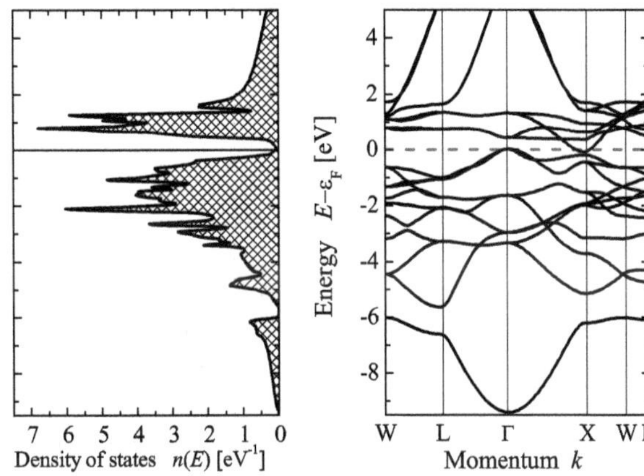

Fig. 7.21 Electronic structure of Fe_2VAl

pounds reported to order in the same space group as Heusler compounds. Both have zero total magnetic moments and thus are half-metallic completely compensated ferrimagnets. It should be firmly emphasised that they are not antiferromagnets due to a completely different symmetry. This phenomenon is discussed in detail in [106].

Inspecting the Fe_2-based compounds, one finds also some cases of half-metallic ferromagnets or at least a behaviour close to it. Overall, the Fe_2 based compounds exhibit in experiments a complicated magnetic behaviour (see References in Table 7.10). The 23 valence electron compound Fe_2TiAl is an interesting case of a ferrimagnet with a near gap-like behaviour. The lowest majority conduction band at the X-point is calculated to be only 300 meV above the highest valence band at Γ and the bands overlap by only 10 meV, what might lead to peculiar transport properties [136].

The magnetic moments of Fe_2TiAl, Fe_2MnAl and Fe_2VSi are in agreement with a Slater–Pauling like behaviour. The Fe_2-based compounds with 24 valence electrons show no magnetic moment, in agreement to the Slater–Pauling rule. The electronically similar $C1_b$ compounds with 18 valence electrons, like CoTiSb [147], are usually semiconductors [148, 149]. However, clear indication for a semiconducting ground state was not indicated in the calculations for the Heusler compounds with 24 valence electrons. Here, only some behaviour close to semi-metallic was found (see Fig. 7.21). Despite the fact of a different symmetry in $L2_1$ and $C1_b$ compounds, this might also point on an underestimation of electron–electron correlation by GGA for $L2_1$ compounds. Some Fe_2-based compounds exhibit at least bands without direct overlap in the Brillouin zone, whereas the iso-electronic compounds Co_2ScC (C = Al, Ga) exhibit touching bands at the Γ and X points. The remaining two Fe_2-based compounds exhibit a gap in the minority band structure. The calculations predict Fe_2MnAl to be a half-metallic ferrimagnet with an antiparallel orientation of the spins at the Fe and Mn sites. Fe_2MnSi is a Type I half-metallic ferromagnet.

A further family of Heusler compounds studied here is based on Ru_2. They all have a gap in the majority states being located above ε_F. Therefore, they are Type III half-metals, at best. Note that the A atoms—here Ru—carry negligible spin moments in the compounds with $N_V = 27$. In all cases, the Mn moment is large and mainly responsible for the total magnetic moment. On the other hand, some of the Ru_2 compounds were recently reported to exhibit antiferromagnetism [137]. In that case the half-metallicity will be lost. The reason for the antiferromagnetic ground state may be found in on-site correlation at the Mn atoms, which was not respected here.

Ir_2MnAl is another $5d$-based Heusler compounds exhibiting half-metallic behaviour in the GGA calculations. The Fermi energy of this material is just at the top of the valence band. In the present study, there was no compound found from other Ni_2, Cu_2, or Pd_2-based Heusler compounds that exhibits half-metallic ferromagnetism. Most of those are paramagnetic or regular ferromagnets without any gap and independent of the valence electron concentration.

It is interesting to note that all Heusler compounds predicted to be half-metallic ferromagnets contain Mn when Co_2 is replaced by other elements. The Mn atoms may replace either the Co atoms on A positions or the second transition metal on the B position. In the latter case, the Mn atoms carry a magnetic moment of about 3 μ_B. The same is true for the Co_2MnC compounds where the Mn atoms also carry a magnetic moment of about 3 μ_B. A very interesting case is the Mn_2CoM ($M = $ Al, Ga, Sb) series that was not mentioned above because it adopts the inverted Li_2AgSb structure and not the regular Heusler Cu_2MnAl structure. Actually, this seem to be the only Heusler-like 2:1:1 compounds that exhibit half-metallic ferrimagnetism in the $F\bar{4}3m$ symmetry [150]. Summarising all these observations, there exist two classes of half-metallic Heusler compounds. First the Co_2-based compounds with the magnetic moment located dominantly in the (100)-like planes and second the Mn compounds with a strongly localised moment at the B site.

7.6 Compensated Ferrimagnets

In the quest for creating magnetic random access memory following the semiconductor roadmap (≈20 nm node in 2010), spin momentum transfer is considered to be the likely mechanism for switching the magnetisation of the so-called free magnetic layers storing the bit information in each storage cell [151]. In 1996, Berger [152] and Slonczewski [153] predicted the possibility of current induced magnetic switching processes. A few years after the theoretical predictions several groups verified the effect experimentally and started initial investigations of those processes in GMR-devices [154–158]. Later, spin transfer switching was also observed in magnetic tunnel junctions (MTJ) [159, 160]. MTJs are suggested to be better suited for devices than GMR-elements based on metallic multilayers because of a large tunneling magnetoresistance (TMR) ratio and a small spin transfer switching current.

One of the critical issues for applications is the current density needed for stable switching using the spin transfer torque mechanism. At present, this current density is order of 2–3 $\mathrm{MA\,cm^{-2}}$ [161, 162]. New materials are needed to achieve a stable switching at significant lower values. Such materials must have a high spin polarisation, a high Curie temperature and a low magnetic moment. Furthermore, the direction of the magnetisation must be thermally stable at room temperature. It is evident that compensated ferrimagnets are suitable materials. They exhibit a high stability against thermal switching at moderate coercitive fields. Ochiai et al. [163] have demonstrated that a synthetic antiferromagnetic layer used for exchange biasing of the free layer is able to reduce the critical current density. In principal, the result is a local (layer-wise) ferrimagnetic order. A reduction of the switching current density by one order of magnitude was also observed if applying exchange biasing to the fixed layer [164]. Both effects suggest that an enhancement of the spin transfer effect may also be reached in real ferrimagnets, although the situation of the spin moments may be more complicated [165].

7.6.1 Half-Metallic Completely Compensated Ferrimagnets

One of the prospective completely compensated half-metallic ferrimagnets (HM-CCF) is $Mn_{2.5}V_{0.5}Si$ or $Mn_2(Mn_{0.5}V_{0.5})Si$. The prediction was based on the supercell approach [104]. It is not clear a priori, however, whether the properties are kept when the Mn and V atoms are randomly distributed on one of the lattice sites.

Here, the electronic structure of $Mn_2(Mn_{0.5}V_{0.5})Si$ was calculated in the local spin density approximation (LSDA) using KKR-CPA. The self-consistent electronic structure calculations were carried out using the spin polarised relativistic Korringa–Kohn–Rostocker method (SPRKKR). The CPA method was used to model the statistic distribution of Mn and V on a common site of the $L2_1$ crystalline structure. The first Mn atom was set on the 8c Wyckoff position, the second Mn atom was placed together with the V atom on the 4a position and Si was finally placed on the 4b position of the cell with $F\,m\bar{3}m$ symmetry. The calculations were performed using the parameterisations of the exchange-correlation functional as given by Vosco, Wilk and Nussair [13] (VWN) as well as the generalised gradient approximation (GGA) in the form given by Perdew et al. [21] (PBE). The CPA tolerance was set to 10^{-4} and the energy convergence criterion to 10^{-5}. f-states were included in the basis of all atoms. 832 irreducible k-points of a $22 \times 22 \times 22$ mesh were used for integration. The density of states was calculated for the same number of k-points from the Greens function by adding a small imaginary part of 0.002 Ry to the energy. For smaller values, the band gaps may become better visible, however, at the same time the DOS becomes much more noisy. The magnetic phase transition temperatures were calculated from the Heisenberg exchange integrals J_{ij} as described in detail by Liechtenstein [166] for LSDA calculations. The cluster size was set to 4.2 times the lattice parameter resulting in overall 1156 ij combinations of the exchange coupling parameter J_{ij} for each atom.

The total density of states of $Mn_2(Mn_{0.5}V_{0.5})Si$—as shown in Fig. 7.22(a)—seems to exhibit not a real gap but only a pseudo gap in one of the spin directions. There are several reasons to explain this observation. First, the KKR method uses Greens functions to calculate the DOS by adding a small imaginary part to the energy. This smears out the density similar to a convolution with a Cauchy Lorenzian

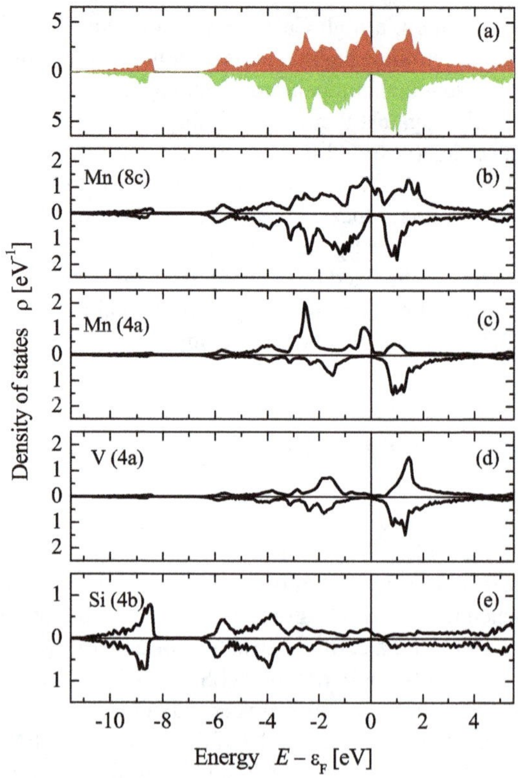

Fig. 7.22 Electronic structure of $Mn_2(Mn_{0.5}V_{0.5})Si$. The calculations were performed using SPRKKR with PBE parameterisation of the exchange-correlation functional. The total DOS (**a**) is given for the primitive cell, the partial densities (**b**)–(**e**) are given per atom

(compare the work of Ebert and Schütz on the half-metallic PtMnSb [167]). Second, the spin–orbit interaction couples the majority and minority spin states. Thus the minority spin states at the Fermi energy are intrinsically coupled in relativistic methods to majority spin states. The mixing of the inseparable spin states results thus always in a small remaining minority DOS at the Fermi energy. Therefore a real gap in only one spin direction cannot exist as solution of the Dirac equation (see also [68]). Finally, the random distribution of Mn and V on the same position destroys the periodicity results also in a continuous spread of the states. In addition, the missing periodicity excludes, of course, the calculation of a band structure.

Nevertheless, the DOS reveals clearly the HMCCF character of $Mn_2(Mn_{0.5}V_{0.5})Si$. This results are in agreement with the findings reported in Ref. [104], even when using a more realistic statistically distribution of Mn and V on the 4a position. Additionally, the partial DOS reveals that the gap in the minority DOS is defined by the states of the Mn atoms located on the 8c position. This coincides with previous calculation for various Heusler compounds, as in most cases, the gap is dominated by the states arising from the atom on the 8c site [25]. The calculated magnetic moments of $Mn_2(Mn_{0.5}V_{0.5})Si$ are listed in Table 7.11. A summation of the site specific magnetic moments yields the result that they sum up to zero total moment, as expected for a completely compensated ferrimagnet. The small part missing in

Table 7.11 Magnetic properties of $Mn_2(Mn_{0.5}V_{0.5})Si$. The calculations were carried out by means of SPRKKR-CPA using VWN and PBE parameterisation of the exchange-correlation potential. All magnetic moments are given in μ_B. The total moments are given per primitive cell. The site specific spin m_S and orbital m_L magnetic moments are given per atom. T_C is the calculated temperature of the magnetic phase transition

Site	VWN		PBE	
	m_S	m_L	m_S	m_L
Mn (8*c*)	−0.7117	−0.01482	−0.8685	−0.01547
Mn (4*a*)	2.5925	0.00728	2.8825	0.00843
V (4*a*)	0.3459	−0.0036	0.4944	−0.00522
m_{tot}	0.0		0.0	
T_C	153 K		262 K	

the table is found at the Si atoms and caused by hybridisation of Si p states with transition metal d-states. The distribution of the magnetic moments on the certain sites in principle follows the distribution as reported in Ref. [104]. Differences are due to the fact that the present work uses space-filling Wigner–Seitz cell integration, whereas the moments in Ref.[104] were calculated for non-overlapping muffin-tin spheres. The partial density of states (see Fig. 7.22(c)) clearly reveals the highly localised electrons at −2.6 eV in the Mn (4a) e_g orbital, in agreement with the Kübler rule.

The calculation of the electronic structure using SPKKR (CPA) revealed a magnetic phase transition temperature of 153 K (see Table 7.11). The calculation of correct magnetic phase transition temperatures is difficult, as there is no expression for the temperature dependence of the exchange correlation functional. Thus magnetism at finite temperatures cannot be expressed with the same accuracy as for $T = 0$ [166], leading to inexactness in the results of calculated magnetic phase transition temperatures. The Heisenberg integrals J_0 found in the calculations amounts to 24.330 meV for Mn (8c), −17.675 meV for Mn (4a) and 14.848 meV for V (4a) if using the VWN exchange-correlation functional. The transition temperatures given in Table 7.11 were, however, calculated from the largest value of the determinant of the J_{ij} matrix. The largest exchange coupling parameter $T_{ij} = J_{ij}/k_B$ was found between Mn atoms on the 4a site with −913 K. This value is reduced due to the nearly vanishing value of the V atoms on the same site, resulting in an average of −223 K. All other integrals between the transition metal atoms have values of about 100 K. The higher T_C found in the PBE calculation emerges mainly from a considerably higher interaction between the Mn atom in 8c position with the Mn and V atoms in the 4a position (average equivalent of 200 K).

In summary, it was found that the results of the electronic structure calculations do not depend upon a particular method for the electronic structure calculations and in particular not on the parameterisation of the exchange-correlation functional. The general shape of the DOS is the same as reported for full symmetry potential calculations with supercells [104]. That is the KKR-CPA calculations also confirm the completely compensated half-metallic ferrimagnetic character of $Mn_2(Mn_{0.5}V_{0.5})Si$ even for the case of Mn-V disorder on the 4b Wyckoff position.

7.7 Summary and Conclusions

Ab-initio calculations have been used to examine the electronic structure of A_2BC Heusler compounds that exhibit half-metallic ferromagnetism. The examination revealed two classes of half-metallic Heusler compounds. First the Co_2-based compounds with the formation of dominant bonds and magnetic moments located in the (100)-like planes containing the Co atoms and second the Mn compounds with a strong localised moment at the B site. This suggests the occurrence of two concurring mechanisms being responsible for the half-metallic ferromagnetism in Heusler compounds.

It has been found that the half-metallic properties in Co_2BC Heusler compounds are dominated by the presence of C atoms. According to the results described here, nearly all Co_2BC compounds will be half-metallic ferromagnets even if no localised magnetic moment is formed. The magnetic moment carried by the Co and B atoms is restricted by the C atoms even though they do not directly contribute to the magnetic properties. The influence of the valence electron concentration on the total and partial magnetic moments was investigated in detail. It is found that the Co_2BC compounds fulfil the Slater–Pauling rule, whereas other compounds may exhibit pronounced deviations from a Slater–Pauling type behaviour. The findings were supported by an analysis of the site resolved occupancy of the d states and their symmetry. Further, it was found that the minority band gap shows a trend to decrease with increasing lattice parameter. It increases with stronger hybridisation as found from an analysis of the Al and Si containing compounds.

The inclusion of electron correlation in the form of LDA+U or LDA+DMFT on top of the LSDA and GGA approximations, does not destroy or produce a half-metallic gap. Its effect is to enlarge the gap if it is already present and to shift the minority bands with respect to majority bands or the Fermi energy. The use of effective Coulomb parameters in the order of distinctly below 2 eV is able to explain the magnetic moments of all compounds if compared to experiments. The remaining question, however, namely whether ε_F is located in reality inside of the gap of the minority states or outside has to be answered by suitable experiments.

In summary, it was shown that state-of-the-art ab-initio calculations are a suitable method to explain the electronic and magnetic structure of Heusler compounds and to predict new half-metallic ferromagnets in this class of compounds.

Acknowledgements The authors are very grateful to P. Blaha (Vienna, Austria) and H. Ebert (Munich) and their groups for development and providing the computer codes (WIEN2k and SPRKKR). Special thanks go to J. Kübler (Darmstadt) for very fruitful discussions about the magnetism in $3d$-intermetallics and in particular Heusler compounds. The authors thank H.C. Kanpal, S.S. Naghavi, D. Schärf and J. Thoene for help with the calculations as well as all members of the group for providing experimental data. This work was financially supported by the Deutsche Forschungsgemeinschaft (project P7 in research unit FOR 559).

References

1. Heusler F (1903) Verh Dtsch Phys Ges 5:219
2. Heusler F, Starck W, Haupt E (1903) Verh Dtsch Phys Ges 5:220
3. Heusler O (1934) Ann Phys (Leipz) 19:155
4. Bradley AJ, Rodgers JW (1934) Proc R Soc A 144:340
5. Webster PJ (1971) J Phys Chem Solids 32:1221
6. Kübler J, Williams AR, Sommers CB (1983) Phys Rev B 28:1745
7. de Groot RA, Müller FM, van Engen PG, Buschow KHJ (1983) Phys Rev Lett 50:2024
8. Coey JMD, Venkatesan M, Bari MA (2002) In: Bertheir C, Levy LP, Mertinez G (eds) Lecture Notes in Physics, vol 595. Springer, Heidelberg
9. Ishida S, Akazawa S, Kubo Y, Ishida J (1982) J Phys F, Met Phys 12:1111
10. Kohn W, Sham LJ (1965) Phys Rev 140:A1133
11. Hedin L, Lundquist BI (1971) J Phys C, Solid State Phys 4:2064
12. von Barth U, Hedin L (1972) J Phys C, Solid State Phys 5:1629
13. Vosko SH, Wilk L, Nusair M (1980) Can J Phys 58:1200
14. Ishida S, Fujii S, Kashiwagi S, Asano S (1995) J Phys Soc Jpn 64:2152
15. Ishida S, Kashiwagi S, Fujii S, Asano S (1995) Physica B 210:140
16. Mohn P, Blaha P, Schwarz K (1995) J Magn Magn Mater 140–144:183
17. Galanakis I, Dederichs PH, Papanikolaou N (2002) Phys Rev B 66:174429
18. Picozzi S, Continenza A, Freeman AJ (2002) Phys Rev B 66:094421
19. Perdew JP, Wang Y (1986) Phys Rev B 33:8800
20. Perdew JP, Wang Y (1992) Phys Rev B 45:13244
21. Perdew JP, Burke K, Ernzerhof M (1996) Phys Rev Lett 77:3865
22. Perdew JP, Burke K, Ernzerhof M (1997) Phys Rev Lett 78:1396
23. Miura Y, Nagao K, Shirai M (2004) Phys Rev B 69:144413
24. Antonov VN, Dürr HA, Kucherenko Y, Bekenov LV, Yaresko AN (2005) Phys Rev B 72:054441
25. Fecher GH, Kandpal HC, Wurmehl S, Morais J, Lin HJ, Elmers HJ, Schönhense G, Felser C (2005) J Phys Condens Matter 17:7237
26. Wurmehl S, Fecher GH, Kandpal HC, Ksenofontov V, Felser C, Lin HJ, Morais J (2005) Phys Rev B 72:184434
27. Kandpal HC, Fecher GH, Felser C, Schönhense G (2006) Phys Rev B 73:094422
28. Anisimov VI, Gunnarson O (1991) Phys Rev B 43:7570
29. Anisimov VI, Zaanen J, Andersen OK (1991) Phys Rev B 44:943
30. Soulen RJ Jr., Byers JM, Osofsky MS, Nadgorny B, Ambrose T, Cheng SF, Broussard PR, Tanaka CT, Nowak J, Moodera JS, Barry A, Coey JMD (1998) Science 282:85
31. Parkin SSP, Kaiser C, Panchula A, Rice PM, Hughes B, Samant M, Yang SH (2004) Nat Mater 862:3
32. Yuasa S, Nagahama T, Fukushima A, Suzuki Y, Ando K (2004) Nat Mater 868:3
33. Felser C, Fecher GH, Balke B (2007) Angew Chem 119:680–713
34. Slater JC (1936) Phys Rev 49:931
35. Pauling L (1938) Phys Rev 54:899
36. Kübler J (2000) Theory of itinerant electron magnetism. Oxford University Press, Oxford
37. Fecher GH, Kandpal HC, Wurmehl S, Felser C, Schönhense G (2006) J Appl Phys 99:08J106
38. Kübler J (1984) Physica B 127:257
39. Balke B, Fecher GH, Kandpal HC, Felser C, Kobayashi K, Ikenaga E, Kim JJ, Ueda S (2006) Phys Rev B 74:104405
40. Fecher GH, Felser C (2007) J Phys D, Appl Phys 40:1582
41. Pauly H, Weiss A, Witte H (1969) Z Metallkd 59(47):414, 554
42. Villars P, Calvert L (1991) Pearson's handbook of crystallographic data for intermetallic phases, 2nd edn. ASM International, Materials Park
43. Puselj M, Ban Z (1969) Croat Chem Acta 41:79

44. Blaha P, Schwarz K, Madsen GKH, Kvasnicka D, Luitz J (2001) WIEN2k, an augmented plane wave + local orbitals program for calculating crystal properties. Karlheinz Schwarz, Techn. Universitaet Wien, Wien, Austria
45. Bacon GE, Plant JS (1971) J Phys F, Met Phys 1:524
46. Webster PJ, Ziebeck KRA (1988) In: Wijn HPJ (ed) Alloys and compounds of d-elements with main group elements. Part 2, Landolt–Börnstein—Group III condensed matter, vol. 19C. Springer, Heidelberg, pp 104–185
47. Ziebeck KRA, Neumann KU (2001) In: Wijn HPJ (ed) Alloys and compounds of d-elements with main group elements. Part 2, Landolt–Börnstein—Group III condensed matter, vol. 32C. Springer, Heidelberg, pp 64–314
48. Gyorffy B (1972) Phys Rev B 5:2382
49. Stocks GM, Winter H (1984) In: Phariseau P, Temmerman WM (eds) The electronic structure of complex systems. Plenum Press, New York, p 463
50. Butler WH (1985) Phys Rev B 31:3260
51. Mankovskyy S, Ebert H (2006) Phys Rev B 74:54414
52. Baudelet F, Odin S, Giorgetti C, Dartyge E, Itie JP, Polian A, Pizzini S, Fontaine A, Kappler JP (1997) J Phys (Paris) 7(C2):441
53. Crisan V, Entel P, Ebert H, Akai A, Johnson DD, Staunton JB (2002) Phys Rev B 66:014416
54. Ebert H (1999) In: Dreysse H (ed) Lecture notes in physics, vol 535. Springer, pp 191–246
55. Ebert H, Minar J, Popescu V (2001) In: Baberschke K, Donath M, Nolting W (eds) Band-ferromagnetism. Lecture notes in physics, vol 580. Springer, Berlin, pp 371–385
56. Czyzyk MT, Sawatzky GA (1994) Phys Rev B 49:14211
57. Anisimov VI, Aryasetiawan F, Lichtenstein AI (1997) J Phys Condens Matter 9:767
58. Cowan RD (1981) The theory of atomic structure and spectra. University of California Press, Berkeley
59. Bandyopadhyay T, Sarma DD (1989) Phys Rev B 39:3517
60. Chadov S, Fecher GH, Felser C, Minár J, Braun J, Ebert H (2009) J Phys Condens Matter 42:084002
61. Kotliar G, Savrasov SY, Haule K, Oudovenko VS, Parcolett O, Marianetti CA (2006) Rev Mod Phys 78:865
62. Lichtenstein AI, Katsnelson MI, Kotliar G (2001) Phys Rev Lett 87:067205
63. Grechnev A, Marco ID, Katsnelson MI, Lichtenstein AI, Wills J, Eriksson O (2006) cond-mat/0610621
64. Perlov A, Chadov S, Ebert H (2003) Phys Rev B 68:245112
65. Minár J, Ebert H, de Nadaï C, Brookes NB, Venturini F, Ghiringhelli G, Chioncel L, Lichtenstein AI, Katsnelson MI (2005) Phys Rev Lett 95:166401
66. Braun J, Minár J, Ebert H, Katsnelson MI, Lichtenstein AI (2006) Phys Rev Lett 97:227601
67. Chadov S, Minár J, Katsnelson MI, Ebert H, Ködderitzsch D, Lichtenstein AI (2008) Europhys Lett 82:37001
68. Mavropoulos P, Sato K, Zeller R, Dederichs PH, Popescu V, Ebert H (2004) Phys Rev B 69:054424
69. Galanakis I (2005) Phys Rev B 71:012413
70. Ishida S, Otsuka Y, Kubo Y, Ishida J (1983) J Phys F, Met Phys 13:1173
71. Buschow KHJ, van Engen PG (1981) J Magn Magn Mater 25:90
72. Bushow KHJ, van Engen PG, Jongebreur R (1983) J Magn Magn Mater 38:1
73. van Engen PG, Buschow KHJ, Erman M (1983) J Magn Magn Mater 30:374
74. Brown PJ, Neumann KU, Webster PJ, Ziebeck KRA (2000) J Phys Condens Matter 12:1827
75. Castelliz L (1953) Monatsh Chem 84:765
76. Gladyshevskii EI (1962) Spvirt powder metallurgy and metal ceramics, translated from Poroshkovaya Metallurgiya, Kiev 1:262
77. Fujita Y, Endo K, Terada M, Kimura R (1972) J Phys Chem Solids 33:1443
78. Szytula A, Kolodziejczyk A, Rzany H, Todorovic J, Wanic A (1972) Phys Status Solidi, a Appl Res 11:57
79. Booth JG, Pritchard RG (1975) J Phys F 5:347

80. Niculescu V, Budnick JI, Hines WA, Rajt K, Pickart S, Skalski S (1979) Phys Rev B 19:452
81. Markiv VY, Gladyshevskii EI, Kuzma AB (1981) J Magn Magn Mater 25:90
82. Soltys J (1983) J Magn Magn Mater 38:1
83. Ooiwa K (1985) J Phys Soc Jpn 54(4):1581
84. Uhl E (1984) Solid State Commun 53:395
85. Ido H (1986) J Magn Magn Mater 54–57:937
86. Sobzcak R (1988) J Phys (Paris), Coll 49:141
87. Kawakami M (1989) Hyperfine Interact 51:993
88. Kido M, Ido H, Kido G (1992) J Magn Magn Mater 104–107:705
89. Carbonari AW, Pendl W Jr., Attili RN, Saxena RN (1993) Hyperfine Interact 80:971
90. Carbonari AW, Saxena RN, Pendl JW, Filho JM, Attili RN, Olzon-Dionysio M, de Souza SD (1996) J Magn Magn Mater 163:313
91. Plogmann S, Schlathölter T, Braun J, Neumann M, Yarmoshenko YM, Yablonskikh MV, Shreder EI, Kurmaev EZ, Wrona A, Slebarski A (1999) Phys Rev B 60:6428
92. Sasaki T, Kanomata T, Narita T, Nishihara H, Note R, Yoshida H, Kaneko T (2001) J Alloys Compd 317–318:406
93. Pierre J, Skolozdra RV, Stadnyk YV (1993) J Magn Magn Mater 128:93
94. Raphael MP, Ravel B, Huang Q, Willard MA, Cheng SF, Das BN, Stroud RM, Bussmann KM, Claassen JH, Harris VG (2002) Phys Rev B 66:104429
95. Yamasaki A, Imada S, Arai R, Utsunomiya H, Suga S, Muro T, Saito Y, Kanomata T, Ishida S (2002) Phys Rev B 65:104410
96. Ritchie L, Xiao G, Ji Y, Chen TY, Chien CL, Zhang M, Chen J, Liu Z, Wu G, Zhang XX (2003) Phys Rev B 68:104330
97. Elmers HJ, Wurmehl S, Fecher GH, Jakob G, Felser C, Schönhense G (2004) Appl Phys A 79:557
98. Zhang M, Brück E, de Boer FR, Li Z, Wu G (2004) J Phys D, Appl Phys 37:2049
99. Kobayashi K, Umetsu RY, Fujita A, Oikawa K, Kainuma R, Fukamichi K, Ishida K (2005) J Alloys Compd 399:60
100. Umetsu RY, Kobayashi K, Fujita A, Oikawa K, Kainuma R, Ishida K, Endo N, Fukamichi K, Sakuma A (2005) Phys Rev B 72:214412
101. Zhang W, Jiko N, Mibu K, Yoshimura K (2005) J Phys Condens Matter 17:6653
102. Kandpal HC, Ksenofontov V, Wojcik M, Seshadri R, Felser C (2006) cond-mat/0609578
103. Wurmehl S, Fecher GH, Felser C (2006) Z Anorg Allg Chem 61b:749
104. Wurmehl S, Fecher GH, Kandpal HC, Ksenofontov V, Felser C, Lin HJ (2006) Appl Phys Lett 88:032503
105. Wurmehl S, Fecher GH, Ksenofontov V, Casper F, Stumm U, Felser C, Lin HJ, Hwu Y (2006) J Appl Phys 99:08J103
106. Wurmehl S, Kandpal HC, Fecher GH, Felser C (2006) J Phys Condens Matter 18:6171
107. Jepsen O, Andersen OK (2000) TB-LMTO-ASA program version 47. MPI für Festkörperforschung, Stuttgart, Germany
108. Öğüt S, Rabe KM (1995) Phys Rev B 51:10443
109. Goerlich EA, Kimec R, Latka K, Matlak T, Ruebenbauer K, Szytula A, Tomala K (1974) J Phys Soc Jpn 36:620
110. Jeitschko W (1983) J Magn Magn Mater 30:374
111. Markiv VY, Voroshilov YV, Kripyakevich PI, Cherkashin EE (1983) J Magn Magn Mater 38:1
112. De Souza SD, Saxena RN (1987) Hyperfine Interact 34:431
113. Slebarski A, Jezierski A, Neumann M, Plogmann S (1999) Eur Phys J B 12:519
114. Neumann KU, Kanomata T, Ouladdiaf B, Ziebeck KRA (2002) J Phys Condens Matter 14:1371
115. Kanomata T, Sasaki T, Nishihara H, Yoshida H, Kaneko T, Hane S, Goto T, Takeishi N, Ishida S (2005) J Alloys Compd 393:26
116. Zhang W, Qian Z, Sui Y, Liu Y, Su W, Zhang M, Liu Z, Liu G, Wu G (2006) J Magn Magn Mater 299:255

117. Wolter AUB, Bosse A, Baabe D, Maksimov I, Mienert D, Klauße HH, Litterst FJ, Niemeier D, Michalak R, Geibel C, Feyerherm R, Hendrikx R, Mydosh JA, Süllow S (2002) Phys Rev B 66:174428
118. Antonov VN, Harmon BN, Bekenov LV, Shpak AP, Yaresko AN (2005) Phys Rev B 71:174428
119. Felser C, Elmers HJ, Fecher GH (2005) In: Galanakis I, Dederichs P (eds) Halfmetallic alloys. Lecture notes in physics, vol 676. Springer, Berlin
120. Mavropoulos P, Galanakis I, Popescu V, Dederichs PH (2004) J Phys Condens Matter 16:S5759
121. Slater JC (1936) Phys Rev 49:537
122. Malozemoff AP, Williams AR, Moruzzi VL (1984) Phys Rev B 29:1620
123. Elmers HJ, Fecher GH, Valdaitsev D, Nepijko SA, Gloskovskii A, Jakob G, Schönhense G, Wurmehl S, Block T, Felser C, Hsu PC, Tsai WL, Cramm S (2003) Phys Rev B 67:104412
124. Yoshimura K, Miazaki A, Vijayaraghavan R, Nakamura Y (1985) J Magn Magn Mater 53:189
125. Sargolzaei M, Richter M, Koepernik K, Opahle I, Eschrig H, Chaplygin I (2006) Phys Rev B 74:224410
126. Jung V, Fecher GH, Balke B, Ksenofontov V, Felser C (2009) J Phys D, Appl Phys 42:084007
127. Galanakis I, Mavropoulos P, Dederichs P (2006) J Phys D, Appl Phys 39:765
128. Kandpal HC, Fecher GH, Felser C (2007) J Phys D, Appl Phys 40:1507
129. Fuji S, Sugimura S, Ishida S, Asano S (1990) J Phys Condens Matter 2:8583
130. Fecher GH, Kandpal HC, Wurmehl S, Felser C, Schönhense G (2006) J Appl Phys 99:08J106
131. Kübler J, Fecher GH, Felser C (2007) Phys Rev B 76:024414
132. Balke B, Wurmehl S, Fecher GH, Felser C, Kübler J (2008) Sci Technol Adv Mater 9:014102
133. Suits JC (1976) Phys Rev B 14
134. Jha S, Seyoum HM, Julian GM, Dunlap RA, Vasquez A, da Cunha JGM, Ramos SMM (1985) Phys Rev B 3:3279
135. Pfleiderer C, Bouf J, von Löhneysen H (2002) Phys Rev B 63:172404
136. Suzuki RO, Kyono T (2004) J All Comp 377
137. Kanomata T, Kikuchi M, Yamauchi H (2006) J Alloys Compd 414:1
138. Gotoh M, Ohashi M, Kanomata T, Yamaguchi Y (1995) Physica B 213–214:306
139. Nagano T, Uwanuyu S, Kawakami M (1995) J Magn Magn Mater 140–144:123
140. Feng Y, Rhee JY, Wiener TA, Lynch DW, Hubbard BE, Sievers AJ, Schlagel DL, Lograsso TA, Miller LL (2001) Phys Rev B 63:165109
141. Jiang C, Venkatesan M, Coey JMD (2001) Solid State Commun 118:513
142. Prakash U, Sauthoff G (2001) Intermetallics 9:107–112
143. Lue CS, Ross JH Jr., Rathnayaka KDD, Naugle DG, Wu SY, Li WH (2001) J Phys Condens Matter 13:1585
144. Lue CS, Li Y, Ross JH Jr., Irwin GM (2003) Phys Rev B 67:224425
145. Nashima O, Kanomata T, Yamaguchi Y, Abe S, Harada T, Suzuki T, Nishihara H, Koyama K, Shishido T, Watanabe K, Kaneko T (2004) J Alloys Compd 383:208
146. Lue CS, Kuo YK, Horng SN, Peng SY, Cheng C (2005) Phys Rev B 71:064202
147. Kroth K, Balke B, Fecher GH, Ksenofontov V, Felser C, Lin HJ (2006) Appl Phys Lett 89:202509
148. Tobola J, Pierre J, Kaprzyk S, Skolozdra RV, Kouacou MA (1998) J Phys Condens Matter 10:1013
149. Tobola J, Pierre J (2000) J Alloys Compd 296:243
150. Dai X, Liu G, Chen L, Chen J, Wu G (2006) Solid State Commun 140:533
151. Rabaey JM (2002) In: CRA Conference
152. Berger L (1996) Phys Rev B 54:9353
153. Slonczewski JC (1996) J Magn Magn Mater 159:L1

154. Tsoi M, Jansen AGM, Bass J, Chiang WC, Seck M, Tsoi V, Wyder P (1998) Phys Rev Lett 80:4281
155. Myers EB, Ralph DC, Katine JA, Louie RN, Buhrman RA (1999) Science 285:867
156. Bussmann K, Prinz GA, Cheng SF, Wang D (1999) Appl Phys Lett 75:2476
157. Katine JA, Albert FJ, Buhrman RA, Myers EB, Ralph DC (2000) Phys Rev Lett 84:3149
158. Grollier J, Cros V, Hamzic A, George JM, Jaffres H, Fert A, Faini G, Youssef JB, Gall HL (2001) Appl Phys Lett 78:3663
159. Huai Y, Albert F, Nguyen P, Valet T (2004) Appl Phys Lett 84:3118
160. Fuchs GD, Emley NC, Krivorotov IN, Braganca PM, Ryan EM, Kiselev SI, Sankey JC, Ralph DC, Buhrman RA, Katine JA (2004) Appl Phys Lett 85:1205
161. Diao Z, Apalkov D, Pakala M, Ding Y, Panchula A, Huai Y (2005) Appl Phys Lett 87:232502
162. Huai Y, Pakala M, Diao Z, Ding Y (2005) Appl Phys Lett 87:222510
163. Ochiai T, Jiang Y, Hirohata A, Tezuka N, Sugimoto S, Inomata K (2005) Appl Phys Lett 86:242506
164. Nguyen HYT, Yi H, Joo SJ, Shina KH, Lee KJ, Dieny B (2006) Appl Phys Lett 89:094103
165. Heide C, Zilberman PE, Elliott RJ (2002) Phys Rev B 63:064424
166. Liechtenstein AI, Katsnelson MI, Antropov VP, Gubanov VA (1986) J Magn Magn Mater 67:65
167. Ebert H, Schütz G (1991) J Appl Phys 69:4627

Chapter 8
Electronic structure of complex oxides

Alexander I. Lichtenstein

Abstract This work discusses the electronic structure magnetic properties and metal–insulator transition in transition metal oxides (TMO). The unique feature of these compounds related to the fact that the spin, charge and orbital degrees of freedom plays an important role in all physical properties. While the local density approximation is quite reasonable for the electronic structure of a metallic oxide, the additional Hubbard-like correlation is important for the energy spectrum of insulating magnetic oxides. The LDA+U method was proven to be a very efficient and reliable tool in calculating the electronic structure of systems where the Coulomb interaction is strong enough to cause localization of the electrons. It works not only for nearly core-like $4f$-orbitals of rare-earth ions, where the separation of the electronic states on the subspaces of the infinitely slow localized orbitals and infinitely fast itinerant ones is valid, but also for such systems as transition metal oxides (NiO). The main advantage of LDA+U method over model approaches is its "first principle" nature with a complete absence of adjustable parameters. At the same time, all the most subtle and interesting many-body effects (such as spectral weight transfer, Kondo resonances, and others) are beyond the LDA+U approach. The LDA+DMFT method seems to be effective and useful to describe the dynamical character using the self-energy instead of the effective exchange-correlation potential acting on the electrons. The results for metal–insulator transition in complex transition metal oxides demonstrate that the dynamical mean field theory gives an opportunity to unify the many-body theory with the practice of first-principle calculations of the electronic structure and properties for real materials.

8.1 Introduction

Complex oxides systems are the most common class of materials, including ironstone and red oxide film, Fe_2O_3 as well as magnetite, Fe_3O_4, which naturally exists in the brain cell of a tun-fish as a half-metallic surface and helps to navigate in the

A.I. Lichtenstein (✉)
Institut für Theoretische Physik, Universität Hamburg, Jungiusstrasse 9, 20355 Hamburg, Germany
e-mail: alichten@physnet.uni-hamburg.de

C. Felser, G.H. Fecher (eds.), *Spintronics*, DOI 10.1007/978-90-481-3832-6_8,

ocean along the magnetic parallel probably using the GMR-effect [1, 2]. Recently, so-called colossal magnetoresistent manganites, $La_{1-x}Sr_xMnO3$, have attracted a lot of attention [3]. Finally, one of the most interesting discoveries of the last decades related with the high-temperature superconductivity in cuprates, $Ls_{1-x}Sr_xCuO_4$, so-called LSO (by J.G. Bednorz and K.A. Müller) [4] was also related with transition metal oxides, and many other oxides such as YBCO ($YBa_2Cu_3O_7$), BISCO ($Bi_2Sr_2CaCu_2O_8$) [5].

The wide class of oxide materials, so-called Mott-insulators, are very important from the physics point of view: they are famous antiferromagnetic insulators, MnO, CoO and NiO, which cannot be described from the standard band theory. The ferromagnetic metallic oxides also exists and EuO represents a classical case of metal–insulator transition under oxygen deficiency. Another famous example of oxide systems with the metal–insulator transition are the well-known V_2O_3 and Ti_2O_3 crystals showing under some conditions an unusual paramagnetic M–I transition without influence of magnetically ordered low-temperature insulating phase.

A common feature of all these transition metal oxides is the existing of well localized $3d$-orbitals with the strong Coulomb interactions between the d-electrons. These electrons are responsible for all unusual electronic properties of oxide systems and strong chemical bonding due to a large hybridization with oxygen $2p$-orbitals. A large variety of interesting properties of complex oxide materials related to the delicate balance between chemical bonding, which try to delocalized the d-electrons and Coulomb interaction which tends to localize magnetic d-electrons.

In this Lecture we will discuss effects of electron–electron interactions on the electronic structure, magnetic behavior and insulating properties of complex transition metal oxides (TMO).

First neutron scattering investigation of magnetic oxide MnO by C.G. Shull and co-workers in 1951 [6] support Néel's idea on the antiferromagnetic ordering. The corresponding theoretical models for insulating and magnetic behaviors of such transition-metal oxides (NiO as a prototype) with partially filled d-shells have been developed later by N.F. Mott, P.W. Anderson, and J.H. van Vleck [7].

Recently, interest has grown in the heterogeneous ferromagnetic materials, such as thin-film transition metals multilayers which display so-called giant magnetoresistance (GMR) [2] and used in the new generation of MR read-head devices. In connection to this discovery, it has become recognized that some $3d$ transition-metal oxide, possess even larger room-temperature magnetoresistivity associated with a paramagnetic–ferromagnetic phase transition in a small magnetic field. Such an effect, called the colossal magnetoresistance (CMR) [3], is the result of a unique type of metal–insulator transition (MIT) in these oxides. The CMR-compounds are the manganite perovskite $T_{1-x}D_xMnO_3$ where T is a trivalent lanthanide cation (e.g. La) and D is a divalent, alkaline-earth cation (e.g. Ca, Sr, Ba). For the end member of the dilution sites, $LaMnO_3$ and $CaMnO_3$, the ground state is antiferromagnetic (AF) as in MnO. In a certain range of doping, $x \approx 0.2$–0.4, the ground state is ferromagnetic (FM), and the paramagnetic-to-ferromagnetic transition is accompanied by a sharp drop in resistivity ρ(T). This phenomenon has been known to exist in manganites since 1950. In addition to the FM-states, there is another type of col-

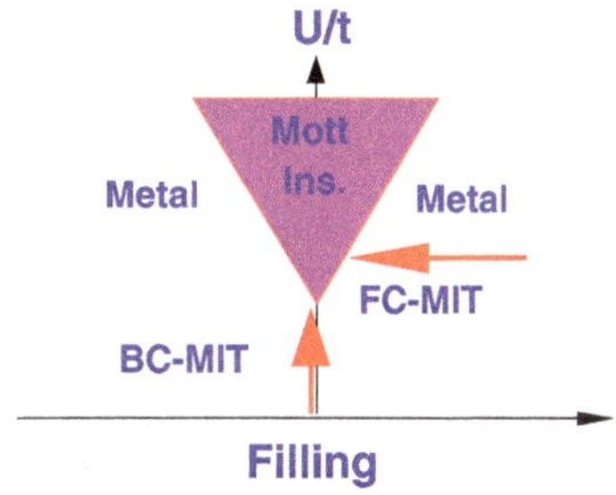

Fig. 8.1 Metal–insulator phase diagram with the two routes for the metal–insulator transition (MIT): The filling-control (FC-MIT) and bandwidth-control (BC-MIT)

lective state with orbital and charge order, typically observed for $x > 0.3$. The new interest come in connection with GMR-success and supported by a rich possibility of the bandwidth-controlling MIT, related with the geometrical factor of lattice distortion (different doping-size cations, which change a so-called electron-correlation strength U/t [8], the average Coulomb energy over effective hopping parameter); as well as filling-control of MIT due to different concentrations of the divalent ions. Such an unique possibility of artificially controlling magnetic and electrical properties (Fig. 8.1) could lead to a new class of artificially designed materials [9].

8.2 Spin, Charge and Orbital Degrees of Freedom

Let us discuss first the important ingredients for the electronic structure of such compounds. The variety of phases with the different magnetic, conducting and lattice properties suggest that both the spin, charge and orbital degrees of freedom play an important roles in their unique physical behavior. There are not many materials in nature where all these different quantum variables are coupled together and could be changed on the very small energy scale.

The spin degrees of freedom define a magnetic ordering in the oxides and are couples by pure quantum exchange interactions. Starting from classical Heisenberg–Dirac–van Vleck theory of exchange interactions, we can write the effective exchange Hamiltonian for non-degenerate case in the Heisenberg form:

$$H_{\text{ex}} = -2\sum_{ij} J_{ij}\vec{S}_i\vec{S}_j \tag{8.1}$$

here $\vec{S}_i$ is the spin operator for site i, and J_{ij} are exchange integrals. A positive sign of J_{ij} corresponds to the ferromagnetic exchange interaction, while negative sign related to the antiferromagnetic one. It is easy to understand the source of ferromagnetic exchange interactions, starting from the total electron–electron Coulomb interactions (in atomic units $e = m = \hbar = 1$):

$$H_{\text{int}} = \frac{1}{2}\sum_{ijkl\sigma\sigma'}\langle ij|\frac{1}{r_{12}}|kl\rangle c^{+}_{i\sigma}c^{+}_{j\sigma'}c_{l\sigma'}c_{k\sigma}$$

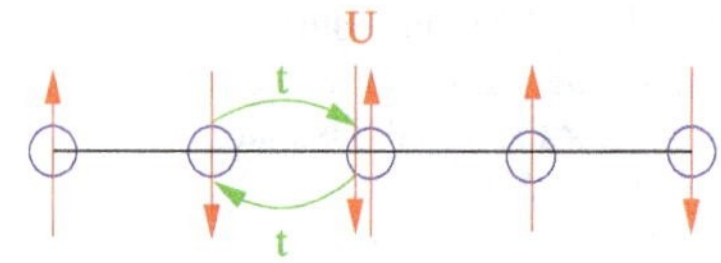

Fig. 8.2 The Anderson kinetic exchange interaction

where $|i\rangle$ is an orthogonal localized basis set for the site i, σ is the spin index and $c^+_{i\sigma}$ ($c_{i\sigma}$) are the fermionic creation (annihilation) operators for electrons with the spin σ on the site i. Taking into account the "exchange" part of the Coulomb interactions with $k = j$ and $l = i$ (the so-called "potential" exchange interactions):

$$J_{ij} = \langle ij|\frac{1}{r_{12}}|ji\rangle$$

and using the following definition of the local spin-operator:

$$\vec{S}_i = \frac{1}{2}\sum_{\sigma,\sigma'} c^+_{i\sigma}\vec{\sigma}_{\sigma\sigma'}c_{i\sigma'}$$

in terms of the Pauli matrices $\vec{\sigma} = (\sigma_x, \sigma_y, \sigma_z)$, one could obtain the Heisenberg exchange Hamiltonian (8.1).

In order to realize the principal mechanism of the antiferrimagnetic exchange interactions we consider a simple non-degenerate half-field Hubbard model (Fig. 8.2) with only two parameters: hopping integral, t, and on-site Coulomb interactions, U:

$$\mathscr{H} = \sum_{i,j,\sigma} t_{ij}\left(c^+_{i\sigma}c_{j\sigma} + c^+_{j\sigma}c_{i\sigma}\right) + U\sum_i n_{i\uparrow}n_{i\downarrow}$$

where the Hubbard parameter U is defined through the total energy differences:

$$U = E\left(d^{n+1}\right) + E\left(d^{n-1}\right) - 2E\left(d^n\right)$$

In the limit of strong Coulomb interaction $U > W = 2zt$ (z is the number of the nearest neighbors) there is one electron per atom in the half-field case. There are additional exchange interactions in the second order perturbation theory with the coupling parameters:

$$J_{ij} = -\frac{2t_{ij}^2}{U} \tag{8.2}$$

After P.W. Anderson this type of interaction is called "kinetic" exchange. It is related to the lowering of the total energy for the antiferromagnetic state due to virtual hopping (Fig. 8.2). These processes are forbidden in the ferromagnetic state due to the Pauli principle.

In a magnetic oxide like MnO, the main mechanism of antiferromagnetic coupling is the so-called superexchange interactions via oxygen $2p$-states, since Mn-atoms are separated by oxygen. The simplest "180-degree" Mn–O–Mn superex-

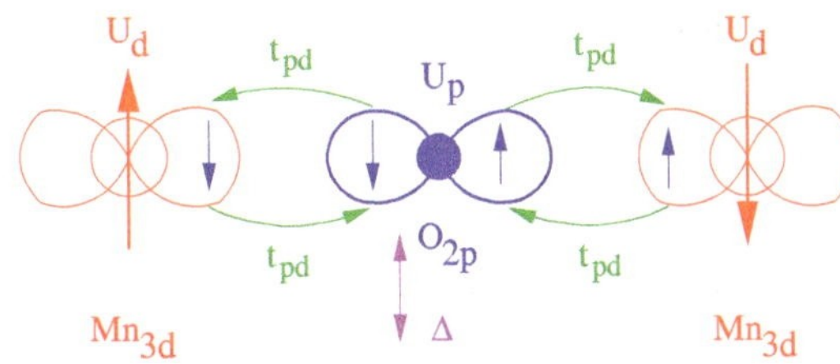

Fig. 8.3 The mechanism of superexchange interactions

change path was shown in Fig. 8.3 and corresponds to the p–d model Hamiltonian:

$$\mathscr{H}_{pd} = \varepsilon_d^0 \sum_{i,\sigma} c_{di\sigma}^+ c_{di\sigma} - t_{pd} \sum_{i,j,\sigma} \left(c_{di\sigma}^+ c_{pj\sigma} + c_{pj\sigma}^+ c_{di\sigma} \right) + U_d \sum_i n_{i\uparrow} n_{i\downarrow} + U_p \sum_j n_{j\uparrow} n_{j\downarrow}$$

where U_p is the oxygen Coulomb energy and the charge-transfer energy defined as

$$\Delta = E\left(d^{n+1}\underline{L}\right) - E\left(d^n\right)$$

here $\underline{L}$ means a hole in the anion band, n is the number of d-electrons.

In this case the exchange interactions appear in the fourth order of perturbation theory:

$$J = -\frac{2t_{pd}^4}{\Delta^2} \left(\frac{1}{U_d} + \frac{2}{2\Delta + U_p} \right) \tag{8.3}$$

if $\Delta \gg U$, we neglect the second term in parentheses and could rewrite this superexchange interaction like the Anderson kinetic exchange (8.2) with the effective d–d hopping integrals via O_{2p} states equal to $t_{dd} = t_{pd}^2/\Delta$.

To understand the diverse physical properties of magnetic oxides from a unified point of view it is important to clarify the systematics of different metallic and insulating compounds as function of the main parameters: Coulomb correlations, U, bandwidth, W, and charge transfer energy, Δ. These parameters depend on the chemical environment and can be calculated from the electronic structure.

The simple ionic picture of $3d$-transition metal oxides $Me^{2+}O^{2-}$ results in the insulating behavior only for MgO and CaOr. For ScO, TiO the metallic behavior is related with the large d-bandwidth: $U < W$. In this situation splitting between d-states due to Coulomb correlation is smaller than the bandwidth due to effective Me-O-Me hopping and the system is metallic for partially filled d-band (Fig. 8.4). On the other hand the oxide like MnO, NiO are insulators and N.F. Mott, was the first who suggested that it is due to the strong correlation: $U > W$. Let us consider the simple lattice model with a single electron orbital on each site. Two electrons sitting on the same site would feel a large Coulomb repulsion (Fig. 8.2). This interaction splits the d-band into two: the lower band is formed from electrons which occupied an empty site and the upper one from electrons which occupied a site already taken

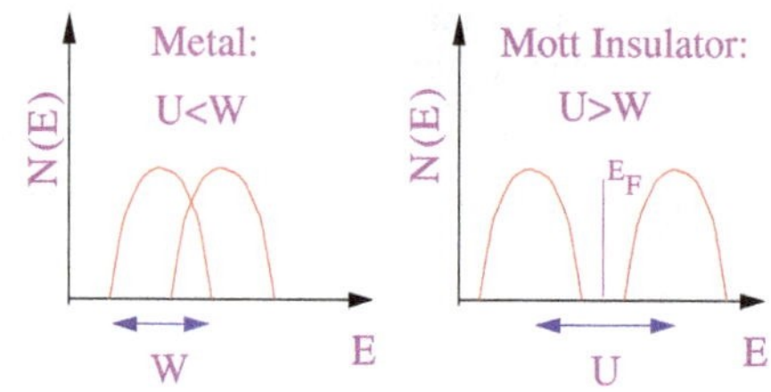

Fig. 8.4 Scheme of metal and correlated Mott-insulator

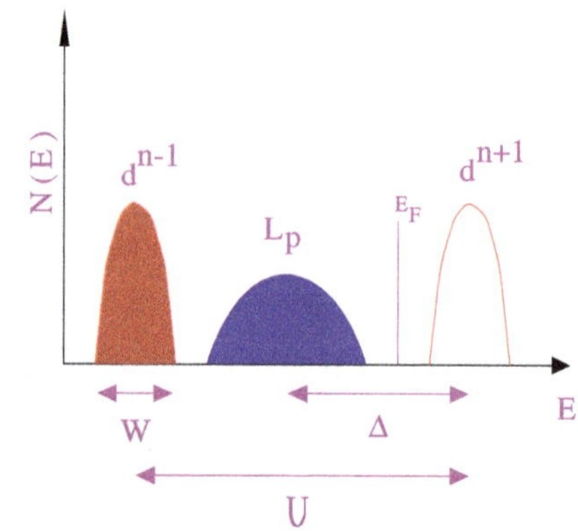

Fig. 8.5 The scheme of charge transfer insulator

by another electron. With the one electron per site the lower band (so-called lower Hubbard band) would be full and upper one (so-called upper Hubbard band) would be empty if the bandwidth is not very large ($U > W$, Fig. 8.4). This type of insulator from the partially occupied d-shell is called Mott-insulator.

In reality, photoemission experiments show that most of the magnetic oxides like MnO, NiO etc. are so-called charge-transfer insulators [10], where top of the valence band is predominantly of O_{2p} character, while the bottom of the empty conducting band has mainly the Me_{3d} character as in the upper Hubbard band. The corresponding theoretical parameter range should be $U > \Delta > W$ and is schematically shown in Fig. 8.5, according to Zaanen–Sawatzky and Allen theory [10]. The general phase diagram in U–Δ–W space (W_M is a metal d-bandwidth, and W_L is a ligand–oxygen bandwidth) is present on Fig. 8.6. We first discuss the insulating part of this diagram. For $U > \Delta$ the band gap is of p–d type and the anion or ligand p-band is located between the lower and upper Hubbard bands. This gap is a charge transfer gap and the corresponding compounds (NiO, FeO, $LaMnO_3$ etc.) are charge transfer insulators. In this case band gap is proportional to Δ. If $U < \Delta$, on other hand, the band gap is of the d–d type and it is called Mott–Hubbard gap while the corresponding compounds (such as V_2O_3) are called Mott–Hubbard insulators. It has a band gap of the magnitude $\sim U$. The straight line $U = \Delta$ separate the Mott–Hubbard and the charge transfer regimes. The diagram also contains a metallic region near the Δ-axis (d-metal as TiO, $YTiO_3$) or near U-axis (p-metal as V_2O_5). This classification scheme is very useful for oxides material science; more examples one could find in the recent review article [9].

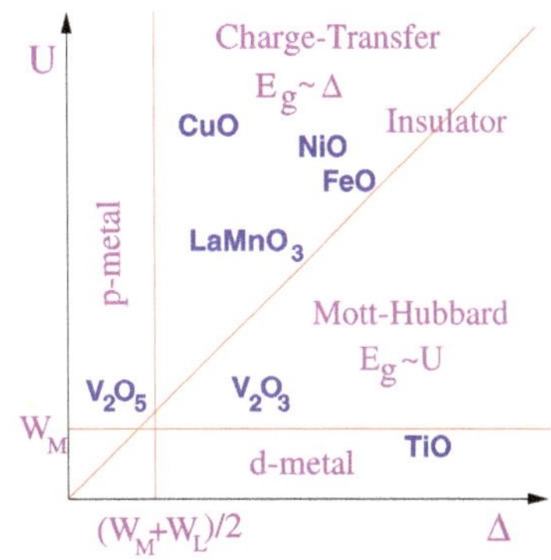

Fig. 8.6 The Zaanen–Sawatzky–Allen phase diagram

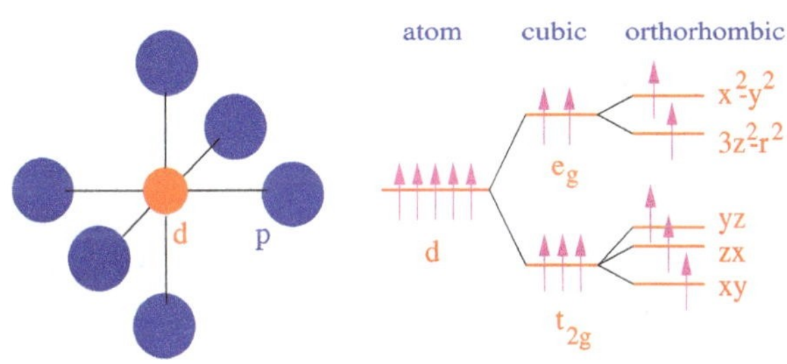

Fig. 8.7 The octahedral ligand MeO_6 complex and scheme of d-states in the crystal field

Since the magnetic oxides contain transition metal d-ions, the orbital degrees of freedom also play an important role in all physical properties. Because the $3d$-orbital has the angular momentum $L = 2$, it has fivefold degeneracy. In the transition metal compounds $3d$-ion is surrounded by octahedral complex of the ligand-oxygen ions (Fig. 8.7). In the cubic crystal field, fivefold atomic d-level splits into the lower threefold degenerate t_{2g} states and twofold degenerate e_g states. In the perovskite structure, the MeO_6 octahedra are a main chemical "block" of the lattice and the e_g orbitals, which pointed in the direction to the ligand atoms, hybridize much better with oxygen than the t_{2g} orbitals and has a larger bandwidth. If the crystal has the orthorhombic distortions like in CMR-compound $LaMnO_3$, then the e_g orbitals split into the $x^2 - y^2$ and $3z^2 - r^2$ states while the t_{2g} orbitals states split into the yz, zx, and xy states.

In general, the relevant electronic orbitals for light transition metals are different from heavy ones. In compounds with the light transition metal elements, such as Ti, V, and Cr, the t_{2g} bands are important, while in magnetic oxides with the heavy transition-metal ions such as Co, Ni, and Cu, the t_{2g} bands are fully occupied and located far below the Fermi level, therefore the most important orbitals are the e_g ones. If the degenerate t_{2g}- or e_g-orbitals are partially filled, it normally leads to the crystal distortion due to cooperative Jahn–Teller effect and results in orbital ordered superstructure.

In magnetic insulators, the spin and orbital degrees of freedom could be coupled via the exchange (superexchange) interactions and therefore, magnetic structure can be strongly dependent on the orbital ordering [11]. We consider the simplest

Fig. 8.8 The Kugel–Khomskii model for double degenerate states

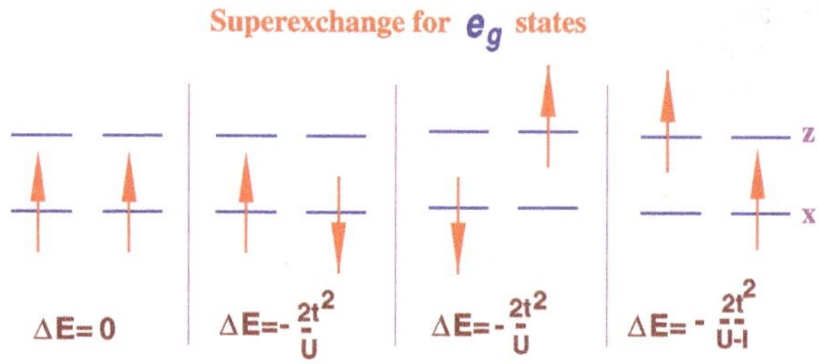

case of the double degenerate e_g^1 configurations or so-called Kugel–Khomskii model (Fig. 8.8). Two e_g orbitals x (which stands for $x^2 - y^2$) and z (for $3z^2 - r^2$) could be described by the spin $s = 1/2$ operator as well as pseudospin $\tau = 1/2$ operator, where $\tau_z = +1/2$ corresponds to z-state and $\tau_z = -1/2$ corresponds to x-state. Different orbital and spin configurations for two atoms with the degenerate e_g orbitals shown on Fig. 8.8, together with the total energy changes due to the kinetic exchange interaction. The most energetically stable configuration is the last one with the ferromagnetic spin and antiferromagnetic orbital ordering. It has the lower transfer energy excitation equal to $(U-I)$, instead of U as for non-degenerate or antiferromagnetic cases. This is a consequence of the inter-atomic Hund exchange interactions (I). The general spin–orbital exchange Hamiltonian for doubly degenerate e_g case can be written as follows [11]:

$$H_{\rm ex} = -2\sum_{ij} J_1 \vec{S_i}\vec{S_j} + J_2 \vec{\tau_i}\vec{\tau_j} + 4J_2(\vec{S_i}\vec{S_j})(\vec{\tau_i}\vec{\tau_j}) \tag{8.4}$$

where $J_1 = -(2t^2/U)(1 - I/U)$ and $J_2 = -(2t^2/U)(1 + I/U)$, for $U \gg I$.

We have discussed some model approaches to the electronic structure of magnetic oxides. In the next section we turn to a more quantitative electronic-structure scheme.

8.3 Correlated Electronic Structure Scheme

8.3.1 LDA+U Method

The proper description of the electronic structure for the complicated materials like $LaMnO_3$ is a hard many-electron problem. The most successful "first-principle" method is the density functional theory [12, 13] within the Local (Spin-) Density Approximation (L(S)DA), where the exchange-correlation potential is approximated by the homogeneous electron gas model. The LDA has proved to be very efficient for the extended systems, such as large molecules and solids [14]. For strongly correlated materials like NiO, application of LDA is problematic. Such systems usually contain transition-metal or rare-earth metal ions with partially filled d- (or f-)

shell. When applying to transition-metal compounds the LDA method with orbital-independent potential one has as a result the partially filled d-band with metallic type of the electronic structure and itinerant d-electrons. This is definitely a wrong answer for the late-transition-metal oxides where d-electrons are well localized and there is a sizable energy separation between occupied and unoccupied subbands (lower Hubbard band and upper Hubbard band in a model Hamiltonian approach).

There were several attempts to improve on the LDA scheme for strongly correlated systems. One of the most popular approaches is the Self Interaction Correction (SIC-LDA) method [15]. It reproduces quite well the localized nature of the d-electrons in transition metal compounds, but SIC one-electron energies are usually in strong disagreement with spectroscopy data and for transition metal oxides occupied d-bands are located too much below the oxygen valence band.

The standard Hartree–Fock (HF) method is also appropriate for describing Mott insulators with spin- and orbital-symmetry broken states. However, a serious problem of the Hartree–Fock approximation is the unscreened nature of the Coulomb interaction used in this method. The bare value of Coulomb interaction parameter U is rather large (15–20 eV) while screening in a solid leads to much smaller values, 8 eV and less [16]. Due to the neglecting of screening, the HF energy gap values are a factor of 2–3 larger than the experimental values.

The best way of addressing this problem is so-called GW-approximation [17], which may be regarded as a Hartree–Fock method with a frequency- and orbital-dependent screening of the Coulomb interaction. Unfortunately GW-method is computationally heavy and even with modern powerful computers any calculations for complex systems are practically impossible. The most successful and computationally simple scheme for magnetic oxides and other correlated materials is the so-called LDA+U method [18], where the frequency-dependent screened GW-Coulomb potential is approximated by statically screened Coulomb parameter U.

The main idea of LDA+U method [17, 18] is to separate electrons into two subsystems: localized d- or f-electrons for which Coulomb d–d correction should be taken into account and delocalized s-, p-electrons which could be described by using LDA orbital-independent one-electron potential. Let us consider a d-ion as an open system with a fluctuating number of d-electrons. If we suggest that the Coulomb energy of d–d interactions as a function of total number of d-electrons $N = \sum n_i$ given by the LDA is a good approximation (but not the orbital energies), then the correct formula for this energy is $E = UN(N-1)/2$. Let us subtract this expression from the LDA total energy functional and add a Hubbard like term (neglecting for a while exchange and non-sphericity). As a result we have the following functional:

$$E = E_{\mathrm{LDA}} - UN(N-1)/2 + \frac{1}{2}U\sum_{i\neq j} n_i n_j \tag{8.5}$$

The orbital energies ε_i are derivatives of (8.5) with respect to orbital occupations n_i:

$$\varepsilon_i = \partial E/\partial n_i = \varepsilon_{\mathrm{LDA}} + U\left(\frac{1}{2} - n_i\right) \tag{8.6}$$

This simple formula shifts the LDA orbital energy by $-U/2$ for occupied orbitals ($n_i = 1$) and by $+U/2$ for unoccupied orbitals ($n_i = 0$) as in the atomic limit of the Hubbard model.

The LDA+U orbital-dependent potential [17] gives upper and lower Hubbard bands with the energy separation between them equal to the Coulomb parameter U, thus reproducing qualitatively the correct physics for Mott–Hubbard insulators. To construct a realistic computational scheme, one needs to define in a more general way an orbital basis set and to take into account properly the direct and exchange Coulomb interactions inside a partially filled d-atomic shell in "rotationally invariant" form [19].

We need to identify the region in space where the atomic characteristics of the electronic states have largely survived ('atomic spheres'), which is not a problem for at least d- or f-electrons. Within these atomic spheres one can expand in a localized orthonormal basis $|inlm\sigma\rangle$ (i denotes the site, n the main quantum number, l the orbital quantum number, m the magnetic number and σ spin index). Although not strictly necessary, let us specialize to the usual situation where only a particular nl shell is partly filled. The density matrix is defined by

$$n^{\sigma}_{mm'} = -\frac{1}{\pi}\int^{E_F} \operatorname{Im} G^{\sigma}_{inlm,inlm'}(E)\, dE \qquad (8.7)$$

where $G^{\sigma}_{inlm,inlm'}(E) = \langle inlm\sigma|(E-\widehat{H})^{-1}|inlm'\sigma\rangle$ are the elements of the Green function matrix in this localized representation, while $\widehat{H}$ will be defined later on. In terms of the elements of this density matrix $\{n^{\sigma}\}$, the generalized LDA+U functional [17] is defined as follows:

$$E^{\mathrm{LDA}+U}\left[\rho^{\sigma}(\mathbf{r}),\{n^{\sigma}\}\right] = E^{\mathrm{LSDA}}\left[\rho^{\sigma}(\mathbf{r})\right] + E^{U}\left[\{n^{\sigma}\}\right] - E_{dc}\left[\{n^{\sigma}\}\right] \qquad (8.8)$$

where $\rho^{\sigma}(\mathbf{r})$ is the charge density for spin-σ electrons and $E^{\mathrm{LSDA}}[\rho^{\sigma}(\mathbf{r})]$ is the standard LSDA functional. Equation (8.8) asserts that the LSDA suffices in the absence of orbital polarizations, while the latter are driven by

$$\begin{aligned} E^{U}\left[\{n\}\right] = \frac{1}{2}\sum_{\{m\},\sigma} \Big\{ &\langle m,m''|V_{ee}|m',m'''\rangle n^{\sigma}_{mm'} n^{-\sigma}_{m''m'''} \\ &+ \left(\langle m,m''|V_{ee}|m',m'''\rangle - \langle m,m''|V_{ee}|m''',m'\rangle\right) n^{\sigma}_{mm'} n^{\sigma}_{m''m'''} \Big\} \qquad (8.9) \end{aligned}$$

where V_{ee} are the screened Coulomb interactions among the nl electrons. Finally, the last term in (8.8) corrects for double counting (in the absence of orbital polarizations, (8.8) should reduce to E^{LSDA}) and is given by

$$E_{dc}\left[\{n^{\sigma}\}\right] = \frac{1}{2}UN(N-1) - \frac{1}{2}J\left[N^{\uparrow}(N^{\uparrow}-1) + N^{\downarrow}(N^{\downarrow}-1)\right] \qquad (8.10)$$

where $N^{\sigma} = \mathrm{Tr}(n^{\sigma}_{mm'})$ and $N = N^{\uparrow} + N^{\downarrow}$. U and J are screened Coulomb and exchange parameters [16].

In addition to the usual LDA potential, we find an effective single particle potentials to be used in the single particle Hamiltonian:

$$\widehat{H} = \widehat{H}_{\rm LSDA} + \sum_{mm'} |inlm\sigma\rangle V^{\sigma}_{mm'} \langle inlm'\sigma| \tag{8.11}$$

$$\begin{aligned} V^{\sigma}_{mm'} = & \sum_{\{m\}} \{\langle m, m''|V_{ee}|m', m'''\rangle n^{-\sigma}_{m''m'''} \\ & + (\langle m, m''|V_{ee}|m', m'''\rangle - \langle m, m''|V_{ee}|m''', m'\rangle) n^{\sigma}_{m''m'''}\} \\ & - U\left(N - \frac{1}{2}\right) + J\left(N^{\sigma} - \frac{1}{2}\right) \end{aligned} \tag{8.12}$$

The V_{ee}'s remain to be determined. We again follow the spirit of LDA+U by assuming that within the atomic spheres these interactions retain largely their atomic nature. Moreover, it is asserted that LSDA itself suffices to determine their values, following the well-tested procedure of the so-called supercell LSDA approach [16]: the elements of the density matrix $n^{\sigma}_{mm'}$ have to be constrained locally and the second derivative of the LSDA energy with respect to the variation of the density matrix yields the wanted interactions. In more detail, the matrix elements can be expressed in terms of complex spherical harmonics and effective Slater integrals F^k as

$$\langle m, m''|V_{ee}|m', m'''\rangle = \sum_{k} a_k(m, m', m'', m''') F^k \tag{8.13}$$

where $0 \leq k \leq 2l$ and

$$a_k(m, m', m'', m''') = \frac{4\pi}{2k+1} \sum_{q=-k}^{k} \langle lm|Y_{kq}|lm'\rangle \langle lm''|Y^{*}_{kq}|lm'''\rangle$$

For d-electrons one needs F^0, F^2, and F^4 and these can be linked to the Coulomb and Stoner parameters U and J obtained from the LSDA-supercell procedures via $U = F^0$ and $J = (F^2 + F^4)/14$, while the ratio F^2/F^4 is to a good accuracy constant $\sim$0.625 for 3d elements [17].

The new Hamiltonian (8.11) contains an orbital-dependent potential (8.12) in the form of a projection operator. This means that the LDA+U method is essentially dependent on the choice of the set of the localized orbitals in this operator. That is a consequence of the basic ideology of the LDA+U approach: the separation of the total variational space into a localized d-orbitals subspace, with Coulomb interaction between them treated with a Hubbard type term in the Hamiltonian, and the subspace of all other states for which local density approximation for Coulomb interaction is regarded as sufficient. The arbitrariness of the choice of the localized orbitals is not as crucial as might be expected. The d-orbitals for which Coulomb correlation effects are important are indeed well localized in direct space and retain

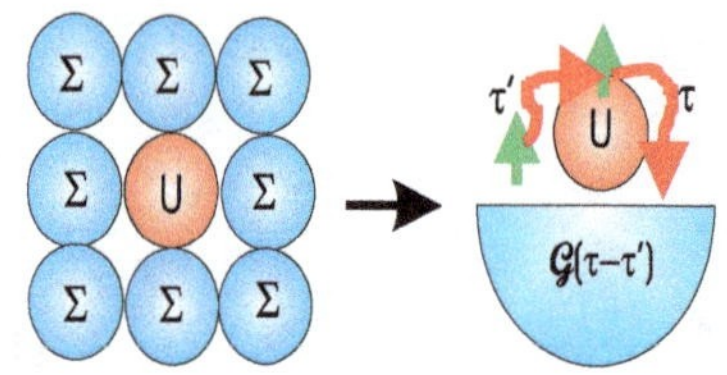

Fig. 8.9 Mapping of the lattice model to the quantum impurity model in the Dynamical Mean Field Theory

their atomic character in a solid. The experience of using the LDA+U approximation in various electronic structure calculation schemes shows that the results are not sensitive to the particular form of the localized orbitals.

Due to the presence of the projection operator in the LDA+U Hamiltonian (8.11) the most straightforward calculation scheme would be to use atomic-orbitals type basis sets, such as the LMTO (Linear Muffin–Tin Orbitals) [20]. However, as soon as localized d-orbitals are defined, the Hamiltonian in (8.11) can be realized even in schemes using plane waves as a basis set, such as pseudopotential methods.

8.3.2 LDA+DMFT: General Considerations

The natural generalization of LDA+U scheme for the local dynamical effects use the recently developed efficient many-body approach—the dynamical mean-field theory (DMFT) [21–23]. The DMFT-scheme map the interaction lattice models onto quantum impurity models (Fig. 8.9) subject to a self-consistency condition [24]). The resulting many-body multi-orbital impurity problem can be solved by various rigorous approaches (Quantum Monte Carlo, exact diagonalization etc.) or by approximated schemes such as Iterated Perturbation Theory (IPT), local Fluctuating-Exchange (FLEX) approximation, or Non-Crossing Approximation (NCA) [24]).

In this section we describe LDA+DMFT approach for the electronic structure calculations. The method was first applied to $La_{1-x}Sr_xTiO_3$ [25] which is a classical example of strongly correlated metal. A general formulation of LDA+DMFT, including the justification of the effective impurity formulation in multi-band case, has been discussed in Ref. [26].

In order to describe the LDA+DMFT scheme, we start from the Hamiltonian of (8.8) where the LDA part was taken from a first-principle LMTO tight-binding method [20, 27] and interaction part is reduced to density–density correlation:

$$H_{\text{LMTO}} = \sum_{ilm,jl'm',\sigma} \left(\delta_{ilm,jl'm'}\,\varepsilon_{il}\,\hat{n}_{ilm\sigma} + t_{ilm,jl'm'}\hat{c}^{\dagger}_{ilm\sigma}\hat{c}_{jl'm'\sigma}\right) \tag{8.14}$$

(i is site index, lm are orbital indices).

As we have mentioned above, the LDA one-electron potential is orbital independent and Coulomb interaction between d-electrons is taken into account in this scheme in an averaged way. We generalize this Hamiltonian for the explicit local

Coulomb correlations with the additional interaction term for correlated il shell:

$$H_{\text{int}} = \frac{1}{2} \sum_{ilmm'\sigma} U^{il}_{mm'} n_{im\sigma} n_{im'-\sigma} + \frac{1}{2} \sum_{ilm \neq m'\sigma} (U^{il}_{mm'} - J^{il}_{mm'}) n_{im\sigma} n_{im'\sigma} \tag{8.15}$$

where i is the site index and m is the orbital quantum numbers; $\sigma = \uparrow, \downarrow$ is the spin projection; c^+, c are the Fermi creation and annihilation operators ($n = c^+ c$); ε and t in (8.14) are effective one-electron energies and hopping parameters obtained from the LDA in the orthogonal LMTO basis set. To avoid the double-counting of electron–electron interactions one must at the same time subtract the averaged Coulomb interaction energy term, which is present in LDA. In the spirit of LDA+U scheme we introduce new ε^0_d where d–d Coulomb interaction is excluded:

$$\varepsilon^0_{d\sigma} = \varepsilon_{d\sigma} - U\left(n_d - \frac{1}{2}\right) + \frac{1}{2} J\left(n^\sigma_d - \frac{1}{2}\right) \tag{8.16}$$

where U and J are the average values of $U_{mm'}$ and $J_{mm'}$ matrices and n_d is the average number of d-electrons.

The screened Coulomb and exchange vertex for the d-electrons are defined as

$$U_{mm'} = \langle mm' | V_{ee}(\mathbf{r} - \mathbf{r}') | mm' \rangle, \qquad J_{mm'} = \langle mm' | V_{ee}(\mathbf{r} - \mathbf{r}') | m'm \rangle$$

Then a new Hamiltonian will have the following form:

$$\begin{aligned} H &= H^0 + H_{\text{int}} \\ H^0 &= \sum_{ilm, jl'm', \sigma} \left(\delta_{ilm, jl'm'} \, \varepsilon^0_{il} \, \hat{n}_{ilm\sigma} + t_{ilm, jl'm'} \hat{c}^\dagger_{ilm\sigma} \hat{c}_{jl'm'\sigma} \right) \end{aligned} \tag{8.17}$$

In the reciprocal space matrix elements of the operator H^0 are

$$H^0_{qlm,q'l'm'}(\mathbf{k}) = H^{\text{LDA}}_{qlm,q'l'm'}(\mathbf{k}) - \delta_{qlm,q'l'm'} \delta_{ql,i_d l_d} \left[U\left(n_d - \frac{1}{2}\right) - \frac{1}{2} J\left(n^\sigma_d - \frac{1}{2}\right) \right] \tag{8.18}$$

(q is an index of the atom in the elementary unit cell).

In the local, frequency dependent dynamical mean-field theory the effect of Coulomb correlation is described by self-energy operator $\Sigma(i\omega)$. The inverse Green function matrix is defined as

$$G^{-1}_{qlm,q'l'm'}(\mathbf{k}, i\omega) = i\omega + \mu - H^0_{qlm,q'l'm'}(\mathbf{k}) - \delta_{ql,q'l'} \delta_{ql,i_d l_d} \Sigma_{m,m'}(i\omega) \tag{8.19}$$

where μ is chemical potential, and the local Green function obtained via integration is over the Brillouin zone:

$$G_{qlm,q'l'm'}(i\omega) = \frac{1}{V_B} \int d\mathbf{k} \, G_{qlm,q'l'm'}(\mathbf{k}, i\omega) \tag{8.20}$$

(V_B is the volume of the Brillouin zone).

A so-called bath Green function which defines a hybridization with the surrounding crystal in the effective Anderson model and preserves the double-counting of the local self-energy is obtained by a solution of the effective impurity problem [24]:

$$\mathcal{G}^{-1}_{m,m'}(i\omega) = G^{-1}_{m,m'}(i\omega) + \Sigma_{m,m'}(i\omega) \tag{8.21}$$

In the simplest case of massive downfolding to the d-orbitals problem, we could incorporate the double counted correction in the chemical potential μ obtained self-consistently from the total number of d-electrons

$$\frac{1}{\beta}\sum_{i\omega_n} e^{i\omega_n 0^+} G(i\omega_n) = N_d \tag{8.22}$$

here $\omega_n = (2n+1)\pi T$ are the Matsubara frequencies for temperature $T \equiv \beta^{-1}$ $(n = 0, \pm 1, \ldots)$. Further, one has to find the self-energy $\Sigma_m(i\omega)$ in terms of the bath Green function $G_{0m}(i\omega)$ and use it in the self-consistent LDA+DMFT loop (8.20), (8.21).

8.3.3 *The Quantum Monte Carlo Solution of Impurity Problem*

Here we describe at first the most rigorous way to solve an effective impurity problem using the multi-band Quantum Monte Carlo (QMC) method [28, 29]. In the framework of LDA+DMFT approach it was used first in Ref. [30] for the case of ferromagnetic iron. In this method the local Green functions is calculated for the imaginary time interval $[0, \beta]$ with the mesh $\tau_l = l\Delta\tau$, $l = 0, \ldots, L-1$, and $\Delta\tau = \beta/L$ ($\beta = \frac{1}{T}$ is the inverse temperature) using the path-integral formalism [24].

The multi-orbital DMFT problem and general cluster DMFT scheme can be reduced to the general impurity action (see Fig. 8.9):

$$S = -\int_0^\beta d\tau \int_0^\beta d\tau' \sum_{i,j} c_i^+(\tau)\mathcal{G}_{ij}(\tau-\tau')c_j(\tau') + \frac{1}{2}\int_0^\beta d\tau \sum_{i,j} n_i(\tau)U_{ij}n_j(\tau)$$

where $i = \{m, \sigma\}$ are orbital (site) and spin. Without spin–orbital coupling we have $\mathcal{G}_{ij} = \mathcal{G}^{\sigma}_{m,m'}\delta_{\sigma\sigma'}$.

The auxiliary fields Green-function QMC use the discrete Hubbard–Stratanovich transformation were introduced by Hirsch [31]

$$\exp\left\{-\Delta\tau U_{ij}\left[n_i n_j - \frac{1}{2}(n_i + n_j)\right]\right\} = \frac{1}{2}\sum_{S_{ij}=\pm 1} \exp\{\lambda_{ij} S_{ij}(n_i - n_j)\}$$

where $S_{ij}(\tau)$ are the auxiliary Ising fields for each pair of orbitals and time slice with the strength

$$\lambda_{ij} = \operatorname{arccosh}\left[\exp\left(\frac{\Delta\tau}{2}U_{ij}\right)\right]$$

Using Hirsch transformation one can integrate out fermionic fields in the path integral [24] and resulting partition function and Green function matrix have the following form:

$$Z = \frac{1}{2^{N_f L}} \sum_{S_{ij}(\tau)} \det\left[\widehat{G}^{-1}(S_{ij})\right]$$

$$\widehat{G} = \frac{1}{Z} \frac{1}{2^{N_f L}} \sum_{S_{ij}(\tau)} \widehat{G}(S_{ij}) \det\left[\widehat{G}^{-1}(S_{ij})\right]$$

where N_f is the number of Ising fields, L is the number of time slices, and $\widehat{G}(S_{ij})$ is the Green function in the auxiliary Ising fields:

$$G_{ij}^{-1}(S) = \mathcal{G}_{ij}^{-1} + \Delta_i \delta_{ij} \delta_{\tau\tau'}$$

$$\Delta_i = \left(e^{V_i} - 1\right)$$

$$V_i(\tau) = \sum_{j(\neq i)} \lambda_{ij} S_{ij}(\tau) \sigma_{ij}$$

here we introduce the generalized Pauli matrix:

$$\sigma_{ij} = \begin{cases} +1, & i < j \\ -1, & i > j \end{cases}$$

For efficient calculation of the Green function in arbitrary configuration of Ising fields $G_{ij}(S)$ we use the following Dyson equation [31]:

$$G' = \left[1 + (1 - G)\left(e^{V' - V} - 1\right)\right]^{-1} G$$

The QMC important sampling scheme allowed us to integrate over the Ising fields with the abs(det[$\widehat{G}^{-1}(S_{ij})$]) as a stochastic weight [24, 31]. For a single spin-flip S_{ij}, the determinant ratio is calculated as follows:

$$\det[\widehat{G}] / \det\left[\widehat{G}'\right] = R_i R_j - R_{ij}$$

$$R_i = 1 + \left[1 - G_{ii}(\tau, \tau)\right] \Delta_i(\tau)$$

$$R_j = 1 + \left[1 - G_{jj}(\tau, \tau)\right] \Delta_j(\tau)$$

$$R_{ij} = G_{ij}(\tau, \tau) \Delta_j(\tau) G_{ji}(\tau, \tau) \Delta_i(\tau)$$

and the Green function matrix updated in the standard manner [24, 31]:

$$G'_{i_1 j_2}(\tau_1, \tau_2) = G_{i_1 j_2}(\tau_1, \tau_2) + \left[G_{i_1 i}(\tau_1, \tau) - \delta_{i_1 i} \delta_{\tau_1, \tau}\right] \Delta_i(\tau) / R_i(\tau) G_{i j_2}(\tau, \tau_2)$$

$$G^{new}_{i_1 j_2}(\tau_1, \tau_2) = G'_{i_1 j_2}(\tau_1, \tau_2) + \left[G'_{i_1 j}(\tau_1, \tau) - \delta_{i_1 j} \delta_{\tau_1, \tau}\right] \Delta_j(\tau) / R_j(\tau) G'_{j j_2}(\tau, \tau_2)$$

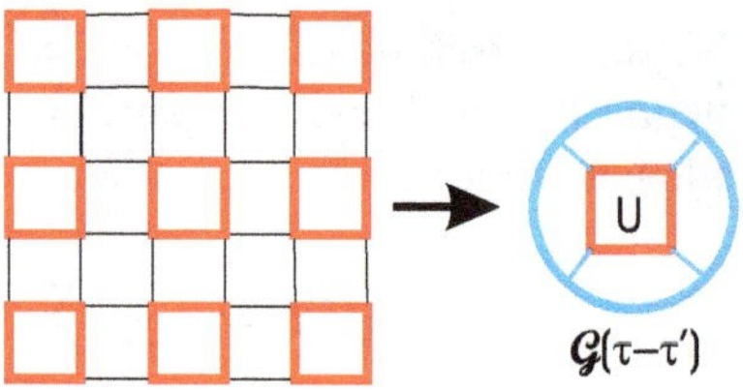

Fig. 8.10 Mapping of the lattice model to the quantum cluster-impurity model in the Cluster Dynamical Mean Field Theory

Using the output local Green function from QMC and input bath Green functions the new self-energy is obtain via (8.21) and the self-consistent loop can be closed through (8.20). The main problem of the multi-band QMC formalism is the large number of the auxiliary fields $s^l_{mm'}$. For each time slice l it is equal to $M(2M-1)$ where M is the total number of the orbitals which gives 45 Ising fields for the d-states case and 91 fields for the f-states. Analytical continuations of the QMC Green functions from the imaginary time to the real energy axis can be done within the maximum entropy method [32].

8.3.4 Cluster LDA+DMFT Scheme

When considering the effects like charge-ordering or d-wave superconductivity which involves explicitly the electronic correlations on *different* sites, a cluster generalization of the LDA+DMFT scheme would be necessary. The most natural way to construct this generalization is to consider the cluster as a "super-site" in an effective medium (for simplicity, consider here the case of two-site cluster). Then the crystal supercell Green function matrix can be written as

$$\mathbf{G}(\mathbf{k}, i\omega) = \left[(i\omega + \mu) * \mathbf{1} - \mathbf{h}(\mathbf{k}) - \Sigma(i\omega)\right]^{-1}$$

where $h_{\alpha\beta}(\mathbf{k})$ is the effective hopping matrix, $\Sigma_{\alpha\beta}(i\omega)$ is the self-energy matrix of the N-site supercell dimension which is assumed to be local, i.e. $\mathbf{k}$-independent, and μ is the chemical potential.

In the cluster version of the DMFT scheme, one can write the matrix equation for a bath Green function matrix $\mathcal{G}$ which describes effective interactions with the rest of the crystal:

$$\mathcal{G}^{-1}(i\omega) = \mathbf{G}^{-1}(i\omega) + \Sigma(i\omega) \tag{8.23}$$

where the local cluster Green function matrix is equal to $G_{\alpha\beta}(i\omega) = \sum_{\mathbf{k}} G_{\alpha\beta}(\mathbf{k}, i\omega)$, and the summation runs over the Brillouin zone of the lattice.

The most efficient ways to solve the cluster-impurity problem is to use the general matrix-QMC scheme described above. This scheme corresponds to so-called "free cluster" approach (Fig. 8.10) [33, 34]. Alternatively, periodic modification of the DMFT in the $\mathbf{k}$-space or so-called "dynamical cluster approximation" [35, 36] can be used.

8.4 Mott–Hubbard Insulators

We will review now the applications of these new methods and computational schemes to real correlated materials.

8.4.1 Electronic Structure of Transition Metal Oxides

The advantages of the simultaneous treatment of the localized and delocalized electrons in the LDA+U method and especially in the LDA+DMFT approaches are seen most clearly for the transition metal compounds, where $3d$-electrons, while remaining localized, hybridize quite strongly with other orbitals. Late-transition metal oxides, for which LSDA results strongly underestimate the energy gap and magnetic moment values (or even give qualitatively wrong metallic ground state for the insulators CoO and $CaCuO_2$), are well described by the LDA+U [17].

As already mentioned, sometimes it is necessary to take into account also the intra-atomic Coulomb interaction on the oxygen sites to have a satisfactory agreement with the experimental data. Here we present the results of such calculations for transition metal oxides [37]. The important part of the LDA+U calculation scheme is the determination of Coulomb interaction parameters U and J in (8.12): Coulomb parameter U_p for p-orbitals of oxygen, U_d for transition metals ion and Hund's parameter J for d-orbitals of transition metals. To get U_d and J one can use the supercell procedure [16] or the constrained LSDA method [38], which are based on calculation of the variation of the total energy as a function of the local occupation of the d-shell. We took the values of Coulomb parameters ($U_d \sim 7$–8 eV and $J \sim 0.9$–1 eV) from the previous LDA+U calculation [18]. The problem is how to determine the Coulomb parameter U_p.

Due to the more extended nature of the O($2p$) Wannier states in comparison with transition metal d states, the constrained occupation calculations cannot be implemented as easy as for the d-shell of transition metals. Nevertheless, several independent and different techniques were used for this purpose previously by different authors. McMahan et al. [39] estimated the value of U_p in high-T_c related compound La_2CuO_4 using the constrained LDA calculation where only atomic-like O($2p$)-orbitals within oxygen atomic spheres were considered instead of the more extended Wannier functions. The corresponding value of the Coulomb interaction parameter U_p was obtained as 7.3 eV. This value can be considered as the upper limit of the exact U_p. The LDA calculations gave the estimation that only 75 % of Wannier function density lies in the oxygen atomic sphere so that renormalized value of Coulomb interaction parameter for oxygen Wannier functions is $U_p = (7.3) \times (0.75)^2 = 4.1$ eV [39].

Later Hybertsen et al. [40] suggested the scheme to calculate U_p, which consists of two steps: (i) via constrained-density-functional approach one can obtain the energy surface $E(N_d, N_p)$ as a function of local charge states and (ii) simultaneously extended Hubbard model was solved in mean-field approximation as a function of

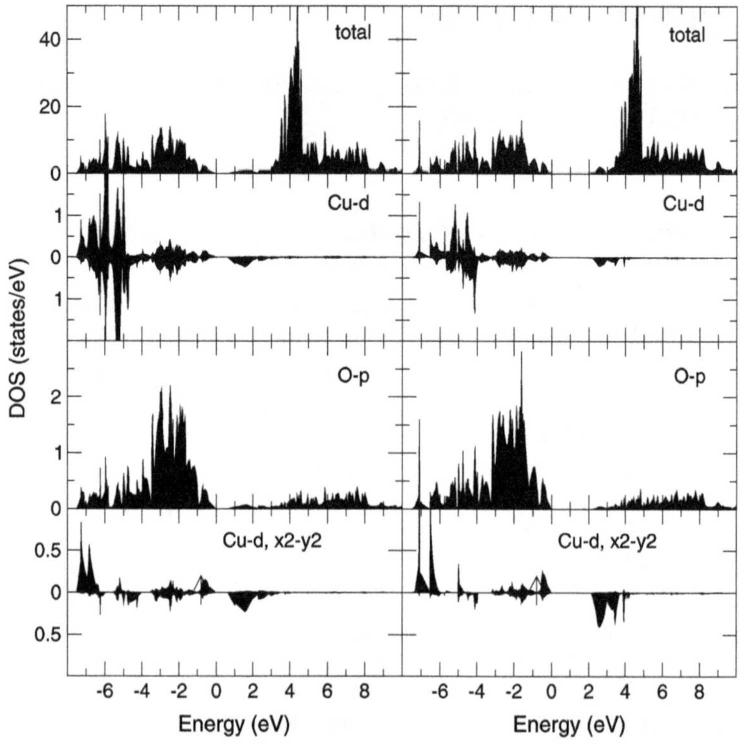

Fig. 8.11 La_2CuO_4 DOS calculated by the LDA+U (*left column*) and the LDA+$U^{(d+p)}$ (*right column*) methods [37]. (On all figures the total DOS is presented per formula unit, the DOS of particular states are per atom. Fermi energy corresponds to zero.)

local charge states N_d and N_p. Corresponding Coulomb interaction parameters were extracted as those which give the energy surface matching the microscopic density-functional calculations results [40]. The obtained values for U_p are 3–8 eV, depending on the parameters of calculations. Another way to estimate U_p is to use Auger spectroscopy data, where two holes in O($2p$)-shell are created in the excitation process. Such fitting to the experimental spectra gave the value of $U_p = 5.9$ eV [41]. In the LDA+$U^{(d+p)}$ calculations [37] the value $U_p = 6$ eV was used.

Comparison between the LDA+U (left column) and the LDA+$U^{(d+p)}$ (right column) calculated density of states (DOS) of NiO, MnO and La_2CuO_4 is presented in Figs. 8.11, 8.12, 8.13. For all compounds one can see that the main difference between the LDA+$U^{(d+p)}$ and the LDA+U calculated densities of states is the increased energy separation between the oxygen $2p$ and transition metal $3d$ bands. The larger value of "charge transfer" energy (O($2p$)–Me($3d$)) (Me = Ni, Mn, Cu) leads to the enhanced ionicity and decreased covalency nature of the electronic structure: the unoccupied bands have more pronounced $3d$ character and the admixture of oxygen states to those bands becomes weaker.

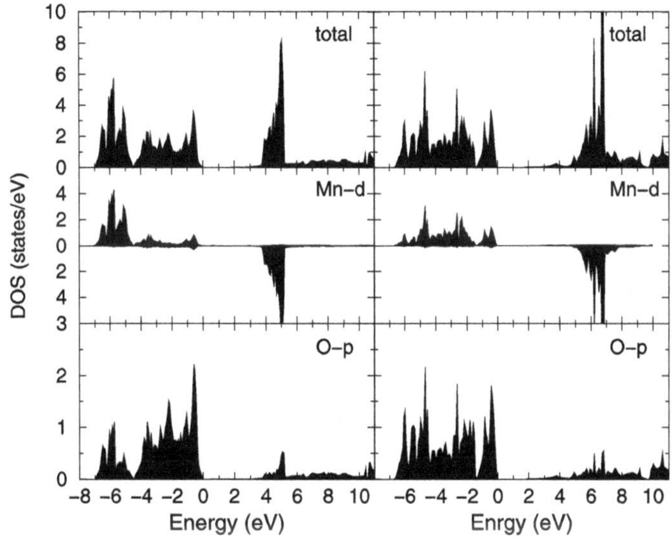

Fig. 8.12 MnO DOS calculated by the LDA+U (*left column*) and the LDA+$U^{(d+p)}$ (*right column*) methods [37]

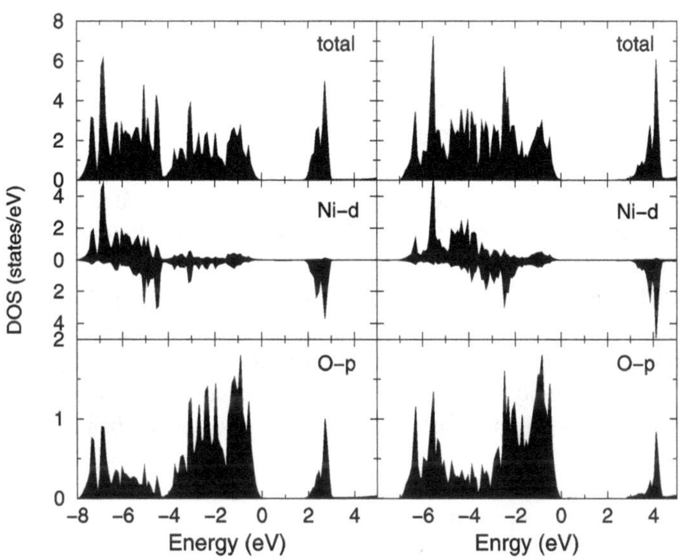

Fig. 8.13 NiO DOS calculated by the LDA+U (*left column*) and the LDA+$U^{(d+p)}$ (*right column*) methods [37]

The ground state is correctly described both by LDA+U and LDA+$U^{(d+p)}$ calculations as antiferromagnetic insulator for all compounds. The values of energy gaps and spin magnetic moments are presented in Table 8.1 (see the discussion of

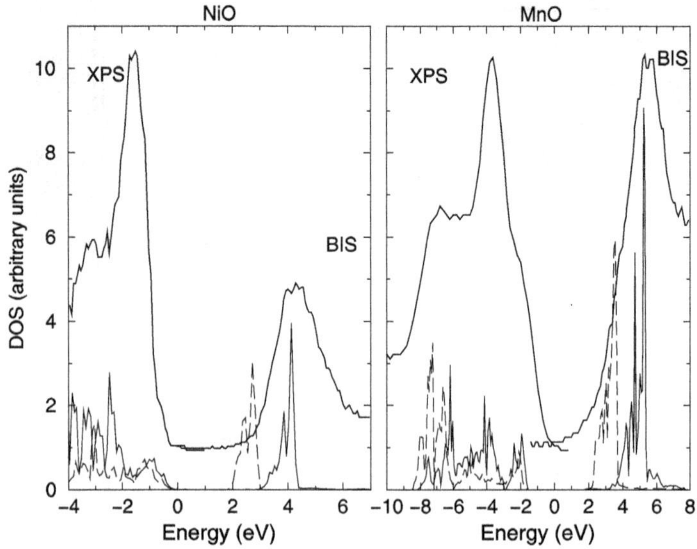

Fig. 8.14 DOS calculated by the LDA+U (*dashed line*) and the LDA+$U^{(d+p)}$ (*solid line*) Ni(3d) and Mn(3d) in comparison with superimposed XPS and BIS spectra [42]

experimental data in [37]). One can see that the values obtained in the LDA+$U^{(d+p)}$ calculations are in general in better agreement with experiment than the LDA+U calculated values. While the increasing of the energy gap values with applying U_p correction was obviously expected with the increasing of "charge transfer" energy in the compounds belonging to the class of "charge transfer" insulators [10], the increasing of the magnetic moments values is a more complicated self-consistency effect due to the increased ionicity in the LDA+$U^{(d+p)}$ calculations compared with the LDA+U results.

In Fig. 8.14 the DOS obtained by LDA+$U^{(d+p)}$ and LDA+U calculations for MnO and NiO compounds are compared with the superimposed XPS and BIS spectra corresponding to the removal of an electron (the occupied bands) and addition of an electron (the empty bands), respectively. The better agreement with the experimental data of position of the main peaks of unoccupied band relative to the occupied one is the direct confirmation of the importance of taking into account Coulomb interactions in oxygen 2p-shell.

It is instructive to compare the results of the LDA+U calculations for NiO with the first-principle "Hubbard I" approach [26]. First of all, to describe Mott insulators in LDA+U approach (as well as in SIC approach) it is necessary to assume magnetic and (or) orbital long-range order [17]. In LDA+DMFT it is possible to consider the *paramagnetic* Mott insulators in the framework of ab initio calculations. Moreover, it is possible to obtain not only the Mott–Hubbard gap in the electron spectrum but also satellites and multiplet structure. The following effective Slater parameters, which define the screened Coulomb interaction in d-shell for NiO, have been used: $F^0 = 8.0$ eV, $F^2 = 8.2$ eV, $F^4 = 5.2$ eV [18]. We have started from the

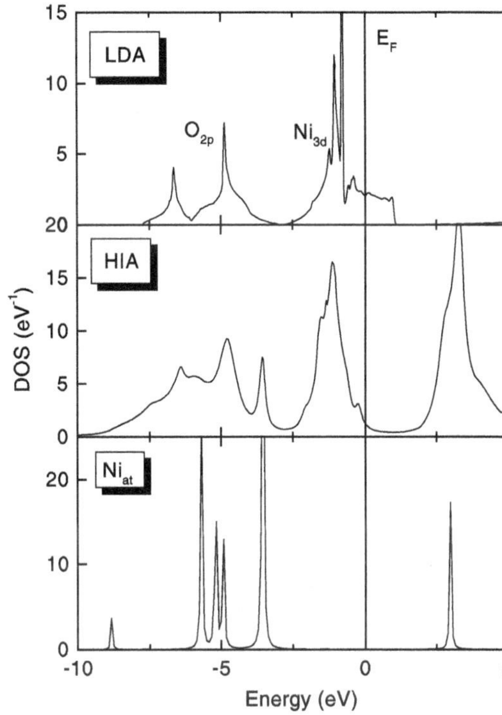

Fig. 8.15 Density of states for paramagnetic nickel oxide in the LDA and HIA approximations as well as Ni-atom Green function. Reprinted figure with permission from Ref. [26]. Copyright 1998 by The American Physical Society

non-magnetic LDA calculations in the LMTO nearly orthogonal representation [27] for experimental crystal structures of NiO. The minimal basis set of s, p, d-orbitals for NiO corresponds to 18×18 matrix of the LDA Hamiltonian $h(\mathbf{k})$. The occupation number for correlated electrons are 8.4 electrons in the d-shell of Ni. Using the corresponding atomic self energy for Ni-atom the total DOS for NiO has been calculated. In Fig. 8.15 we compare the paramagnetic LDA results with HIA scheme. It is well known that paramagnetic LDA calculations cannot produce the insulating gap in nickel oxide: the Fermi level is located in the middle of the half-filled e_g bands [17]. In the HIA approximation there is a gap (or pseudogap in Fig. 8.15 due to temperature broadening) of the order of 3.5 eV even in this "nonmagnetic" state. This gap and the satellites at -5 and -8 eV are related to the structure of atomic Green function shown in the lower panel of Fig. 8.15.

8.4.2 Exchange Interactions in Transition-Metal Oxides

The values of the intersite exchange interaction parameters J_{ex} depend on the parameters of the electronic structure in a rather indirect implicit way. The developing of the good calculating scheme for exchange parameters is very important because the ab-initio calculation is often the only way to describe the magnetic properties

Table 8.1 Calculated and experimental values of energy gaps (eV), spin magnetic moments (μ_B) and intersite exchange interaction parameters J_{ex} (meV)

		LDA+U	LDA+$U^{(d+p)}$	Experiment
La_2CuO_4	E_g	0.7	2.0	2.0
	μ_B	0.45	0.68	0.60
	J_{ex}	−82.9	−100.9	−136
MnO	E_g	3.8	4.5	3.6, 3.8
	μ_B	4.51	4.59	4.58, 4.79
	J_{ex}^1	5.4	−5.4	−4.8, −5.4
	J_{ex}^2	−9.3	−5.1	−5.6, −5.9
NiO	E_g	1.8	2.8	4.0, 4.3
	μ_B	1.50	1.64	1.64, 1.77
	J_{ex}^1	−0.8	−0.2	−1.4
	J_{ex}^2	−23.2	−19.4	−19.0

of complicated compounds such as, for example, "spin-gap" systems [43]. Recently Solovyev and Terakura [44] did a very through analysis of the exchange interaction parameters for MnO calculated using different methods of electronic structure calculations. They used the positions of the Mn(3d)-spin-up and Mn(3d)-spin-down bands relative to the oxygen 2p states as adjustable parameters to fit the values of exchange interaction for the nearest and second Mn-Mn neighbors. Their results gave nearly the same splitting between Mn(3d)-spin-up and Mn(3d)-spin-down states as in standard LDA+U calculations (10.6 eV) but the position of those states relative to the oxygen band was shifted approximately on 3 eV up relative to the LDA+U case. It is practically the same as we have in our LDA+$U^{(d+p)}$ calculations, because with $U_p = 6$ eV the shift of the position Me(3d)-band relative to the oxygen O(2p)-band is equal to $U_p/2 = 3$ eV.

Comparison between LDA+U and LDA+$U^{(d+p)}$ calculated J_{ex} parameters and experimental data is presented in Table 8.1. J_{ex} were calculated from the Green function method as second derivatives of the ground state energy with respect to the magnetic moment rotation angle [19, 30, 45–48] as was described above. Again one can see that in general the LDA+$U^{(d+p)}$ gives better results than the LDA+U, especially for the MnO compound.

The change in the electronic structure, due to the U-corrections affect the superexchange interactions (see e.g. (8.3)). It is useful to compare the spin-wave spectrum for different U with experimental one for NiO (Fig. 8.16). We can see the improvement of theoretical spin-wave dispersion in magnetic oxide compare with the standard LSDA calculations, which overestimate exchange excitations by factor of three [49]. For reasonable value of $U = 9$–13 eV, theoretical spin-wave spectrum in the LDA+U scheme agree quite well with the experimental one.

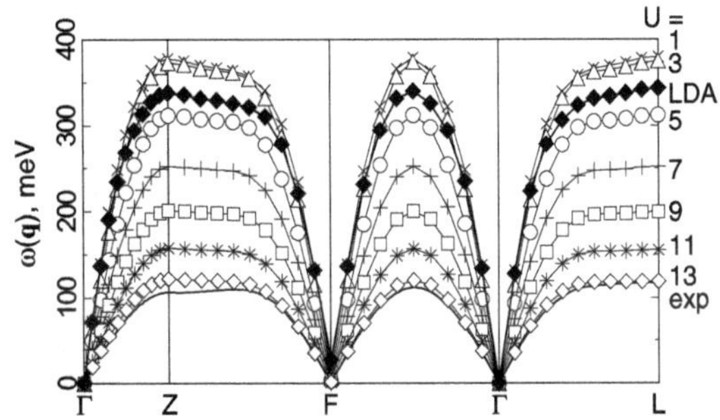

Fig. 8.16 The spin-wave spectrum of NiO as a function of U compared with LDA result and experiment. Reprinted figure with permission from Ref. [49]. Copyright (1998) by the American Physical Society

8.4.3 Orbital Magnetism: CoO

As we mention already at the beginning, novel phenomena caused by strong coupling among spin, orbital and lattice degrees of freedom are the central issue in the physics of transition-metal compounds for the last few years. One of the modes, when this coupling is mediated by the relativistic spin–orbit (S–O) interaction leads to the orbital magnetism, which is manifested in the magnetocrystalline anisotropy, magneto-optical effects, magnetic X-ray circular dichroism, etc. Due to the quenching effects in the crystal field, the orbital moments are expected to be well localized in the spherical potential region near atomic nuclei, and well described in terms of site-diagonal elements of the one-particle 10×10 density matrix in the basis of atomic-like ($3d$) orbitals $n_{\gamma_1\gamma_2} = \langle\gamma_1|\hat{n}(\mathbf{r},\mathbf{r}')|\gamma_2\rangle$ as $\langle\widehat{\mathbf{L}}\rangle = \mathrm{Tr}_{SL}(\widehat{\mathbf{L}}\hat{n})$, where $\widehat{\mathbf{L}}$ is the orbital angular momentum operator, $\gamma \equiv \{s, m\}$ is the joint index including spin (s) and azimuthal (m) counterparts, and Tr_{SL} denotes the trace over all s and m. The matrix $\hat{n} = \|n_{\gamma_1\gamma_2}\|$ generally consists of both spin-diagonal and spin-non-diagonal elements. The latter can be due to the S–O interactions or a non-collinear magnetic order.

In the rotationally invariant LDA+U scheme, one needs to include S–O interaction in the LDA functional and change correlated terms to the spin-density matrix form [50]:

$$E_U[\hat{n}] = \frac{1}{2}\sum_{\{\gamma\}}(U_{\gamma_1\gamma_3\gamma_2\gamma_4} - U_{\gamma_1\gamma_3\gamma_4\gamma_2})n_{\gamma_1\gamma_2}n_{\gamma_3\gamma_4} \tag{8.24}$$

It was shown that the renormalization can be described by retaining the atomic-like form for the electron–electron (e–e) interactions $U_{\gamma_1\gamma_3\gamma_2\gamma_4} = \langle m_1 m_3|\frac{1}{r_{12}}|m_2 m_4\rangle \times \delta_{s_1s_2}\delta_{s_3s_4}$ with the screened effective Slater integrals F^0, F^2, and F^4. If the orbital populations are integer (0 or 1), $\hat{n}^2 = \hat{n}$ holds. Then, an analog of two Hund's rules can be derived from $E_{\mathrm{HF}}[\hat{n}]$: first, the s-dependent occupation is driven by J; second, the m-dependent occupation is driven by $B = \frac{1}{441}(9F^2 - 5F^4)$. This is the atomic picture. In solids, however, the local orbital populations are fractional and shall be treated as independent variational degrees of freedom.

Let us illustrate this scheme for the rock-salt oxide CoO, where the orbital moment is not necessarily quenched in the $\frac{2}{3}$ filled t_{2g} manifold [51]. The antiferro-

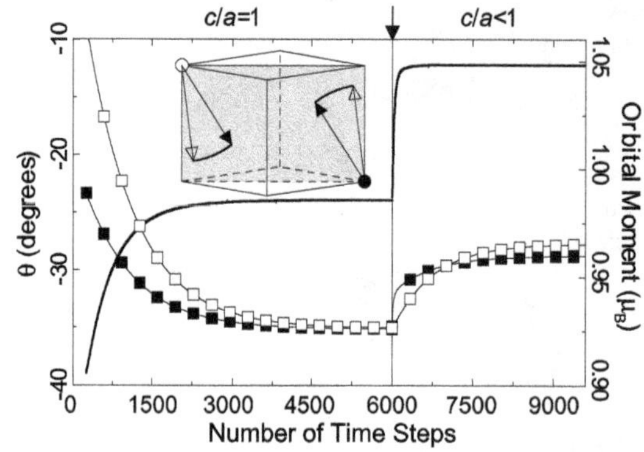

Fig. 8.17 Relaxation to the new magnetic equilibrium after turning on the S–O interaction in CoO: Orbital moment (*full line*) and deviations of spin (*white squares*) and orbital (*black squares*) magnetic moments from the [001] axis. The *inset* shows trajectories of the spins attached to magnetically different Co sites in the plane $(1\bar{1}0)$. *Open* and *filled arrows* correspond to the initial and final states. After reaching the equilibrium, a small tetragonal distortion $c/a = 0.988$ has been turned on at the point shown by the *arrow*. Reprinted figure with permission from Ref. [50]. Copyright (1998) by the American Physical Society

magnetic spin order additionally lowers the cubic symmetry of CoO to the trigonal one, resulting in complicated anisotropy effects.

The results of our numerical calculations are shown on Fig. 8.17. We start with a self-consistent LDA+U solution without S–O interaction where spins can take an arbitrary direction and there is no orbital moment. With turning on the S–O interaction, a typical relaxation process to the new equilibrium state as a function of iteration steps is shown in Fig. 8.17, where we used $U = 8$ eV, $J = 1$ eV and $B = 0.1J$, suggested by the constraint-LSDA calculations. On approaching the equilibrium, the orbital moment grows at the Co site and is stabilized between two high-symmetry directions [001] and $[\overline{11}1]$, causing a similar reorientation of the spin counterpart. The orbital instability is directly related with the appearance of the band gap in CoO. Once the band gap opens when U varies in the wide range from 2.2 to 8 eV, the orbital moment becomes well localized and the angle θ is stabilized between $-29°$ and $-35°$. On the contrary, $U = 0$ closes the band gap, and aligns magnetic moments parallel to the cube diagonal. Finally, the magnetostriction is responsible for the tetragonal deformation in CoO in the direction $c/a < 1$ which further enhances the orbital magnetic moment (Fig. 8.17).

8.4.4 Transition Metal Perovskite: $LaMeO_3$

The transition metal perovskite $LaMeO_3$ presents an interesting group of magnetic oxide with very rich properties accompanying the metal–insulator transition and related to CMR-phenomena. Traditionally ferromagnetism in the mixed manganite system $La_{1-x}D_xMnO3$ (D = Ca, Sr, and Ba) for $0.2 < x < 0.4$ is related to the

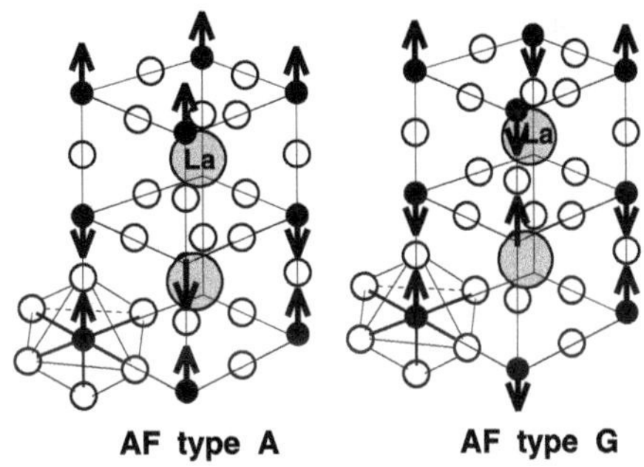

Fig. 8.18 The crystal and antiferromagnetic structure of perovskite $LaMnO_3$

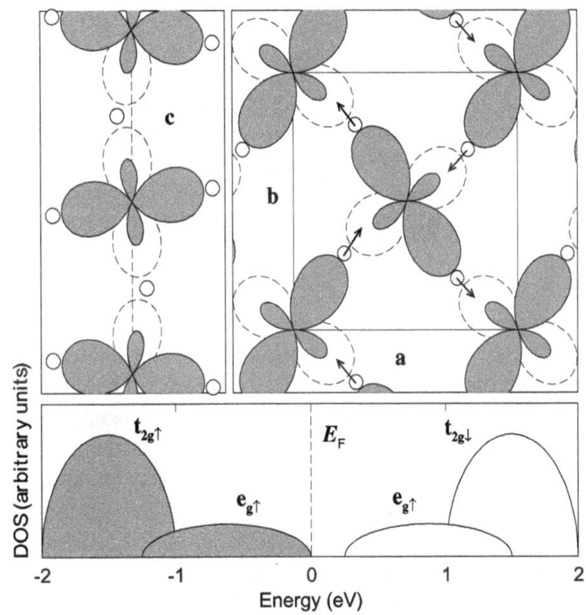

Fig. 8.19 *Upper panel*: Orbital ordering along **c** axis and in **a**–**b** plane of orthorhombic $LaMnO_3$ lattice. *Shaded* orbits denote occupied e_g states of $3x^2 - r^2$ and $3z^2 - r^2$ symmetry, and orbits with *broken lines* denote empty e_g states of the $y^2 - z^2$ and $x^2 - z^2$ symmetry. Local displacements of the oxygen atoms are shown by *arrows*. *Lower panel*: Schematic position of the $t_{2g\uparrow}$, $t_{2g\downarrow}$ and $e_{g\uparrow}$ band split by Jahn–Teller distortion in the LDA calculations. Reprinted figure with permission from Ref. [52]. Copyright (1996) by the American Physical Society

"double-exchange" model of Zener: Mn^{3+}–O–Mn^{4+}, while antiferromagnetism of the undoped system, $x = 0$, corresponds to Anderson superexchange (Fig. 8.18). The ferromagnetic sign of the double-exchange interactions is easy to understand in connection with Kugel–Khomskii exchange in degenerate e_g state (Fig. 8.8), since Mn^{3+} ion is exactly double degenerate $t_{2g}^3 e_g^1$ case. The recent investigation shows that the crystal distortion (Table 8.2) also plays an important role in magnetic and electronic properties of manganites. In order to see how the crystal distortion (due to cooperative Jahn–Teller effect for orbital degenerate e_g^1 of Mn^{3+} ion) related to insulating properties, we show in Fig. 8.19 the scheme of the LDA-energy band in orbital ordered state [52]. The crystal structure of $LaMnO_3$ has two principal types of distortion: the local tetragonal distortion of the oxygen atoms around each Mn-site (the Jahn–Teller distortion) and small tilting of MnO_6 octahedra resulting in the orthorhombic superstructure with four formula unit in the primitive cell. The

Table 8.2 Crystal and magnetic structure of some complex oxides

Compound	Crystal structure	Magnetic structure
$LaTiO_3$	orthorhombic	AFM–G
$LaVO_3$	monoclinic	canted AFM
$LaCrO_3$	orthorhombic	AFM–G
$LaMnO_3$	orthorhombic	AFM–A
$LaFeO_3$	orthorhombic	AFM–G
$LaCoO_3$	rhombohedral	paramagnetic
$LaNiO_3$	rhombohedral	paramagnetic
$LaCuO_3$	tetragonal	paramagnetic

strength of the crystal distortion appears to be sufficient to split double degenerate e_g states around the Fermi level and create a band gap and the orbital ordering. Moreover the calculated exchange interactions change the sign from ferromagnetic in the undistorted lattice, to the proper antiferromagnetic in the orbital ordered orthorhombic crystal (Table 8.2).

Finally we discuss the general trends in the electronic structure of magnetic perovskite oxides. On Fig. 8.20 we present the band structure of $LaMeO_3$ series (Me = Ti, V, Cr, Mn, and Fe) calculated in the LSDA and LDA+U approaches with two type of the U-corrections. The left panel corresponds to the U-corrections added only to the localized t_{2g} states of Me^{3+}-ion, since the e_g states in the perovskite structure are much more delocalized [53]. The right panel corresponds to the standard LDA+U scheme with the U-corrections applied to all d-states. Since the U-corrections to the t_{2g} states are much smaller due to the effective screening from e_g states: $U_{t2g} \approx (1 + \delta n_{e_g}/\delta n_{t_{2g}})U$ and general tendency of the local charge conservation: $\delta n_{e_g}/\delta n_{t_{2g}} < 0$, this LDA+$U$ scheme (left panel in Fig. 8.20) is much closer to the LSDA results. For $LaMnO_3$, on the other hand, both LDA+U methods give similar electronic structure, because Mn^{3+} ion has a large Hund splitting. Nevertheless, the LSDA method and the "$U_{t_{2g}}$ scheme" underestimate the value of the band gap in comparison with the experimental gap value 1.1 eV.

8.5 Highly Correlated Metallic Oxides

8.5.1 Doped Mott Insulators

The LDA+DMFT approach was successfully applied to study electronic structure of correlated oxide with perovskite structure [54]. Transition metal perovskites have been studied for decades because of their unusual electronic and magnetic properties arising from narrow $3d$ bands and strong Coulomb correlations. The $3d^1$ perovskites are particularly interesting since, despite their lack of multiplet structure, similar materials have very different properties: $SrVO_3$ and $CaVO_3$ are correlated metals, while $LaTiO_3$ and $YTiO_3$ are Mott insulators with gaps of, respectively, 0.2 and 1 eV [9].

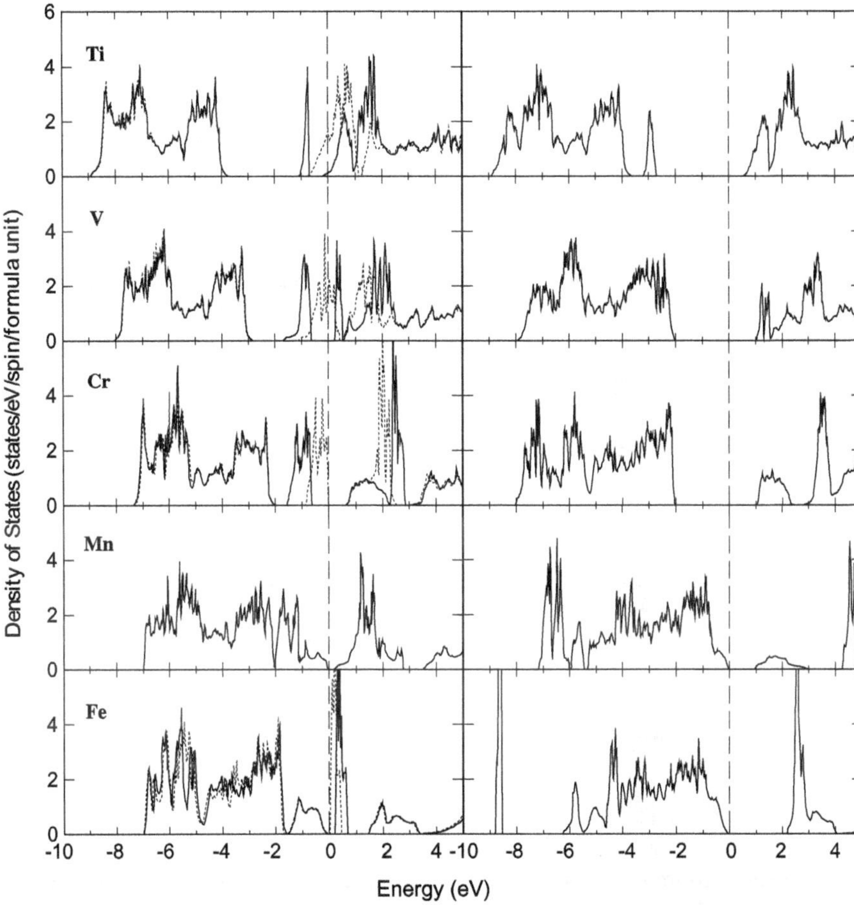

Fig. 8.20 Density of states for $LaMeO_3$ perovskite obtained with LDA+$U_{t_{2g}}$ (*left panel*) and with LDA+U (*right panel*) method. *Dotted line* in the *left panel* corresponds to LDA results. Position of the Fermi level shown by *vertical dashed line*. Reprinted figure with permission from Ref. [53]. Copyright (1996) by the American Physical Society

In Fig. 8.21 we show the DMFT spectral functions together with the LDA total DOS. For cubic $SrVO_3$ we reproduce the results of previous calculations [55]: the lower Hubbard band (LHB) is around -1.8 eV and the upper Hubbard band (UHB) around 3 eV. Going to $CaVO_3$, the quasiparticle peak loses weight to the LHB, which remains -1.8 eV, while the UHB moves down to 2.5 eV. These results are in good agreement with photoemission data [9] and show that $SrVO_3$ and $CaVO_3$ are rather similar, with the latter slightly more correlated. From the linear regime of the self-energy at small Matsubara frequencies we estimate the quasiparticle weight to be $Z \simeq 0.45$ for $SrVO_3$ and $Z \simeq 0.29$ for $CaVO_3$. For a **k**-independent self-energy as assumed in DMFT, this yields $\frac{m^*}{m} = \frac{1}{Z} \simeq 2.2$ for $SrVO_3$ and $\simeq 3.5$ for $CaVO_3$, in reasonable agreement with experimental values obtained from optical conductivity

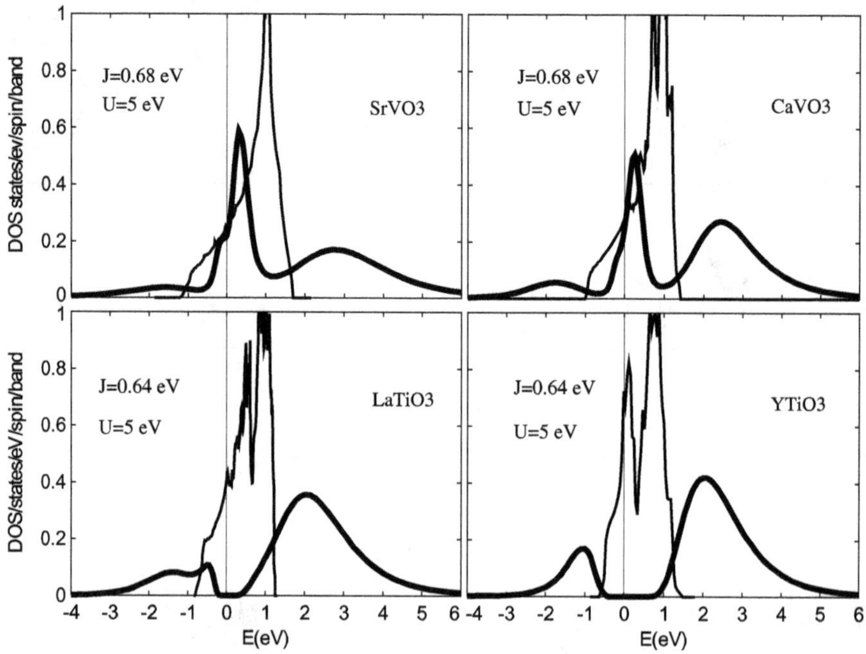

Fig. 8.21 LDA+DMFT spectral function for different perovskite compounds at $T = 770$ K (*thick line*) and LDA DOS (*thin line*). Reprinted with permission from Ref. [54]. Copyright 1998 by the American Physical Society

(2.7 and 3.6). For $LaTiO_3$ and $YTiO_3$ the LHB is around -1.5 eV, in accord with photoemission [9], but despite very similar bandwidths ($W = 2.1$ and 1.9 eV), the gaps are very different, 0.3 and 1 eV, and this agrees with experiments. This result shows that in $3d^1$ systems the Mott transition and the gap-size depend not only on U/W, but on the full band structure.

Doping of $LaTiO_3$ by a very small value of Sr (few percent) leads to the transition to a paramagnetic metal with a large effective mass. As photoemission spectra of this system also show a strong deviation from the noninteracting electrons picture, $La_{1-x}Sr_xTiO_3$ is regarded as an example of strongly correlated metal. Photoemission spectroscopy of the early transition metal oxides provides a direct tool for the study of the electronic structure of strongly correlated materials. A comparison of the experimental photoemission spectra [56, 57] with the results obtained from LDA and LDA+DMFT(QMC) [55] at 1000 K are shown in Fig. 8.22. To take into account the uncertainty in U, we present the results for $U = 3.2$, 4.25 and 5 eV. All spectra are multiplied with the Fermi step function and Gaussian-broadened with a broadening parameter of 0.3 eV to simulate the experimental resolution [56, 57]. The LDA band structure calculation clearly fails to reproduce the broad band observed in the experiment at 1–2 eV below the Fermi energy [56, 57]. Taking the correlations between the electrons into account, this lower band is easily identified as the lower Hubbard band whose spectral weight originates from the quasiparticle band

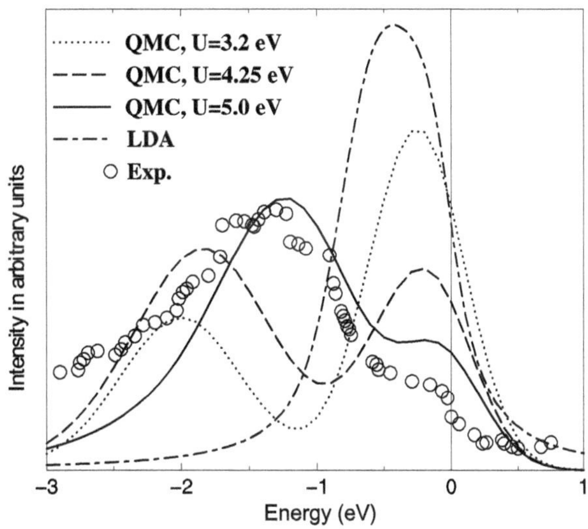

Fig. 8.22 Comparison of the experimental photoemission spectrum [56, 57], the LDA result, and the LDA+DMFT(QMC) calculation for $LaTiO_3$ with 6 % hole doping and different Coulomb interactions $U = 3.2, 4.25$, and 5 eV. Reprinted from Ref. [55] with kind permission from Springer Science+Business Media

at the Fermi energy and increases with U. The best agreement with experiment concerning the relative intensities of the Hubbard band and the quasiparticle peak and, also, the position of the Hubbard band is found for $U = 5$ eV. The value $U = 5$ eV is still compatible with the ab-initio calculation of this parameter. One should also note that the photoemission experiment is sensitive to surface properties. Due to the reduced coordination number at the surface, the bandwidth is likely to be smaller and the Coulomb interaction to be less screened, i.e., larger. Both effects make the system more correlated and, thus, might also explain why better agreement is found for $U = 5$ eV. Besides, the polycrystalline nature of the sample and, also, spin and orbital [58] fluctuation, not taking into account in the LDA+DMFT approach, could further reduce the quasiparticle weight.

The LDA+DMFT approach not only explains the existence of the lower Hubbard band in doped $LaTiO_3$, but also, in contrast to LDA, reproduces the qualitative picture of the spectral weight transfer from the quasiparticle band to the lower Hubbard band, the position of the lower Hubbard band, and the narrowing of the quasiparticle band.

8.5.2 Metal–Insulator Transition in Ti_2O_3

The complicated electronic properties of TMO closely related with delicate balance between electron–electron interactions and chemical bonding. Let us discuss the general trends in competition correlation effects and hybridization (Fig. 8.23). In the limit of weak Coulomb interactions there is strong bonding–antibonding splitting among the d-orbitals. In the opposite case, where the on-site electron–electron interactions U is much stronger than d–d hybridization t_{ij} the wave-function will be localized with large renormalization of intersite hybridization and reduced bonding–

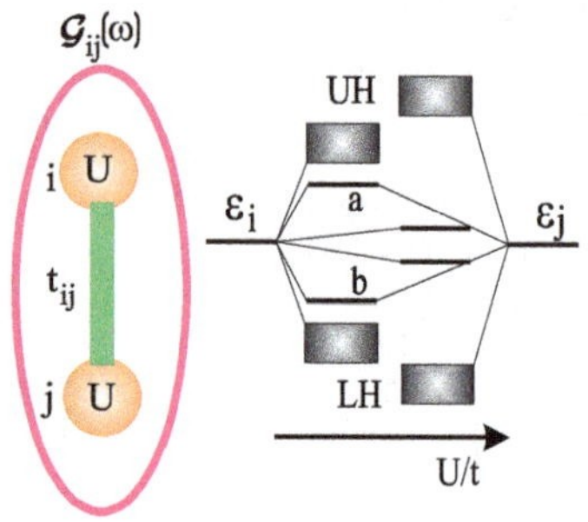

Fig. 8.23 General trends in correlated electronic materials with the strong chemical d–d bonds as function on interacting strength U/t

antibonding splitting among the quasiparticle bands. In addition, there are non-quasiparticle, so-called Hubbard bands: Lower-Hubbard (LH) and Upper-Hubbard (UH) bands with the splitting of the order of Coulomb interaction U. Therefore the strong interactions reduced the chemical bonding and otherwise in the limit of strong bonding–antibonding splitting, the correlation effects largely reduced. Here we discuss this competition of correlation effects and chemical bonding for the example of Ti_2O_3 [59].

The complicated electronic structure and the nature of the metal–insulator transition (MIT) in Ti_2O_3 and V_2O_3 has been the object of intensive experimental and theoretical investigation over the past half century [9]. Recent progress in high-energy photoemission spectroscopy [60] and correlated electrons dynamical-mean field theory (DMFT) [24] has shed new light on the MIT in V_2O_3. It has been shown that a realistic description of the metallic and insulating phases of V_2O_3 can be obtained from the combination of a band structure scheme with the local electron–electron interaction given from DMFT [61]. The correlation effects in Ti_2O_3 are less clear but angle resolved photoemission experiment [62] shows a strong reduction of the Ti $3d$-bandwidth compared to band structure calculations. The important question is related to the mechanism of the small, about 0.1 eV, semiconductor band-gap formation. The generally accepted view is that the MIT is related to the decrease of the c/a ratio in rhombohedral Ti_2O_3 and the formation of a Ti–Ti pair along z-axis [63]. Below the broad (almost 250 K in width) MIT at around 470 K the Ti–Ti pair distance is seen to decrease without any change of the rhombohedral structure or the formation of long-range antiferromagnetic order [64]. This is in contrast to the case of V_2O_3 where the V–V pair distance increases within a monoclinic distortion in the antiferromagnetic phase [9].

Ti_2O_3 has an α-Al_2O_3 corundum structure (Fig. 8.24) in the metallic and insulating phases with two formula units per rhombohedral cell [65, 66]. Each Ti atom is surrounded by the octahedron of oxygens leading to the large t_{2g}–e_g^σ splitting. The trigonal distortion gives an additional splitting of t_{2g} bands into e_g^π–a_{1g} states and a_{1g} subbands of Ti–Ti pair form strong bonding–antibonding counterparts (Fig. 8.24). In principle, the large decrease of the Ti–Ti distance could split further an occupied single-degenerate a_{1g} states from a double-degenerate e_g^π states of t_{2g} subband and form the insulating d^1 configuration of this Ti compound. Nevertheless, state of the art LDA calculations have shown that for reasonable Ti–Ti pair distances Ti_2O_3 will stay metallic [67].

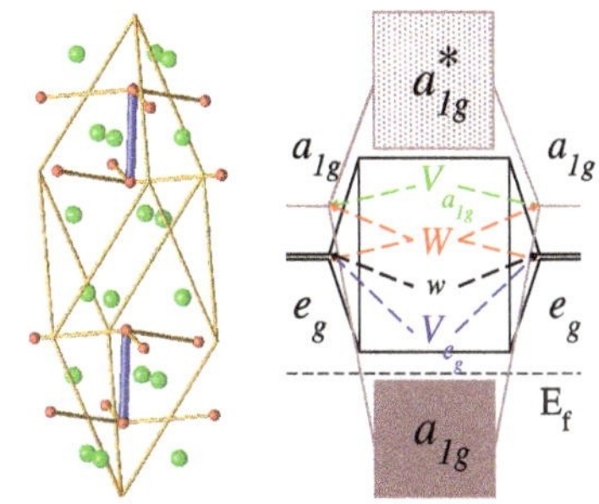

Fig. 8.24 *Left*: Rhombohedral unit cell of Ti_2O_3 corundum structure. Titanium ions are indicated by the *red color*, oxygens by *green*, and the pair of Ti atoms in z direction by *blue*. *Right* is a schematic representation of the t_{2g} splitting in Ti_2O_3 (*top part*)

In order to investigate the role of electron–electron interactions in the formation of this insulating low-temperature phase one needs an accurate estimation of the a_{1g} and e_g bandwidths in this complex structure [68]. For example a simple free $[Ti_2O_9]^{12-}$ cluster mean-field investigation can easily produce a gap due to drastic underestimation of the a_{1g} and e_g bandwidths [69]. On the other hand a more accurate band structure calculation within the unrestricted Hartree–Fock approximation results in a large gap antiferromagnetic state [70]. Thus it is crucial to use both the correct Green-function embedding of the Ti–Ti pairs as well as a more accurate treatment of the electron–electron interaction.

The role of metal–metal pair formation and the "molecular" versus band pictures of the electronic structure have attracted much attention in these compounds [71]. The combination of a strong on-site Coulomb interaction and the large anisotropy between the hopping parameters in and perpendicular to the pair direction can favor a localized molecular-orbital picture of the insulating phase. However, realistic tight-binding calculations for V_2O_3 have shown the importance of long-range hopping parameters [72]. It is also unclear how good an on-site approximation is for the electron–electron interaction. Since the pair forms a natural "molecular like" element in the corundum-type Ti_2O_3 structure it might be expected that non-local electron correlations are important in this system. Thus an approach which combines pair and beyond pair hopping with non-local electron interactions would seem to be ideal for this problem.

We apply the cluster DMFT (CDMFT) scheme [33, 34], which contains all the physics of correlated pairs in crystals to determine the origin of the insulating phase and the MIT in Ti_2O_3. A numerically exact multi-orbital Quantum Monte-Carlo (QMC) scheme is deployed for the solution of the CDMFT problem and an accurate first principles tight-binding parametrization used for the one electronic structure. Our strategy here is to investigate the gap formation using single site [25, 26] and cluster LDA+DMFT with only local correlations included. We then deploy the full non-local CDMFT and in this way are able to directly elucidate the impact non-local Coulomb interactions have on the physics. We show that the competition between strong bonding within the Ti–Ti pair and localization from correlation effects leads to the unique situation of the small semiconducting gap structure in Ti_2O_3 oxide and that non-local Coulomb correlations are of crucial importance for the physics of these small gap insulators.

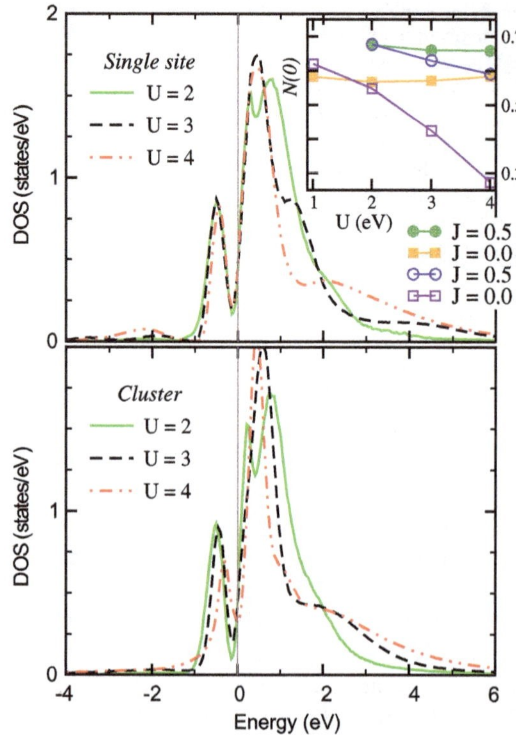

Fig. 8.25 DOS for the single site (*upper panel*) and cluster (*lower panel*) DMFT calculations with different values of the Coulomb repulsion U and $J = 0.5$ eV. *Inset*: $N(0)$ *versus* Coulomb parameter. *Filled dark green circles*: DMFT results with $J = 0.5$ eV, *filled orange squares*: DMFT with $J = 0$ eV. *Open blue circles* and *violet squares* are CDMFT with $J = 0.5$ and 0 eV, respectively

We start with the orthogonal LDA Hamiltonian $H^{\mathrm{LDA}}_{mm'}(\mathbf{k})$ in the massively downfolded Nth order muffin-tin orbital representation [73] (m corresponds to the 12 t_{2g} orbitals of two Ti–Ti pairs in rhombohedral unit cell) and include different Coulomb interactions (see Fig. 8.24). DMFT results for the local and non-local Coulomb interactions are presented in Figs. 8.25, 8.26.

The bare LDA density of states (DOS) is shown in Fig. 8.26 by the dashed lines for the low temperature structure (LTS, ~300 K [65, 66]) and high temperature structural (HTS, ~870 K [65, 66]) data on the upper and lower panels, respectively. Both LTS and HTS electronic structures are metallic within the LDA scheme. The a_{1g} subband (green dashed line in Fig. 8.26) has a strong bonding–antibonding splitting in contrast to the e_g^{π} subbands (red dashed line). The bandwidth of the HTS is approximately 2.8 eV and smaller than the bandwidth of the LTS (3.2 eV) due to the reduction of the $t_{a_{1g},a_{1g}}$ hopping from −0.85 to −0.63 eV.

The Hamiltonian and the self-energy matrix have the following super-matrix form corresponding to the symmetry of two Ti–Ti pairs in the unit cell:

$$\begin{pmatrix} \mathbf{H}_{11}+\Sigma_{11} & \mathbf{H}_{12}+\Sigma_{12} & \mathbf{H}_{13} & \mathbf{H}_{14} \\ \mathbf{H}_{21}+\Sigma_{21} & \mathbf{H}_{22}+\Sigma_{11} & \mathbf{H}_{23} & \mathbf{H}_{24} \\ \mathbf{H}_{31} & \mathbf{H}_{32} & \mathbf{H}_{33}+\Sigma_{11} & \mathbf{H}_{34}+\Sigma_{12} \\ \mathbf{H}_{41} & \mathbf{H}_{42} & \mathbf{H}_{43}+\Sigma_{21} & \mathbf{H}_{44}+\Sigma_{11} \end{pmatrix} \tag{8.25}$$

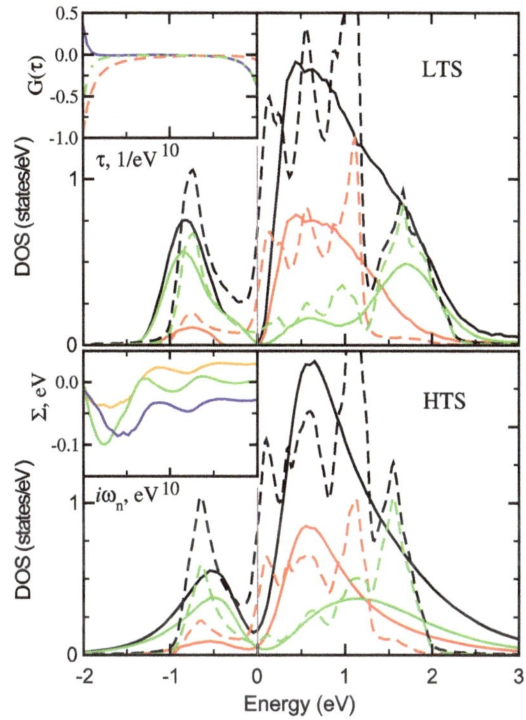

Fig. 8.26 Partial and total CDMFT (*solid line*) compared to the LDA (*dashed*) DOS with $W = 0.5$ eV and $V_{a_{1g}} = V_{e_g} = 0$. Total DOS are shown by *black*, the e_g states by *red*, and a_{1g} states by *green*. On the *upper panel* the low temperature structure and $\beta = 20$ eV^{-1} are used. For the *lower panel* the high temperature structure and $\beta = 10$ eV^{-1} are used. The diagonal $G_{a_{1g}}$, G_{e_g} and the largest $G_{a_{1g}-a_{1g}}$ off-diagonal Green functions are shown in the *upper inset* by the *green*, *red* and *blue colors*, respectively. In the *lower inset* the Re $\Sigma_{a_{1g},a_{1g}}$ with intersite Coulomb interactions are shown by *blue* (we have added 0.25 eV to this quantity to make it visible in the same region), Re $\Sigma'_{a_{1g},a_{1g}}$ without intersite Coulomb interactions by *orange* and Im $\Sigma_{a_{1g}}$ are shown by *green*

where $\mathbf{H}_{ij}(\mathbf{k})$ and $\Sigma_{ij}(\omega_n)$ are 3×3 matrices for the t_{2g} states and Σ_{11} and Σ_{12} correspond to the intrasite and intersite contributions to the self-energy, respectively.

Firstly, in Fig. 8.25 we show the total DOS for both conventional single site and cluster DMFT where only local electron correlations have been included. The QMC simulation has been carried out for $\beta = 20$ eV^{-1}, which corresponds to a temperature of $T \simeq 580$ K which is on the border of the MIT. In the upper panel of Fig. 8.25 are shown the DMFT results with $U = 2, 3, 4$ eV and exchange parameter $J = 0.5$ eV. For all values of Coulomb interactions there is a peak below the Fermi level at around -0.5 eV, predominantly of a_{1g} character with in all cases the same intensity. Above the Fermi level there are two peaks. The first is at 0.5 eV and has e_g character while the other peak is strongly dependent on the Coulomb parameter and can be associated with an upper Hubbard band. A lower Hubbard band can be seen at around -2 eV. We see that for all values of U the shape of the pseudogap is unchanged and the system remains metallic. On the lower part of Fig. 8.25 the results of the CDMFT calculation are shown for the same values of the Coulomb and exchange parameters. The general structure of the DOS is seen to be similar to the single site calculation, however, one may note interesting differences. The lower a_{1g} quasiparticle band is decreased in intensity and shifted towards the Fermi level from -0.6 eV to -0.3 eV on increasing U from 2 to 4 eV. This has the result that for $U = 4$ eV the pseudogap is now located directly at Fermi level, whereas for other U-values and for all DMFT results it lies on the slope of the quasiparticle peak.

Using the temperature DOS at the Fermi level, defined as $N(0) \equiv -\operatorname{Im} G(\omega_0)/\pi$ with $\omega_0 = \pi T$ we are able to estimate the critical value of U. This is indicated in the inset in the upper panel of Fig. 8.25. We see that for the single site calculations $N(0)$ depends weakly on U and the system will remain metallic up to very large values, about 8 eV, of the Coulomb parameter. On the other hand for the cluster calculation $N(0)$ is seen to decrease strongly as a function of U for both values of exchange parameter, and the critical value for an insulating solution is now lower at $U \sim 5$–6 eV. As expected for the d^1 configuration the finite value of the exchange parameter effectively decreases the Coulomb interaction matrix. We see the single site results are in greater contradiction to the experiment as compared to LDA (see Fig. 8.26): the local Coulomb interaction leads to the reduction of the bonding–antibonding splitting of the a_{1g} subband and this acts to suppress gap formation. On the other hand in the cluster case a small semiconducting gap is developed for large U due to dynamical antiferromagnetic correlation [74] in the Ti–Ti pair.

Nevertheless, using the DMFT scheme with only local correlations there remains a dramatic absence of gap formation in Ti_2O_3. We now deploy the full non-local correlation in CDMFT which is normally regarded to the effect of non-local correlations on low and high temperature electronic structure [33, 34]. We have used different values of the non-local Coulomb parameters and found that the most important interactions correspond to non-diagonal W terms [75]. For both structures we have chosen values of $U = 2$ eV and $J = 0.5$ eV which are close to those from constrained LDA estimations [53], while the off diagonal Coulomb parameters have been chosen as $V_{e_g} = 0.6$, $V_{a_{1g}} = 1.0$ and $W = 0.5$ eV. On the upper panel of Fig. 8.26 is shown the total and partial DOS for $\beta = 20$ eV^{-1}. Shown also is the LDA result. One can see that for the reasonable parameters chosen we can reproduce the correct value of the semiconducting gap ~0.1 eV while keeping the bonding–antibonding splitting on the LDA level. In the lower panel the high temperature metallic solution corresponding to $\beta = 10$ eV^{-1} is shown. Here we emphasize that the proper inclusion of the structural effect [76] on the LDA level is important as evinced by the fact that for $\beta = 20$ eV^{-1} and high temperature Hamiltonian we again obtain a metallic solution. The e_g states are similar for both LTS and HTS calculations with a small shift of occupied part and clearly seen gap in LTS case. However, the difference between the LTS and HTS phases is more pronounced for the a_{1g} states. The bonding–antibonding splitting in the LTS is about 2.5 eV whereas in the HTS case it is only 1.5 eV. The occupied a_{1g} states in the LTS phase are shifted down opening the insulating gap. The important difference between the large U and small U plus non-local Coulomb interaction is the absence of well defined Hubbard bands. This absence makes possible a critical test of the theory proposed here, and thus it would be very interesting for photoemission experiments to check the existence or not of a lower Hubbard band at around -2 eV.

We have shown that the cluster LDA+DMFT calculation with a moderate Coulomb repulsion among the a_{1g} orbitals is essential to produce the high temperature semimetallic state and the low temperature insulating state. To understand the role play of the intersite Coulomb interaction we focus on the quantity $t_{a_{1g},a_{1g}} + \mathrm{Re}\,\Sigma_{a_{1g},a_{1g}}(i\omega)$ which we can interpret as a frequency dependent "effective $a_{1g} - a_{1g}$ hopping" which describes the hopping matrix element in the titanium pair. We find that this quantity is surprisingly frequency dependent (see lower inset of Fig. 8.26).

The main role of the intersite Coulomb interaction is dynamic and results in the effective a_{1g}–a_{1g} hopping that *changes* with the frequency. This enhancement produces a strong level repulsion of the bonding–antibonding a_{1g} levels, lowering the a_{1g} level relative to the e_g level at the low frequency. This effect combined with a small narrowing of the a_{1g} band opens the e_g–a_{1g} band gap which results in the insulating state. We checked that this enhancement of the effective hopping as frequency is decreased is absent if we turned off the intersite Coulomb repulsion.

8.6 Conclusions

We have discussed the electronic structure magnetic properties and metal–insulator transition in TMO. The unique feature of these compounds related to the fact that the spin, charge and orbital degrees of freedom plays an important role in all physical properties. While the local density approximation is quite reasonable for the electronic structure of metallic oxide, the additional Hubbard-like correlation is important for energy spectrum of insulating magnetic oxides. The most difficult problem is the doping dependence of the electronic structure for the charge-transfer insulators and the joint efforts of ab initio calculations and many-body model approaches remains the most efficient way for electronic structure of magnetic oxides [9].

The LDA+U method was proven to be a very efficient and reliable tool in calculating the electronic structure of systems where the Coulomb interaction is strong enough to cause localization of the electrons. It works not only for nearly core-like $4f$-orbitals of rare-earth ions, where the separation of the electronic states on the subspaces of the infinitely slow localized orbitals and infinitely fast itinerant ones is valid, but also for such systems as transition metal oxides (NiO), where $3d$-orbitals hybridize quite strongly with oxygen $2p$-orbitals. In spite of the fact that the LDA+U is a mean-field approximation which is in general insufficient for the description of the metal–insulator transition and strongly correlated metals, in some cases, such as the metal–insulator transition in FeSi and $LaCoO_3$, LDA+U calculations gave valuable information by giving insight into the nature of these transitions. However, in general LDA+U overestimates the tendency to localization as is well known for Hartree–Fock type methods. The main advantage of LDA+U method over model approaches is its "first principle" nature with a complete absence of adjustable parameters. Another asset is the fully preserved ability of LDA-based methods to address the intricate interplay of the electronic and lattice degrees of freedom by computing total energy as a function of lattice distortions. When the localized nature of the electronic states with Coulomb interaction between them is properly taken into account, this ability allows to describe such effects as polaron formation and orbital polarization. As the spin and charge density of the electrons is calculated self-consistently in the LDA+U method, the resulting diagonal and off-diagonal matrix elements of one-electron Hamiltonian could be used in more complicated calculations where many-electron effects are treated beyond mean-field approximation.

At the same time, all the most subtle and interesting many-body effects (such as spectral weight transfer, Kondo resonances, and others) are beyond the LDA+U approach. To describe these effects a dynamical character of the effective potential acting on the electrons should be taken into account, or, in other words, we have to work with the Green function instead of the density matrix and with the self-energy instead of the effective exchange-correlation potential. The LDA+DMFT method seems to be effective and useful form of such approaches. In particular, in contrast with the LDA+U method it is not necessary to consider only the magnetically or orbitally ordered phases to describe the Mott insulator states, spectral weight effects are taken into account. These results for the metal–insulator transition in complex transition metal oxides demonstrate that the dynamical mean field theory does give us an opportunity to unify the many-body theory with the practice of first-principle calculations of the electronic structure and properties for real materials.

References

1. Bazylinski DA, Frankel RB (2004) Nat Rev Microbiol 2:217
2. Grünberg P et al. (1996) Phys Rev Lett 57:2442
3. Ramires AP (1997) J Phys Condens Matter 9:8171
4. Bednorz JG, Müller KA (1986) Z Phys B 64:189
5. Damascelli A, Hussain Z, Shen Z-X (2003) Rev Mod Phys 75:473
6. Shull CG, Strauser WA, Wollan EO (1951) Phys Rev 83:333
7. Mott NF (1974) Metal–insulator transition. Taylor & Francis, London
8. Scalapino DJ (1995) Phys Rep 250:329
9. Imada M, Fujimori A, Tokura Y (1998) Rev Mod Phys 70:1039
10. Zaanen J, Sawatzky GA, Allen JW (1985) Phys Rev Lett 55:418
11. Kugel KI, Khomskii DI (1982) Usp Fiz Nauk 136:621 [(1982) Sov Phys Usp 25:231]
12. Hohenberg P, Kohn W (1964) Phys Rev 136:B864.
13. Kohn W, Sham LJ (1965) Phys Rev 140:A1133
14. Jones RO, Gunnarsson O (1989) Rev Mod Phys 61:689
15. Svane A, Gunnarsson O (1990) Phys Rev Lett 65:1148
16. Anisimov VI, Gunnarson O (1991) Phys Rev B 43:7570
17. Anisimov VI, Aryasetiawan F, Lichtenstein AI (1997) J Phys Condens Matter 9:767
18. Anisimov VI, Zaanen J, Andersen OK (1991) Phys Rev B 44:943
19. Liechtenstein AI, Anisimov VI, Zaanen J (1995) Phys Rev B 52:R5467
20. Andersen OK (1975) Phys Rev B 12:3060

21. Metzner W, Vollhardt D (1989) Phys Rev Lett 62:324.
22. Georges A, Kotliar G (1992) Phys Rev B 45:6479
23. Jarrell M (1992) Phys Rev Lett 69:168
24. Georges A, Kotliar G, Krauth W, Rosenberg MJ (1996) Rev Mod Phys 68:13
25. Anisimov VI, Poteryaev AI, Korotin MA, Anokhin AO, Kotliar G (1997) J Phys Condens Matter 9:7359
26. Lichtenstein AI, Katsnelson MI (1998) Phys Rev B 57:6884
27. Andersen OK, Jepsen O (1984) Phys Rev Lett 53:2571
28. Takegahara K (1992) J Phys Soc Jpn 62:1736.
29. Rozenberg MJ (1997) Phys Rev B 55:R4855
30. Katsnelson MI, Lichtenstein AI (2000) Phys Rev B 61:8906
31. Hirsch JE, Fye RM (1986) Phys Rev Lett 25:2521
32. Jarrell M, Gubernatis JE (1996) Phys Rep 269:133
33. Lichtenstein AI, Katsnelson MI (2000) Phys Rev B 62:R9283.
34. Kotliar G, Savrasov SY, Palsson G, Biroli G (2001) Phys Rev Lett 87:186401
35. Hettler MH, Tahvildar-Zadeh AN, Jarrell M et al. (1998) Phys Rev B 58:7475.
36. Hettler MH, Mukherjee M, Jarrell M et al. (2000) Phys Rev B 61:12739
37. Nekrasov IA, Korotin MA, Anisimov VI. cond-mat/0009107
38. Dederichs PH, Blügel S, Zeller R, Akai H (1984) Phys Rev Lett 53:2512
39. McMahan AK, Martin RM, Satpathy S (1988) Phys Rev B 38:6650
40. Hybertsen MS, Schlüter M, Christensen NE (1989) Phys Rev B 39:9028
41. Knotek ML, Feibelman PJ (1978) Phys Rev Lett 40:964
42. Sawatzky GA, Allen JW (1984) Phys Rev Lett 53:2239
43. Korotin MA, Elfimov IS, Anisimov VI, Troyer M, Khomskii DI (1999) Phys Rev Lett 83:1387
44. Solovyev IV, Terakura K (1998) Phys Rev B 58:15496
45. Liechtenstein AI, Katsnelson MI, Gubanov VA (1984) J Phys F 14:L125
46. Liechtenstein AI, Katsnelson MI, Gubanov VA (1985) Solid State Commun 54:327
47. Liechtenstein AI, Katsnelson MI, Antropov VP, Gubanov VA (1987) J Magn Magn Mater 67:65
48. Gubanov VA, Liechtenstein AI, Postnikov AV (1992) Magnetism and the electronic structure of crystals. Springer series in solid-state sciences, vol 98
49. Solovyev IV, Terakura K (1998) Phys Rev B 58:15496
50. Solovyev IV, Liechtenstein AI, Terakura K (1998) Phys Rev Lett 80:5758
51. Brandow B (1977) Adv Phys 26:651
52. Solovyev IV, Hamada N, Terakura K (1996) Phys Rev Lett 76:4825
53. Solovyev IV, Hamada N, Terakura K (1996) Phys Rev B 53:7158
54. Pavarini E, Biermann S, Poteryaev A, Lichtenstein AI, Georges A, Andersen OK (2004) Phys Rev Lett 92:176403
55. Nekrasov IA, Held K, Blümer N, Poteryaev AI, Anisimov VI, Vollhardt D (2000) Eur Phys J B 18:55
56. Fujimori A et al. (1992) Phys Rev Lett 69:1796.
57. Fujimori A et al. (1992) Phys Rev B 46:9841
58. Keimer B et al. (2000) Phys Rev Lett 85:3946
59. Poteryaev AI, Lichtenstein AI, Kotliar G (2004) Phys Rev Lett 93:086401
60. Mo S-K et al. (2003) Phys Rev Lett 90:186403
61. Held K et al. (2001) Phys Rev Lett 86:5345
62. Smith KE, Henrich VE (1988) Phys Rev B 38:5965
63. Goodenough JB (1972) Prog Solid State Chem 5:145
64. Moon RM, Riste T, Koehler WC, Abrahams SC (1969) J Appl Phys 40:1445
65. Abrahams SC (1963) Phys Rev 130:2230.
66. Rice CE, Robinson WR (1977) Acta Cryst B 33:1342
67. Mattheiss L (1996) J Phys Condens Matter 8:5987
68. Zeiger HJ (1975) Phys Rev B 11:5132
69. Nakatsugawa H, Iguchi E (1997) Phys Rev B 56:12931

70. Catti M, Sandrone G, Dovesi R (1997) Phys Rev B 55:16122
71. Shiina R, Mila F, Zhang F-C, Rice TM (2001) Phys Rev B 63:144422
72. Elfimov IS, Saha-Dasgupta T, Korotin MA (2003) Phys Rev B 68:113105
73. Andersen OK, Saha-Dasgupta T (2000) Phys Rev B 62:16219
74. Haule K, Rosch A, Kroha J, Wölfle P (2003) Phys Rev B 68:155119
75. Frésard R, Kotliar G (1997) Phys Rev B 56:2909
76. Lucovsky G, Sladek RJ, Allen JW (1977) Phys Rev B 16:4716

Chapter 9
Local Structure of Highly Spin Polarised Heusler Compounds Revealed by Nuclear Magnetic Resonance Spectroscopy

Sabine Wurmehl and Jürgen T. Kohlhepp

Abstract A key tool in the rational design of spin polarised materials is the precise control of the relationships between structure and physical properties, such as between structure and magnetism or transport properties. Thus, a sophisticated and comprehensive characterisation is required in order to understand, tune and control the macroscopic properties of spin polarised materials towards optimised performance in spintronics devices. Nuclear magnetic resonance spectroscopy (NMR) probes the local environments of the active nuclei and is based on the interaction of the spin of a nucleus with the effective field present at the nucleus. The local character of NMR arises from local contributions to the hyperfine field, namely the transferred field which depends on the nearest neighbour atoms and their magnetic moments. This enables NMR to study structural properties of bulk samples as well as of thin films of spin polarised materials. Recent results confirmed that NMR is a very suitable tool to reveal structural contributions and foreign phases in spin polarised materials which are very difficult to detect with other methods like, e.g., conventional X-ray diffraction. In this chapter, recent NMR studies of the local structure of various Heusler compounds will be presented and the impact of the NMR results on their potential for spintronics will be discussed.

9.1 Why Is the Knowledge of the Local Structure Crucial for Highly Spin Polarised Materials

Materials with a high spin polarisation at the Fermi energy are interesting objects of a new field: spintronics. Spintronics combines the use of the charge and the spin of electrons as information carriers. This leads to an additional degree of freedom and to potential advantages over purely charge-based electronics like, e.g., non-

S. Wurmehl (✉)
Leibniz Institute for Solid State and Materials Research, 01171 Dresden, Germany
e-mail: s.wurmehl@ifw-dresden.de

J.T. Kohlhepp
Eindhoven University of Technology, 5600 MB Eindhoven, The Netherlands
e-mail: J.T.Kohlhepp@tue.nl

C. Felser, G.H. Fecher (eds.), *Spintronics*, DOI 10.1007/978-90-481-3832-6_9,

volatility of the data, large storage density and lower energy consumption [1–4]. Potential applications for spintronics are magnetic hard discs, magnetic random access memories (MRAMS), storage devices based on solitons in magnetic nanostructures, magnetic domain-wall race track memory, as well as angle and field dependent sensors (see e.g. [1–5]). Materials with a high spin polarisation at the Fermi energy are interesting for spintronics applications due their potential to realise devices with improved magneto-resistance (MR) and low signal-to-noise ratio. Particularly, half-metallic ferromagnets (HMFs) are currently of scientific interest, as they are predicted to be completely spin polarised at the Fermi level. Suitable half-metallic ferromagnets are found among the Heusler compounds [6, 7].

Heusler compounds are ternary X_2YZ intermetallics where X and Y are transition metals and Z is a main group element. The electrical and magnetic properties of Heusler compounds range widely from metals to semiconductors and from ferrimagnets to half-metallic ferromagnets (Co_2FeSi [8, 9]). The cobalt-based Heusler compounds, crystallising in the $L2_1$ structure (space group $Fm\overline{3}m$), show some of the highest Curie temperatures (1100 K), high magnetic moments (6 μ_B/f.u.) and a complete spin polarisation at the Fermi edge, leading to their description as HMFs [6, 7].

The physical properties such as magnetisation and transport behaviour are strongly affected by the (local) structure. Thus, also the performance of such a spin polarised material in a spintronics device will depend on the (structural) quality of the material itself. In particular, it is well known that the relation between structure and the preferred high spin polarisation is essential [8, 10–14]. The Heusler compounds are known to exhibit different structure types and band structure calculations revealed a decrease in spin polarisation and magnetic moment if some types of disorder occur on certain atomic positions. In many cases, when the main group element is from the same period of the periodic system as the transition metals, conventional X-ray or neutron diffraction may not provide enough information to determine the correct structure unambiguously [15].

The optimisation of spin polarised materials for spintronics devices demands a precise control of the relations between structure and properties, in particular of the magnetic and the structural properties. Thus, a sophisticated and comprehensive characterisation is required in order to understand, tune and control the macroscopic properties of spin polarised materials towards optimised performance in spintronics devices.

Nuclear magnetic resonance (NMR) probes the local hyperfine fields of the active atoms, which strongly depend on the local environment. This spectroscopic technique is able to reveal the next neighbouring shells of the NMR active nuclei. This sensitivity to small changes in the local (magnetic and electronic) environment makes NMR an ideal method to determine the local modifications upon e.g. disorder. In particular, ferromagnetic materials are ideal to be studied by NMR, though the application of this spectroscopic method to spin polarised materials is quite unique. This renders NMR an ideal method to study the local order in Heusler compounds.

9.2 The Nuclear Magnetic Resonance (NMR) Experiment

9.2.1 The Basics of NMR

Nucleons have a spin S and orbital angular momentum L, which couple to an overall nuclear angular momentum I. If I of a certain nucleus differs from zero, this nucleus is a magnetic dipole with a spin magnetic moment μ. NMR is based on the interaction of the spin nuclear magnetic moment with an effective field B_{eff} present at the nucleus [16–22]. The presence of a magnetic field $\overrightarrow{B_0} = (0, 0, B_0)$ causes a precession of the spin around the $\overrightarrow{B_0}$ axis. In a quantum mechanical formalism, the energy of the nuclear precession is quantised and given by

$$E = -m\gamma\hbar B_0, \tag{9.1}$$

with γ the gyromagnetic ratio and m the magnetic quantum number which can have the values $m = +I, +(I-1), \ldots, -(I-1), -I$. Without a magnetic field, these energy levels are degenerated, but the presence of a magnetic field causes a splitting into $2I + 1$ equidistant energy levels (neglecting quadrupole interaction). This is the nuclear Zeeman splitting.[1] The energy difference between adjacent levels at the nucleus is resonantly probed by magnetic pulses in the radio frequency range (so-called RF pulses, typically in the range of 10 MHz–1 GHz). To do so, the RF pulses are applied, usually using a small coil wrapped around the sample. In order to avoid a dephasing of the NMR signal, a Hahn- or spin-echo sequence of RF pulses is used. For details of pulsed NMR and the appropriate pulse sequences see e.g. [23]. The intensity of the NMR spin-echo signal is measured inductively. The relative numbers of nuclei with a certain resonance frequency and, thus, in a certain local environment, can be determined from the intensity S of the resonance signal which is given by

$$S \propto \frac{\gamma\hbar^2 I(I+1)\omega_L^2}{3k_B T} \cdot N \tag{9.2}$$

with ω_L the so-called Larmor frequency, T the temperature and N the number of resonantly excited nuclei [18]. Equation 9.2 demonstrates that the intensity of the resonance signal becomes larger with decreasing temperature T. This is in particularly important for the investigation of ultra-thin films, where only a relatively small number of nuclei (corresponding to a small number N in Eq. 9.2) is involved. Consequently, most NMR experiments are performed at 4.2 K or at even lower temperatures (if possible). Equation 9.2 also states that nuclei with a high abundance and high gyromagnetic ratio (e.g. hydrogen ^{1}H or ^{59}Co) are preferable. Furthermore, in order to determine the correct number of nuclei at a certain resonance frequency, the measured data have to be corrected for the square of the resonance frequency and the inverse of the temperature.

---

[1] Note that in ferromagnetic materials, these energy levels are not degenerated even without application of a magnetic field due to the exchange interaction.

9.2.2 NMR on Ferromagnets

Contrary to paramagnetic and diamagnetic materials, where the externally applied magnetic field is the determining factor for the effective field at the nucleus, the hyperfine field is the dominating contribution in ferromagnetic (spin polarised) materials. The hyperfine field $B_{\rm hf}$ [18] refers to the exchange interaction of the magnetic moment of the nucleus itself and directly surrounding electronic and orbital moments. An important contribution to the hyperfine field is related to the conduction electron polarisation caused by the surrounding neighbour atoms and their magnetic moments. This term is usually referred to as the transferred hyperfine field and explains, why the hyperfine field $B_{\rm hf}$ is highly affected by the influence of the directly neighbouring atoms [18–20]. In general, the contribution from the next nearest neighbours is the most dominating contribution, but higher shell effects have been observed [24, 25]. This enables the NMR technique to study structural properties of bulk samples as well as thin films on a very local scale (see for instance [26–33]).

The short range dependence of the hyperfine field is precisely what gives NMR of ferromagnets its analytical strength, specifically its excellent sensitivity to the local magnetic environment of a nucleus. The presence of foreign atoms, disorder, the existence of a different crystal structure (fcc, bcc, etc.), even lattice distortions due to strain (e.g. in thin films) all cause variations in the local field at a nucleus. These variations are sensed by the NMR technique.

The considerable magnitude of the hyperfine field in ferromagnets leads to very high resonance frequencies, which are typically in the order of hundreds of MHz. Also, unlike in standard NMR where the applied magnetising field B_0 is well defined and can be made very homogeneous, the distribution of the hyperfine field in ferromagnetic materials is usually very wide due to its environmental dependence. Correspondingly, ferromagnetic line widths are typically very broad (MHz, Teslas) and the free induction decay (FID) of the NMR signals very fast. This fast decay can be overcome by the use of spin echo techniques.

When Portis and Gossard [34] carried out the first NMR experiments on ferromagnetic materials, the RF power required to observe resonance was found to be much less than expected. At the same time, the intensity of the resulting signal was also much larger. It was found that rather than directly exciting the nucleus, the applied RF field B_1 was interacting primarily with the electronic magnetisation. The oscillations induced in the local electronic magnetisation ($M_{\rm elec}$) due to this interaction are subsequently passed on to the nucleus greatly amplified, via the hyperfine interaction. The extent of this amplification is defined by $\eta = B_1^*/B_1$, where B_1^* is the enhanced RF field experienced by the nucleus.

There are two types of enhancement effect in ferromagnetic materials, the transmitting and the receiving enhancement. The RF field induces oscillations in the electronic magnetisation $M_{\rm elec}$ of the $3d$ metals and therefore the electronic magnetisation $M_{\rm elec}$ undergoes a rotation θ dependent on three fields: the RF field B_1, an effective field accounting for any intrinsic magnetic anisotropy $\mathrm{B}_{\rm aniso}$ and any

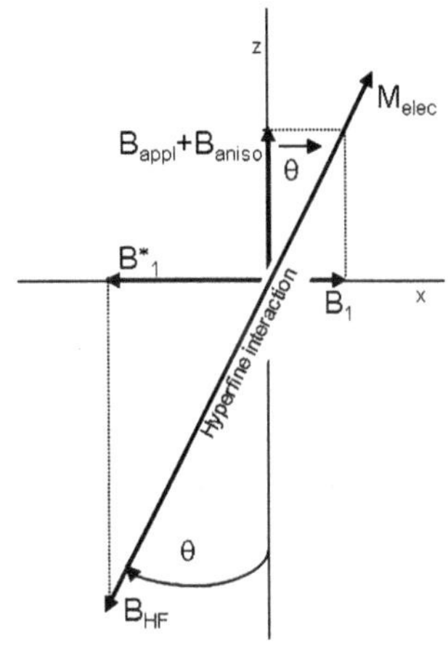

Fig. 9.1 Illustration of the B_1 enhancement

external field applied, B_{appl}. These moment oscillations lead to changes in the hyperfine field due to their coupling via the contact interaction, resulting in an RF component of the hyperfine field and therefore to a similar rotation of the hyperfine field B_{hf} (see Fig. 9.1). The actual RF field experienced by the nucleus $\overrightarrow{B}_{1,\text{eff}}$ is given by the sum of the applied RF field $\overrightarrow{B}_{\text{rf,applied}}$ and the RF field induced by the oscillating electronic moments (transmitting enhancement). The receiving enhancement describes the amplification of the NMR signal by the hyperfine interaction. The precession of a nucleus in resonance leads to a corresponding coherent motion of the electronic moments, leading to an enhancement of the signal. In cobalt, a B_1 amplification (η) by the enhancement was predicted to be of the order of 10–100, however, again, significantly larger effects were originally observed. This increase has been explained in terms of domain wall motion [35]. For a single domain, the application of a weak alternating field such as B_1 simply induces a coherent rotation of the local (electronic) magnetisation of the atoms in a single domain. In multidomain samples, such a field causes an expansion and contraction of the individual domains. Correspondingly, domain walls undergo an oscillating displacement, and so does the magnetisation of the atoms contained within the walls, but to a much greater extent than in the domains themselves. In line with Fig. 9.1, larger rotations of M_e lead to a larger enhancement.

The enhancement of the RF signal in ferromagnetic materials enables the measurement of magnetic thin films and the possibility to measure at elevated temperatures. In order to correct NMR data for the enhancement, the optimal duration of the RF pulse for the desired rotation of the nuclear moments during the NMR experiment has to be chosen. Usually, the rotation angle leading to a maximum spin-echo signal is preferred. The optimum field strength and the enhancement factor is found experimentally by monitoring the spin-echo intensity as a function of the RF field strength on a logarithmic scale. This dependence follows a Gaussian line and can be fitted. With this procedure, the optimum RF power is found, for which the enhancement factor can be determined. Afterwards, the experimental data can be corrected using this enhancement factor.

9.3 Selected NMR Studies of Spin Polarised Heusler Compounds

In an $L2_1$ ordered Heusler compound, each atom has one local environment and therefore one hyperfine field should be observed. If atoms are mixed on crystallographic sites due to atomic disorder or due to partial substitution, different local environments become possible, leading to a distribution of hyperfine fields and therefore to line broadening or line splitting.

Quite a few Heusler compounds have already been subject to NMR studies since the 1960s (see e.g. the studies of Niculescu et al. [36, 37] and LeDang Khoi et al. [38]). Most NMR studies of Heusler compounds focus on the structural characterisation of bulk materials like e.g. in Co_2TiSn [39], Co_2NbSn [40], or the $C1_b$ Heusler compounds CoVSb [41], or CoMnSb [42] as well as of Heusler thin films like e.g. NiMnSb [26, 27], Co_2FeAl [43, 44] or Co_2FeSi [45, 46]. The NMR studies reveal in most cases distinct disorder contributions and might even be leading to a completely new structural model, like e.g. in CoMnSb [42], but are also used to study the electronic structure like e.g. in Fe_2VGa [47]. Here, a study of the Knight shift and the spin-lattice relaxation was used to confirm a two-band model for Fe_2VGa, one overlapping with the Fermi level and one having an energy gap, leading to the predicted semi-metallic characteristics [47]. Moreover, the presence of paramagnetic Fe antisite impurities were confirmed by analysis of the temperature dependence of the NMR line width [47]. Importantly, Fe at an antisite position, as revealed by NMR, provokes the observed giant magnetoresistance effect and the pure compound can be regarded as non-magnetic, in line with the calculations [48]. NMR spectroscopy provides also an indirect tool to measure the densities of states of spin polarised materials [49], which are different for majority and minority electrons, via a measurement of the temperature dependence of the spin-lattice relaxation. In 2009, a study of the spin lattice relaxation rate of CoTiSb was used to investigate the change in the DOS upon annealing [50].

In the following, we will present some selected NMR studied with respect to Heusler materials for spintronics in more detail.

9.3.1 $Co_2Mn_{1-x}Fe_xSi$

Members of the substitutional series $Co_2Mn_{1-x}Fe_xSi$ with $x \approx 0.5$ are expected to be particular stable half-metals [51]. The $L2_1$ structure of the complete substitutional series $Co_2Mn_{1-x}Fe_xSi$ is observed by XRD and EXAFS [8, 15, 51] independent of the Fe concentration x. From a crystallographic point of view, the $L2_1$ structure type requires a random distribution of Mn and Fe at the $4b$ Wyckoff position. Note that the investigation of such a local distribution of atoms requires a sophisticated local characterisation method such as NMR.

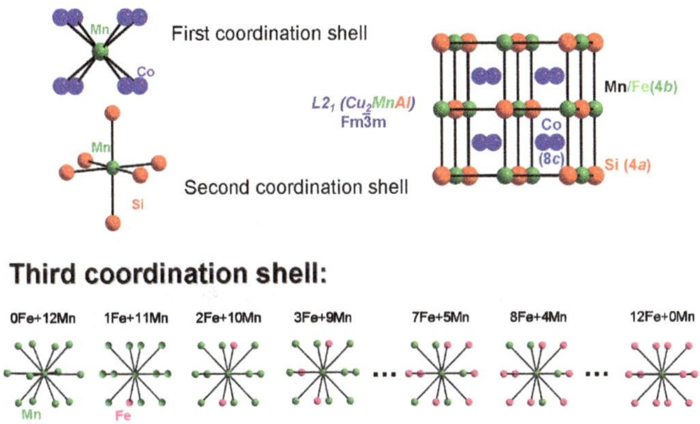

Fig. 9.2 First, second and third coordination sphere of Mn in the $Co_2Mn_{1-x}Fe_xSi$ Heusler compound and the corresponding different local environments with different numbers of Mn and Fe next neighbours

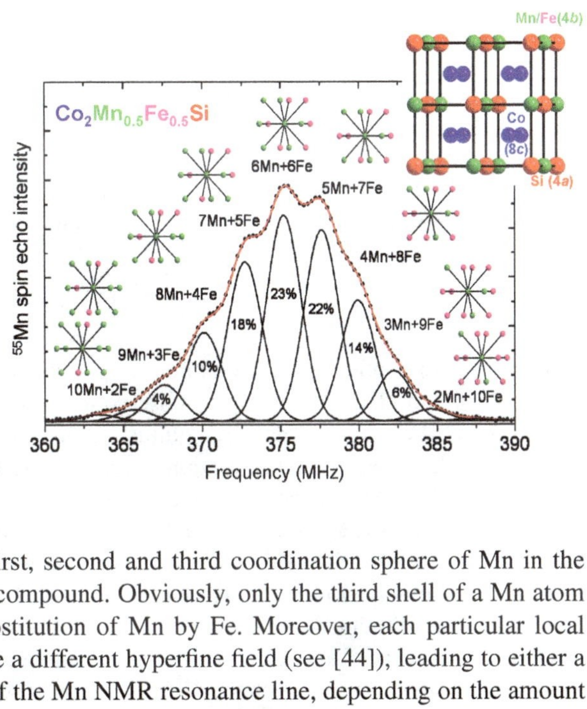

Fig. 9.3 The spin echo intensity as a function of frequency for ^{55}Mn in $Co_2Mn_{0.5}Fe_{0.5}Si$ (*black dots*) and the fit (*red line*). The resonance lines were fitted using a sum of Gaussian lines (*black lines*). Their widths were restrained within each compound to be identical. The distribution of Mn and Fe atoms in the third coordination shell of the ^{55}Mn are given for each line as well as the relative areas obtained from Gaussian lines [44]

Figure 9.2 shows the first, second and third coordination sphere of Mn in the $Co_2Mn_{1-x}Fe_xSi$ Heusler compound. Obviously, only the third shell of a Mn atom will be altered by the substitution of Mn by Fe. Moreover, each particular local environment will also have a different hyperfine field (see [44]), leading to either a broadening or a splitting of the Mn NMR resonance line, depending on the amount of the difference of the corresponding hyperfine fields.

Figure 9.3 exemplarily shows the ^{55}Mn NMR spectrum of $Co_2Mn_{0.5}Fe_{0.5}Si$. The spectrum exhibits a splitting into several resonance lines with a spacing between adjacent resonances of about 2.35 MHz. Thus, every line corresponds to a specific local environment of Mn experiencing a different distribution of Mn and Fe atoms in the third coordination shell (see Fig. 9.3).

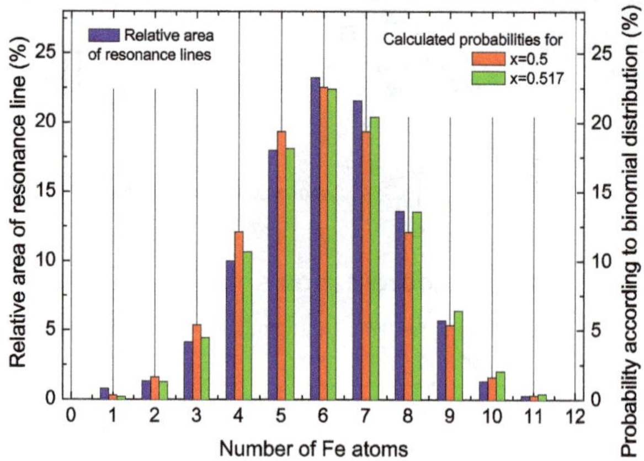

Fig. 9.4 Relative areas of the resonance lines as a function of the number of Fe atoms in the third coordination shell of ^{55}Mn. The *blue bars* represent the fitting results of the relative line areas. Shown are the results for the $Co_2Mn_{0.5}Fe_{0.5}Si$ compound with ideal stoichiometry $x = 0.5$ (*red*) and optimised stoichiometry $x = 0.517$ (*green*)

High frequency lines correspond to Fe-rich environments, while low frequencies lines correspond to Mn-rich environments, in agreement with the distribution of local environments of the complete substitutional series.

The resonance lines were fitted using a sum of Gaussian lines (grey line in Fig. 9.3). Their widths were restrained within each compound to be identical. From this fit with Gaussian lines, the relative areas of the resonance lines are obtained. In order to further confirm the validity of a random distribution of Mn and Fe for the complete substitutional series, the experimentally found relative areas of the resonance lines as a function of the number of Fe atoms are compared with the relative probabilities of a random atom model as obtained by a binomial distribution:

$$P(n,x) = \frac{N!}{(N-n)!n!}(1-x)^{N-n}x^n \tag{9.3}$$

with $N = 12$ corresponding to the number of possible crystallographic sites in the third shell of Mn. The probability to observe a specific environment depends of course on the Mn:Fe ratio. Thus, by optimising the Mn:Fe fraction x using a least square method, the real Fe concentration is derived. Figure 9.4 compares the probabilities using Eq. 9.3 with nominal (red bars) and the optimised Fe concentration (green bars) with respect to the measured relative lines areas (red). The good agreement between the experimental results and the random atom model (Eq. 9.3) confirms the predominantly random distribution of Fe and Mn on the 4*b* position for the complete substitutional series.

The ^{55}Mn spectra of the quaternary alloys unambiguously demonstrate the random distribution of Mn and Fe on the 4*b* Wyckoff position and a very high degree of order, even with regards to the next neighbouring shells. The high crystallographic order of $Co_2Mn_{1-x}Fe_xSi$ demonstrated by the ^{55}Mn NMR results will have strong impact on the half-metallic character and the high degree of spin polarisation of $Co_2Mn_{1-x}Fe_xSi$, including the expectation of high TMR ratios. Thus, these NMR studies confirm that the Heusler compounds $Co_2Mn_{1-x}Fe_xSi$ have the potential to be ideal candidates for spintronics applications [24, 44].

9.3.2 Co_2TiSn

Powder X-ray diffraction and Mößbauer spectroscopy revealed the need to further analyse the structure and local ordering in Co_2TiSn bulk samples [39]. The ^{59}Co NMR spectrum of Co_2TiSn consists of a main line at 21.1 MHz and satellite lines on the low and high frequency side. The main line arises from Co atoms in the ordered stoichiometric environment (4 Ti + 4 Sn atoms). The additional lines arise from Co atoms with excessive Co atoms in the first coordination shell and consequently excessive Ti atoms in the second shell. The advantage of the NMR technique over non-local methods is that it allows that the Co-Ti antisite disorder, the so-called DO_3 type order, can be quantified. In the present case, the contribution of DO_3 type order is found to be about 9 %. Eventually, the band structure calculations reveal that the half-metallic character is conserved even if taking the 9 % of DO_3 type structure contributions into account. This renders Co_2TiSn to be a good candidate for spintronics applications [39].

9.3.3 Ni_2MnGa

The Heusler compound Ni_2MnGa is reported to be a magnetic shape memory compound [52]. It undergoes a transition from a cubic to a tetragonal distorted structure at 195 K. The bulk sample of Ni_2MnGa was measured using ^{55}Mn NMR (see Fig. 9.5). At room temperature, the spectrum is centred at 230 MHz, while the spectrum is shifted to higher frequencies at 4 K (centre at 310 MHz). The resonance line at room temperature represents the cubic ordered Ni_2MnGa, while at 4 K, mainly the tetragonal structure is obtained. The shift of the main resonance line is in agreement with the measurement of the temperature dependence of the magnetic moment as shown in reference [53].

9.3.4 Co_2FeAl

The Heusler compound Co_2FeAl is reported to occur in various structures ranging from the completely ordered $L2_1$ to the completely disordered $A2$ structure type. The different hyperfine fields in the local environment in $B2$ and $A2$ ordered Co_2FeAl thin films was first reported by Inomata et al. [43] using NMR. Inomata et al. observed a line splitting in thin films of Co_2FeAl due to Co atoms experiencing a random distribution of Fe and Al in the $B2$ and of Co, Fe and Al in the $A2$ type structure. In a more recent NMR study, the local structure of Co_2FeAl bulk samples was probed [25]. The ^{59}Co NMR measurements reveal a distribution of Fe and Al not only in the first neighbouring shells of the ^{59}Co nuclei but also in more distant shells.

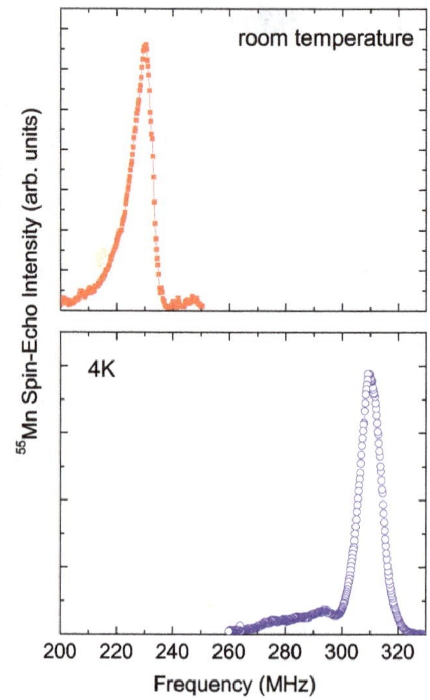

Fig. 9.5 The spin-echo intensity as a function of frequency for ^{55}Mn in Ni_2MnGa at (**a**) room temperature and (**b**) at 4 K

As shown in Figs. 9.6(a), (b), the ^{59}Co NMR spectrum of the bulk Co_2FeAl samples (as-melted) revealed a splitting into main and sub-lines. These main lines originate from the distribution of Fe and Al in the neighbouring shells of the ^{59}Co nuclei. Each main resonance line can be attributed to a certain distribution of Fe and Al in the first shell of the ^{59}Co nuclei in agreement with the results of Inomata et al. [43]. A complete intermixing of the Fe and Al atoms leads to a structure type different from L21, the $B2$ structure. From a crystallographic point of view, a basic requirement of the $B2$ structure is an entire random distribution of Fe and Al of the Wyckoff 1a position. The random distribution of Fe and Al ($B2$ type) is described by a binomial distribution, revealing the probability $P(n, x)$ for a particular environment of the form n Fe atoms $+ (8 - n)$Al atoms in the first shell of the Co central atom. A comparison between the fitted relative areas of the resonance lines (grey line in Fig. 9.6) with the probabilities given by a random distribution of Fe and Al demonstrates that as-melted Co_2FeAl bulk samples mainly exhibit a $B2$ type structure with contribution of an $L2_1$ type structure of about 10 %. This $L2_1$ contribution was not apparent in the NMR spectrum of Co_2FeAl thin films [43].

The structure of the bulk samples can be changed by annealing (Figs. 9.6(c), (d)). The NMR spectrum of a bulk sample annealed at 1300 K consists of several main lines with an underlying sub-structure similar to the NMR spectrum of an as-melted sample. However, the main lines are less resolved after annealing and seem to merge. It was not possible to sufficiently fit the main lines of the spectrum of the annealed sample using only the Gaussian lines expected for a pure $B2$ type ordering or a mixture of $B2$ and $L2_1$. These observations point to a loss of the high degree

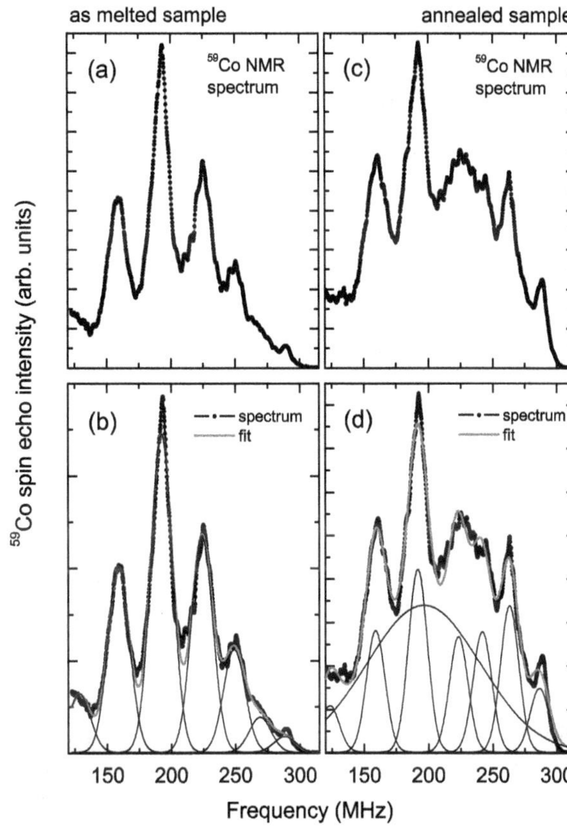

Fig. 9.6 The ^{59}Co spin echo intensity in Co_2FeAl bulk samples (**a**), (**c**) as a function of frequency in comparison with a fit (*grey line*) (**b**), (**d**) for the as-melted (**a**), (**b**) and annealed sample (**c**), (**d**)

of order as shown for the as-melted sample. A satisfactory fit was obtained by implementation of an additional broad resonance line. Interestingly, the additional very broad line exhibits the characteristic resonance frequency and width of a completely $A2$ ordered Co_2FeAl sample as shown by Inomata et al. [43]. It seems that annealing at 1300 K leads to a mixture of $B2$ and $A2$ contributions. This result is in perfect agreement with the results of Balke et al. who found that this annealing temperature leads to samples with a mixture of $A2$ and $B2$ contributions [54], thus supporting the NMR results.

Additional contributions of structural phases as $L2_1$ and $A2$ can be excluded to cause the sub-lines. However, a random distribution of Fe and Al may not only affect the composition of the first coordination shell of the ^{59}Co nuclei but may also affect the composition of higher shells. Thus, the sub-structure is related to different higher shell environments of the ^{59}Co nuclei. This sub-structure of the resonance lines was not resolved in ^{59}Co NMR spectrum of thin Co_2FeAl films [43], pointing to a more pronounced long range order in the bulk samples than in thin film samples. These NMR studies demonstrates the possibility to further improve the degree of order in Co_2FeAl, which may lead to high TMR ratios when implemented in magnetic tunnel junctions.

9.3.5 CoFe–Al

The magnetotransport properties of current-perpendicular-to-the plane giant magneto-resistance (CPP-GMR) devices consisting of ferromagnetic Co–Fe alloys have recently been shown to be significantly improved by the addition of up to 28 % Al [55]. In order to further optimise CoFe-Al spin-valves, it is important to understand the impact of the Al alloying on the local and electronic structure. NMR is able to reveal the next neighbouring shells of the ^{59}Co nuclei in the Co–Fe–Al magnetic films. Using NMR spectroscopy, the local formation of a Heusler-like structure by addition of Al to the Co–Fe alloy in CPP-GMR multilayers was shown. The observed local formation of a highly spin polarised Heusler compound may be correlated to the observed enhancement of the GMR effect [56].

9.3.6 Co_2FeSi

Co_2FeSi is predicted to be a half-metallic ferromagnet with an extraordinarily high magnetic moment and Curie temperature. In the case of complete $L2_1$ type ordering, there is only one possible way to distribute the atoms on the crystal lattice, leading to one hyperfine field for the ^{59}Co nuclei and therefore to one narrow resonance line. Thus, in the case of the $L2_1$ ordered Co_2FeSi poly- and single crystalline bulk samples, a very sharp line at 139 MHz is observed [45, 57]. The magnetic moment of the $L2_1$ ordered bulk samples follow the Slater Pauling rule and therefore have a magnetic moment of 6 μ_B/f.u., as expected. However, a low tunnel magneto-resistance ratio, a lower spin polarisation than predicted, and a lower magnetic moment were experimentally observed in thin film samples. Therefore, NMR was used to investigate the local structure of Co_2FeSi films. Interestingly, similar ^{59}Co NMR spectra of Co_2FeSi thin films have been presented and discussed by different groups. However, the interpretation of the data leads to the same results and indicates Fe off-stoichiometry in Co_2FeSi films caused by selective sputtering. Figure 9.7 shows the ^{59}Co spectrum of a thin film of Co_2FeSi. The spectrum exhibits a main resonance line roughly at 155 MHz (slightly shifted to higher frequencies compared to measurements of an ordered bulk sample of Co_2FeSi [45, 57]) and satellite lines in the high frequency site of the main line only. The spacing between the main and satellite lines is comparable to the spacing found in $B2$ ordered Co_2FeAl. In contrast to the ^{59}Co-NMR spectrum of Co_2FeAl, the distribution of main resonance lines is not symmetric around the main resonance line and reveals only high frequency satellites lines. Thus, a $B2$ disorder is excluded for the origin of the line splitting. In line with reference [37], the main lines are interpreted to arise from Fe excess atoms (and, thus, to a deficit of Si atoms) in the first shell of the ^{59}Co nuclei. These Fe excess lines correspond to the high frequency lines of a $B2$ ordered compound, which explains the observed resemblance in the spectra of $B2$ ordered Co_2FeAl and the thin films of Co_2FeSi. The observed sub-lines are attributed to a corresponding off-stoichiometry in higher shells. In summary,

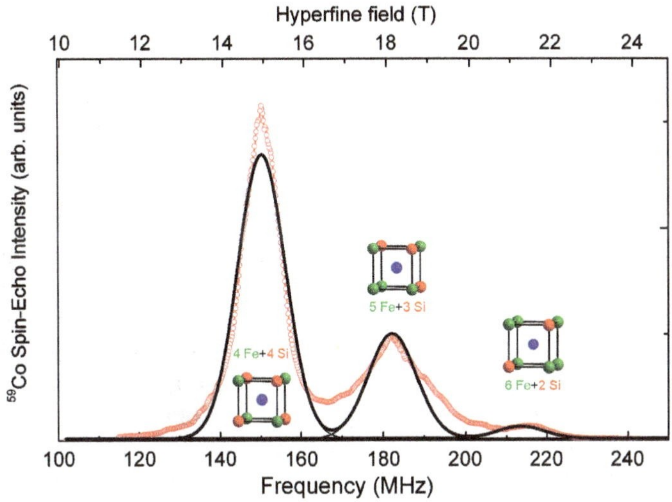

Fig. 9.7 The ^{59}Co spin echo intensity in thin film of Co_2FeSi thin films as a function of frequency demonstrating the splitting into the main lines with a spacing in the order of 30 MHz. The line splitting is attributed to Fe-rich environments (see the corresponding local environments of Co with different numbers of Fe and Si next neighbours) caused by selective sputtering

films made from a stoichiometric target reveal an off-stoichiometric composition caused by selective sputtering. However, Sakuraba et al. [58] report stoichiometric Co_2MnSi thin films from off-stoichiometric targets. Remarkably, thin Co_2FeSi films prepared in different groups show quite similar results [45, 46]. The off-stoichiometry revealed by NMR might explain the observed deviations from the expected behaviour [46].

9.3.7 NiMnSb

A NMR study by van Roy et al. [27] of the half-metallic $C1_b$ Heusler compound NiMnSb showed that thin epitaxial films of NiMnSb on GaAs(111)B substrates exhibit a very low degree of disorder. Only very few Mn atoms are involved in planar defects including stacking faults, only 0.5 % of Sb and As (from the interdiffusion from the GaAs buffer layer) interchange is detected, and only 0.2 % of Sb atoms have an altered environment compared to the $C1_b$ structure. These results demonstrate the capability of NMR spectroscopy to detect chemical disorder even at interfaces. However, NMR reveals a significant local separation into MnSb (about 14 %) in films on GaAs(001) and a very low degree (less than 4 %) of Mn and Sb antisite disorder [26]. Using these NMR results, bandstructure calculations show that NiMnSb maintains a half-metallic band structure even at the ferromagnet/semiconductor interface [27].

9.3.8 CoMnSb

For quite a long time, the $C1_b$ Heusler compound CoMnSb was expected to belong to the class of half-metallic ferromagnets [59]. Interestingly, the experimentally observed magnetic moment significantly differs from the theoretically calculated moment [60] and powder X-ray diffraction data show a superstructure. In order to investigate the local structure, NMR measurements were performed. The NMR spectrum consists of a superposition of many lines separated into three groups, which are related to the ^{59}Co, ^{55}Mn, ^{121}Sb and ^{123}Sb nuclei [42]. Assuming $C1_b$ ordering, one would expect the intensity ratio for the group to be 1:1:1. An analysis of the NMR data reveals two crystallographic sites for the ^{121}Sb and ^{123}Sb nuclei, one for the ^{59}Co nuclei and three for ^{55}Mn. This excludes the $C1_b$ structure and leads to a new structural model. As a result, the crystal structure of CoMnSb is described by alternation of Co_2MnSb and MnSb structural units, with three different Mn and two Sb positions in the unit cell. This new structural model was used for band structure calculations, which finally confirms the experimentally observed magnetic moment and dispels CoMnSb from the class of half-metallic ferromagnets.

9.4 Summary

All NMR studies presented in this review demonstrate that NMR is indeed helpful to interpret the measurements of physical properties. This renders NMR spectroscopy a very useful tool to characterise highly spin polarised materials, which is also reflected by the rising number of publications. The understanding and the control of the local structure of e.g. Heusler compounds will advance the whole class of materials.

Acknowledgements The authors thank W. de Jonge, B. Koopmans, H.J.M. Swagten, T. Ellis, P.J. Jacobs, M. Wojcik, V. Ksenofontov, G.H. Fecher, C. Felser, C.G.F. Blum, G. Jakob, H. Schneider, D. Ebke, G. Reiss, B. Balke, S. Maat, M.J. Carey and J.R. Childress for fruitful discussions.

References

1. Baibich MN, Brote JM, Fert A, Nyugen van Dau N, Petroff F, Eitenne P, Creuzet G, Friederich A, Chazelas J (1988) Phys Rev Lett 61:2472
2. Binasch G, Grünberg P, Saurenbach F, Zinn W (1989) Phys Rev B 39:4828
3. Coey JMD, Venkatesan M, Bari MA (2002). In: Berthier C, Levy LP, Martinez G (eds) Lecture notes in physics, vol 595. Springer, Heidelberg, p 377
4. Prinz GA (1998) Science 282:1660
5. Parkin LTSSP, Hayashi M (2008) Science 320:190
6. Groot RAd, Müller FM, Engen PGv, Buschow KHJ (1983) Phys Rev Lett 50:2024
7. Felser C, Fecher GH, Balke B (2007) Angew Chem, Int Ed Engl 46:668
8. Wurmehl S, Fecher GH, Kandpal HC, Ksenofontov V, Felser C, Lin HJ, Morais J (2005) Phys Rev B 72:1844341

9. Wurmehl S, Fecher GH, Kandpal HC, Ksenofontov V, Felser C, Lin HJ (2006) Appl Phys Lett 88:032503
10. Orgassa D, Fujiwara H, Schulthess TC, Butler WH (1999) Phys Rev B 60:13237
11. Block T, Carey MJ, Gurney BA, Jepsen O (2004) Phys Rev B 70:205114/1
12. Miura Y, Nagao K, Shirai M (2004) Phys Rev B 69:144413
13. Miura Y, Nagao K, Shirai M (2004) J Appl Phys 95:7225
14. Fecher GH, Kandpal HC, Wurmehl S, Morais J, Lin HJ, Elmers HJ, Schönhense G (2005) J Phys Condens Matter 17:7237
15. Balke B, Wurmehl S, Fecher GH, Felser C, Alves MCM, Bernardi F, Morais J (2007) Appl Phys Lett 90:172501
16. Abragam A (1996) Principles of nuclear magnetism. Oxford University Press, Oxford
17. Schlichter CP (1963) Principles of magnetic resonance. Springer, Berlin
18. Panissod P (1986) In: Gonser U (ed) Nuclear magnetic resonance in topics in current physics: microscopic models in physics. Springer, Berlin
19. Panissod P (1997) In: Baryakhtar VG, Wigen PE, Lesnik NA (eds) NATO ASI series high tech, vol 48. Kluwer Academic, Dordrecht, p 225
20. Riedi PC, Thomson T, Tomka GJ (1999) In: Buschow KHJ (ed) Handbook of magnetic materials, vol 12. North-Holland, Amsterdam, p 97
21. Gladden LF (1994) Chem Eng Sci 49:3339
22. Veeman WS (1997) Geoderma 80:225
23. Wurmehl S, Kohlhepp J (2008) J Phys D, Appl Phys 41:173002 (Topical review)
24. Wurmehl S, Kohlhepp J, Swagten H, Koopmans B, Wojcik M, Balke B, Blum C, Ksenofontov V, Fecher G, Felser C (2008) J Appl Phys 103:07D706
25. Wurmehl S, Kohlhepp J, Swagten H, Koopmans B (2008) J Phys D, Appl Phys 41:115007
26. Wojcik M, van Roy W, Jedryka E, Nadolski S, Borghs G, De Boeck J (2002) J Magn Magn Mater 240:414
27. van Roy W, Wojcik M, Jedryka E, Nadolski S, Jalabert D, Brijs B, Borghs G, De Boeck J (2003) Appl Phys Lett 83:4214
28. Wojcik M, Jedryka E, Nadolski S, Navarro J, Rubi D, Fontcuberta J (2004) Phys Rev B 69:100407
29. Wojcik M, Jedryka E, Nadolski S, Rubi D, Frontera C, Fontcuberta J, Jurca B, Dragoe N, Berthet P (2005) Phys Rev B 71:104410
30. Wieldraaijer H, de Jonge WJM, Kohlhepp JT (2005) Phys Rev B 72:155409
31. Wieldraaijer H, de Jonge WJM, Kohlhepp JT (2005) J Magn Magn Mater 286:390
32. Wojcik M, Jedryka E, Skorvanek I, Svec P (2005) J Magn Magn Mater 290–291:1431
33. Wojcik M, Jedryka E, Skorvanek I, Marcin J, Svec P (2006) J Magn Magn Mater 304:e712
34. Gossard AC, Portis AM (1959) Phys Rev Lett 3:164
35. Portis AM, Gossard AC (1960) J Appl Phys 31:205
36. Niculescu VA, Budnick JI, Hines W, Raj K, Pickart S, Skalski S (1979) Phys Rev B 19:452
37. Niculescu VA, Burch TJ, Budnick JI (1983) J Magn Magn Mater 39:223
38. Khoi LD, Veillet P, Campell IA (1978) J Phys F, Met Phys 8:1811
39. Kandpal HC, Ksenofontov V, Wojcik M, Seshadri R, Felser C (2006) J Phys D, Appl Phys 40:1587
40. Wolter AUB, Bosse A, Baabe D, Maksimov I, Mienert D, Klauß HH, Litterst FJ, Niemeier D, Michalak R, Geibel C, Feyerherm R, Hendrikx R, Mydosh JA, Süllow S (2002) Phys Rev B 66:174428
41. Nishihara H, Kanomata T, Furutani Y, Igarashi T, Koyama K, Goto T (2006) Phys Status Solidi 3:2779
42. Ksenofontov V, Melnyk G, Wojcik M, Wurmehl S, Kroth K, Reimann S, Blaha P, Felser C (2006) Phys Rev B 74:134426
43. Inomata K, Okamura S, Miyazaki A, Tezuka N, Wojcik M, Jedryka E (2006) J Phys D, Appl Phys 39:816
44. Wurmehl S, Kohlhepp J, Swagten H, Koopmans B, Wojcik M, Balke B, Blum C, Ksenofontov V, Fecher G, Felser C (2007) Appl Phys Lett 91:052506

45. Inomata K, Wojcik M, Jedryka E, Ikeda N, Tezuka N (2008) Phys Rev B 77:214425
46. Wurmehl S, Kohlhepp J, Swagten H, Koopmans B, Blum C, Ksenofontov V, Schneider H, Jakob G, Ebke D, Reiss G (2009) J Phys D, Appl Phys 42:084017
47. Lue CS, Ross JHR Jr. (2001) Phys Rev B 63:054420
48. Lue CS, Ross JH Jr., Rathnayaka KDD, Naugle DG, Wu SY, Li WH (2001) J Phys Condens Matter 13:17968
49. Kuhns P, Hoch MJR, Reyes AP, Moulton WG, Wang L, Leighton C (2006) Phys Rev Lett 96:167208
50. Lue CS, Chen CF, Chiang FK, Chu MW (2009) Phys Rev B 80:174202
51. Balke B, Fecher GH, Kandpal HC, Felser C, Kobayashi K, Ikenaga E, Kim JJ, Ueda S (2006) Phys Rev B 74:104405
52. Ullakko K, Huang JK, Kantner C, O'Handley RC, Kokorin VV (1996) Appl Phys Lett 69:1966
53. Ooiwa K, Endo K, Shinogi A (1992) J Magn Magn Mater 104–107:2011
54. Balke B, Fecher GH, Felser C (2007) Appl Phys Lett 90:242503
55. Maat S, Carey MJ, Childress JR (2007) J Appl Phys 101:093905
56. Wurmehl S, Jacobs PJ, Kohlhepp JT, Swagten HJM, Koopmans B, Maat S, Carey MJ, Childress JR (2011) Appl Phys Lett 98:012506
57. Blum CGF, Jenkins C, Barth J, Felser C, Wurmehl S, Friemel G, Hess C, Behr G, Büchner B, Reller A, Riegg S, Ebbinghaus SG, Ellis T, Jacobs PJ, Kohlhepp JT, Swagten HJM (2009) Appl Phys Lett 95:161903
58. Sakuraba Y, Miyakoshi T, Oogane M, Ando Y, Sakuma A, Miyazaki T, Kubota H (2006) Appl Phys Lett 89:052508
59. Galanakis I (2002) J Phys Condens Matter 14:6329
60. Tobola J, Piere J (2000) J Alloys Compd 296:243

Chapter 10
New Materials with High Spin Polarization Investigated by X-Ray Magnetic Circular Dichroism

Hans-Joachim Elmers, Michael Kallmayer, and Peter Klaer

Abstract We investigate element-specific spin and orbital magnetic moments of polycrystalline bulk Heusler alloys that are predicted to be half-metallic with composition Co_2YZ (Y = Ti, Cr, Mn, Fe and Z = Al, Ga, Si, Ge, Sn, Sb) using magnetic circular dichroism in X-ray absorption spectroscopy (XAS/XMCD). In addition to stoichiometric compounds we also investigate composition series with partly replaced elements on the Y-site ($Co_2Fe_xCr_{1-x}Si$, $Co_2Mn_xTi_{1-x}Si$ and $Co_2Mn_xTi_{1-x}Ge$) and on the Z-site ($Co_2MnGa_{1-x}Ge_x$) promising a tailoring of the Fermi level with respect to the minority band gap. We compare experimental results with theoretical predictions elucidating the influence of local disorder in the experimental samples. Moreover, we demonstrate that a consideration of electron correlation in local density approximation theories is necessary for reproducing experimental results. Increased orbital magnetic moments in respect to theoretical predictions put forward the role of spin–orbit coupling for half-metallic properties. For the case of single crystalline thin films we developed a method of simultaneous measurement of bulk and surface sensitive magnetic properties including those at the crucial interface to a tunneling barrier. Exploiting the comparison of bulk and interface information, film growth can be improved for specific applications. In order to directly determine the spin-resolved density-of-states function we present a calculation scheme for extracting this information from the XMCD spectra considering final-state electron correlations. We investigate the electronic properties of epitaxial $Co_2(Fe_xMn_{1-x})Si$, $Co_2Fe(Al_{1-x}Si_x)$, and $Co_2(Cr_{0.6}Fe_{0.4})Al$ films on MgO(100) substrates and for the case of several bulk samples including Co_2TiSb as a reference sample for normal metallic ferromagnetism. The experimental results, revealing the distribution of magnetic moments and the relative position of the Fermi energy as a function of the number of valence electrons, confirm the predicted possibility of

H.-J. Elmers (✉) · M. Kallmayer · P. Klaer
Institut für Physik, Johannes Gutenberg – Universität, 55099 Mainz, Germany
e-mail: elmers@uni-mainz.de

M. Kallmayer
e-mail: kallmaye@uni-mainz.de

P. Klaer
e-mail: klaer@uni-mainz.de

C. Felser, G.H. Fecher (eds.), *Spintronics*, DOI 10.1007/978-90-481-3832-6_10,

tailoring the minority band gap using substitutional quaternary Heusler compounds. These findings may be of general importance for the understanding of the electronic structures in complex intermetallic compounds.

10.1 Introduction

Currently, Heusler alloys are in the focus of great scientific interest due to their potential for applications in spin valves and other spintronics devices [1–5]. For these purposes, materials with a high spin polarization, i.e. half-metallic ferromagnets, are desirable and theory suggests that many Heusler alloys show this property [6]. Using local density approximations (LDA), calculations of the band structure predict half-metallic properties for half- and full-Heusler compounds [7]. The consideration of electron correlation effects, e.g. local spin density approximation with added correlation potential (LDA+U), led to half-metallic Heusler alloys that were previously considered as normal metals [8]. Experimental attempts, however, to directly detect a high spin polarization in Heusler alloys using spin-polarized photoemission have failed [9, 10]. Recent experimental progress in the fabrication of tunneling magnetoresistance (TMR) devices provides indirect evidence for HMF in Heusler alloys because of the large TMR values [2, 3, 5, 11–13] observed for magnetic tunneling junctions using Heusler electrodes.

For most applications, the high spin polarization potentially offered by Heusler alloys must stay constant up to the topmost atomic layers, even at room temperature. Magnetization and spin–orbit coupling are strongly related to the spin polarization and therefore an understanding of the magnetic interface properties is of great help for further optimization of half-metallic materials. The simultaneous determination of bulk and surface properties of Heusler films has shown that in many cases the magnetization decreases at the tunneling barrier interface [14] but this can be avoided by optimizing deposition procedures [15]. Hence, analyzing buried interfaces is indispensable for further improving application devices. Investigation of free surfaces provides an alternative way of material characterization. However, the preparation of an ideal free surface is difficult. Cleaning by sputtering and annealing procedures easily lead to surface structures that strongly deviate from the Heusler structures.

Magnetic circular dichroism in X-ray absorption spectroscopy (XAS/XMCD) is a powerful tool for studying the element-specific electronic structure even at buried interfaces [16, 17] thus solving the above mentioned problems. Due to the limited mean free path of 2.5 nm for photo emitted electrons XAS provides surface sensitive information through thin capping layers. Element-specific information on spin and orbital magnetic moments derived from the spectra allow a detailed test of theoretical predictions and can be used for optimizing sample preparation. As an example, the antiferromagnetic exchange coupling between nearest-neighbor Mn/Cr atoms at antisite defects causes an antiparallel alignment of moments [18]. Since in this case the Mn/Cr moment counts twice, this type of defect will drastically reduce the Mn/Cr moment.

More spectroscopic information can be extracted from the XAS/XMCD measurements. In principle, the L-edge absorption spectra for left and right circularly polarized X-ray light reflects the spin-resolved partial density of states (PDOS) at the $3d$ transition metal atoms [19, 20]. For strongly localized states, e.g. in an oxide, the strong interaction between the core–hole and the conduction band in the final state leads to an additional splitting of the spectra, often denoted as multiplet effects [21]. These multiplet effects may effectively mask the band structure and in this case it is impossible to disentangle the PDOS and multiplet contributions. Telling et al. [22] pointed out that the existence of local moments at the Y site in Heusler compounds also give rise to a pronounced multiplet structure in the absorption spectra. However, previous investigations at the Co and Ni L-edge in intermetallic compounds clearly revealed PDOS related features in the absorption spectra of Heusler alloys [23, 24]. For Co-based Heusler compounds a proper consideration of the final state effects partly recovers the spin-resolved PDOS for unoccupied Co states from the XMCD spectra [15, 25]. This procedure enables a direct quantitative comparison of ab-initio calculations with experimental results in respect to the existence and position of the minority band gap relative to the Fermi level.

An important aspect is the temperature dependence of the electronic states in half-metals because a temperature induced change directly affects the device performance. A recent high energy photo emission study for a half-metallic Co_2MnSi/MgO interface [26] found no significant temperature dependence of the electronic states. Instead, the temperature dependence of the spin-resolved unoccupied density of states derived from XAS/XMCD data shows an unexpected strong temperature dependence of the DOS beyond the change expected from the Fermi distribution function in the case of Mn_2VAl and Co_2FeAl [27].

Results for element-specific spin and orbital moments of bulk samples are summarized in Sect. 10.2. A comparison of simultaneously measured interface and bulk information on epitaxial Heusler films with potential half-metallic properties is presented in Sect. 10.3. Section 10.4 outlines results for the spin-resolved PDOS derived from XAS/XMCD.

10.2 Bulk Samples

Magnetization values that follow the Slater–Pauling rule are a substantial precondition for half-metallic properties. Since deviations from the predicted local magnetization values at different lattice sites may be positive or negative, these deviations may compensate each other. Therefore, it is important to measure element-specific magnetic moments in addition to the mean magnetization. Bulk samples of the Heusler alloys $Co_2Y_xY'_{1-x}Z_yZ'_{1-y}$ (Y, Y' = Ti, Cr, Mn, Fe and Z, Z' = Al, Ga, Si, Ge, Sn) predicted to be half-metallic ferromagnets were investigated and compared to theoretical results. Leading ideas for the search of promising materials are doping series that have a constant number of valence electrons but varying composition elucidating the distribution of the constant magnetization among the constituents. A second approach includes doping series with constant composition on the X and Y sites but varying valence electron number for optimizing the level of the Fermi energy relative to the energy gap of the minority band.

A further important aspect is the value of the magnetic orbital moment contribution. Strictly speaking half-metallic properties only exist if electron states can be classified in pure spin up and spin down states which is not the case for non-vanishing orbital moment resulting from spin–orbit coupling. Directly related to spin–orbit coupling is the magnetic anisotropy, which is the most prominent macroscopic impact of spin–orbit coupling. It determines the shape of the magnetization curve, and the information on magnetic anisotropies is therefore crucial for the development of thin film devices based on Heusler alloys. On the other hand, spin–orbit coupling is expected to be quite small in Heusler alloys as long as the alloy comprises light elements, and therefore the coupling leads to only subtle variations in the band structure. In contrast, it is expected that Heusler alloys with a heavy element at the Co lattice site will show a large orbital magnetic moment.

XMCD measurements based on acquiring the total electron yield from bulk samples require a careful preparation of a clean surface. For the first experiments the surfaces were scraped in situ in ultrahigh vacuum before the XMCD measurements in order to remove the surface oxide layer. The success of the scraping was confirmed by the vanishing of the O 1s absorption edge. It turned out that the scraping leads to a comparatively large orbital magnetic moment most probably caused by mechanical stress in the surface region. For the $Co_2Cr_{1-x}Fe_xAl$ and Co_2FeAl samples which were scraped before the measurement (Table 10.1) orbital moments of 0.5 μ_B and 0.7 μ_B per formula unit marked the maximum values observed for all compounds. A better cleaning technique is an in situ cleaving of the samples, which can be done easily in the case of brittle samples. All other materials noted in Table 10.1 were cleaved by a newly developed in-situ pair of nippers. Although the surfaces are far from smooth, the advantage of cleaving bulk samples is certainly a clean surface with the best approximation to bulk composition.

XAS/XMCD spectra obtained for cleaved $Co_2Ti_{1-x}Mn_xGe$ samples are shown in Fig. 10.1. All measurements were performed at 110 K except for $x = 0$ which was done at 300 K. For consideration of the thermal magnetization decrease, we multiplied the room temperature XMCD signal of the sample with $x = 0$ by a factor of 1.66. The Co XAS spectra in Fig. 10.1(e) reveal the largest change with composition. Similar changes occur at the L_3-edge and L_2-edge. Replacing Ti by Mn causes a vanishing of the second maximum A ($\sim$1 eV above L_3 maximum) and an arising satellite peak B ($\sim$4 eV above). Peak A represents a dominant Co–Ti d-band hybridization state. The satellite peak B corresponds to a Co-Ge sd-band hybridization state and was described as characteristic for highly ordered Heusler compounds [28]. In the quaternary Heusler alloy Mn has also some effects on the Ti states and the corresponding XMCD spectra. A systematic change of the peak height and an additional feature at 470 eV occurs in the Ti XMCD spectra (marked in Fig. 10.1(b)).

In order to quantitatively discuss spin μ_{spin} and orbital μ_{orb} moment contribution, we apply the magneto-optical sum rules as discussed in Refs. [29, 30], neglecting the

Table 10.1 Comparison of element-specific magnetic moments derived from the sum rule analysis, SQUID magnetometry (E) and values calculated by a standard LDA (T) or LDA+U (U) scheme for the indicated $X_2Y_{1-x}Y'_xZ$ Heusler compounds. Experimental values are measured at low temperatures compared to T_C. In some cases (indicated by $*$) XMCD values measured at higher temperatures are scaled to the corresponding low-temperature SQUID value. The sum moment μ_{XMCD} results from a weighted sum of the atomic moments. Values are given in μ_B per atom for element-specific moments and μ_B per formula unit for the magnetization, μ_{sum}

		μ_{spin} X	μ_{orb} X	μ_{spin} Y	μ_{orb} Y	μ_{spin} Y'	μ_{orb} Y'	μ_{sum} XMCD	μ_{sum} SQUID	Ref.
Co_2CrAl	E	0.49	0.06	0.15	0.04	–	–	1.30*	1.3	[31]
	T	0.76	0.01	1.53	0.00	–	–		3.0	[31]
$Co_2Cr_{0.6}Fe_{0.4}Al$	E	0.96	0.12	2.37	0.33	0.40	0.04	3.50*	3.5	[32]
	T	0.86	0.04	2.59	0.08	1.47	0.00		3.8	[33]
Co_2FeAl	E	1.35	0.22	1.89	0.27	–	–	5.30*	5.3	[32]
	T	1.10	0.04	2.75	0.06	–	–		5.0	[31]
	U	1.22	–	2.97	–	–	–		5.0	[34]
Co_2MnGa	E	0.86	0.03	2.56	0.04	–	–	4.38	4.0	[25]
	T	0.72	0.02	2.70	0.02	–	–		4.1	[35]
$Co_2MnGa_{0.4}Ge_{0.6}$	E	0.82	0.07	2.83	0.11			4.71	4.6	[25]
Co_2FeGa	E	1.22	0.08	2.96	0.06	–	–	5.61	5.2	[36]
	T	1.11	0.07	2.74	0.09	–	–		5.1	[35]
Co_2TiSi	E	0.94	0.05	−0.05	0.00	–	–	1.96*	1.96	[25]
	T	1.03	–	−0.02	–	–	–		2.0	[37]
$Co_2Mn_{0.6}Ti_{0.4}Si$	E	0.90	0.05	3.08	0.12	−0.01	0.00	3.82	3.8	[25]
$Co_2Mn_{0.5}Ti_{0.5}Ge$	E	0.79	0.08	3.84	0.11	−0.05	0.00	3.69	3.4	[38]
	T	1.01	–	3.19	–	−0.10	–		3.5	[38]
Co_2TiGe	E	0.94	0.08	−0.04	0.00	–	–	2.0*	2.0	[25]
	T	1.05	–	−0.04	–	–	–		2.0	[37]
Co_2TiSn	E	0.94	0.08	−0.04	0.00	–	–	2.0*	2.0	[25]
	T	1.07	–	−0.05	–	–	–		2.0	[37]
Co_2TiSb	E	0.67	0.05	0.00	0.00	–	–	1.44	1.62	[25]
	T	0.85	–	0.05	–	–	–		1.73	[37]
Co_2MnSi	E	0.72	0.05	3.34	0.11	–	–	4.99	5.0	[25]
	T	1.02	0.05	3.03	0.02	–	–		5.10	[35]
	U	0.98	–	3.29	–	–	–		5.0	[34]
Co_2FeSi	E	1.44	0.12	3.12	0.06	–	–	5.97*	5.97	[8]
	T	1.20	0.06	2.67	0.16	–	–		5.3	[35]
	U	1.56	0.08	3.24	0.07	–	–		6.0	[39]
Co_2MnGe	E	0.98	0.02	3.12	0.08	–	–	5.20	5.10	[25]
	E	1.04	0.07	3.17	0.03	–	–	5.42		[40]
	T	0.98	0.05	3.09	0.02	–	–		5.07	[35]
Co_2MnSn	E	0.76	0.09	3.52	0.12	–	–	5.32	4.94	[41]
	T	0.70	0.10			–	–		5.10	[42]
Rh_2MnGe	E	0.15	0.03	3.29	−0.02	–	–	3.63	3.5	[43]
	T	0.40	–	3.68	–	–	–		4.50	[43]

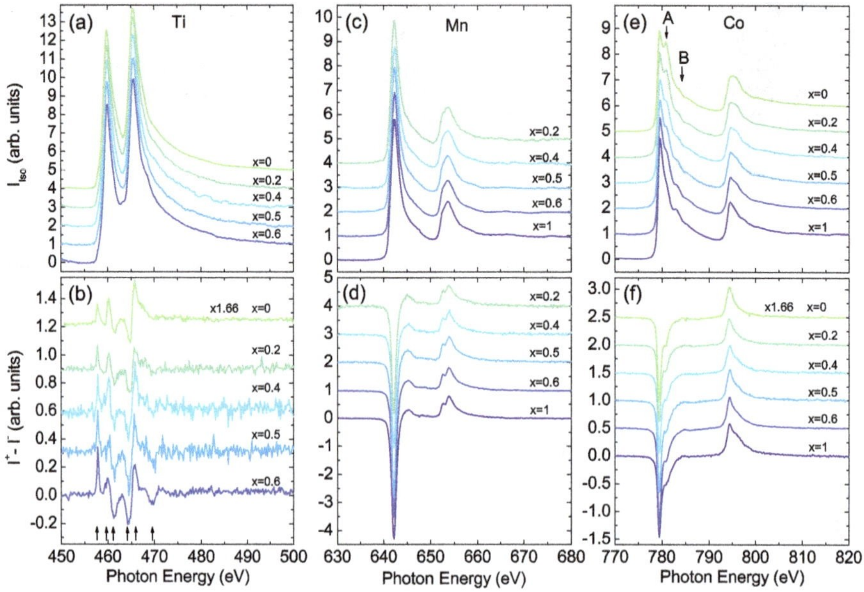

Fig. 10.1 (**a**), (**c**), (**e**) X-ray absorption spectra for $Co_2Ti_{1-x}Mn_xGe$ measured at the $L_{3,2}$ edge averaged from total electron yield intensities I^+ and I^- for magnetization direction parallel and antiparallel to the X-ray polarization. (**b**), (**d**), (**f**) corresponding XMCD spectra $I^+–I^-$. Reproduced from P. Klaer, E.A. Jorge, M. Jourdan, W.H. Wang, H. Sukegawa, K. Inomata, H.J. Elmers, *Phys. Rev. B* 82, 024418 (2010). DOI 10.1103/PhysRevB.82.024418. Copyright (2010) by the American Physical Society

magnetic dipole operator. The omitting of the dipole operator was later confirmed by angle resolved measurements of single crystals. Both orbital and spin magnetic moments are proportional to the number of *d* holes N_h which is different for each element. Its measurement is a principal problem for alloys as it cannot be uniquely determined by bulk magnetization measurements. N_h can in principle be determined quantitatively from the XAS spectra with, however, a quite large error [17]. Instead one may use theoretically calculated values. Fortunately, it turned out that the variation for a certain element in a Heusler alloy depends only weakly on the composition. Accepting an error of 10 % one may use values of $N_h(\mathrm{Ti}) = 8$, $N_h(\mathrm{V}) = 7$, $N_h(\mathrm{Cr}) = 6$, $N_h(\mathrm{Mn}) = 4.5$, $N_h(\mathrm{Fe}) = 3.5$, $N_h(\mathrm{Co}) = 2.5$, and $N_h(\mathrm{Ni}) = 1.5$ for the 3*d* elements. In addition a correction for the jj mixing for the light 3*d* elements must be considered. The correction factors are $c_{jj}(\mathrm{Ti}) = 5$, $c_{jj}(\mathrm{V}) = 3.6$, $c_{jj}(\mathrm{Cr}) = 2$, $c_{jj}(\mathrm{Mn}) = 1.4$ [44, 45]. Considering these uncertainties the relative error of element-specific moments is at least 10 %. However, results evaluated in a similar way show a considerably better reproducibility and despite a large common error the discussion of trends is useful.

For Co_2CrAl and $Co_2Cr_{0.6}Fe_{0.4}Al$ samples that were quenched after melting we obtained spin magnetic moments in agreement with the band structure calculations for Co and Fe atoms, whereas Cr atoms possess a much smaller magnetic moment. The comparatively large orbital magnetic moments are attributed to the surface scraping before the XMCD measurement. Extensive annealing of Co_2CrAl and $Co_2Cr_{0.6}Fe_{0.4}Al$ causing severe deviations of the $L2_1$ order reduces magnetic moments at Cr atoms drastically. Therefore, a small Cr moment is indicative for a reduced atomic order [32]. For Co_2FeAl the magnetization depends weakly on the degree of local order with a trend to an increased magnetization for reduced order. For this compound we also find a quite large orbital moment anisotropy [46] in combination with a large absolute orbital moment.

The Ti moment in the $Co_2Ti_{1-x}Mn_xZ$ compounds is in all samples small and oriented antiparallel to the mean magnetization direction, indicating that Ti containing Heusler compounds are ferrimagnets. The Mn moments are comparatively large, consisting mainly of a spin moment as the orbital to spin ratio is less than 0.03. The Co moment shows values of 0.8 μ_B–1.0 μ_B. Co_2TiSb is an exception with a vanishing Ti moment and a smaller Co moment. For the parent compounds with $x = 0$ and $x = 1$, the total magnetization and the element-specific moments are in agreement with theoretical results.

The substitutional series of quaternary Heusler compounds show a linearly increasing magnetization with increasing number of valence electrons obeying the generalized Slater–Pauling rule. Slightly exceeding values compared to the Slater–Pauling rule were reported in the case of $Co_2Mn_{1-x}Fe_xAl$ [47] and attributed to partial disorder.

Two complementary series are investigated with transition-metal substitution on the Y site for $Co_2Ti_{1-x}Mn_xGe$ and main group element substitution on the Z site for $Co_2MnGa_{1-x}Ge_x$ as shown in Fig. 10.2. We find that the distribution of the magnetization on the X and Y sublattices varies with the number of valence electrons N_v in a similar way for the two complementary series. For $Co_2Ti_{1-x}Mn_xGe$, the element-specific moments remain almost constant and the increase in the magnetization with increasing x is directly related to the increasing Mn content. Thus, the Y sublattice covers the total magnetization increase. For $Co_2MnGa_{1-x}Ge_x$, we find an almost constant Co moment and an increasing Mn moment on the Y sublattice. The almost constant Co moment reflects the fact that the distribution of the Co d electrons in the majority and minority states does not change with N_v. In order to maximize the number of unpaired electrons, the minority d states should be filled by eight electrons to have a gap and the rest of the electrons fill the majority states [48]. As the Co minority d states are occupied by roughly 3.5 electrons at each atom, only one minority d electron stems from the Y atom. This explains the increasing moment on the Y site going from Ti to Mn. The increase in the Mn moment for $Co_2Ti_{1-x}Mn_xGe$ cannot be explained by this general behavior of d electrons. In this case, the hybridization of electrons at the Z site causes a redistribution of majority and minority electrons when the Ga is replaced by Ge. Because Ge has a higher electronegativity compared to Ga, the bonding interaction between Co and Ge is stronger than for Co and Ga. This causes an electron transfer to delocalized states in the interstitial volume. In particular, the t_{2g} minority states at the Y site are emptied. At the same time, the majority t_{2g} states are filled resulting in a larger moment.

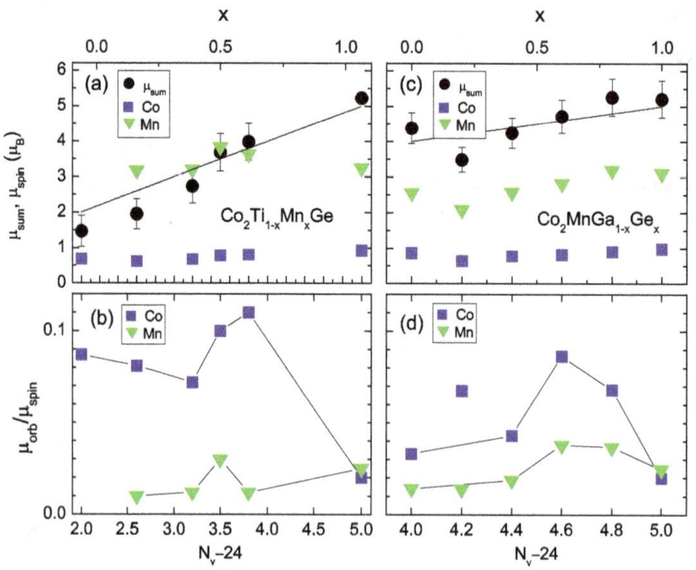

Fig. 10.2 (**a**) Total magnetic moments (*circles*) and element-specific spin moments for $Co_2Ti_{1-x}Mn_xGe$ as a function of the Mn concentration. The *solid line* indicates the Slater–Pauling rule. (**b**) Ratio of the orbital magnetic moment over the spin magnetic moment. (**c**) Total magnetic moments (*circles*) and element-specific spin moments for $Co_2MnGa_{1-x}Ge_x$ as a function of the Ge concentration. The *solid line* indicates the Slater–Pauling rule. (**d**) Ratio of the orbital magnetic moment over the spin magnetic moment. Reproduced from P. Klaer, M. Kallmayer, C.G.F. Blum, T. Graf, J. Barth, B. Balke, G.H. Fecher, C. Felser, H.J. Elmers, *Phys. Rev. B* 80, 144405. DOI 10.1103/PhysRevB.80.144405. Copyright (2009) by the American Physical Society and from P. Klaer, E.A. Jorge, M. Jourdan, W.H. Wang, H. Sukegawa, K. Inomata, H.J. Elmers, *Phys.Rev. B* 82, 024418 (2010). DOI 10.1103/PhysRevB.82.024418. Copyright (2010) by the American Physical Society

Although the orbital moment contributes less than 10 % to the total moment, its variation with composition is an important indication for local deviations from cubic symmetry. Large orbital moments may also be a hint to an increased magnetic crystal anisotropy and to an increased Gilbert damping. The orbital magnetic moments are small for the parent Heusler compounds but increased values are found in the case of intermediate compounds. For $Co_2Ti_{1-x}Mn_xGe$ the maximum orbital moment appears at $x = 0.4$ and for $Co_2MnGa_{1-x}Ge_x$ at $x = 0.6$. We attribute this observation to a local distortion from the cubic symmetry caused by a random occupation of the Y and Z sites, respectively.

Not all quaternary Heusler compounds form homogeneous alloys. Examples for decomposing alloys are $Co_2Ti_{1-x}Mn_xSi$ and $Co_2Ti_{1-x}Mn_xSn$ [38]. Interestingly, these compounds decompose nearly completely into the two parent Heusler compounds Co_2MnZ and Co_2TiZ, which could be interesting for increasing the ratio between electrical and thermal conductivity for thermoelectric applications [49].

XAS/XMCD spectra at the Rh $M_{3,2}$ edges and the Mn $L_{3,2}$ edges of the ferromagnetic Heusler alloy Rh_2MnGe were compared with ab-initio calculations using a single-electron model for the description of the X-ray absorption and with an atomic multiplet calculation [43]. The overall line shapes of the Mn $2p$ XAS and XMCD spectra are better reproduced by a multiplet calculation than for other Mn containing Heusler alloys. The Rh spectra show no multiplet features and can be reproduced by LDA calculations. From this observation we conclude that the Mn moment is strongly localized. The Rh moment is an order of magnitude smaller than the Mn moment. The temperature dependences of the Mn and Rh moments are almost equal to each other in contrast to the prediction by a Heisenberg model. This might be attributed to the remaining itinerant character of the Rh moment. The contribution of the orbital magnetic moment is very small for both elements with orbital-to-spin moment ratios of 0.01 ± 0.02 for Mn and 0.20 ± 0.15 for Rh. The magnetization agrees within error limits with bulk magnetometry at low temperatures but shows a stronger decrease with increasing temperature. We attribute this reduction in surface magnetization to the surface effect induced by the broken symmetry.

10.3 Epitaxial Films

For most application devices the relevant property is the spin polarization at the surface of the material instead of its bulk properties. For tunneling magnetoresistance devices the topmost atomic layer at the interface exclusively determines the TMR value. Therefore, the evaluation of magnetic properties at the surface is important for the search of materials with high spin polarization and consequently high TMR values at the interface.

Quantitative information on element-specific spin and orbital magnetic moment obtained by XMCD comprises uncertainties because of the experimentally unaccessible number of d-holes N_h as discussed in the previous section. Additional problems may arise for total electron yield measurements. For thin films with a common in-plane easy axis a favorable geometry is provided by the X-ray light shining at grazing incidence onto the sample thus requiring only small fields to reach saturation. In this case the sum rule analysis leads to large errors caused by the X-ray self absorption.

In order to avoid these problems we simultaneously measure the XMCD signal in transmission (TM) and by total electron yield (TEY) and apply similar evaluation parameters for the sum rule analysis in both cases [14, 50]. An external magnetic field of 1.6 Tesla saturates the sample magnetization perpendicular to the film surface and parallel to the incident X-ray beam. The transmission signal was measured via the X-ray luminescence within the substrate, thus allowing the investigation of epitaxially grown films, in contrast to previous approaches using parylene foils or SiN membranes. Common errors including finite polarization of X-ray light, unknown number of unoccupied d-states, background subtraction for transition into s-states, and errors due to separation of L_3 and L_2 transitions cancel for comparison of TEY and TM. As TEY is surface sensitive while the transmission signal averages along the complete film normal we can infer a tendency for the depth dependence of the magnetization.

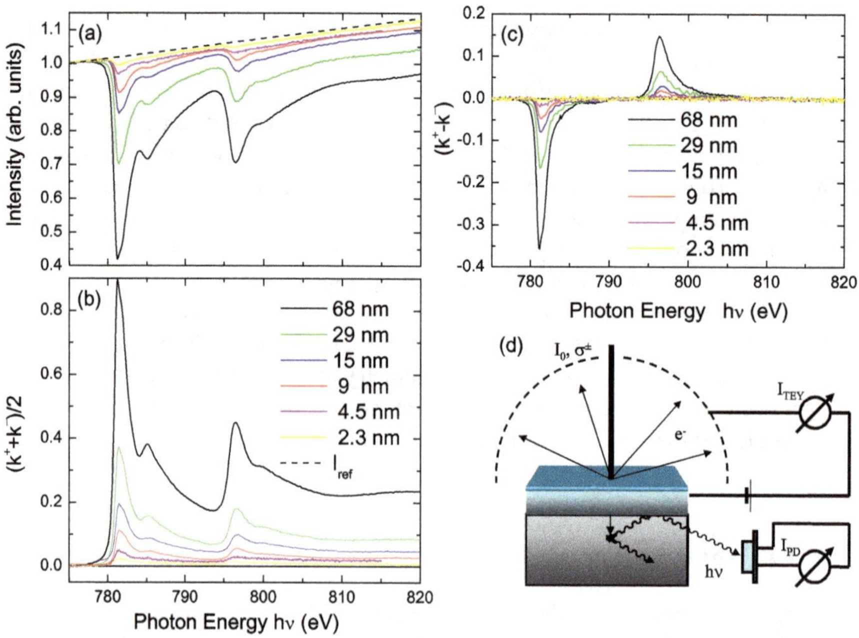

Fig. 10.3 (**a**) Intensity of the luminescence light emerging from the substrate measured at the Co $L_{2,3}$-edge of Co_2FeSi films on Al_2O_3. Thicknesses of the films are indicated in the figure. The intensity was normalized at the pre-edge. (**b**) Corresponding absorption coefficient $k = -\ln(I/I_{ref})$ for various film thicknesses. (**c**) Circular magnetic dichroism signal $(k^+ - k^-)$ for thicknesses indicated in the figure. (**d**) Cartoon of the experimental setup used for the X-ray absorption experiment in transmission and TEY. Reproduced with permission from M. Kallmayer, H. Schneider, G. Jakob, H.J. Elmers, B. Balke, S. Cramm, *J. Phys. D* 40, 1552. DOI 10.1088/0022-3727/40/6/S07. Copyright (2007), IOP Publishing

For illustration Fig. 10.3(a) shows the incident-photon-flux-normalized transmission XAS spectra of Co_2FeSi/Al_2O_3 Heusler alloy films for thicknesses $d = 2.3$ nm to $d = 68$ nm averaged over the two magnetization directions. For the thickest film the transmission signal at the L_3 edge is less than half the intensity measured at the pre-edge. Assuming that the luminescence signal of the substrate $I^{\pm}_{lum}$ is proportional to the transmitted X-ray intensity, the X-ray absorption coefficient k can be calculated using the equation $k^{\pm}(h\nu) = -\ln[I^{\pm}(h\nu)/I_{ref}(h\nu)]$ (Fig. 10.3(a)). The reference spectra $I_{ref}(h\nu)$ was measured by the bare substrate crystal and found to increase linearly with the photon energy. $I_{ref}(h\nu)$ was then normalized at the pre-edge region of the corresponding element. This normalization corresponds to an infinitely large penetration depth at the pre-edge. XAS and XMCD signals are very similar to measurements of bulk samples.

Experiments on $Co_2Cr_{0.6}Fe_{0.4}Al$ films that were deposited in a not fully optimized process show that the magnetic moments calculated from the surface sensitive TEY are reduced by 17 % averaged over the electron escape depth of about 25 Å with respect to the corresponding moments determined from the transmission

Table 10.2 Comparison of element-specific magnetic moments derived from the transmission spectra (TM) and from the surface sensitive total electron yield spectra (TEY) for the indicated $X_2Y_{1-x}Y'_xZ$ epitaxial Heusler films. In some cases (indicated by $*$) XMCD values measured at higher temperatures are scaled to the corresponding low-temperature SQUID value. The sum moment μ_{XMCD} results from a weighted sum of the atomic moments. Values are given in μ_B per atom for element-specific moments and μ_B per formula unit for the magnetization

		μ_{spin} X	μ_{orb} X	μ_{spin} Y	μ_{orb} Y	μ_{spin} Y'	μ_{orb} Y'	μ_{sum} XMCD	μ_{sum} SQUID	Ref.
Co_2CrAl	TM	0.70	0.07	0.79	0.00	–	–	2.33	1.3	[51]
	TEY	0.68	0.07	0.46	0.00	–	–	1.96		
$Co_2Cr_{0.6}Fe_{0.4}Al$	TM	0.89	0.07	2.64	0.05	0.20	0.00	3.12	3.4	[14]
	TEY	0.74	0.08	1.90	0.10	0.10	0.00	2.50		
$Co_2Cr_{0.6}Fe_{0.4}Al$	TM	0.96	0.09	2.50	0.08	0.80	0.10	3.67	3.5	[51]
	TEY	0.99	0.06	2.51	0.10	0.96	0.11	3.79		
$/Al_2O_3$	TEY	0.99	0.06	2.42	0.10	0.10	0.00	3.17		
Co_2FeAl	TM	1.36	0.14	2.28	0.12	–	–	5.40	5.0	[52]
	TEY	1.27	0.12	2.52	0.04	–	–	5.34		
Co_2MnAl	TM	0.84	0.03	3.02	0.00	–	–	4.76	4.2	[53]
	TEY	0.73	0.03	2.81	0.00	–	–	4.33		
$Co_2FeAl_{0.3}Si_{0.7}$	TM	1.12	0.08	2.45	0.10	–	–	4.75	4.9	[54]
	TEY	1.26	0.12	2.21	0.10	–	–	4.92		
$Co_2FeAl_{0.5}Si_{0.5}$	TM	1.41	0.09	2.43	0.10	–	–	5.5*	5.5	[55]
	TEY	1.36	0.10	2.56	0.10	–	–	5.5*		
Co_2MnSi	TM	0.87	0.04	3.20	0.00	–	–	5.02	5.0	[15]
	TEY	0.80	0.05	3.04	0.11	–	–	4.85		
$Co_2Fe_{0.5}Mn_{0.5}Si$	TM	1.20	0.08	2.85	0.10	3.00	0.00	5.56	5.5	[15]
	TEY	1.20	0.08	2.80	0.12	2.80	0.06	5.46		
Co_2FeSi	TM	1.37	0.14	2.66	0.08	–	–	5.75	6.0	[15]
	TEY	1.35	0.07	2.80	0.12	–	–	5.54		
$Co_2FeGa_{0.5}Ge_{0.5}$	TM	1.25	0.07	2.50	0.06	–	–	5.2	5.3	[53]
	TEY	1.14	0.06	2.64	0.10	–	–	5.14		

signal (see Table 10.2) [14]. Since a homogeneous magnetization reduction at the surface is unlikely the reduction infers an even larger reduction for the topmost layer and may explain the observed low TMR values and spin polarization at the Fermi level of these films. Depositing in chambers with lower base pressure and adjusting the stoichiometry of the compounds carefully, these problems could partly be avoided [51].

A similar reduction of surface magnetization was detected for $Co_2FeSi(110)$ films grown by RF magnetron sputtering on Al_2O_3 $(11\bar{2}0)$ substrates [50, 56]. From

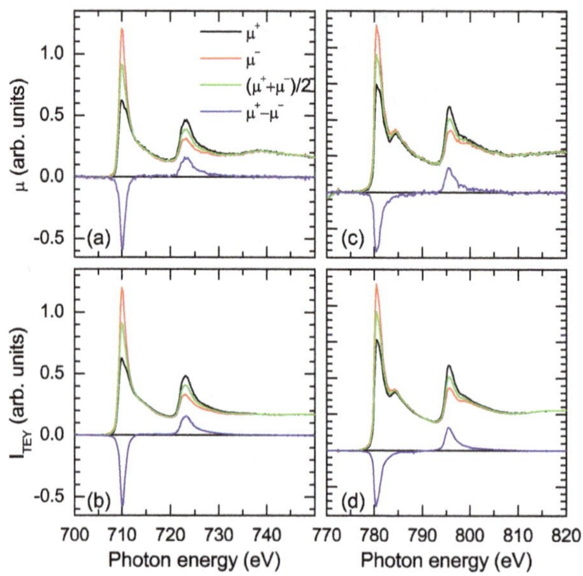

Fig. 10.4 (**a**) Absorption coefficient at the Fe $L_{2,3}$ edge for magnetization direction parallel (antiparallel) to the X-ray polarization, $\mu^{+(-)} \propto -\ln(I^{+(-)}/I_{ref})$, for a Co_2FeSi film on MgO. The XMCD spectrum, $\mu^+ - \mu^-$ is shown on the same scale. (**b**) Simultaneously measured total electron yield spectra $I_{TEY}^{+(-)}$ for the same sample. (**c**)–(**d**) Corresponding data measured at the Co $L_{2,3}$ edge. Reproduced with permission from M. Kallmayer, P. Klaer, H. Schneider, G. Jakob, H.J. Elmers, D. Legut, and P.M. Oppeneer, Element-specific magnetic properties of $Co_2(Mn_{1-x}Fe_x)Si$ films probed by X-ray magnetic circular/linear dichroism, *Physical Review B* 84, 054448. Copyright (2011) by the American Physical Society

a comparison of the two experimental techniques TEY and TM we found that the magnetization at the interface between Heusler film and Al capping is reduced by 8 % integrated over the electron penetration depth of 2.5 nm. Combining this result with the total interface reduction evaluated from the thickness dependence of the magnetic moment, we obtain a second interesting information. At the interface between the Heusler film and the substrate the magnetization is reduced, too, corresponding to 0.55 nm of magnetically dead layers.

Epitaxial Co_2FeAl [52] and $Co_2FeAl_{0.5}Si_{0.5}$ [55] films for which large TMR values have been measured and a large spin polarization was inferred from other methods show a negligible difference between bulk and interface magnetization. Interestingly, in these films the Co moment is slightly reduced and the Fe moment slightly increased at the interface (TEY) compared to their bulk (TM) values. Hence, the interface moments are even closer to the theoretical prediction for the fully $L2_1$-ordered compound.

Similar favorable trends were observed for laser ablated films consisting of the compounds $Co_2Fe_xMn_{1-x}Si$ and $Co_2FeGa_{1-x}Ge_x$ (see Table 10.2) [53, 57]. Transmission and TEY Co XAS signals (Figs. 10.4(c)–(d)) show the typical features, i.e. a satellite peak at 4 eV above the L_3 absorption edge, observed already for bulk samples, which is indicative for a high degree of local order [28, 32]. The extra peak was explained by a hybridization of Co d-band states with sp-states of the main group element. For all samples of this series the element-specific spectra do not show a prominent deviation between TEY and TM data indicating the high quality of the interface. The element specific moments (see Table 10.2) do not show large differences except for the orbital moments. The Fe and Mn surface orbital moments are larger compared to the corresponding bulk values. For the Co orbital moments we observe a smaller bulk orbital moment in the case of Co_2MnSi and a larger value in the case of Co_2FeSi compared to the surface orbital moment. This observation might indicate a deviation from the cubic crystal symmetry in the surface region invoked by the deposition process or by the capping layer.

10.4 Spin-Resolved Density of States

The band gap in the minority-spin states arises from the hybridization of Co and Y $3d$ orbitals. The width of the gap is determined by the Co-Co interaction because these states are closest to the Fermi energy. In order to overcome the thermally induced suppression of high spin polarization further band structure tailoring through doping of the Heusler alloys has been proposed [6, 58, 59]. An especially interesting example is $Co_2FeAl_{1-x}Si_x$ [59] for which ab-initio calculations predict a Fermi energy E_F in the center of the minority gap for $x = 0.5$, in contrast to $x = 0$ and $x = 1$ with E_F positioned close to the upper or lower boundary of the gap [60, 61]. Therefore, a direct study of the band gap is of particular importance. Although spin-resolved photoemission [9, 62] or scanning tunneling spectroscopy can directly probe the spin polarization at a half-metal surface, these methods have no access to the crucial buried interfaces in spintronics devices.

In contrast, XMCD/XAS is capable to resolve the element-specific electronic structure at buried interfaces [17]. In principle, the L-edge absorption spectra for left and right circularly polarized X-ray light reflects the spin-resolved partial density of states (PDOS) at the $3d$ transition metal atoms [19, 20]. Previous investigations at the Co and Ni L-edge in intermetallic compounds clearly revealed PDOS related features in the absorption spectra of Heusler alloys [23, 24].

The experimental spin-resolved unoccupied Co PDOS is shown in Fig. 10.5 derived from the Co L_3-edge absorption data for series of $Co_2Mn_xTi_{1-x}Ge$ and $Co_2MnGa_xGe_{1-x}$ Heusler compounds. For the calculation we used the spin-resolved unoccupied PDOS function (Fermi function f_F) following from the XAS spectra I^+ and I^-

$$D^{\uparrow(\downarrow)}(1 - f_F) \propto I_{\text{iso}} - s + (-)\frac{1}{P_j}\frac{I^+ - I^-}{2}, \tag{10.1}$$

where I_{iso} denotes the isotropic absorption coefficient $(I^+ + I^-)/2$, s is the step function and P_j is the spin polarization of the excited photoelectrons, i.e., $P_{L3} = 0.25$ and $P_{L2} = -0.5$. An electron correlation effect causes a shift of the majority

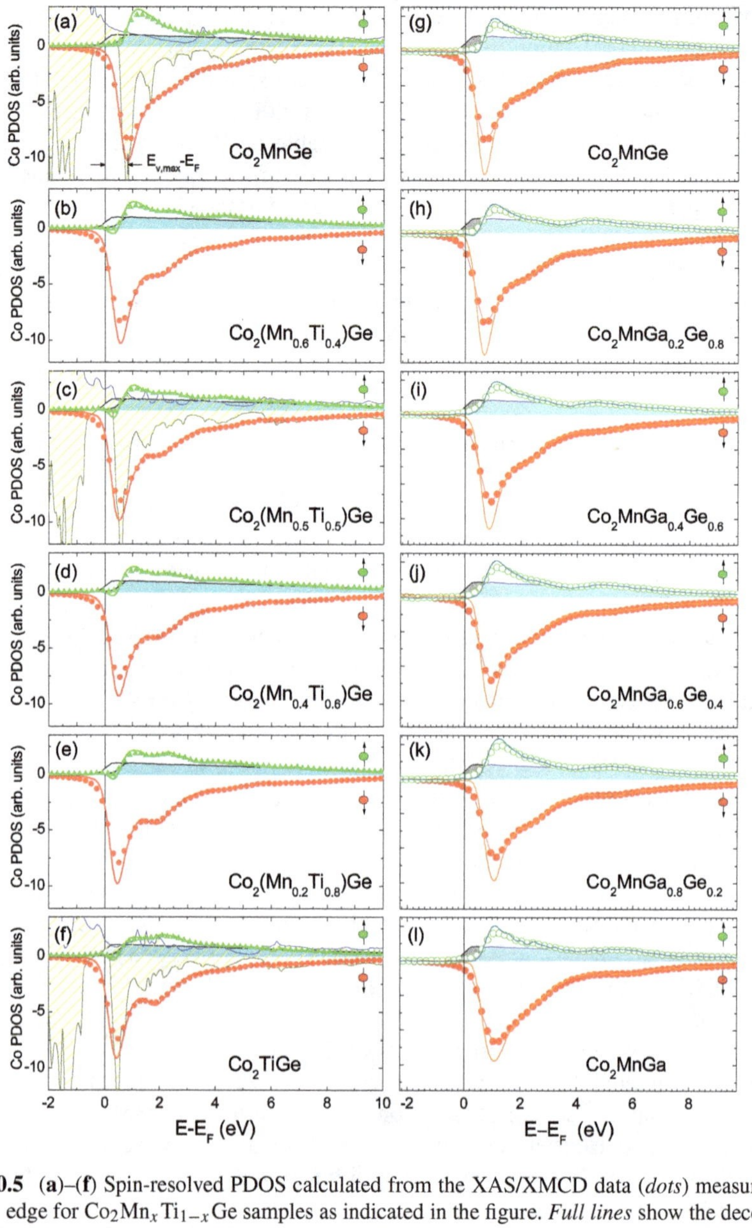

Fig. 10.5 (**a**)–(**f**) Spin-resolved PDOS calculated from the XAS/XMCD data (*dots*) measured at the L_3 edge for $Co_2Mn_xTi_{1-x}Ge$ samples as indicated in the figure. *Full lines* show the deconvoluted data. *Full filled areas* indicate the function used for approximation of the itinerant band as measured (*blue*) and shifted by ΔE_c (*light blue*). Data from theoretical ab initio calculations are indicated by a *yellow patterned area* in (**a**), (**c**) and (**f**). (**g**)–(**l**) Similar data for a $Co_2MnGa_xGe_{1-x}$ samples series. Reproduced from P. Klaer, M. Kallmayer, C.G.F. Blum, T. Graf, J. Barth, B. Balke, G.H. Fecher, C. Felser, H.J. Elmers, *Phys. Rev. B* 80, 144405. DOI 10.1103/PhysRevB.80.144405. Copyright (2009) by the American Physical Society and from P. Klaer, T. Bos, M. Kallmayer, C.G.F. Blum, T. Graf, J. Barth, B. Balke, G.H. Fecher, C. Felser, H.J. Elmers, *Phys. Rev. B* 82, 104410. DOI 10.1103/PhysRevB.82.104410. Copyright (2010) by the American Physical Society

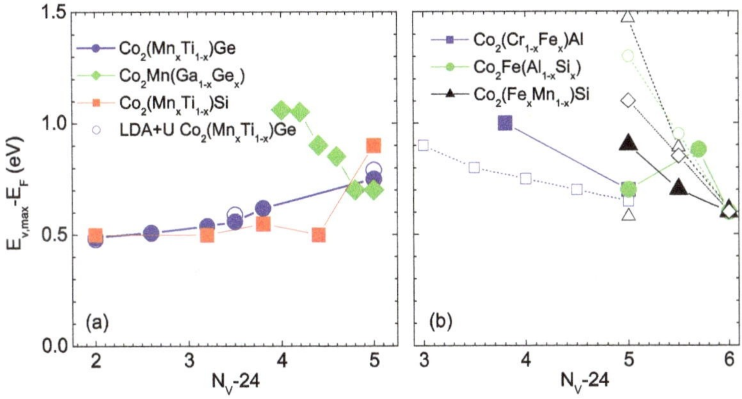

Fig. 10.6 (**a**) Difference of the minority PDOS maximum and the Fermi energy $E_{v,\max} - E_F$ for the indicated Heusler alloy bulk samples (*full symbols*). N_v indicates the number of valence electrons per formula unit. (**b**) Difference of the minority PDOS maximum and the Fermi energy $E_{v,\max} - E_F$ for the epitaxial Heusler films (*full symbols*) compared with calculations (*open symbols*) using LDA (*squares* [31], *triangle* [63]), LDA+U (*triangles* [64], *circles* [60]) and LDA+DMFT (*diamonds* [65]). Reproduced from M. Kallmayer, P. Klaer, H. Schneider, E. Arbelo-Jorge, C. Herbort, G. Jakob, M. Jourdan, H.J. Elmers, *Phys. Rev. B* 80, 020406(R). DOI 10.1103/PhysRevB.80.020406. Copyright (2009) by the American Physical Society and from P. Klaer, T. Bos, M. Kallmayer, C.G.F. Blum, T. Graf, J. Barth, B. Balke, G.H. Fecher, C. Felser, H.J. Elmers, *Phys. Rev. B* 82, 104410. DOI 10.1103/PhysRevB.82.104410. Copyright (2010) by the American Physical Society

PDOS of $\Delta E_c = 0.5$ eV with respect to the minority states as explained in Refs. [15, 25]. The position of the Fermi level was derived from the initial increase in the majority states considering the energy shift of ΔE_c.

Due to the lifetime broadening of the XAS spectra we have deconvoluted the experimental spectra with a Lorentzian function with a width of $\Gamma = 0.35$ eV. Note that the Co minority PDOS is close to zero in a finite energy interval around E_F. The observation of a vanishing PDOS for $E < E_F$ is a consequence of a near coincidence of the Lorentzian function and the experimental data. Thus, the experimental data support the theoretical prediction of half-metal behavior particular for Co_2MnGe. In comparison with the LDA+U calculations we can find a nice coincidence (see Figs. 10.5(a), (c), (f)). Only Co_2MnGe shows a deviation in the majority states which has not been understood yet. Co–Ti d hybridization states are well reproduced in the minority states (1.8 eV above E_F), even though from the experimental results one can assume that the Co–Ti hybridization state should be more pronounced in the calculations.

Replacing Ti by Mn reduces the difference $E_{v,\max} - E_F$ between the prominent maximum $E_{v,\max}$ in the minority PDOS and E_F as indicated in Fig. 10.5(a). Results for $E_{v,\max} - E_F$ are summarized in Fig. 10.6 and compared to ab initio calculations. The theoretical results show good agreement with the experimental results. The linear increase in $E_{v,\max} - E_F$ with increasing valence electron number N_v is quantitatively reproduced. This indicates that the previously predicted half-metallic property for this series indeed exists in the samples. Figure 10.6 also shows the results for isoelectronic $Co_2Mn_xTi_{1-x}Ge$ and $Co_2MnGa_{1-x}Ge_x$ series from Ref. [25] for comparison. Instead of an increasing value of $E_{v,\max} - E_F$, $Co_2Mn_xTi_{1-x}Si$ shows only one step from $x = 0.8$ to $x = 1$. This can be explained by the phase inhomogeneity discussed before. Co_2TiSi impurities in the quaternary $Co_2Mn_xTi_{1-x}Si$ compound dominate the energetic position of $E_{v,\max} - E_F$. A completely different behavior is observed for $Co_2MnGa_{1-x}Ge_x$. In this case we observe a linear decrease in $E_{v,\max} - E_F$ with increasing N_v. The decrease in $E_{v,\max} - E_F$ is a consequence of filling the majority Co states with electrons. This shifts the Fermi energy to the right side of the band gap as expected from the rigid-band model. For $Co_2Mn_xTi_{1-x}Ge$ the opposite behavior is observed. This unexpected behavior can only be explained by a charge transfer from Ti to Co and Mn due to the small electronegativity of Ti in combination with the half-metallic properties of this series.

We also investigated the spin-resolved DOS of epitaxial (100) oriented films of $Co_2Fe_xMn_{1-x}Si$, $Co_2FeAl_{1-x}Si_x$ and $Co_2Cr_{0.6}Fe_{0.4}Al$ films grown on MgO(100) substrates using XMCD data [15]. The experimental results corroborate the predicted half-metallic ferromagnetic properties of these alloys. In particular, data for Co_2MnSi films provide a solid base for half-metallic properties. We observed a variation of the position of E_F within the minority band gap with the substitution of the transition metal element or the main group element in agreement with earlier predictions. For the $Co_2(Fe_xMn_{1-x})Si$ films the spin-resolved Co PDOS reveals a minority maximum at $E_{v,\max}$ approaching E_F with increasing x as depicted in Fig. 10.6. Ab-initio calculations have predicted this trend, although absolute values varied depending on the model assumptions and calculation schemes. LDA+U [39] fits better to the experiment, however, the value of U appears overestimated for Co_2MnSi. The best agreement with experiment is achieved for a recent LDA+DMFT calculation [65]. The comparison with theory allows an estimation of theoretical parameters, e.g. the electron–electron correlation potential. In addition one can observe a correlation of local order and electronic structure. For $Co_2FeAl_{1-x}Si_x$ films a broadening of the minority PDOS with increased local disorder was observed.

The temperature dependence of the electronic states in half-metals might contribute to the strong decrease of the spin polarization and its experimental investigation is highly interesting. In a single-particle model changes of the electronic structure are only expected in the vicinity of the Fermi level due to thermally excited occupation of previously unoccupied states which is in agreement with a recent high energy photo emission study for a half-metallic Co_2MnSi/MgO interface [26]. Instead, a model beyond the single-particle picture of the local density approximation considering finite temperature many-body effects [58, 66] results in a considerable temperature dependence of the DOS.

Using XAS/XMCD spectroscopic measurements of temperature dependent data of the spin-resolved DOS were examined in two different Heusler alloys, which were predicted to be half-metallic comprising band gaps in the majority spin band (Mn_2VAl) [67] and in the minority spin band (Co_2FeAl) [31], respectively.

The result for the Mn related PDOS in Mn_2VAl is shown in Figs. 10.7(a), (b) and compared to theoretical data. The minority states clearly reproduce the ab-initio

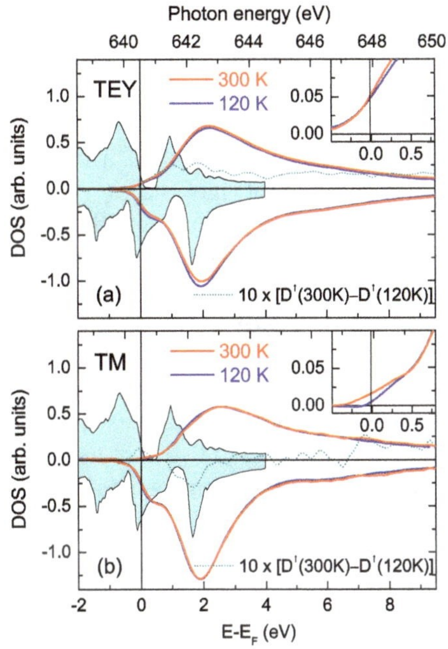

Fig. 10.7 (**a**) Spin-resolved unoccupied PDOS calculated from the XAS/XMCD TEY data measured at the Mn L_3-edge of the Mn_2VAl film. The majority and minority PDOS is shown on a positive and negative scale, respectively. The difference of the low and high temperature majority PDOS is ten-fold enhanced, emphasizing the temperature induced spectral changes near the majority band gap. The theoretical data from Ref. [66] are shown as *shaded areas*. The *inset* shows the magnified energy region around E_F for majority states. (**b**) Spin-resolved PDOS derived from the transmission data. Reproduced from P. Klaer, E.A. Jorge, M. Jourdan, W.H. Wang, H. Sukegawa, K. Inomata, H.J. Elmers, *Phys. Rev. B* 82, 024418 (2010). DOI 10.1103/PhysRevB.82.024418. Copyright (2010) by the American Physical Society

calculation considering the lifetime broadening of the spectral features. The large maximum at $E - E_F = 1.9$ eV corresponds to the peak in the theoretical data at 1.7 eV. The pronounced shoulder closer to E_F is the remainder of the maximum in the theoretical data just below E_F. The pronounced maximum of the majority states observed at 2.2 eV above E_F (see Figs. 10.7(a), (b)) is at variance with theory. This might be explained by static electron correlation effects that are not properly considered by the evaluation procedure or by the theoretical calculation.

The majority state data are different for surface and bulk sensitive measurements. While the TEY data [Fig. 10.7(a)] reveals a pronounced non-zero PDOS at E_F, the transmission data [Fig. 10.7(b)] is almost zero at E_F for low temperatures and shows a small positive value at room temperature. The difference $D^{\uparrow}(300\text{K}) - D^{\uparrow}(120\text{K})$ has clearly a maximum below E_F. This temperature dependence cannot be explained by a thermal broadening of the Fermi function, which is much smaller at 300 K and would show up antisymmetrically below and above E_F. Similar temperature dependence is clearly absent for the case of the TEY related data. This fact coincides with the strong decrease of the Mn and V interface magnetic moments by a factor of two. The origin of the reduced interface magnetization is unclear. One might speculate that it is caused by a reduced atomic order at the interface since films deposited at room temperature without annealing do show a strongly reduced atomic and magnetic order in the bulk of the films. Therefore we suggest that the interface region is metallic instead of half-metallic.

Our experimental observation of the increase of spectral weight just below the Fermi level for the case of the TM data of Mn_2VAl with a majority band gap is in qualitative agreement with the model of non-quasiparticle (NQP) states, calculated for real systems using local density approximation plus dynamical mean-field theory (LDA+DMFT) [66]. For the case of Co_2FeAl a similar unexpected change of the PDOS was observed with increasing temperature in this case both for interface and bulk related data. For the minority states an extra peak appears slightly above E_F as predicted by the non-quasiparticle model [58]. These temperature induced changes of the spin-resolved DOS of two Heusler compounds cannot be explained by thermal excitations within a single-particle model. The increase of the DOS within the band gap possibly contributes to the observed strong temperature dependence of the TMR effect.

10.5 Summary

Element-specific spin and orbital magnetic moments of polycrystalline bulk Heusler alloys and epitaxial Heusler alloy films with potential half-metallic properties were investigated using magnetic circular dichroism in X-ray absorption spectroscopy (XAS/XMCD). In addition to stoichiometric compounds composition series with partly replaced elements on the Y-site and on the Z-site with the aim of tailoring the band structure for specific applications were measured. Comparison of experimental results with theoretical predictions elucidate the influence of local disorder and the importance of electron correlation. Increased orbital magnetic moments in respect to theoretical predictions put forward the role of spin–orbit coupling for half-metallic properties. For the case of single crystalline thin films a method of simultaneous measurement of bulk and surface sensitive magnetic properties including those at the crucial interface to a tunneling barrier was presented. Exploiting the comparison of bulk and interface information improves film growth processes. The spin-resolved DOS was extracted from XAS/XMCD spectra considering final-state electron correlations. This information allows a direct comparison with theoretical predictions. The experimental results, revealing the distribution of magnetic moments and the relative position of the Fermi energy as a function of the number of valence electrons, confirm the predicted possibility of tailoring the minority band gap using substitutional quaternary Heusler compounds and may be of general importance for the understanding of the electronic structures in complex intermetallic compounds. Temperature dependent measurements shed light on many-body correlation effects in half-metals beyond the single-particle model. Thus, XAS/XMCD experiments have revealed that Heusler alloys form a very attractive class of new materials with high spin polarization and tunable band structure providing unexpected and interesting new correlation physics.

Acknowledgements We wish to thank S. Cramm for many useful discussions and support at the Beamline UE56/1-SGM at the German Light source BESSY II. This work was financially supported by the Deutsche Forschungsgemeinschaft (project P5 in research unit FOR 559).

References

1. Inomata K, Okamura S, Goto R, Tezuka N (2003) Jpn J Appl Phys 42:L419. doi:10.1143/JJAP.42.L419
2. Kämmerer S, Thomas A, Hütten A, Reiss G (2004) Appl Phys Lett 85:79. doi:10.1063/1.1769082
3. Sakuraba Y, Hattori M, Oogane M, Ando Y, Kato H, Sakuma A, Miyazaki T, Kubota H (2006) Appl Phys Lett 88:192508. doi:10.1063/1.2202724
4. Wang W, Sukegawa H, Shan R, Inomata K (2008) Appl Phys Lett 93:122506. doi:10.1063/1.2988649
5. Tsunegi S, Sakuraba Y, Oogane M, Takanashi K, Ando Y (2008) Appl Phys Lett 93:112506. doi:10.1063/1.2987516
6. Felser C, Fecher GH, Balke B (2007) Angew Chem, Int Ed Engl 46:668. doi:10.1002/anie.200601815
7. Galanakis I, Dederichs PH, Papanikolaou N (2002) Phys Rev B 66:174429. doi:10.1103/PhysRevB.66.174429
8. Wurmehl S, Fecher GH, Kandpal HC, Ksenofontov V, Felser C, Lin HJ, Morais J (2005) Phys Rev B 72:184434. doi:10.1103/PhysRevB.72.184434
9. Wang WH, Przybylski M, Kuch W, Chelaru LI, Wang J, Lu YF, Meyerheim JBHL, Kirschner J (2005) Phys Rev B 71:144416. doi:10.1103/PhysRevB.71.144416
10. Donath M, Rangelov G, Braun J, Grentz W (2005) Lect Notes Phys 678:261
11. Tezuka N, Ikeda N, Mitsuhashi F, Sugimoto S (2009) Appl Phys Lett 94:162504. doi:10.1063/1.3116717
12. Taira T, Ishikawa T, Itabashi N, Matsuda KI, Uemura T, Yamamoto M (2009) J Phys D 42:084015. doi:10.1088/0022-3727/42/8/084015
13. Tezuka N, Ikeda N, Sugimoto S, Inomata K (2007) Jpn J Appl Phys 46:L454. doi:10.1143/JJAP.46.L454
14. Kallmayer M, Schneider H, Jakob G, Elmers HJ, Kroth K, Kandpal HC, Stumm U, Cramm S (2006) Appl Phys Lett 88:072506. doi:10.1063/1.2175486
15. Kallmayer M, Klaer P, Schneider H, Arbelo-Jorge E, Herbort C, Jakob G, Jourdan M, Elmers HJ (2009) Phys Rev B 80:020406(R). doi:10.1103/PhysRevB.80.020406
16. Stöhr J (1995) J Electron Spectrosc Relat Phenom 75:253. doi:10.1016/0368-2048(95)02537-5
17. Stöhr J (1999) J Magn Magn Mater 200:470. doi:10.1016/S0304-8853(99)00407-2
18. Miura Y, Nagao K, Shirai M (2004) Phys Rev B 69:144413. doi:10.1103/PhysRevB.69.144413
19. Ebert H (1996) Rep Prog Phys 59:1665. doi:10.1088/0034-4885/59/12/003
20. Antonov VN, Shapak AP, Yaresko AN (2008) J Low Temp Phys 34:1. doi:10.1063/1.2832352
21. de Groot FMF, Fuggle JC, Thole BT, Sawatzky GA (1990) Phys Rev B 42:5459. doi:10.1103/PhysRevB.42.5459
22. Telling ND, Keatley PS, van der Laan G, Hicken RJ, Arenholz E, Sakuraba Y, Oogane M, Ando Y, Takanashi K, Sakuma A, Miyazaki T (2008) Phys Rev B 78:184438. doi:10.1103/PhysRevB.78.184438

23. Jakob G, Eichhorn T, Kallmayer M, Elmers HJ (2007) Phys Rev B 76:174407. doi:10.1103/PhysRevB.76.174407
24. Kallmayer M, Hild K, Eichhorn T, Schneider H, Jakob G, Conca A, Jourdan M, Elmers HJ (2007) Appl Phys Lett 91:192501. doi:10.1063/1.2805812
25. Klaer P, Kallmayer M, Blum CGF, Graf T, Barth J, Balke B, Fecher GH, Felser C, Elmers HJ (2009) Phys Rev B 80:144405. doi:10.1103/PhysRevB.80.144405
26. Miyamoto K, Kimura A, Miura Y, Shirai M, Ye M, Cui Y, Shimada K, Namatame H, Taniguchi M, Takeda Y, Saitoh Y, Ikenaga E, Ueda S, Kobayashi K, Kanomata T (2009) Phys Rev B 79:100405. doi:10.1103/PhysRevB.79.100405
27. Klaer P, Jorge EA, Jourdan M, Wang WH, Sukegawa H, Inomata K, Elmers HJ (2010) Phys Rev B 82:024418. doi:10.1103/PhysRevB.82.024418
28. Telling ND, Keatley PS, van der Laan G, Hicken RJ, Arenholz E, Sakuraba Y, Oogane M, Ando Y, Miyazaki T (2006) Phys Rev B 74:224439. doi:10.1103/PhysRevB.74.224439
29. Thole BT, Carra P, Sette F, van der Laan G (1992) Phys Rev Lett 68:1943. doi:10.1103/PhysRevLett.68.1943
30. Carra P, Thole BT, Altarelli M, Wang X (1993) Phys Rev Lett 70:694. doi:10.1103/PhysRevLett.70.694
31. Fecher GH, Kandpal HC, Wurmehl S, Morais J, Lin HJ, Elmers HJ, Schönhense G, Felser C (2005) J Phys Condens Matter 17:7237. doi:10.1088/0953-8984/17/46/008
32. Elmers HJ, Fecher GH, Valdaitsev D, Nepijko SA, Gloskovskii A, Jakob G, Schönhense G, Wurmehl S, Block T, Felser C, Hsu PC, Tsai WL, Cramm S (2003) Phys Rev B 67:104412. doi:10.1103/PhysRevB.67.104412
33. Wurmehl S, Fecher GH, Kroth K, Kronast F, Dürr HA, Takeda Y, Saitoh Y, Kobayashi K, Lin HJ, Schönhense G, Felser C (2006) J Phys D 39:803. doi:10.1088/0022-3727/39/5/S06
34. Kandpal HC, Fecher GH, Felser C (2007) J Phys D 40:1507. doi:10.1088/0022-3727/40/6/S01
35. Sargolzaei M, Richter M, Koepernik K, Opahle I, Eschrig H, Chaplygin I (2006) Phys Rev B 74:224410. doi:10.1103/PhysRevB.74.224410
36. Umetsu RY, Nakamura T, Kobayashi K, Kainuma R, Sakuma A, Fukamichi K, Ishida K (2010) J Phys D 43:105001. doi:10.1088/0022-3727/43/10/105001
37. Lee SC, Lee TD, Blaha P, Schwarz K (2005) J Appl Phys 97:10C307. doi:10.1063/1.1853899
38. Klaer P, Bos T, Kallmayer M, Blum CGF, Graf T, Barth J, Balke B, Fecher GH, Felser C, Elmers HJ (2010) Phys Rev B 82:104410. doi:10.1103/PhysRevB.82.104410
39. Kandpal HC, Fecher GH, Felser C (2006) Phys Rev B 73:094422. doi:10.1103/PhysRevB.73.094422
40. Miyamoto K, Kimura A, Iori K, Sakamoto K, Xie T, Moko T, Qiao S, Taniguchi M, Tsuchiya K (2004) J Phys Condens Matter 16:5797. doi:10.1088/0953-8984/16/48/051
41. Paudel MR, Wolfe CS, Ali N, Stadler S, Christodoulides JA, Ederer DL, Li Y, Callcott TA, Freeland JW (2009) J Appl Phys 105:103907. doi:10.1063/1.3126502
42. Ishida S, Otsuka Y, Kubo Y, Ishida J (1983) J Phys F 13:1173. doi:10.1088/0305-4608/13/6/015
43. Klaer P, Kallmayer M, Elmers HJ, Basit L, Thöne J, Chadov S, Felser C (2009) J Phys D 42:084001. doi:10.1088/0022-3727/42/8/084001
44. Scherz A, Wende H, Baberschke K (2004) Appl Phys A 78:843. doi:10.1007/s00339-003-2440-x
45. Goering E (2005) Philos Mag 85:2895. doi:10.1080/14786430500155221
46. Elmers HJ, Wurmehl S, Fecher GH, Jakob G, Felser C, Schönhense G (2004) Appl Phys A 79:557. doi:10.1007/s00339-003-2366-3
47. Jung V, Fecher GH, Balke B, Ksenofontov V, Felser C (2009) J Phys D 42:084007. doi:10.1088/0022-3727/42/8/084007
48. Kübler J (1983) Phys Rev B 28:1745. doi:10.1103/PhysRevB.28.1745
49. Barth J, Fecher GH, Balke B, Ouardi S, Graf T, Felser C, Shkabko A, Weidenkaff A, Klaer P, Elmers HJ, Yoshikawa H, Ueda S, Kobayashi K (2010) Phys Rev B 81:064404. doi:10.1103/PhysRevB.81.064404

50. Kallmayer M, Schneider H, Jakob G, Elmers HJ, Balke B, Cramm S (2007) J Phys D 40:1552. doi:10.1088/0022-3727/40/6/S07
51. Jourdan M, Jorge EA, Herbort C, Kallmayer M, Klaer P, Elmers HJ (2009) Appl Phys Lett 95:172504. doi:10.1063/1.3254252
52. Hamrle J, Pistora J, Wolf G, Trudel S, Hillebrands B, Klaer P, Kallmayer M, Elmers HJ, Sukegawa H, Wang W, Inomata K (2010) unpublished
53. Vilanova E, Jakob G, Klaer P, Kallmayer M, Elmers HJ (2010) unpublished
54. Arbelo-Jorge E, Jourdan M, Kallmayer M, Klaer P, Elmers HJ (2010) J Phys Conf Ser 200:072006. doi:10.1088/1742-6596/200/7/072006
55. Trudel S, Wolf G, Hamrle J, Hillebrands B, Klaer P, Kallmayer M, Elmers HJ, Sukegawa H, Wang W, Inomata K (2011) Phys Rev B 83:104412
56. Kallmayer M (2011) Röntgenabsorptionsspektroskopie und zirkulardichroismus von heusler – verbindungen. Ph.D. thesis, Johannes Gutenberg – Universität Mainz
57. Kallmayer M, Klaer P, Schneider H, Jakob G, Elmers HJ, Legut D, Oppeneer PM (2011) Phys Rev B 84:054448
58. Chioncel L, Sakuraba Y, Arrigoni E, Katsnelson MI, Oogane M, Ando Y, Miyazaki T, Burzo E, Lichtenstein AI (2008) Phys Rev Lett 100:086402. doi:10.1103/PhysRevLett.100.086402
59. Wang W, Sukegawa H, Shan R, Furubayashi T, Inomata K (2008) Appl Phys Lett 92:221912. doi:10.1063/1.2940595
60. Fecher GH, Felser C (2007) J Phys D 40:1582. doi:10.1088/0022-3727/40/6/S12
61. Gercsi Z, Hono K (2007) J Phys Condens Matter 19:326216. doi:10.1088/0953-8984/19/32/326216
62. Schneider H, Jakob G, Kallmayer M, Elmers HJ, Cinchetti M, Balke B, Wurmehl S, Felser C, Aeschlimann M, Adrian H (2006) Phys Rev B 74:174426. doi:10.1103/PhysRevB.74.174426
63. Özdogan K, Aktas B, Galanakis I, Sasioglu E (2007) J Appl Phys 101:073910. doi:10.1063/1.2714502
64. Balke B, Fecher GH, Kandpal HC, Felser C (2006) Phys Rev B 74:104405. doi:10.1103/PhysRevB.74.104405
65. Chadov S, Fecher GH, Felser C, Minar J, Braun J, Ebert H (2009) J Phys D 42:084002. doi:10.1088/0022-3727/42/8/084002
66. Chioncel L, Arrigoni E, Katsnelson MI, Lichtenstein AI (2009) Phys Rev B 79:125123. doi:10.1103/PhysRevB.79.125123
67. Galanakis I, Ozdogan K, Sasioglu E, Aktas B (2007) Phys Rev B 75:092407. doi:10.1103/PhysRevB.75.092407

Chapter 11
Hard X-Ray Photoelectron Spectroscopy of New Materials for Spintronics

Gerhard H. Fecher and Claudia Felser

Abstract In this work, results of hard X-ray photoelectron spectroscopy (HAXPES) of Heusler compounds and new materials for spintronics are presented. The class of Heusler materials includes some interesting half-metallic and ferromagnetic properties that were predicted by theory. HAXPES allows a direct comparison of the measured and the calculated electronic structure. Valence band spectroscopy of bulk materials by HAXPES is illustrated for the case of the half-metallic ferromagnet Co_2MnGe. The feasibility of HAXPES to explore the valence band electronic structure in deeply buried metallic layers is demonstrated for buried Co_2MnSi films. The films exhibit the same valence density of states as bulk samples and confirm the promise of an epitaxial, single-crystalline Co_2-based Heusler compound film as a ferromagnetic electrode for spintronics devices. The study of complete CoFe(B)/MgO/CoFe(B) tunneling junctions demonstrates the capability of HAXPES to explore the electronic structure in deeply buried layers in a non-destructive way. The improvement of the TMR by annealing of the junction is explained by an improvement of the structure together with a change of the composition in the CoFeB layers.

11.1 Introduction

Since the pioneering works of Kai Siegbahn and his co-workers (see [1] and references therein), X-ray photoemission spectroscopy has grown to be one of the most universal and powerful tools available for the investigation of chemical states and electronic structures of materials. The first high resolution hard X-ray photoemis-

G.H. Fecher · C. Felser
Institut für Anorganische Chemie und Analytische Chemie, Johannes Gutenberg – Universität, 55099 Mainz, Germany

G.H. Fecher (✉) · C. Felser
Max-Planck-Institut für Chemische Physik fester Stoffe, 01187 Dresden, Germany
e-mail: fecher@cpfs.mpg.de

C. Felser
e-mail: felser@cpfs.mpg.de

C. Felser, G.H. Fecher (eds.), *Spintronics*, DOI 10.1007/978-90-481-3832-6_11,

sion spectroscopy experiments using synchrotron radiation were performed by Lindau et al. [2] in 1974, who used X-rays from a bending magnet at SPEAR (Stanford Synchrotron Radiation Laboratory) to measure the intrinsic linewidth of the Au $4f$ core levels with high energy resolution. The observed peak count rate was very low, thus no further trials aiming at high resolution hard X-ray spectroscopy for practical studies were attempted until the first report of HAXPES using a third-generation undulator X-ray source in 2003 [3].

Photoelectron spectroscopy is of great importance in many fields of research because of its numerous advantageous properties. Among its various applications, in particular, it is used to investigate the symmetries of the electronic structure of various materials. However, thus far, such studies have been restricted to atoms, molecules, adsorbates, and surfaces because low-energy (<1 keV) electrons have limited probing depths. This is disadvantageous in that three-dimensional (3D) bulk states cannot be studied. In contrast, the use of hard X-rays for excitation results in the emission of electrons having high kinetic energies, in turn leading to a high probing depth because of the increased electron mean free path. Recently, hard X-ray photoelectron spectroscopy (HAXPES) has emerged as a powerful tool to investigate the electronic structure of solids [4] as well as multilayer systems [5, 6] and the valence band of buried thin films [7]. The valence transitions of several bulk systems were successfully investigated by means of HAXPES [8–10] and very recently, the polarization dependence of the emission from Au and Ag samples was reported [11].

HAXPES is the method of choice to perform bulk sensitive (at excitation energies of about 8 keV an escape depth greater than 90 Å is attained) photoemission spectroscopy on correlated systems in thin films, multilayer and devices because it overcomes the shortcomings of a low electron mean free path in solids. At high kinetic energies of the electrons as used in HAXPES the valence band as well as core level can be investigated spectroscopically with real bulk sensitivity. In standing wave methods this is used to distinguish the valence band signals emerging from bulk or interfaces. The wide energy range easily allows a chemical analysis and moreover the details of the core level spectra allow to investigate satellites that appear in magnetic and correlated systems. Magnetic dichroism in the angular distribution (MDAD) can be excited by any kind of photon polarization and is accessible by laboratory sources. It makes use of the break of symmetry implied by the magnetization of the sample or by orientation of correlated electron states.

At the present time, the high energy excitation and analysis of electrons has become easily feasible due to the development of high intensity sources (insertion devices at synchrotron facilities) and multi-channel electron detection. Thus, HAXPES was recently introduced by several groups (for a complete overview see [5, 12] and references therein) as a bulk sensitive probe of the electronic structure of complex materials. The resolution of HAXPES with down to below 50 meV photon band-width (BL47XU and BL29XU at SPring-8) is competitive to low energy solid state photoemission that is often governed by life time broadening. Even higher resolution (20 meV) seems to be possible after recent improvements concerning the intensity at the SPring-8 beamlines and using a micro/nano focused beam [5].

The feasibility of HAXPES for the investigation of thin films was reported by Kim et al. [13]. The authors show that HAXPES is well suited for the characterization of electronic and chemical states of thin solid films for which conventional photoelectron spectroscopy is not applicable. Nowadays it is possible to grow thin films and devices (for the later use in spintronics applications) from a huge range of materials including a lot different Heusler compounds (for a recent review see Ref. [14]). The detailed investigation of these materials is more complicated. To prevent the freshly prepared thin film or device from contamination while transferring the sample from the preparation chamber to the analysis chamber very often a cap layer is used. This cap layer disturbs a huge variety of surface analysis methods and has to be removed again before starting the investigation [15]. This surface preparation may change the state of the sample from a well known and defined state to an undefined state so one is never sure if the investigated sample has something in common with the prepared one. There are no in-situ investigations of the electronic structure available that were carried out during the growth of layered devices. Indeed, ex-situ investigations of devices by depth profiling might lead to wrong conclusions (see [16] and Sect. 11.3.3).

Several investigations of bulk Heusler compounds by HAXPES were performed. Photoemission data for $Co_2Cr_{0.6}Fe_{0.4}Al$ were reported in several publications [17, 18]. In addition to the effect of disorder, it was found, by using resonant and high energy photoemission methods, that there may be a strong influence of the electron–electron correlation in Heusler compounds. The HAXPES results reported in Ref. [18] showed the strong deviation of the structure of the valence band from the structure that is expected from the calculated density of states. High resolution photoelectron spectroscopy measurements of the valence band of polycrystalline $Co_2Mn_{1-x}Fe_xSi$ ($x = 0, 0.5, 1$), excited by photons having an energy of approximately 8 keV, was reported in [19, 20]. It was shown that the high energy spectra indicate the bulk electronic structure better than low energy XPS spectra. High resolution measurements of the valence band close to the Fermi energy indicate the existence of a gap in the minority states for all three alloys.

Further, HAXPES investigations were also performed on thin films fabricated from Heusler compounds. High energy photoelectron spectroscopy from the valence band of buried Heusler thin films (Co_2MnSi and $Co_2FeAl_{0.5}Si_{0.5}$) excited by photons of about 6 keV energy was carried out and reported in [7, 21]. The measurements were performed on thin films covered by MgO and SiO_x with different thickness from 1 nm to 20 nm of the insulating layer and additional AlO_x or Ru protective layers. It was shown that the insulating layer does not affect the high energy spectra of the Heusler compound close to the Fermi energy. The high resolution measurements of the valence band close to the Fermi energy indicated a very large electron mean free path of the electrons through the insulating layer. The spectra of the buried thin films agreed well with previous measurements from bulk samples. The valence band spectra of the two different Heusler compounds exhibit clear differences in the low lying *s* bands as well as close to the Fermi energy. The valence band spectra have been used to estimate the mean free path of the electrons through the MgO layer to be 17 nm at kinetic energies of about 6 keV. The electron inelastic mean free path was experimentally determined for Co_2MnSi in a wide energy range [22]. As expected for materials with unfilled d-shells, its value is slightly smaller than the calculated one. The buried Co_2MnSi thin film resembles the valence band of the bulk sample that confirms its promise as an electrode for spintronics devices. The electronic structure of the buried thin films at 20 K does not differ from the one measured at 300 K. Accounting for the bulk sensitivity of HAXPES, this illustrates that the electronic structure of the thin Co_2MnSi film itself does not depend on the temperature for $T \leq 300$ K. This fact rules out the possibility that the temperature dependence of the TMR is related to the changes in the bulk electronic structure of the electrodes. The observed temperature dependence of the TMR has to be directly related to the properties of the Co_2MnSi–MgO interface.

A systematic examination of the electronic structure of Heusler compounds was carried out in this work. The results from photoelectron spectroscopy are compared to electronic structure calculations to study the effect of the different kind of atoms and valence electron concentration on the properties of the compounds.

11.2 Experimental Details

Hard X-ray photoelectron spectroscopy was used as a bulk-sensitive probe [7] to study the valence band and core level of the samples. The HAXPES experiments were performed using the beamlines BL15XU and BL47XU at SPring-8 (Japan) [5]. In the regular set-up, photons with energies of 5.9468 keV (BL15XU, BL47XU) or 7.938 keV (BL47XU) that were linearly p polarized in the horizontal direction were used for selective excitation. The p polarized light was obtained from undulator sources without the use of any polarization optics. At BL15XU, the photons were monochromized using a Si(111) double crystal monochromator and the (333) reflection of a Si channel-cut post monochromator. At BL47XU, the (333) or (444) reflections of the channel-cut post monochromator following the Si(111) double crystal monochromator were used to fix the energy.

At both beamlines, the kinetic energy of the photoemitted electrons was analyzed using hemispherical analyzers (VG-Scienta R4000-HV). The overall energy resolution was set to 150 or 250 meV, as verified by the spectra obtained at the Fermi edge of a Au sample. The angular resolution was set to 2°. The angle between the electron spectrometer and photon propagation was fixed at 90°. The detection angle was set to $\theta = 2°$ in order to reach a near-normal emission mode. The angle of incidence of $\alpha = 88°$ ensures that the polarization vector of the p polarized photons is either nearly parallel to the surface normal. The experimental set-up is sketched in Fig. 11.1.

For polarization dependent HAXPES at BL47XU, an in-vacuum phase retarder based on a 200-μm-thick diamond crystal with (220) orientation [23] was used to rotate the linear polarization into the vertical plane or to produce circularly polarized photons. The direct beam is linearly polarized with $P_p = 0.99$. The degrees of polarization were $P_L = (I_h - I_v)/(I_h + I_v) = 0.99$ for p polarization and $-0.7 \cdots - 0.9$

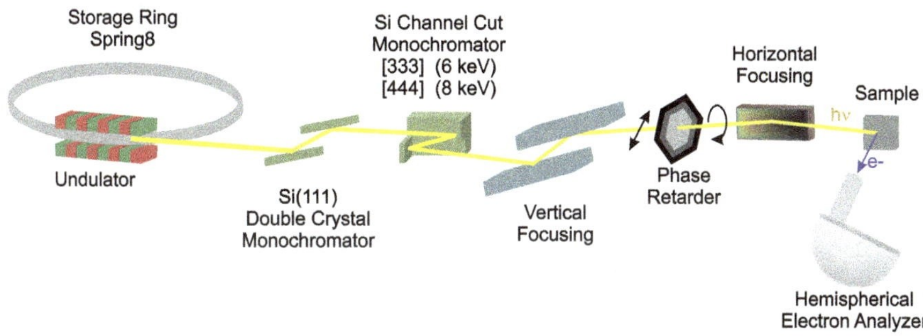

Fig. 11.1 Set-up of the HAXPES experiment (beamline BL47XU at SPring-8, Japan)

for s polarization, where I_h and I_v denote the X-ray intensities of the horizontal and vertical polarization components, respectively. The nearly grazing incidence ensures that the polarization vector of the linearly polarized photons is either nearly parallel (p) or perpendicular (s) to the surface normal. The use of the phase retarder for s polarized light considerably decreases the overall intensity because of the absorption by the diamond crystal. The integration of the count rate over the valence band of NiTiSn results in a ratio of $I^p/I^s \approx 20$ when p polarization without the phase retarder is compared with s polarization with the retarder. For circularly polarized photons the polarization vector is nearly parallel (σ^+) or antiparallel (σ^-) to the in-plane magnetization M^+. The sign of the magnetization was varied by mounting samples with opposite directions of magnetization (M^+, M^-). Using the phase retarder, the degree of circular polarization is set such that $P_c > \pm 0.9$. The circular dichroism is characterized by an asymmetry that is defined as the ratio of the difference between the intensities I^+ and I^- and their sum, $A = (I^+ - I^-)/(I^+ + I^-)$, where I^+ corresponds to σ^+-polarized light and I^-, to σ^--type helicity. Magnetic dichroism may be defined in a similar manner using the differences in the intensities if the direction of the magnetization is changed keeping the polarization of the photons fixed.

11.3 Results and Discussions

11.3.1 Inelastic Mean Free Path and Cross Sections at High Energies

A short description of the effects of the inelastic mean free path and cross sections on the high energy photoelectron spectra will be given before proceeding with results obtained from particular materials. As mentioned in the introduction, one of the most important properties of HAXPES is its high bulk sensitivity. The bulk sensitivity is based on the large electron mean free path at high kinetic energies resulting in a large escape depth of the electrons. A reduction of the photon intensity can be

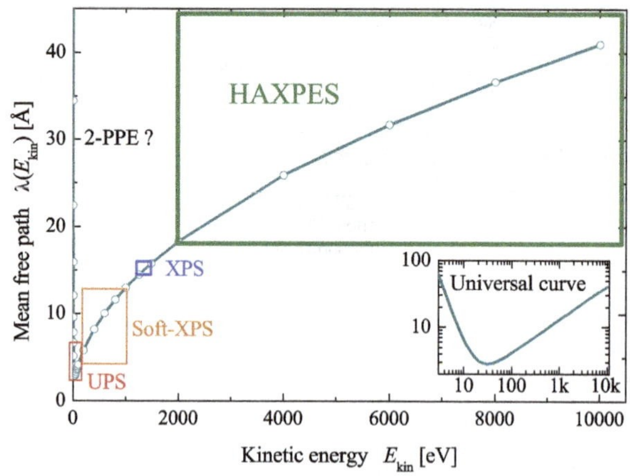

Fig. 11.2 *Universal curve* of the electron mean free path $\lambda(E)$. Shown is the variation of the inelastic mean free path as function of the photoelectron kinetic energy. The *inset* shows $\lambda(E)$ on the frequently used log-log scale. Typical ranges of different methods are marked by *boxes*

safely neglected because the mean free path of X-rays is by orders of magnitude larger than the one of the electrons, anyway.

Figure 11.2 illustrates the general behavior of the electron mean free path with variation of the kinetic energy of the photoelectrons. The shape of this dependence is rather independent of the material wherefore it became named *universal curve*. A closer inspection reveals that the mean free path, indeed, is material dependent (see [24]). From Fig. 11.2 it is clear that HAXPES covers the widest range of spectroscopic possibilities. Ultraviolet photoelectron spectroscopy (UPS) covers the energy region of valence band but is most surface sensitive. The escape depth becomes larger for XPS using soft X-rays or regular laboratory sources (Al–K_α, Mg–K_α) but is still beyond true bulk sensitivity. The role of threshold or two-photon photoemission (2-PPE) with very low kinetic energies (<5 eV) is ambiguous. On the one hand one expects a high escape depth whereas on the other hand the spectroscopic information is limited to the Fermi energy and work function, with the latter being again a typical surface property.

The effective probing depth $z_{\mathrm{eff}} = l\cos(\theta)$ is given by the path l of the electrons through the solid with θ being the angle between the direction of electron emission and the surface normal. The effect of varying θ is a high surface sensitivity at large emission angles, rather independent on energy. Therefore, normal emission ($\theta = 0$, $z = l$) will be assumed in the following. The intensity of the electrons passing through a layer of thickness z is reduced according to an exponential law:

$$I(z) = I_0\, e^{-\frac{z}{\lambda(E)}}, \qquad (11.1)$$

where $\lambda(E)$ is the energy (E) dependent mean free path. The effect of intensity reduction is illustrated in Fig. 11.3 for the example of Co_2MnSi. It is seen that the intensity drops down very fast at low energies. At 80 eV not much more than the topmost cell is probed by the electrons. The situation changes drastically at 8 keV, even after 20 cells the intensity is reduced by only $1/e \approx 0.368$. This also means that still more than 60 % of the signal emerge from deeper layers.

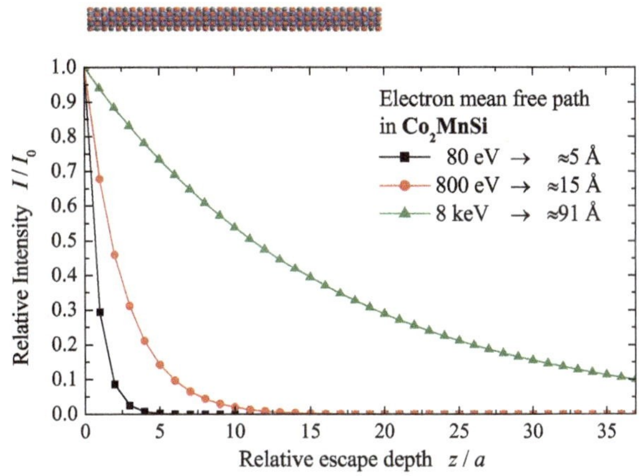

Fig. 11.3 Electron mean free path in Co_2MnSi. Shown is the intensity as function of the number of cubic cells ($a = 5.654$ Å) for different kinetic energies

From Fig. 11.3 and Eq. 11.1 it is also expected that intensity is gained when using high kinetic energies as the integrated intensity normalized to the lattice parameter a

$$I_{\mathrm{eff}} = \frac{1}{a}\int_0^\infty I(E,z)\,dz = I_0\frac{\lambda(E)}{a} \tag{11.2}$$

increases with increasing electron mean free path. Unfortunately, this is not the complete truth. Reason is the decrease of the photoelectron cross sections with increasing photon energy. The partial cross sections (σ_{nl}) of the valence states of the elements contributing in Co_2MnSi are shown in Fig. 11.4. The cross sections for core level (not shown here) are reduced in a similar way. A strong reduction of σ_{nl} by up to three orders of magnitude between 1 keV and 8 keV is obtained. Further, it is obvious that the cross sections for excitation of d states decrease faster than for s states. The effective cross section ($\sigma_{\mathrm{eff}} = \sigma_{nl}\lambda/a$) respecting the gain of intensity by the increasing mean free path is shown in Fig. 11.4(d) for the states contributing most to the valence band of Co_2MnSi.

11.3.2 Valence Band HAXPES of Heusler Compounds

The photoelectron spectroscopy of the valence band at high kinetic energies is not straightforward. A wide energy scan photoelectron spectrum of a Co_2MnGe/MgO (2 nm)/1 nm AlO_x thin film system is shown in Fig. 11.5. It demonstrates the demanding task of high energy valence band spectroscopy. The intensity of the valence band is about 200 times lower compared to the dominant emission from the Co $2p$ states and becomes only weakly visible in the spectrum taken with reduced energy and intensity scales.

In the following hard X-ray photoelectron spectroscopy on the valence bands of the Heusler compound Co_2MnGe is reported.

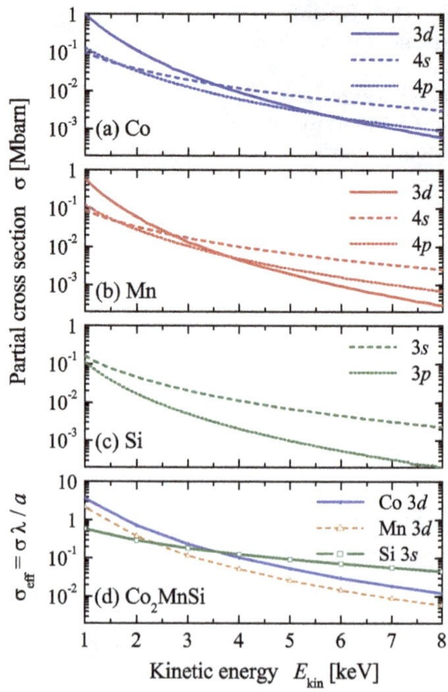

Fig. 11.4 Partial cross sections of Co_2MnSi. Shown are the cross sections of the valence states of Co (**a**), Mn (**b**), and Si (**c**). Panel (**d**) shows the effective cross section respecting the intensity gain by the increased mean free path for the states that contributes most to the valence bands of Co_2MnSi

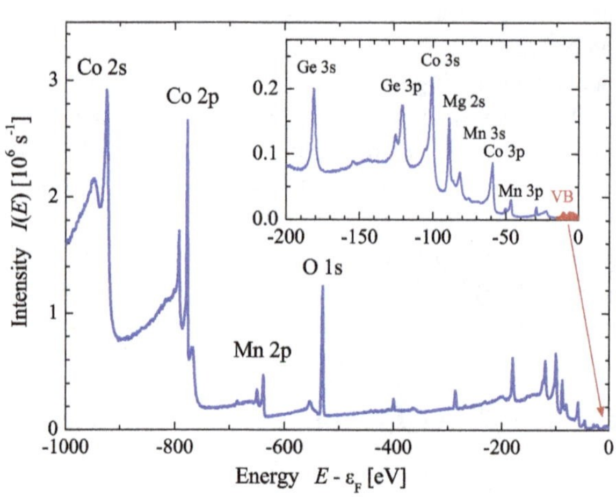

Fig. 11.5 High kinetic energy survey spectrum of Co_2MnGe/2 nm MgO/1 nm AlO_x. The *inset* shows the energy range of the semi-core level and the valence band. The excitation energy was 7.938 keV resulting in a kinetic energy of about 8 keV for the electrons emitted from the valence band

Figure 11.6(a) compares the valence band spectra of Co_2MnGe taken at low (16 K) and high (300 K) temperature at an excitation energy of about 6 keV. The changes in the spectra at different temperatures are unremarkable. The only effect is the expected broadening of the spectra and the slight smearing of the states at the Fermi edge caused by the change of the Fermi–Dirac distribution while going from 16 K to 300 K. This observation is in accordance to the work on Co_2MnSi, where also no temperature dependent changes were detected [21, 25].

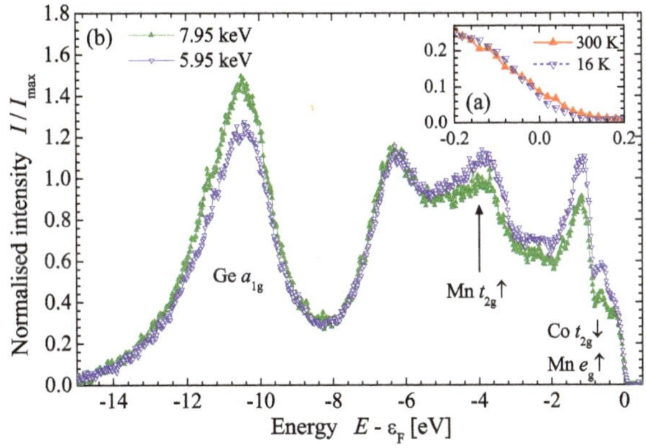

Fig. 11.6 Valence band spectra of Co_2MnGe. Compared are the spectra taken at 16 K excited by different photon energies (5.9468 keV or 7.9392 keV). The *inset* compares the spectra in the vicinity of the Fermi energy at 16 K and 300 K taken with $h\nu = 5.9468$ keV

Figure 11.6(b) compares the valence band spectra taken with different photon energies of about 6 keV and 8 keV. The differences in the spectra are caused by the different weights of the partial cross sections for s, p, and d electrons at different excitation energies. With increasing energies the cross sections for d electron excitation decrease faster than those for s or p electrons (see [20, 26]). Therefore, the contribution from d electrons is more pronounced at lower energy.

The maximum of the emission from the a_{1g} states appears at -10.5 eV what is about 1 eV lower compared to the calculated density of states. The maximum at about -3.9 eV corresponds to an excitation of the Mn t_{2g} minority states, its center is about 0.9 eV lower compared to the center of the states in the calculated DOS. The calculated site resolved DOS exhibits pronounced Co and Mn e_g majority states as well as Co t_{2g} minority states at about -1 eV where a strong maximum is observed in the spectrum. A pronounced influence of a correlation energy can thus be excluded what is proven by the small energy shifts between the spectrum and calculated density of states. The observed energy shifts increase when going further away from the Fermi energy. This points clearly on lifetime effects where the holes at ε_F have a longer lifetime compared to those at lower energies. Accounting for the lifetime effects by an increasing imaginary part in the self-energy of the photoexcited electrons has the result that the observed states do not only have an increased width when going away from ε_F but also exhibit an increase of the energy shift. States that are farer away from ε_F are stronger influenced because of the nonlinearity of the imaginary part of the self-energy, it is Zero at ε_F and rises within a small energy range (few eV) to its final value. In summary, it is concluded that the observed shifts are due to the photoemission process and are not directly related to an electron–electron interaction in the ground state.

11.3.3 HAXPES of Thin Films and Devices

The following reports on the detection of the valence band of buried Heusler compounds in thin films. The measurements have been performed on so-called "*half*"

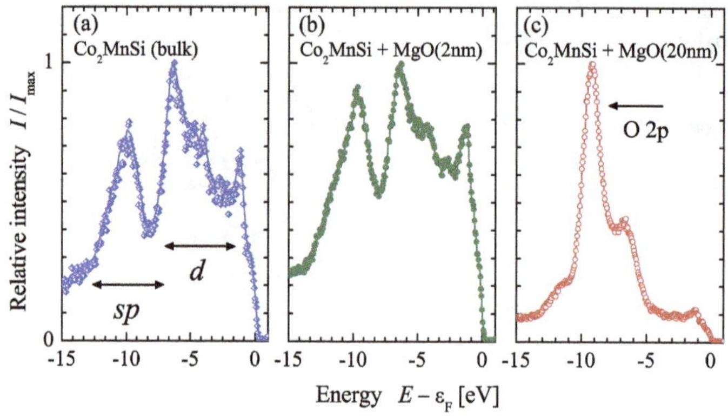

Fig. 11.7 High kinetic energy photoemission spectra of buried Co_2MnSi. The spectrum bulk material (**a**) is compared to the spectra with different thickness 2 nm (**b**), 20 nm (**c**) of the MgO top layer ($h\nu = 5.95$ keV). States of the Co_2MnSi valence band and oxygen p-states are indicated by *arrows* in (**a**) and (**c**), respectively

tunnel junctions that are thin films of Co_2MnSi underneath MgO. To develop high performance spintronics devices with Co_2-based Heusler compound thin films as ferromagnetic electrodes, it is essential to clarify the electronic structures of these films in terms of the half-metallic nature. The purpose of the present study was to investigate the valence band electronic structure of Co_2MnSi thin films buried underneath a MgO barrier by means of hard X-ray photo emission spectroscopy (HAXPES).

For the study of the valence band of thin Heusler films, special multicomponent arrangements were produced that correspond to half of a magnetic tunnel junction as used in TMR devices (see [27–29]). In particular the free electrode was modeled. The fabricated sample layer structure was as follows: MgO(001) substrate/MgO buffer layer (10 nm)/Co_2MnSi (50 nm)/MgO barrier (t_{MgO})/AlO_x (t_{AlO_x}) cap. $t_{\mathrm{MgO}} = 2$ nm, 20 nm and $t_{\mathrm{AlO}_x} = 1$ nm, 2 nm were chosen for the thickness of the MgO and AlO_x cap, respectively. The topmost AlO_x was used for protection of the MgO layer. The fabrication procedure is reported in detail in references [27, 30, 31]. Al $1s$ core level spectra (not shown here) revealed that the 1 nm thick Al layer was completely oxidized to Al_2O_3, whereas the 2 nm layer had a remaining, small contribution of metallic Al. Polycrystalline bulk material was used for comparison. It was produced, prepared, and characterized as described in [19].

Figure 11.7 compares the valence band spectra from the 50 nm thick Co_2MnSi layer buried underneath MgO/AlO_x with different thickness of the MgO interlayer with spectra from bulk material. The valence band spectra of the thin films agree well with those reported from $Co_2Mn_{1-x}Fe_xSi$ bulk samples [19, 20], even though the emission is strongly suppressed in the film with the 20 nm thick MgO layer. In particular, the peak in the density of states at -1.3 eV below ε_F is clearly resolved (see Fig. 11.8). The low lying s-band below -8 eV is only seen in Figs. 11.7(a) and (c). This part, in addition to the lower parts of the p and d-bands, is covered by

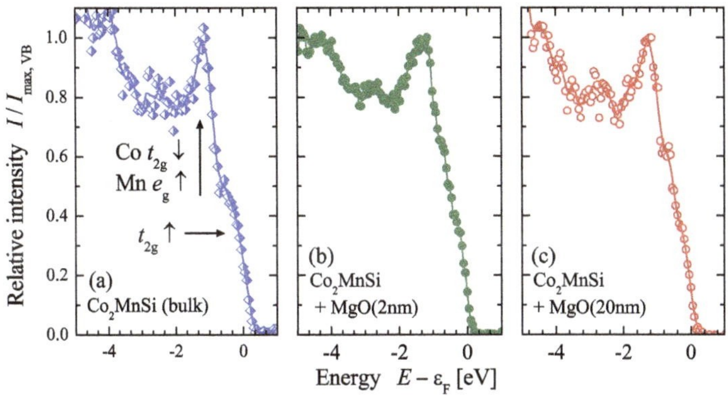

Fig. 11.8 Comparison of the Co_2MnSi valence band spectra close to ε_F. Co_2MnSi d-states with different character are indicated by *arrows*. ((**a**)–(**c**) like in Fig. 11.7)

the emission from oxygen $2p$ states in MgO in Fig. 11.7(b) where the MgO film is 20 nm thick. The similar structure of the valence band close to ε_F in the thin films and the bulk material is obvious from Fig. 11.8.

The electron mean free path (λ) in MgO is expected to be about 9 nm at 6 keV energy, as calculated from the TPP2M-equation (Tanuma–Powell–Penn [24]). In Al_2O_3, one has a mean free path of 11 nm at the same kinetic energy, which should be much smaller in Co_2MnSi (7 nm). From those values, one expects that the electrons emitted from the Fermi energy easily penetrate the thin MgO and AlO_x films but not through the 50 nm thick Co_2MnSi layer. The electron mean free path is easily extracted from the experiment by comparing the intensities for different thickness of the MgO layer. Figure 11.9(b) displays the relative intensities $I_{2\text{ nm}}/I_{20\text{ nm}}$. It is seen that the factor $I_{2\text{ nm}}/I_{20\text{ nm}}$ is constant for energies between -4 eV and ε_F. The scaling factor amounts to $I_{2\text{ nm}}/I_{20\text{ nm}} = 2.88 \pm 0.02$ on average. The constant intensity ratio is expected as MgO does not contribute to the emission close to the Fermi energy due to its wide band gap. From the exponential decay of the intensity $I(z) \propto \exp\{-z/\lambda\}$ with the layer thickness z, one finds the electron mean free path in the MgO layer from $\lambda = \Delta z / \ln(I_{2\text{ nm}}/I_{20\text{ nm}})$ to be $\lambda_{\text{MgO}} = 17$ nm.

In Fig. 11.9(a), the valence band spectra for different MgO thicknesses are directly compared by scaling up the intensity of the 20 nm layer. The identical structure of the spectra, in particular close to the Fermi energy, is easily recognized. The high intensity with a center at about -1.3 eV is due to emission from flat d-bands (see also Fig. 11.8) belonging to minority states localized in the Co planes and the highly localized Mn d majority states (e_g), as already described in detail for the emission from bulk samples [19, 20]. The Co t_{2g} like states mainly define the upper energy of the minority valence bands, whereas the lower intensity at the Fermi energy arises from steep delocalized majority d-bands with t_{2g} character at Γ [32, 33].

The value observed here for the electron mean free path in MgO is about twice as large compared to the one calculated from the TPP2M-equation. This is explained

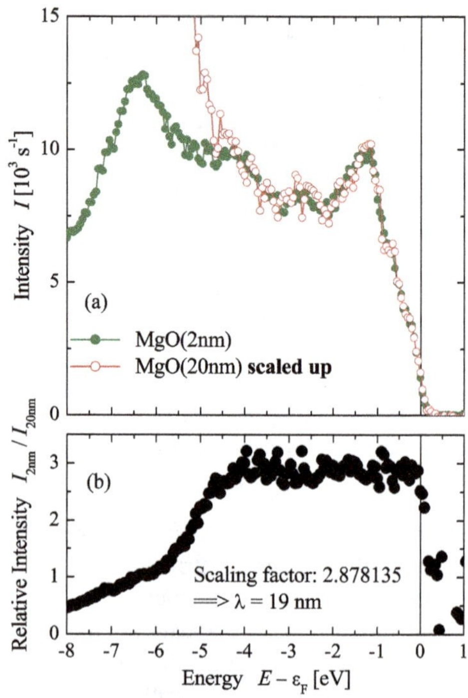

Fig. 11.9 Comparison of the spectra of the buried Co_2MnSi for different MgO thickness. (**a**) compares the spectrum of the 2 nm MgO covered and the scaled-up spectrum of the 20 nm MgO covered Co_2MnSi film, (**b**) shows the relative intensities of the spectra for different MgO thickness

by the high crystalline order of the MgO film and its large optical gap which prevents inelastic scattering mediated by interband transitions. The large mean free path makes it possible to look through the material of the tunnel barrier as seen in Figs. 11.8(b) and 11.9. In summary, the present study demonstrates the feasibility of HAXPES to explore the valence band electronic structure in deeply buried metallic layers. In particular it was shown that buried Co_2MnSi films exhibit the same valence density of states like bulk samples that is typical for half-metallic ferromagnets for spintronics applications. This result confirms the promise of an epitaxial, single-crystalline Co_2-based Heusler compound film as a ferromagnetic electrode for spintronics devices.

The following part of the study is devoted to a non-destructive analysis of the chemical composition of annealed CoFeB tunnel junctions using high resolution hard X-ray photoelectron spectroscopy (HAXPES). Aim is to explain the role of B content in the magnetic electrodes and B diffusion for the observed improvement of the TMR with increasing annealing temperature. Special multicomponent thin film arrangements were produced that correspond to half of a magnetic tunnel junction as well as to the full junction as used in TMR devices. The structures of the layer stacks are sketched in Fig. 11.10. The conditions of MTJ fabrication and TMR measurement have been described elsewhere [34–36]. The film structures were investigated by X-ray diffraction and high resolution transmission electron microscopy with energy dispersive X-ray spectroscopy. For similar junctions as used here, a TMR ratio of 604 % at 300 K was found for junctions annealed to 798 K [36].

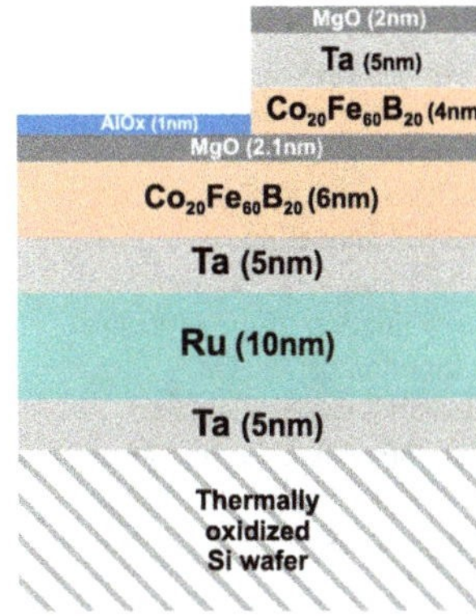

Fig. 11.10 Sketch of the film structure used for the photoemission experiments. The layer structure corresponds to the lower part (*left*) of a tunnel junction and a full junction (*right*), respectively

The Fe and Co $2p$ core level spectra taken from samples annealed to 523 K and 773 K are shown in Fig. 11.11. The difference spectra of both constituents exhibit slightly smaller width of the $2p_{3/2}$ emission peaks for the sample annealed to the optimized annealing temperature of $T_a = 773$ K where the highest tunneling magneto resistance ratio is obtained. This points at an improved crystalline structure in the sample annealed to 773 K. Besides the change of the peak width, the difference in the spectra may give a hint on a very small chemical shift. Changes in the intensity are below the order of ± 0.3 %. This observation includes spectra from samples with other annealing temperatures. It is concluded that the Fe and Co content in the CoFeB layer is rather stable against variation of the annealing temperature. Traces of oxidation are found neither at Fe nor at Co.

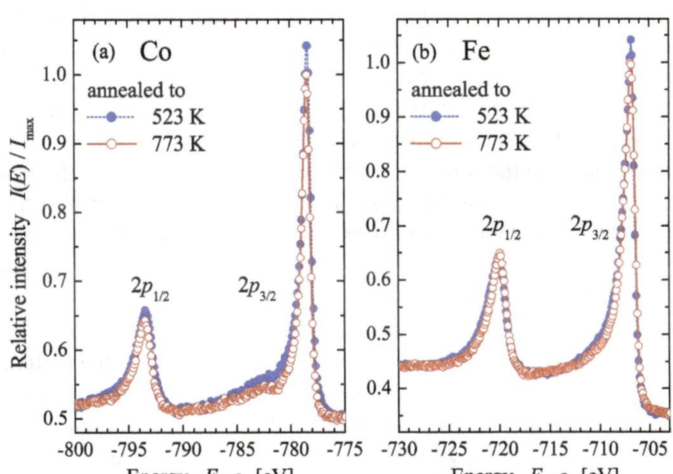

Fig. 11.11 $2p$ core level spectra of Co and Fe. The emission from the $2p$ states of Co and Fe is displayed in (**a**) and (**b**), respectively. The spectra were taken from samples annealed to 523 K and 773 K

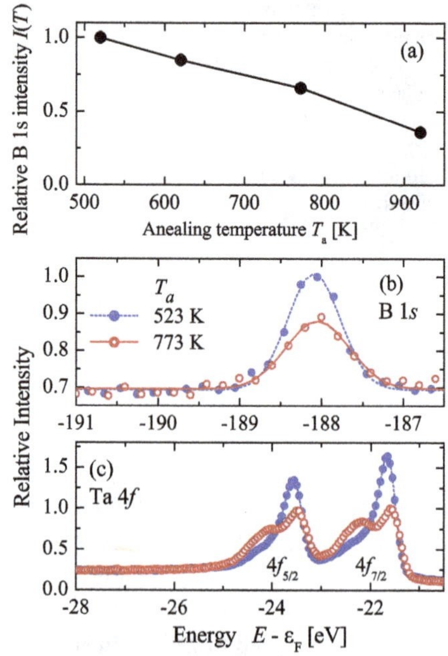

Fig. 11.12 B $1s$ and Ta $4f$ core level spectra of the half tunnel junction. Shown are the spectra of the B $1s$ (**b**) and Ta $4f$ (**c**) states for selected annealing temperatures T_a. (**a**) shows the evaluation of the intensity as function of T_a. The intensities were calculated from the peak areas and are given with respect to the area of the peak at $T_a = 523$ K

More interesting are the observations made at the boron core level. Figure 11.12 shows the B $1s$ core level spectra from samples that were annealed to different temperatures. Obviously the B $1s$ signal is very low and exceeds the secondary background by only 30 %. The signal is drastically reduced in the sample annealed to 923 K. Due to the low intensity, a chemical shift cannot be detected. In the sample with the optimized annealing temperature, the B $1s$ intensity is about 60 % compared to the sample with the lowest annealing temperature. Overall, the decrease of intensity points to a decrease of the B content in the CoFeB layer with increasing annealing temperature. From the reduction of the signal it is obvious that a diffusion of B to the lower parts of the samples takes place.

The diffusion of B to the lower parts of the junction becomes more clear if considering the emission from the tantalum $4f$ states. It is well known that the binding energy of those states is changed in different chemical environment leading to pronounced chemical shifts. The dependence of the Ta $4f$ emission is displayed in Fig. 11.12 for different annealing temperatures. The appearance of a state with a chemical shift of about -0.7 eV at higher annealing temperatures is evident. This shift is in agreement to the formation of TaB_x. The intensity of the shifted component (TaB) increases on cost of the intensity of the initially non-shifted state (Ta^0). Assuming a replacement of B by Ta in the CoFeB layer, one would expect a strong increase of the overall intensity of the Ta $4f$ emission opposite to the decrease of the B $1s$ intensity. From Fig. 11.12 it is found that the total Ta $4f$ intensity does not change much up to 773 K. Therefore, a complete exchange of B by Ta in the CoFeB layer can be excluded. At the highest annealing temperature (923 K) a slight increase of the overall intensity is observed and a long tail becomes visible at the

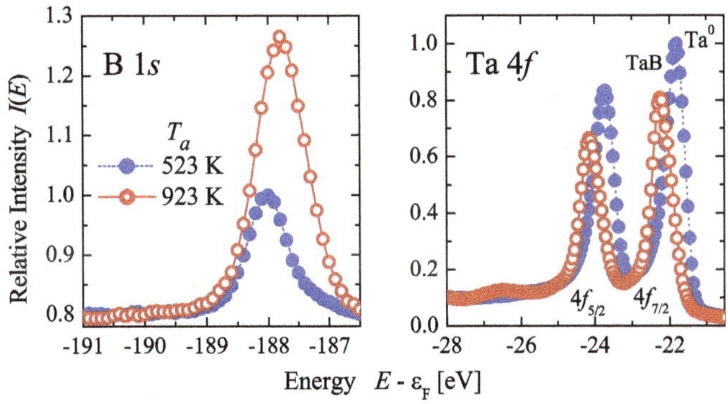

Fig. 11.13 B 1*s* and Ta 4*f* core level spectra of the full tunnel junction. Displayed are spectra for selected annealing temperatures T_a

low energy side of the $4f_{5/2}$ state. This might be a result of Ta diffusion and an additional degradation of the Ta layer in the over annealed sample.

The B 1*s* and Ta 4*f* core level spectra of the samples that refer to the upper electrode of the full tunnel junction revealed the same tendency of diffusion of B atoms into the neighboring Ta layer. From Fig. 11.13 the increase of the B signal with increasing annealing temperature is obvious. At the overheated sample (923 K) the Ta 4*f* states are completely shifted to the TaB binding energy, similar to the observation of the lower electrode. A BO_x signal is not detected. From this observation it is clear that B is always removed from the CoFeB layers and moves into the Ta layers.

In summary, an enforced diffusion of B from the CoFeB into the Ta layers of a magnetic tunnel junction was detected. The B 1*s* and Ta 4*f* core level spectra reveal that Boron is rapidly diffusing into the contiguous tantalum layers upon annealing of the tunnel junction. In the lower half of the junction the B signal decreases with higher annealing temperatures, whereas it increases in the upper part of the junction. At the same time the chemically shifted Ta-B part of the Ta 4*f* signal increases in both layers. The improvement of the TMR is thus not only caused by an improvement of the structure but also by a change of the composition in the CoFeB layers. The degradation of the (001) orientation caused by Ta diffusion into the CoFe(B) layer [36], i.e. the lowering of tunneling spin polarization, explains the sudden drop of the TMR when the sample is overheated.

The results derived here by non-destructive HAXPES are much different from those using destructive depth profiling [16]. Those authors found BO_x in both electrodes and particularly a higher oxygen concentration in the lower electrode. Most probably this BO_x was produced by the knocking effect of the Ar^+ bombardment where the O is pushed from the MgO layer into the lower electrode. The comparison of the non-destructive HAXPES results with destructive investigation of thin films makes it clear that non-destructive techniques in combination with in-situ film preparation are necessary to find reliable results about the electronic structure if using photoemission.

11.3.4 Magnetic Circular Dichroism in Core Level HAXPES

This section is devoted to MCDAD experiments in the HAXPES range on different types of exchange-biased structure with epitaxially grown ferromagnetic layers of $Co_{50}Fe_{50}$ (CoFe) and Co_2FeAl, these being typical materials used in tunnel magnetoresistive devices. The on-top approach multilayers were deposited in the sequence MgO (100) substrate/MgO buffer layer (10 nm)/$Ir_{78}Mn_{22}$ (IrMn) (10 nm)/CoFe (3 nm)/MgO barrier (2 nm)/AlO_x (1 nm) that corresponds to the lower exchange-biased electrode of an MTJ [37]. The on-bottom configuration was realized in the multilayer sequence MgO(100) substrate/Cr buffer layer (40 nm)/Co_2FeAl (30 nm)/IrMn (10 nm)/AlO_x (1 nm) [38]. In both cases, the topmost AlO_x layers served as a protective coating. All metal layers were deposited by magnetron sputtering and electron beam evaporation was used to epitaxially grow the MgO barrier. IrMn serves as an exchange-biasing layer that keeps CoFe or Co_2FeAl magnetized in preset directions. Remanent magnetization was achieved by annealing the multilayer stacks in a magnetic field.

Figure 11.14 shows the $2p$ core-level spectra of Co and Fe that were taken from an exchange-biased CoFe film that was covered by oxide films. Pronounced differences were observed in the spectra taken with photons having opposite helicity for a fixed direction of magnetization. The MCDAD asymmetry was determined after subtracting a Shirley-type background. The largest obtained asymmetry values are $-29\ \%$ at Co $2p_{3/2}$ and $-46\ \%$ at Fe $2p_{3/2}$. The dichroism vanished in an identical manner in spectra taken in the geometry in which the projection of the photon vector is perpendicular to the magnetization, independently of whether the photon helicity or the magnetization was reversed. This indicates that the films are perfectly magnetized in the direction forced by the exchange bias layer magnetization.

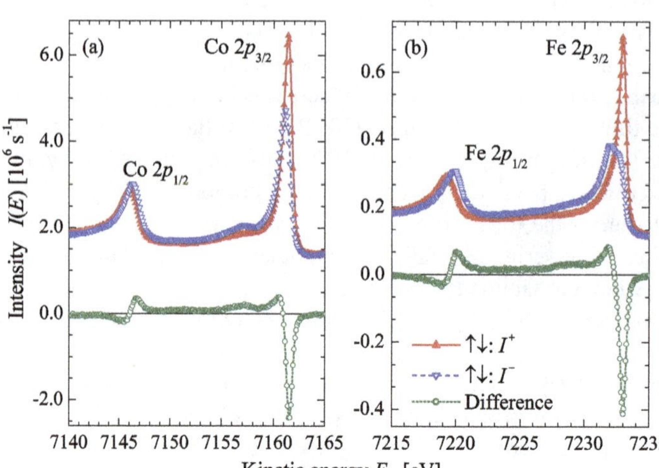

Fig. 11.14 Polarization-dependent photoelectron spectra of Co $2p$ and Fe $2p$ core-level emission from CoFe on top of IrMn exchange-biasing layer. The film is covered with 3-nm-thick oxide layers (MgO/AlO_x) (resolution: 150 meV)

For both Co and Fe, the spin–orbit splitting of the $2p$ states is clearly resolved, as expected. When going from $p_{3/2}$ to $p_{1/2}$, the dichroism changes its sign across the $2p$ spectra in the sequence $- + + -$; this appears characteristic of Zeeman-type m_j sub-level ordering. The details of the MCDAD reveal, however, that the situation is more complicated. In particular, the dichroism in the Fe $2p$ spectra does not vanish between the spin–orbit doublet.

Theoretical atomic single-particle models were quite successful in describing, explaining, and predicting many aspects of magnetic dichroism. Cherepkov et al. elaborated the general formalism for the dichroism in photoemission excited by circularly, linearly, and unpolarized radiation [39]. They showed that MCDAD is very sensitive to the geometry of the experiment and depends strongly on the relative orientation among the magnetization, helicity, and momentum of the excited electrons. The maximum effect is obtained when the magnetization and helicity vectors are parallel; the effect decreases with an increase in the angle between these vectors. MCDAD has previously been used to investigate the itinerant magnetism of ferromagnetic elements such as Co, Fe, and Ni, where it was explained in terms of such models [40–43]. As demonstrated in the case of Ni, however, the single-particle approach poorly describes all the peculiarities of the complex spectra. Van der Laan and Thole considered the MCDAD phenomenon by taking into account the influence of electron correlation effects in the frame of atomic many-particle models that were successfully used to describe both localized and itinerant magnetism phenomena [44–47]. Many-body effects play an important role when using polarized incident photons. The correlation among spin and orbital moments, $2p$ core-hole, and spin-polarized valence band results in a rich multiplet structure that spreads out over a wide energy range of a spectrum [48]. In the case of strongly correlated systems, the bulk magnetic and electronic properties are quite different from the surface ones. However, as observed previously, MCDAD with radiation in the soft X-ray range is highly sensitive to the surface where the dichroism is influenced by symmetry breaking [49]. Because of the strong inelastic electron scattering in this energy range, the escape depth of the photoemitted electrons of a few Å becomes comparable to the thickness of a monolayer. The tuning of the excitation energy also affects the photoionization cross sections. At high energies, the intensities from the d states of transition metals are reduced as compared to the partial cross sections of the s and p states [3, 10, 20]. The shape and magnitude of the asymmetry depend on the partial bulk to surface spectral weights; hence, only at high energies, the dichroism effects appear to be related to the bulk properties.

Figure 11.15 shows the polarization dependence of the CoFe valence band spectra together with the resulting magnetic dichroism. The MCDAD observed for the valence band is much smaller as compared to the core-level photoemission. The largest asymmetry is approximately -1.5 % at approximately -0.1 eV below the Fermi energy. Such low asymmetry values were also observed when using low photon and kinetic energies [50]. Only in the threshold region, higher asymmetries arise in the case of one- [51] and two-photon photoemission [52]. In the range of the valence states, the detection is further complicated by the signal from the underlying IrMn layer that does not contribute to the dichroism. Because of the thin layer of

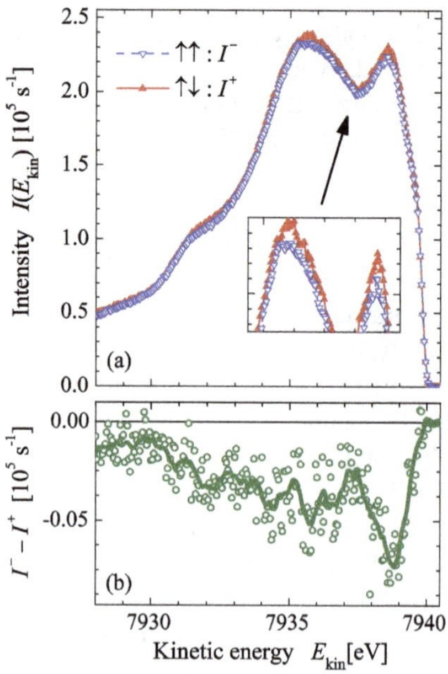

Fig. 11.15 MCDAD in valence band of CoFe on top of IrMn. Same sample as that shown in Fig. 11.14 (resolution: 250 meV)

CoFe and the large escape depth of the nearly 8 keV fast electrons, the two layers cannot be distinguished in the valence band. It is worthwhile to note that the dichroic signal itself arises exclusively from the buried, ferromagnetic CoFe layer.

For studies aimed toward the development of novel devices, it is necessary to also detect the magnetic signal from deeply buried layers. To prove the reliability of the proposed method, experiments were also performed on samples in which the IrMn exchange-biasing layer was on top of the layer structure. Figure 11.16(a) shows the MCDAD results for the shallow core levels. For such complex multilayer structures, the situation becomes complicated in that the signals from all the elements contained in the system are detected. Strong signals can be detected from the buried elements even though the ferromagnetic Co_2FeAl layer lies beneath the antiferromagnetic IrMn layer. As expected, no MCDAD is observed for the Ir and Mn states. A non-vanishing asymmetry is clearly observed only for Co and Fe signals, and these are the only ones responsible for the ferromagnetic properties of the system. The asymmetry of 50 % in the Fe $3p$ signal is quite large. In Co $3p$, it is well detected even though the direct spectra overlap with the Ir states.

Figure 11.16(b) shows the Fe $2p$ MCDAD from a deeply buried layer of Co_2FeAl beneath a 10-nm-thick IrMn film. The Fe $2p$ emission has a much lower intensity and the resolution was therefore reduced to 250 meV in order to keep the counting rates comparable to those of the CoFe measurements. The multiplet splitting at Fe $2p_{3/2}$ is still well resolved and the MCDAD is well detected. It was shown that linear magnetic dichroism (LMDAD) along with the circular one can be successfully applied to investigate the electronic and magnetic properties of surfaces and interfaces. The LMDAD asymmetry observed at Fe $2p_{3/2}$, however, was only at most

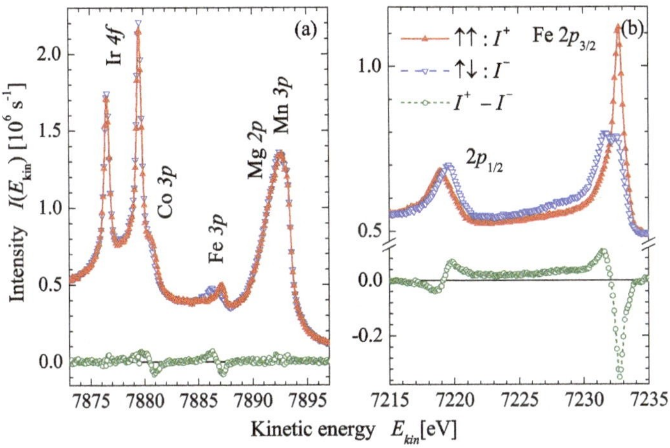

Fig. 11.16 Core-level spectra and MCDAD from Co_2FeAl beneath a 10-nm-thick IrMn film. (**a**) shallow core levels; (**b**) Fe $2p$ (resolution: 250 meV)

-9 % for a low excitation energy [53]. Here, the maximum asymmetry is -40 % at Fe $2p_{3/2}$, and this is ideal for the analysis of the magnetic properties.

There is a striking distinction between the Fe $2p$ spectra of the two layer systems. Even though taken with a lower resolution, the multiplet splitting of the Fe $2p_{3/2}$ emission from Co_2FeAl is in the I^- spectrum and is better resolved as compared to the corresponding spectrum from CoFe. The mean splitting ΔE_{exch} of the Fe $2p_{3/2}$ states is 0.8 eV and 1.0 eV for CoFe and Co_2FeAl, respectively. Co_2FeAl is supposed to be a half-metallic ferromagnet with a magnetic moment of around 1.25 μ_B per atom, whereas CoFe is a regular band ferromagnet with a very high magnetic moment ($>2.2\ \mu_B$). One of the major differences is the localized magnetic moment of Fe in Co_2FeAl that is caused by a strong localization of the t_{2g} bands. In the ordered case of both compounds, the Fe atoms are in a cubic environment and are surrounded by eight Co atoms. Co_2FeAl forms a perfect 2^3 CsCl supercell with every second Fe atom of CoFe replaced by Al. This causes additional Co–Al bonds that reduce the Co–Fe d-state overlap. The result is localized moment at the Fe sites. From this viewpoint, Fe in Co_2FeAl is closer to an ionic than to a metallic state. For the Fe atoms, this causes a more pronounced interaction of the core hole at the ionized $2p$ shell with the partially filled $3d$ valence shell.

11.3.5 Linear Dichroism in Valence Band HAXPES

The linear dichroism in the angular distribution (LDAD) of photoelectrons has attracted considerable interest as a powerful probe to study the symmetry, orientation, and alignment phenomena in photoionization. The prerequisites for this effect are the orientation of the photons and the alignment of the electronic states. The first is realized by the linear polarization of the photons whereas the second might be produced by the alignment of the molecular axes or by the symmetry of the bands in a solid. The main difference between linear and circular dichroism in the angular distribution is that the former does not require a chirality or magnetization of the investigated system because the electric field vector of linearly polarized photons is a real vector whereas the helicity of circularly polarized photons is a pseudovector. Angle-resolved photoelectron spectra indicated that the emission from adsorbates exhibits strong polarization dependent variations [54]. Many theoretical [55–57] and experimental [9, 58, 59] studies of the LDAD have been carried out. The nonmagnetic and nonchiral system 1T-TaS_2 is an example of LDAD from solids [60].

In the present work, a combination of the LDAD with the bulk-sensitive HAXPES technique was used to investigate the symmetry of the valence states of nonmagnetic, polycrystalline Heusler compounds. The compounds NiTiSn and NiMnSb were selected in this study. NiMnSb was selected because it was suggested to be a half-metallic ferromagnet [61]. NiTiSn is supposed to be a semiconductor [62]; therefore, this sample was heavily hole doped by Ti↔Sc substitution with a composition of $NiTi_{0.9}Sc_{0.1}Sn$. This ensures that the Fermi energy clearly lies in the valence band and it suppresses unwanted *in-gap* states that were reported previously [63]. The details of the sample production and characterization are reported in Ref. [64]. Polycrystalline samples were cut to discs and the surface was polished. The sample surfaces were prepared by Ar^+ bombardment to remove traces of oxidation. Their cleanness was checked by regular low-energy XPS (Al K_α).

Figure 11.17 compares the polarization dependent valence band spectra of NiTiSn and NiMnSb. The corresponding density of states are shown in Fig. 11.18. The density of states (DOS) of NiTiSn exhibits a typical 4-peak structure in the energy range of the d states as well as the split-off s band with a_1 symmetry. These structures can be clearly observed in the photoelectron spectra. The sum of the polarization-resolved spectra corresponds to a spectrum with unpolarized photons. The width of the d-part of the density of states $(0 \cdots -5$ eV) corresponds to the width of the measured spectra. The spectrum exhibits a rather high intensity in the s-part below (-7 eV) that is caused by the higher cross section for s states as compared to that for d states at a high excitation energy. The intensity of the d states is governed by the states localized at the Ni atoms. Dirac–Fock calculations for the photoexcitation of Ni, Ti, and Sc valence electrons indicate that at $h\nu = 8$ keV, the cross section for Ni $3d$ states is 20 times higher than that for light $3d$ transition metals.

It is evident that striking differences appear in the spectra if the polarization is changed from p to s. The spectra shown in Fig. 11.17 indicate that the s states have a higher intensity for p polarization and the s polarization result in a higher intensity in the energy range of the d states. In particular, the intensity of the Ni $3d$ states at -2 eV is enhanced and the structure of the d states becomes better resolved under illumination with s polarized photons.

With a change from NiTiSn to NiMnSb, and thus, from Ti $3d^2$ to Mn $3d^5$, one expects a change in the spectra in the energy range of the d states. Figures 11.18(b) and 11.17(b) show a comparison of the valence band spectra and the electronic structure of NiMnSb. The DOS exhibits a five-peak structure in the range of the d states

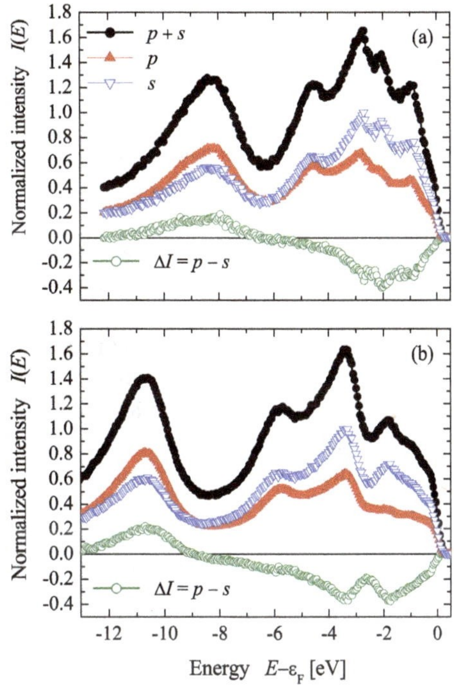

Fig. 11.17 Polarization-dependent photoelectron spectra of NiTiSn (**a**) and NiMnSb (**b**). Spectra obtained with s and p polarized light together with the difference and sum (resolution: 150 meV for NiTiSn (**a**) and 250 meV for NiMnSb (**b**))

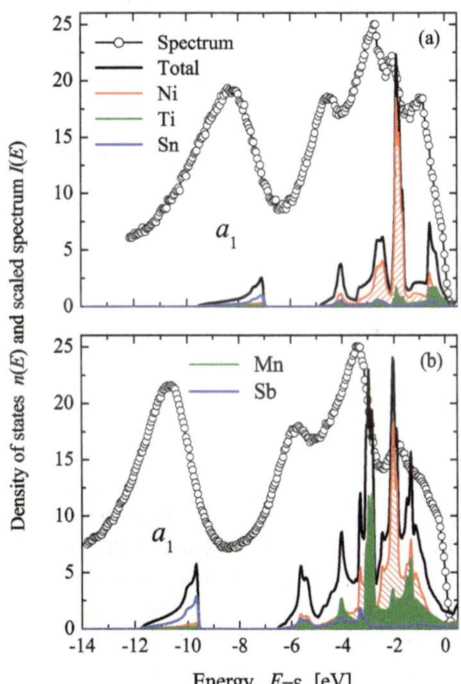

Fig. 11.18 Electronic structure of NiTiSn (**a**) and NiMnSb (**b**). Shown is the total and partial density of states together with the sum spectra ($I^s + I^p$)

that is not completely resolved in the spectra. A pronounced maximum appears at approximately -3 eV in the DOS and spectra. It emerges from the Mn t_2 states that are responsible for the localized magnetic moment in NiMnSb. The changes in the spectra while changing the polarization are the same as those observed for NiTiSn. Again, the intensity of the Ni 3d states at -2 eV is noticeably enhanced for s polarized photons. The intensity of the Mn 3d states is enhanced in a similar manner.

The polarization-dependent spectra can be most easily analyzed for direct emission by neglecting all final state effects in the first step (for example: electron diffraction [65]). In the dipole approximation, the observed intensity ($I \propto \frac{d\sigma}{d\Omega}$) depends on the partial cross section (σ_κ) and the angular asymmetry parameter (β_κ) of the photoexcited states. κ assigns a complete set of quantum numbers describing the initial state. For atomic-like states in a single electron description, these are $\kappa = n, l$ with main quantum number n and orbital angular momentum l. In the case of linearly polarized photons, the angular dependence of the intensity in the dipole approximation is given as follows [66]:

$$I_\kappa^{E_1}(\theta) = \sigma_\kappa \left[1 + \beta_\kappa P_2(\theta)\right],$$

where $P_2(\theta) = (3\cos^2\theta - 1)/2$ is the second Legendre polynomial and θ is measured with respect to the electric field vector. The angular parameter is limited to the range $-1 \le \beta \le 2$. Both parameters, σ_κ and β_κ, depend on the kinetic energy of the emitted electrons. In addition, higher-order nondipole terms may appear (M_1 and E_2 transitions) in addition to the E_1 transitions. The first-order nondipole approximation results in an additional term

$$I_\kappa^{M_1,E_2} \propto \sigma_\kappa \left[\left(\delta_\kappa + \gamma_\kappa \cos^2\theta\right) \sin\theta \cos\phi\right]$$

in the angular distribution [67], with the angular parameters γ_κ and δ_κ that depend on the magnetic dipole and electric quadrupole matrix elements. Note that δ has to vanish if $\beta = 2$ or -1.

In the present case, where the emitted electrons are either parallel (p) or perpendicular (s) to the electric field vector, the intensities are given by

$$\begin{aligned} I_\kappa^p &\propto \sigma_\kappa (1 + \beta_\kappa), \\ I_\kappa^s &\propto \sigma_\kappa \left(1 + \delta_\kappa - \frac{\beta_\kappa}{2}\right). \end{aligned} \tag{11.3}$$

For s polarization, positive values of δ enhance and negative values attenuate the intensity. For atoms or linear molecules, the dipole asymmetry parameter for s electrons is $\beta_{n0} = 2$ in the non-relativistic approximation, implying $\delta_{n0} = 0$.

Neglecting an incomplete polarization, in the geometry used, $\beta_0 = 2$ should lead to a vanishing intensity for excitation of the s states by s polarized photons; however, this is clearly not the case. In the solid with T_d symmetry, however, the photoexcitation of the s band is better described by an $a_1 \to t_2$ transition as compared to an atom in which one simply has transitions into final p states [68, 69]. Restricting the partial waves of the initial a_1 state to $l = 0$ will allow already direct transitions into partial waves with $l' = 1, 2$ of the final t_2 state and the β parameter becomes [70]

$$\beta_{a_1,l=0} = \frac{2D^2_{t_2,l'=1} - 6/7D^2_{t_2,l'=2}}{D^2_{t_2,l'=1} + D^2_{t_2,l'=2}}, \tag{11.4}$$

where D are the radial matrix elements. The additional $D_{t_2,l'=2}$ term immediately allows for $\beta_{a_1,l=0} < 2$.

Further differences are explained by photoelectron diffraction [69, 71–73] because not only the directly emitted electrons but also the electrons scattered in the atoms of the solid approach the detector independently of their initial direction. Generally, this implies that the angular dependence of the final state contains contributions from much higher l' values that are compatible with the irreducible representations of the T_d symmetry group. Indeed, because of polycrystalline samples without preferential crystallographic directions, a rather uniform contribution is expected instead of strong forward scattering effects. This affects not only the angular distribution of the emission from the initial a_1 states but also that of the e and t_2 states. In the latter case, the backscattering terms may be additionally modified by different terms of the angular distribution of the direct channel.

From the enhancement of the intensity from the d states with s polarized light, one would expect negative β_{3d} parameters because the restriction on the nondipole parameter to $\delta \leq 1 - \beta/2$ (for $\beta > 0$) [74] does not allow for I^s_{3d} to become larger than I^p_{3d}, at least for direct emission. The calculated atomic-type β_{3d} parameters for Ni and Mn are in the range of 1.16 to 1.28, but they are not negative. Assuming more complete $e \rightarrow (t_1, t_2)$ and $t_2 \rightarrow (a_1, e, t_1, t_2)$ dipole transitions, the differences become obvious. The equation for $\beta_{e,l=2}$ and $\beta_{t_2,l=2}$, however, cannot be given as easily as above for $\beta_{a_1,l=0}$ without accounting for the contributions of one other than the two that are present in the initial e and t_2 states. In NiMnSb, the Mn states at -3 eV have a nearly pure t_2 character, whereas the Ni d states of both compounds at -2 eV have t_2 as well as e characters. A detailed analysis reveals that the relative change in the intensity in NiMnSb is clearly higher at -2 eV ($\approx$30 %) as compared to that at -3 eV ($\approx$20 %), even without any background subtraction. Similarly, in NiTiSn, the relative change is slightly higher at approximately -1 eV. At -1 eV, most states have e character whereas at -3 eV, the states mainly have t_2 character.

The variation of the relative changes for both compounds with differences observed over the entire valence band clearly indicates a symmetry dependence of the angular asymmetry parameters, in addition to the above mentioned diffraction effects.

11.4 Summary and Conclusions

Hard X-ray photoelectron spectroscopy was used to examine the electronic structure of Heusler compounds, materials for spintronics applications and complete devices.

It was shown that the strongly reduced cross sections of the photoionization process at high kinetic energies are compensated by the high brilliance of the synchrotron sources and that the resolution of HAXPES comes into the same order as is known from low energy spectroscopy. The advantages of the method are the high bulk sensitivity on one hand and the nearly equal differential cross sections for states of different angular momentum on the other. The latter point, in particular, makes HAXPES superior for the study of the valence density of states of solids.

The study of MgO(001)/Co_2MnSi/MgO barrier/AlO_x thin films revealed a high electron mean free path in the oxide layers. The large mean free path makes it possible to look through the material of the tunnel barrier. The feasibility of HAXPES to explore the valence band electronic structure in deeply buried metallic layers was demonstrated. It was shown that buried Co_2MnSi films exhibit the same valence density of states like bulk samples that is typical for half-metallic ferromagnets for spintronics applications. The study of a complete CoFe(B)/MgO/CoFe(B) tunneling device demonstrates the feasibility of HAXPES to explore the electronic structure in deeply buried layers in a non-destructive way. An enforced diffusion of B from the CoFeB into the Ta layers of a magnetic tunnel junction was detected. The improvement of the TMR by annealing of the junction is thus not only caused by an improvement of the structure but also by a change of the composition in the CoFeB layers.

MCDAD in hard X-ray photoelectron spectroscopy was used to study the magnetic response of the core level of buried, remanently magnetized layers. Using bulk-sensitive HAXPES-MCDAD, it was shown that IrMn exchange-biasing layers keep thin films of CoFe or Co_2FeAl remanently magnetized in a well-defined direction. The magnetic dichroism from core levels of CoFe and buried Co_2FeAl multilayer has asymmetries up to above 50 % when it is excited by circularly polarized hard X-rays and is thus much larger compared to that in the case of excitation by soft X-rays. As a noteworthy result, the differences in the Fe $2p$ emission from a regular ferromagnet (CoFe) and a suggested half-metallic ferromagnet (Co_2FeAl) were demonstrated.

LDAD in hard X-ray photoelectron spectroscopy was used to study the angular asymmetry in photoemission from the valence states of not magnetized, polycrystalline Heusler compounds. The polarization-dependent measurements reveal that even high energies and polycrystalline samples do not yield a one-to-one correspondence of photoemission spectra and density of states. This is mainly because there not only exist differences in the partial cross sections but there also exists the symmetry dependence of the angular asymmetry parameters. It was shown that the change in the linear polarization allows for different states of the valence band of complex materials to be easily distinguished.

The polarization-dependent HAXPES experiments can clearly be applied not only to valence states but also to core states. Overall, the high bulk sensitivity of HAXPES combined with linearly polarized photons will have a major impact on the study of the electronic structure of bulk materials, thin films, deeply buried materials, and interfaces. The combination with circularly polarized photons will have a major impact on the study of the magnetic phenomena of deeply buried magnetic materials. The combination of polarization-dependent HAXPES with recently proposed standing wave methods [75, 76] will allow a layer-resolved, element-specific study of buried interfaces.

Acknowledgements This work was financially supported by the Deutsche Forschungsgemeinschaft (project P 7 in DfG research unit FOR 559). The authors are very grateful to K. Kobayashi, S. Ueda, Y. Yamashita, H. Yoshikawa (NIMS, SPring-8, Japan), E. Ikenaga, T. Sugiyama, N. Kawamura, and M. Suzuki (JASRI, SPring-8, Japan) for help with the synchrotron experiments. Thin film samples were provided by T. Ishikawa, T. Taira, T. Uemura, M. Yamamoto (Hokkaido University, Sapporo, Japan), S. Ikeda, H. Ohno (Tohoku University, Sendai, Japan), K. Inomata, H. Sukegawa, and W. Wang (National Institute for Materials Science, Tsukuba, Japan). The authors thank B. Balke, A. Gloskowski, X. Kozina, S. Ouardi, and G. Stryhanyuk (JoGU Mainz) for help with the experiments, for providing bulk samples as well as experimental XRD and magnetization data. Further financial support for the set-up of the phase retarder at BL47XU was provided by DfG-JST (FE633/6-1). The synchrotron radiation HAXPES measurements were performed at BL47XU with the approval of the Japan Synchrotron Radiation Research Institute (JASRI) (Proposal Nos. 2008B0017, 2009B0017) and at BL15XU under the approval of the NIMS beamline station (Proposal Nos. 2007A4903, 2007B4907).

References

1. Siegbahn K (2005) Nucl Instrum Methods Phys Res A 547:2005
2. Lindau I, Pianetta P, Doniach S, Spicer WE (1974) Nature 250:214
3. Kobayashi K, Yabashi M, Takata Y, Tokushima T, Shin S, Tamasaku K, Miwa D, Ishikawa T, Nohira H, Hattori T, Sugita Y, Nakatsuka O, Sakai A, Zaima S (2003) Appl Phys Lett 83:1005
4. Various authors (2005) Nucl Instr Methods Phys Res A 547:1–238
5. Kobayashi K (2009) Nucl Instr Methods Phys Res A 601:32
6. Ueda S, Tanaka H, Ikenaga E, Kim JJ, Ishikawa T, Kawai T, Kobayashi K (2009) Phys Rev B 80:092402
7. Fecher GH, Balke B, Gloskowskii A, Ouardi S, Felser C, Ishikawa T, Yamamoto M, Yamashita Y, Yoshikawa H, Ueda S, Kobayashi K (2008) Appl Phys Lett 92:193513
8. Sato H, Shimada K, Arita M, Takeda Y, Sawada M, Nakatake M, Yoshikawa K, Namatame H, Takata Y, Kobayashi K, Ikenaga E, Shin S, Yabashi M, Miwa D, Nishino Y, Tamasaku K, Ishikawa T, Hiraoka K, Kojima K, Taniguchi M (2004) Physica B 351:298
9. Kuznetsov VV, Cherepkov NA, Fecher GH, Schönhense G (1999) J Chem Phys 110:9997
10. Panaccione G, Cautero G, Cautero M, Fondacaro A, Grioni M, Lacovig P, Monaco G, Offi F, Paolicelli G, Sacchi M, Stojic N, Stefani G, Tommasini R, Torelli P (2005) J Phys Condens Matter 17:2671
11. Sekiyama A, Yamaguchi J, Higashia A, Obara M, Sugiyama H, Kimura MY, Suga S, Imada S, Nekrasov IA, Yabashi M, Tamasaku K, Ishikawa T (2010) New J Phys 12:043045
12. Kobayashi K (2005) Nucl Instrum Methods Phys Res A 547:98
13. Kim J, Ikenaga E, Kobata M, Takeuchi A, Awaji M, Makino H, Chen PP, Yamamoto A, Matsuoka T, Miwa D, Nishino Y, Yamamoto T, Yao T, Kobayashi K (2006) Surf Sci 252:5602
14. Reiss G, Schmalhorst J, Thomas A, Hütten A, Yuasa S (2007) Springer Tracts Mod Phys 227:291
15. Wüstenberg J, Fischer J, Herbort C, Jourdan M, Aeschlimann M, Cinchetti M (2007) J Phys D, Appl Phys 42:084016
16. Jang Y, Nam C, Lee KS, Choa BK, Cho YJ, Kim KS, Kim KW (2007) Appl Phys Lett 91:102104
17. Felser C, Elmers HJ, Fecher GH (2005) Lect Notes Phys 676:113

18. Wurmehl S, Fecher GH, Kroth K, Kronast F, Dürr HA, Takeda Y, Saitoh Y, Kobayashi K, Lin HJ, Schönhense G, Felser C (2006) J Phys D, Appl Phys 39:803
19. Balke B, Fecher GH, Kandpal HC, Felser C, Kobayashi K, Ikenaga E, Kim JJ, Ueda S (2006) Phys Rev B 74:104405
20. Fecher GH, Balke B, Ouardi S, Felser C, Schönhense G, Ikenaga E, Kim JJ, Ueda S, Kobayashi K (2007) J Phys D, Appl Phys 40:1576
21. Ouardi S, Balke B, Gloskovskii A, Fecher GH, Felser C, Schönhense G, Ishikawa T, Uemura T, Yamamoto M, Sukegawa H, Wang W, Inomata K, Yamashita Y, Yoshikawa H, Ueda S, Kobayashi K (2009) J Phys D, Appl Phys 42:084010
22. Ouardi S, Gloskovskii A, Balke B, Jenkins CA, Barth J, Fecher GH, Felser C, Gorgoi M, Mertin M, Schäfers F, Ikenaga E, Yang K, Kobayashi K, Kubota T, Oogane M, Ando Y (2009) J Phys D, Appl Phys 42:084011
23. Suzuki M, Kawamura N, Mizukami M, Urata A, Maruyama H, Goto S, Ishikawa T (1998) Jpn J Appl Phys 37:L1488
24. Tanuma S, Powell CJ, Penn DR (1993) Surf Interface Anal 21:165
25. Miyamoto K, Kimura A, Miura Y, Shirai M, Ye M, Cui Y, Shimada K, Namatame H, Taniguchi M, Takeda Y, Saitoh Y, Ikenaga E, Ueda S, Kobayashi K, Kanomata T (2009) Phys Rev B 79:1004205(R)
26. Fecher GH, Gloskovskii A, Kroth K, Barth J, Balke B, Felser C, Schäfers F, Mertin M, Eberhardt W, Mähl S, Schaff O (2007) J Electron Spectrosc Relat Phenom 156–158:97
27. Kijima H, Ishikawa T, Marukame T, Koyama H, Matsuda K, Uemura T, Yamamoto M (2006) IEEE Trans Magn 42:2688
28. Kijima H, Ishikawa T, Marukame T, Matsuda KI, Uemura T, Yamamoto M (2007) J Magn Magn Mater 310:2006
29. Marukame T, Kijima H, Ishikawa T, Matsuda KI, Uemura T, Yamamoto M (2007) J Magn Magn Mater 310:1946
30. Ishikawa T, Marukame T, Kijima H, Matsuda Ki, Uemura T, Yamamoto M (2006) Appl Phys Lett 89:192505
31. Ishikawa T, Hakamata S, Matsuda KI, Uemura T, Yamamoto M (2008) J Appl Phys 103:07A919
32. Kandpal HC, Fecher GH, Felser C, Schönhense G (2006) Phys Rev B 73:094422
33. Kandpal HC, Fecher GH, Felser C (2007) J Phys D, Appl Phys 40:1507
34. Lee YM, Hayakawa J, Ikeda S, Matsukura F, Ohno H (2007) Appl Phys Lett 90:212507
35. Lee YM, Hayakawa J, Ikeda S, Matsukura F, Ohno H (2006) Appl Phys Lett 89:042506
36. Ikeda S, Hayakawa J, Ashizawa Y, Lee YM, Miura K, Hasegawa H, Tsunoda M, Matsukura F, Ohno H (2008) Appl Phys Lett 93:082508
37. Masuda M, Uemura T, Matsuda Ki, Yamamoto M (2008) IEEE Trans Magn 44:3996
38. Wang WH, Sukegawa H, Shan R, Mitani S, Inomata K (2009) Appl Phys Lett 95:182502
39. Cherepkov NA, Kuznetsov VV, Verbitskii VA (1995) J Phys B, At Mol Opt Phys 28:1221
40. Hillebrecht FU, Roth C, Rose HB, Park WG, Kisker E, Cherepkov NA (1996) Phys Rev B 53:12182
41. Tobin JG, Goodman KW, Schumann FO, Willis RF, Kortright JB, Denlinger JD, Rotenberg E, Warwick A, Smith NV (1998) Surf Sci 395:L227
42. Henk J, Niklasson AMN, Johansson B (1999) Phys Rev B 59:13986
43. Bansmann J, Lu L, Meiwes-Broer KH, Schlathölter T, Braun J (1999) Phys Rev B 60:13860
44. Thole BT, van der Laan G (1991) Phys Rev B 44:12424
45. Thole BT, van der Laan G (1993) Phys Rev B 48:210
46. Thole BT, van der Laan G (1994) Phys Rev B 49:9613
47. van der Laan G, Thole BT (1994) Phys Rev B 52:15355
48. van der Laan G, Dhesi SS, Dudzik E (2000) Phys Rev B 61:12277
49. Fecher GH (1995) Europhys Lett 29:605
50. Venus D (1997) Phys Rev B 56:2661
51. Nakagawa T, Yokoyama T (2006) Phys Rev Lett 96:237402

52. Hild K, Maul J, Schönhense G, Elmers HJ, Amft M, Oppeneer PM (2009) Phys Rev Lett 102:057207
53. Rossi G, Panaccione G, Sirotti F, Lizzit S, Baraldi A, Paolucci G (1997) Phys Rev B 55:11488
54. Smith RJ, Anderson J, Lapeyre GJ (1976) Phys Rev Lett 37:1081
55. Cherepkov NA, Schönhense G (1993) Europhys Lett 24:79
56. Cherepkov NA, Raseev G (1995) J Chem Phys 103:8238
57. Kabachnik NM (1996) J Electron Spectrosc Relat Phenom 79:269
58. Ostertag C, Bansmann J, Grünewald C, Jentzsch T, Oelsner A, Fecher GH, Schönhense G (1995) Surf Sci 331–333:1197
59. Fecher GH, Oelsner A, Schicketanz M, Schönhense G (1998) J Electron Spectrosc Relat Phenom 88–91:185
60. Matsushita T, Imada S, Daimon H, Okuda T, Yamaguchi K, Miyagi H, Suga S (1997) Phys Rev B 56:7687
61. de Groot RA, Müller FM, van Engen PG, Buschow KHJ (1983) Phys Rev Lett 50:2024
62. Aliev FG (1991) Physica B 171:199
63. Miyamoto K, Kimura K, Sakamoto K, Ye M, Cui Y, Shimada K, Namatame H, Taniguchi M, Fujimori SI, Saitoh Y, Ikenaga E, Kobayashi K, Tadano J, Kanomata T (2008) Appl Phys Express 1:081901
64. Ouardi S, Fecher GH, Balke B, Kozina X, Stryganyuk G, Felser C, Lowitzer S, Ködderitzsch D, Ebert H, Ikenaga E (2010) Phys Rev B 085108
65. Fadley CS (2005) Nucl Instrum Methods Phys Res A 647:24
66. Cooper J, Zare RN (1968) J Chem Phys 48:942
67. Cooper JW (1990) Phys Rev A 42:6942
68. Kuznetsov VV, Cherepkov NA, Fecher GH, Schönhense G (2002) J Chem Phys 117:7180
69. Fecher GH, Kuznetsov VV, Cherepkov NA, Schönhense G (2002) J Electron Spectrosc Relat Phenom 122:157
70. Chandra N (1987) J Phys B, At Mol Phys 20:3417
71. Daimon H, Imada S, Suga S (2001) Surf Sci 471:143
72. Oelsner A, Schicketanz M, Fecher GH, Schönhense G, Morais J (2002) Surf Rev Lett 9:889
73. Rennert P (2001) J Electron Spectrosc Relat Phenom 119:1
74. Hemmers O, Guillemin R, Lindle DW (2004) Radiat Phys Chem 70:123
75. Fadley CS (2010) J Electron Spectrosc Relat Phenom 178–179:2
76. Zegenhagen J, Detlefs B, Lee TL, Thiess S, Isern H, Petit L, Andre L, Roy J, Mi Y, Joumard I (2010) J Electron Spectrosc Relat Phenom 178–179:258

Chapter 12
Characterization of the Surface Electronic Properties of $Co_2Cr_{1-x}Fe_xAl$

Jan-Peter Wüstenberg, Martin Aeschlimann, and Mirko Cinchetti

Abstract Due to their half metallic properties and high Curie temperatures Heusler alloys are potential key materials for future application in spin-based devices. One crucial aspect for device performance is the electron spin polarization at surfaces and interfaces, which in general deviates from bulk values due to the different bonding environment. Here, we present spin-resolved photoemission data on surfaces of $Co_2Cr_{1-x}Fe_xAl$ thin films. The influence of bulk and surface contributions on the spin-resolved spectra is discussed for $x = 0$, $x = 0.4$ and $x = 1$, yielding information on disorder, surface states and calculation schemes that are consistent with our spin-resolved photoemission experiments.

12.1 Introduction

Half metallic ferromagnets have received a great scientific interest due to their potential application as spin injection layer in spintronics devices. Among this class of materials, Co-based Heusler alloys are of particular technological interest since they exhibit Curie temperatures high above room temperature and a wide electronic tunability through the exchange or the substitution of atoms within a sublattice of the alloy [1]. Such Heusler alloys display Co_2YZ composition, with Y being a transition metal and Z being a main group element. While the majority spins have metallic character at the Fermi energy ε_F, a Co related hybridization gap has been proposed for the minority electrons [2], leading to a full spin polarization of the conduction electrons. These peculiar electronic properties are strongly connected to the bulk $L2_1$ crystal structure. Even small deviations in bulk stoichiometry or sublattice

J.-P. Wüstenberg · M. Aeschlimann · M. Cinchetti (✉)
Department of Physics and Research Center OPTIMAS, University of Kaiserslautern, Erwin-Schrödingerstr. 46, 67663 Kaiserslautern, Germany
e-mail: cinchett@rhrk.uni-kl.de

J.-P. Wüstenberg
e-mail: jpwuest@physik.uni-kl.de

M. Aeschlimann
e-mail: ma@physik.uni-kl.de

C. Felser, G.H. Fecher (eds.), *Spintronics*, DOI 10.1007/978-90-481-3832-6_12,

Table 12.1 Nomenclature of the investigated $Co_2Cr_{1-x}Fe_xAl$ samples

Fe content x	Alloy formula	Short name
0	Co_2CrAl	CCA
0.4	$Co_2Cr_{0.6}Fe_{0.4}Al$	CCFA
1	Co_2FeAl	CFA

atomic purity (such as antisites, swaps or impurities) can induce local electronic states in the bulk minority spin gap and consequently reduce or even invert the spin polarization of the electrons at the Fermi energy [3, 4].

Here we report on the surface electronic structure of three prominent members of the $Co_2Cr_{1-x}Fe_xAl$ family of quaternary Heusler alloys, where the Y lattice site is shared by Cr and Fe atoms (see Table 12.1). This alloy is one of the most thoroughly investigated Heusler alloys to date [5].

Co_2CrAl has been predicted to be a half metal in the $L2_1$ structure, showing a large Cr-induced majority density of states. However, magnetic moments have been determined to be significantly below the predicted values. This is attributed to site disorder, in particular swaps between the Co and the Cr sublattice [6]. The latter leads to an antiferromagnetic coupling between nearest neighbor Cr atoms which is responsible for the reduction of the Cr magnetic moment. Recently, Hoshino calculated the formation energies of various defects in Co_2CrAl, showing that these values are lower than the corresponding formation energies in Co_2MnSi [7]. To our knowledge, Co_2CrAl has not been employed in magnetic tunneling junctions so far, even though Miura et al. predicted a half metallic interface between a CrAl terminated Co_2CrAl and a MgO tunneling barrier [8]. CCFA has been widely investigated as an electrode material for magnetic tunnel junctions [9]. The influence of interface morphology was studied by Herbort and coworkers [10, 11]. A maximum magnetoresistance effect was found for an annealing temperature of 550 °C. For these films, a B2 structure was found with 19 % disorder on the Co sites. Using STM it was shown that the annealed films showed flat terraces, possibly supporting a defined surface termination. However, such flat terraces did not yield the maximum magnetoresistance values. They proposed that the good surface quality might give rise to surface states that reduce the tunneling spin polarization. Such surface and interface states have indeed been predicted recently for a CCFA/GaAs interface [12].

In contrast to early GGA calculations [13], LDA+U calculations predict Co_2FeAl to be half metallic, with ε_F being situated close to minority valence band edge [14]. The crystal structure of bulk samples is B2, with up to 10 % of $L2_1$ order [15, 16]. In thin films A2 or B2 type disorder was observed, depending on the thermal treatment of the samples [17]. Using a MgO tunneling barrier and a CoFe counterelectrode Wang et al. found a very large magnetoresistive effect of 340 % at room temperature. Recently Xu and coworkers calculated the surface electronic structure of Co_2FeAl and found that the band gap was closed only in a very limited region close to the surface [18].

Measuring the spin polarization of Co-based Heusler alloys remains a nontrivial task, in bulk systems as well as at surfaces and interfaces. While theoretical calculations often predict a full spin polarization at the Fermi energy using the bulk density of states values, an unambiguous experimental verification is still missing. For magnetic tunneling junctions using amorphous barriers, the Jullière formula $TMR = \frac{2P_1P_2}{1-P_1P_2}$ can be applied to extract an effective tunneling spin polarization, provided that both electrode materials are equal ($P_1 = P_2$) or one of the spin polarization values is known (e.g., $P_2 = 50$ % in $Co_{70}Fe_{30}$) [19]. A value of $P_1 = 67$ % at low temperature has been reported by Herbort et al. for a magnetic tunnel junction (MTJ) based on CCFA and $Co_{70}Fe_{30}$ [11]. Similar values have been obtained by Andreev reflection using a superconducting point contact [15]. Only recently, the spin-resolved bulk half metallic valence band density of unoccupied states could be revealed using circular dichroism in X-ray absorption [20–22].

The lack of convergence between first principle calculations and experiments is attributed to three main effects: bulk, surface and interface, and method-related effects. Bulk effects, such as dynamic correlation effects [23, 24] or crystallographic disorder may destroy half metallicity from the very beginning. However, the effects are not restricted to a particular energy such as the Fermi energy. Instead, the band structure is modified for all energies. The electronic structure and effects of disorder in $Co_2(Cr_{1-x}Fe_x)Al$ have been studied theoretically by Miura and coworkers [4]. For low Fe concentrations, B2 disorder between the Cr/Fe and Al sublattices does not affect half metallicity. However, with increasing Fe content, the calculated spin polarization was found to decrease due to a shift of the Fermi energy out of the minority gap. For CCFA ($x = 0.4$) and Co_2FeAl ($x = 1$) a spin polarization of 90 % (77 % in the B2 structure) and 30 % were predicted, respectively. More severely, disorder between Co and Cr sites (in particular Co antisites) was shown to induce a strong peak in the minority density of states at the Fermi energy. The formation energy of such defects was found to be almost one order of magnitude higher than for B2 disorder. In alloys, phase separation has to be considered as well. Karthik and coworkers used dark field transmission electron microscopy to reveal bulk phase separations into A2 and B2 regions for Co_2CrAl, A2 and $L2_1$ regions for CCFA and a single B2 phase in Co_2FeAl [15]. However, the spin polarization values for these samples obtained by point contact Andreev reflection were very similar (62 %, 54 %, and 56 % respectively).

Surface and interface states are relevant for real devices such as MTJs. Such states may be intrinsic properties of the specific surface or interface termination and appear at very specific energetic positions on top of the surface projected bulk band structure. They have been predicted by first principle calculations for several Heusler surfaces [12, 25].

Nearly no spin-resolving measurement technique yields the true bulk spin polarization as given by the theoretical density of states calculations. Instead, the measured spin asymmetry is altered by additional spin dependent filtering properties of the instruments transmission function, including all effects from the electronic structure to the final measurement signal. This may also include, for example, a strong surface sensitivity, spin dependent scattering effects or symmetry selection rules within the excitation process.

One of the most direct ways to probe the electronic properties of a magnetic material is spin-resolved photoelectron spectroscopy (SR-PES). By varying the photon energy and polarization, and the electron emission angle, the electronic band structure and wave function symmetry can be determined with high energy and momentum accuracy [26]. The addition of a spin polarimeter in principle gives access to the spin-resolved band structure [27]. The spin polarization at the Fermi level of Heusler alloys determined by SR-PES is typically smaller than the values reported above using other techniques such as the tunneling magnetoresistance or point contact Andreev reflection. For Co_2MnSi films, a spin polarization of only 12 % was observed using 70 eV photons, and attributed to disorder effects and a possible small degree of sample oxidation [28]. Using a UHV cleaning recipe, Cinchetti et al. obtained $P = 45$ % for an Fe buffered $Co_2Cr_{0.6}Fe_{0.4}Al$ film and a photon energy of 5.9 eV [29, 30].

In order to explain these low values, effects intrinsic to the photoemission process have to be considered as well. For example, the surface sensitivity depends strongly on the electron kinetic energy and thus on the wavelength of the incident photons, being at its minimum of about 5 Å in the VUV photon energy range (30–100 eV) [31]. Hence, surface states can contribute strongly to the photoelectron yield in photoemission spectra of Heusler alloy surfaces and interfaces. Since surface states are localized in the direction of the surface normal, they appear as energetically relatively narrow features in the photoelectron spectrum. Due to the spin dependent band structure, such surface states affect majority and minority electrons in different ways, leading to significant changes also in the measured spin polarization. Especially if the majority density of states at the Fermi energy is low (e.g., due to strongly dispersive bands with sp character) a surface state located in the minority spin gap of a half metal will dominate the spin polarization at the Fermi level [32].

In addition to the increased surface sensitivity, dipole selection rules may alter the spin dependent photoemission intensity, depending on photon energy and polarization. This effect turns out to be particularly important at low photon energies. Also spin-dependent lifetime effects in the final state may change the measured polarization [33].

In this work, we investigate the spin-resolved electronic structure of three important members of the $Co_2Cr_{1-x}Fe_xAl$ Heusler family. By means of spin-resolved photoelectron spectroscopy (SR-PES), LEED and Auger analysis we study 80 nm-thick films of the Heusler alloys listed in Table 12.1. The experimental spin-resolved photoelectron spectra are compared to calculated bulk electronic structures. Considering the peculiar properties of the photoemission mechanism the observed features are discussed in terms of bulk origin, additional surface states and disorder effects.

12.2 Experimental Details

All Heusler films have been deposited on MgO(001) substrates by rf sputtering from stoichiometric targets. In order to improve epitaxial growth of the Heusler films, an additional MgO buffer layer of 10 nm was evaporated prior to deposition. The thickness of the films was 80 nm. The CCFA film was not capped and transported in vacuo to the experimental chamber using a well-degassed vacuum suitcase. The Co_2CrAl and Co_2FeAl samples were covered with a 4 nm Al protection layer for ex situ transport. Further details considering the sample fabrication process can be found in [10, 34]. The experiments are performed in a separate UHV chamber with a base pressure of 10^{-10} mbar, equipped with sputter and heating facilities for sample cleaning. The chemical composition and structure was checked using Auger spectroscopy and low energy electron diffraction (LEED). Sample preparation was accomplished by sputtering and heating cycles. For the capped samples Co_2CrAl and Co_2FeAl, the Al protection layer was first removed by 500 eV Ar^+ etching, until the Co to Al peak ratio was stable in the Auger spectra. After exposing the Heusler surface all films have been sputtered gently using 500 eV Ar^+ ions in order to remove residual carbon and oxygen surface contaminations. Surface ordering was restored by careful annealing up to 550 °C for 10 minutes and checked by LEED. The films were reported to show B2 order after thermal annealing [17, 35], which is in agreement with the LEED patterns obtained in our analysis chamber as described in more detail in the next section.

The photoemission spectra were obtained by irradiating the sample with the 4th harmonic (photon energy 5.9 eV) of a 100 fs Ti:Sapphire oscillator (Spectra Physics Tsunami). The photon angle of incidence onto the samples was set to 45°, and the spectra were taken in normal electron emission (ΓX momentum direction). However, due to the large unit cell size of CCFA and the angular acceptance angle of the analyzer of $\pm 13°$ along the sample [110] direction (ΓK momentum direction), the photoemission spectrum integrates the electron momentum over almost 2/3 of the Brillouin zone of $Co_2Cr_{1-x}Fe_xAl$. Spin-resolved photoemission spectra were recorded by means of a commercial 90° cylindrical sector energy analyzer (Focus CSA 300) equipped with an additional spin detector based on spin-polarized low-energy electron diffraction at a W(100) crystal (Focus SPLEED). Due to the detection geometry, the spin projections along the $Co_2Cr_{1-x}Fe_xAl$ [110] direction (in plane) and along the surface normal (out of plane) can be determined. A Sherman factor of 0.2 was used to determine the spin polarization from the measured intensity asymmetry of opposite diffraction spots. Due to uncertainties of the Sherman factor a systematic relative spin polarization error of up to 25 % apart from statistical errors [27] cannot be excluded. The achieved effective energy resolution is 210 meV as determined from the work function cutoff. For the spin-resolved measurements the films are remanently magnetized by an external in-plane magnetic field of $\mu_0 H = 15$ mT, which is sufficient to saturate the magnetization of the investigated samples along the [110] direction. Detector related asymmetries are canceled by taking separate measurements with reversed sample magnetization. With this setup an in-plane spin polarization of 30 % at the Fermi energy was achieved on a 200 nm thick film of polycrystalline $Co_{70}Fe_{30}$.

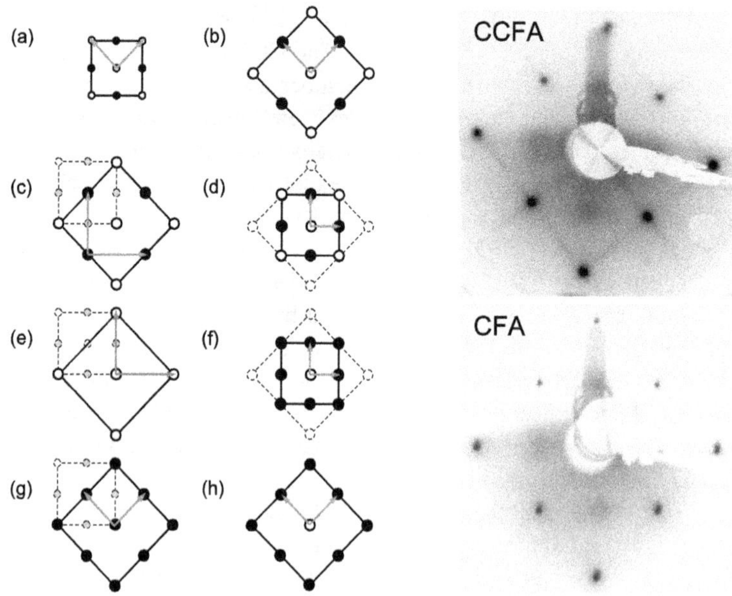

Fig. 12.1 *Left*: Common $L2_1$ surface configurations (*left*) and expected LEED patterns (*right*) within the kinematic LEED approximation. (**a**) and (**b**) MgO (001) surface, (**c**) and (**d**) YZ termination, (**e**) and (**f**) vacancy-Z or Y-vacancy termination, (**g**) and (**h**) X_2 or disordered YZ (B2) termination. The MgO reference is shown with *dashed lines*. In real space different circles represent different atom species, in reciprocal space *open* (*full*) *circles* correspond to high (low) intensities, as determined by the values of the structure factor. *Right*: LEED images of CCFA and Co_2FeAl, taken at $E = 57$ eV primary energy

12.3 Results and Discussion

Before investigating the spin polarization, we examine the structural quality of the crystal surface. As an example, we present the LEED pattern of CCFA in the right panel of Fig. 12.1. The pattern has been obtained after heating the film to 550 °C. The MgO substrate [100] axis is directed horizontally and corresponds to the Heusler [110] axis.

We can observe a clear diamond shaped spot pattern, indicating a high degree of crystallographic ordering. The spots are connected by diagonal lines along the [100] and [010] directions of CCFA, which are missing for Co_2FeAl and Co_2CrAl (not shown). A pattern of parallel lines with constant reciprocal spacing implies a real space structure that is irregular parallel to the direction of the line and periodic in the normal direction. Since these properties are mutually exclusive, the presence of line patterns intersecting at an angle of 90° indicate the presence of two different domain patterns that are rotated by 90° with respect to each other. The periodic spacing matches the lattice constant of the B2 ordered surface which is half the lattice constant of the corresponding $L2_1$ structure (compare Figs. 12.1c and g). The aperiodicity along the line directions might be caused by irregularly sized terraces or domains with otherwise good crystallographic order. Such features have, for example, been observed also by Herbort et al. in STM images of rf-sputtered CCFA films [10]. We have observed such lines for CCFA samples both with and without capping layer [36].

The electronic spectrum of the surface depends crucially on the outermost atom layer of the crystal, i.e., the termination. For Co-based Heusler alloy (001) surfaces in the ideal $L2_1$ structure, the outermost layer consists either of a layer of Co atoms in a square lattice or Y and Z atoms occupying the corners and the center of a centered square lattice (Co or YZ-termination). In this report we consider three defect types. B2 disorder leads to a random occupation of the Y and Z sites and thus to a reduction of the effective surface unit cell size. The same effect occurs for A2 disorder in which all four sublattices are occupied randomly by the constituting atoms. A regular vacancy on any surface lattice (Co, Y or Z) site leads to a larger surface unit cell as compared to the ideal structure, and thus to a constricted LEED pattern.

In order to learn more about the surface termination we have constructed kinematic LEED patterns of several possible terminations of the (001) Heusler surface (left panel of Fig. 12.1). As a reference the pattern of MgO (001) is used. The primitive surface basis vectors are marked by arrows, and the reciprocal lattice basis vectors indicate the possible positions of diffraction spots. If different atomic species are present in the surface unit cell, the associated kinematic structure factor leads to an intensity modification of a regular set of reflexes, as known, for example, from the fcc lattice extinction rules in X-ray diffraction. The spot intensity modulation depends on the difference in the complex atomic scattering amplitudes of the alloy elements (a vacancy has a zero atomic scattering factor). Unfortunately, we cannot say whether the surface has B2 or A2 disorder without a detailed intensity analysis, since the A2, B2 and Co termination have the same surface unit cell geometry. Even very small domains of $L2_1$ order are possible considering the weak and broad intensity contribution at intermediate spot positions. For CCFA and Co_2FeAl, calculations predict a Cr-Al and Fe-Al termination, respectively [12, 18]. Thus we conclude that the surfaces investigated in this report are well ordered and most probably of dominant YZ-B2 or A2 order.

The spin polarization of Co_2CrAl, CCFA and Co_2FeAl is shown in Fig. 12.2. Co_2CrAl shows no sign of spin polarization within the statistical error in the whole spectral range. At first sight, a strong reduction of the Co_2CrAl spin polarization might not seem surprising since the Curie temperature of Co_2CrAl is only in the range of 334 K [5]. However, no change of the spin-resolved spectrum was observed upon cooling of the sample to 120 K (not shown here).

In contrast to Co_2CrAl, CCFA and Co_2FeAl show a clear and structured spin polarization spectrum. A plateau with positive spin polarization at low energy is followed by a characteristic minimum. Towards the Fermi energy, the spin polarization rises towards positive values again. Only in CCFA, this trend is interrupted by a sudden drop as the energy approaches the Fermi level, leading to a slightly negative value.

As a first step to interpret the spectra we have calculated artificially broadened spectra using the GGA density of states (taken from [13]) and a gaussian resolution

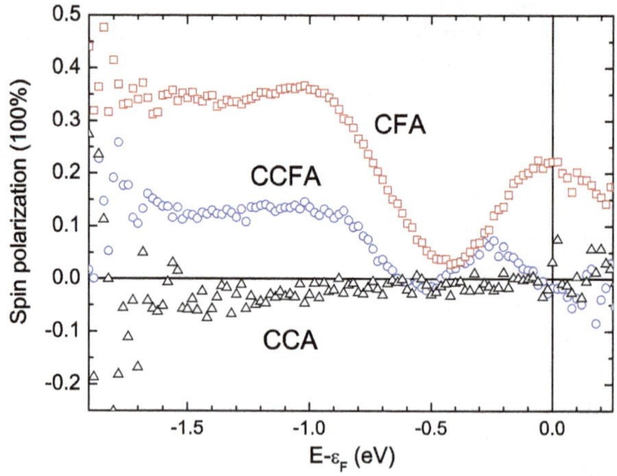

Fig. 12.2 Energy-resolved surface spin polarization of $Co_2Cr_{1-x}Fe_xAl$ for $x = 0$ (Co_2CrAl, *triangles*), $x = 0.4$ (CCFA, *circles*) and $x = 1$ (Co_2FeAl, *squares*) measured at room temperature. The photon energy is 5.9 eV

function of 210 meV full width at half maximum (FWHM), corresponding to our experimental facilities. The result is plotted and compared to the spectra obtained experimentally (Fig. 12.3).

We first observe that in contrast to Co_2CrAl the common trend of the experimental spin polarization spectra of CCFA and Co_2FeAl is reproduced in the calculated spin polarization spectra. For a rising Fe/Cr ratio the theoretical spin polarization spectra suggest two trends. First, the spin polarization at the Fermi energy is expected to drop from 100 % for Co_2CrAl to about 40 % for Co_2FeAl. Second, the minimum of the spin polarization is expected to get broader and to move towards the Fermi energy (from $E = -0.9$ eV for Co_2CrAl and CCFA to $E = -0.6$ eV for Co_2FeAl).

The shift of the calculated spin polarization minimum towards the Fermi energy is confirmed in the measured spin polarization spectra of CCFA ($E_{min} = -0.53$ eV) and Co_2FeAl ($E_{min} = -0.46$ eV). The measured spin polarization value of Co_2FeAl at the Fermi energy is reduced to half of the calculated value and even negative for CCFA.

Comparing the measured majority and minority spectra of each compound we observe a further interesting feature. The minority electrons of the compounds Co_2CrAl and CCFA, which are predicted to show a band gap at the Fermi energy by GGA calculations, display a clear metallic behavior at the Fermi energy (within the resolution limits). Co_2FeAl minority electrons are predicted to behave like a metal in GGA calculations. Instead, a Fermi–Dirac cutoff is barely visible in the measured minority spectra of Co_2FeAl, indicating poor minority conduction properties.

In order to distinguish between matrix element related and surface state related spectral features we have to consider also the band structure of the compounds, including initial and final states. Here we only take into account the energy- and momentum-allowed transitions. Symmetry effects, reducing the number of possible transitions, are not considered. In Figs. 12.4 and 12.5 overlays of the ideal $L2_1$ occupied band structure and a replica of the unoccupied states, shifted downwards by the used photon energy of $\hbar\omega = 5.9$ eV, are shown for the borderline cases Co_2CrAl and

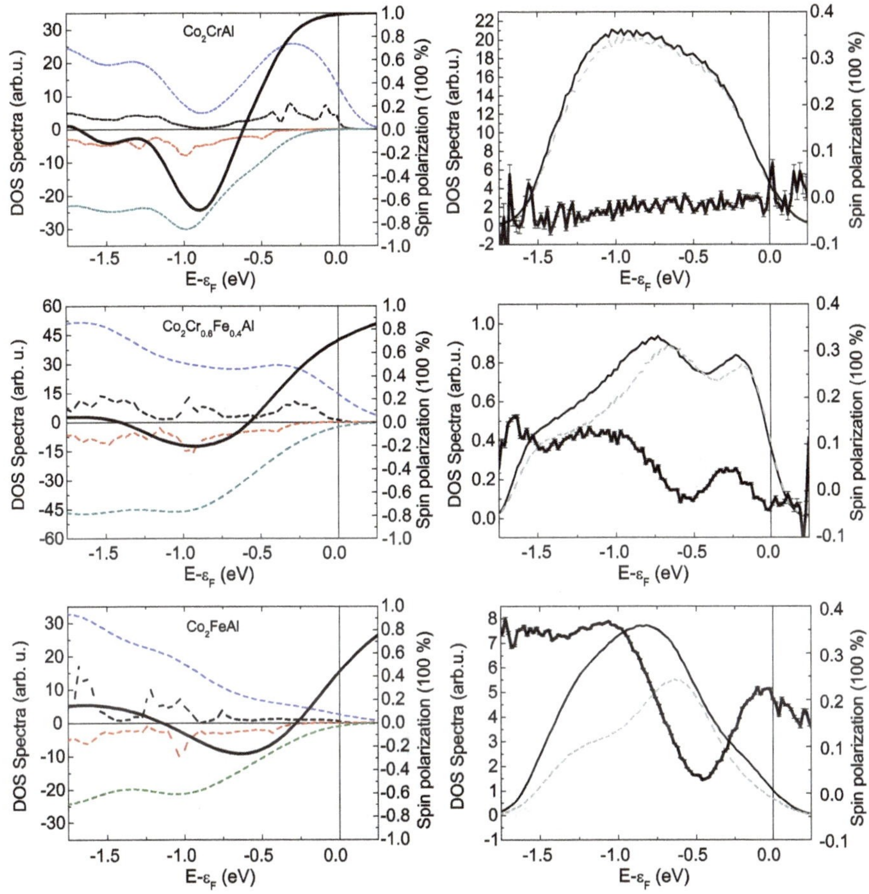

Fig. 12.3 *Left*: Majority (positive values) and minority (negative values) density of states for Co_2CrAl, CCFA (using $x = 0.5$) and Co_2FeAl without disorder. The spectra were truncated by a Fermi–Dirac distribution at room temperature and convoluted with a resolution function of FWHM = 210 meV. The resulting spin polarization is indicated by the *thick black line*. GGA supercell calculations adapted from [13]. *Right*: Measured majority (*black line*), minority (*gray dashed*) and spin polarization (*thick black line*)

Co_2FeAl, respectively. Possible bulk resonant interband transitions appear as crossing points between the bands and can directly be compared to the binding energy of measured spectral features.

We start the discussion with Co_2CrAl (see Figs. 12.3 and 12.4). In principle, the spin polarization of Co_2CrAl should be robust against surface related influences, since the majority density of states is high compared to CCFA and Co_2FeAl. The latter is caused by the influence of localized bands in the majority spin band structure at the point (Γ, ε_F). Clearly, the upper edges of the occupied majority and minority bands are connected by interband transitions to final state bulk bands and thus should appear as pronounced features in the measured spectra due to resonant

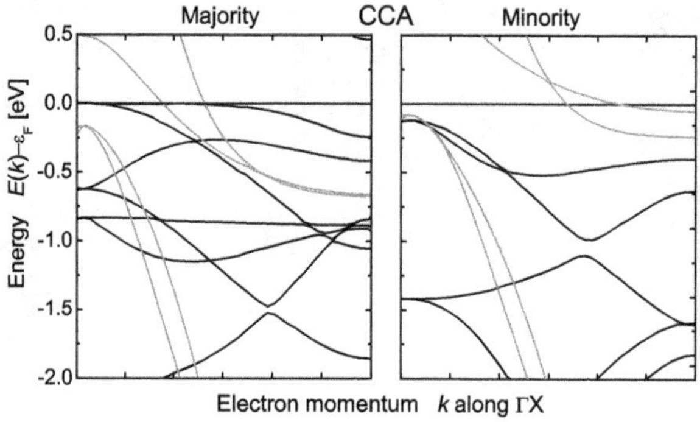

Fig. 12.4 Energy-allowed optical transitions along the surface normal direction for majority and minority electrons in Co_2CrAl as obtained within the GGA scheme: Possible bulk transitions exist at the crossing point of occupied initial states (*black bands*) and unoccupied final states (*gray bands*). Reproduced from G. Fecher, H. Kandpal, S. Wurmehl, J. Morais, H.J. Lin, H.J. Elmers, G. Schonhense, C. Felser, *J. Phys.: Cond. Mat.* 17, 7237. Copyright (2005) by IOP Publishing

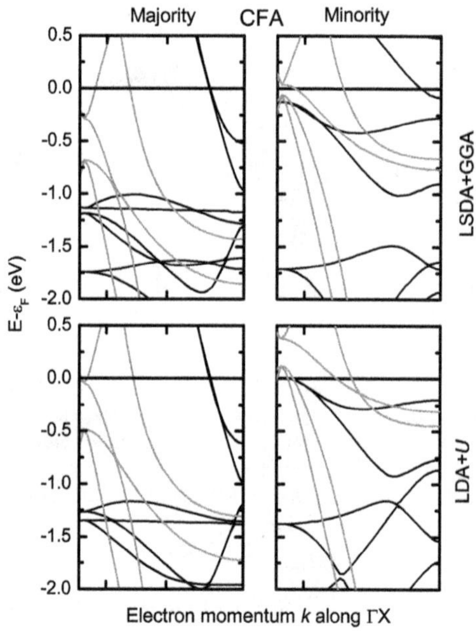

Fig. 12.5 Energy-allowed optical transitions along the surface normal direction for majority and minority electrons in Co_2FeAl as obtained by the GGA (*top*) and LDA+U scheme (*bottom*): Possible bulk transitions exist at the crossing point of occupied initial states (*black bands*) and unoccupied final states (*gray bands*). Reproduced from G. Fecher, C. Felser, *J. Phys. D: Appl. Phys.* 40, 1582. Copyright (2007) by IOP Publishing

coupling of the bulk final states and outgoing inverse LEED states. The measured majority and minority spectra are identical, but only the measured majority electron spectrum can qualitatively be explained using the calculated majority bulk transition scheme of $L2_1$ ordered Co_2CrAl. The strong deviation of the minority spectrum from the calculated band structure might be caused by Co-Cr disorder. Co-Cr disorder results in a strong increase in the minority density of states at ε_F which is related to Co d-states [4]. We conclude that for Co_2CrAl bulk optical transitions cannot explain the vanishing of the spin polarization for all energies, implying that the calculated bulk Co_2CrAl band structure is not conserved within the probing depth of the electrons, probably as a consequence of surface disorder or of bulk atomic disorder beyond a B2 type disorder. Since the surface showed a clear LEED pattern we exclude surface roughness as origin of the vanishing spin polarization. Thus we conclude that the investigated sample still showed a substantial degree of bulk Co-Cr disorder that changes the band structure and leads to a canceling of the spin polarization. This discrepancy between measurements and calculations is in line with the experimental low values for the bulk magnetic moments of Co_2CrAl, which are strongly reduced as compared to the theoretical values calculated for the ordered $L2_1$ structure [35].

Turning our attention to the spectra of CCFA and Co_2FeAl in Fig 12.3 we observe much more pronounced features. The spin polarization of Co_2FeAl resembles most clearly the calculated spin polarization as obtained from the density of states. Its value of 22 % at ε_F corresponds to approximately half of the numerical value. Interestingly, the surface spin polarization is also highest for Co_2FeAl, which has the lowest majority density of states of all investigated compounds and thus should be particularly sensitive to the presence of minority surface states. However, in a recent publication, Xu et al. have shown that although one indeed has to expect surface states at Fe-Al terminated surfaces of Co_2FeAl, these states are localized in the first three layers and the bulk band structure is quickly recovered [18]. Thus the bulk spin polarization could be at least partly captured by our measurements, provided that (a) the escape depth of the electrons is larger than the electronically disturbed surface region and that (b) the bulk electronic structure is not significantly disturbed by disorder effects.

It is interesting to relate the observed spectra with the band structures that are displayed in Fig. 12.5, obtained for Co_2FeAl in the GGA and the LDA+U calculation scheme, respectively [13, 14]. The differences between the two calculation schemes show up mainly in the minority band structure. The dispersive band that crosses the Fermi energy in the LSDA calculations of Co_2FeAl is lifted above the Fermi energy by the inclusion of static correlation effects in the LDA+U scheme, opening a large gap in the minority band structure. However, at the same time the valence band maximum is shifted upwards and touches the Fermi energy [14]. Compared to the Co_2CrAl calculations, the minority sp-bands in Co_2FeAl are located at higher energies. We also observe a remarkable difference to the occupied majority bands in Co_2CrAl. The strong contribution of localized majority states at the Γ point of Co_2CrAl is replaced by dispersive sp-bands that cross the Fermi energy far away from the Brillouin zone center.

Considering possible bulk majority transitions for Co_2FeAl within the energy range of our spectra we observe that no bulk majority interband transitions take place down to a binding energy of at least 1 eV, since the sp-bands close to ε_F are not crossed by bulk final state bands. Bulk related transitions can only stem from the sp-bands, coupling directly into inverse LEED final states at the surface. This scenario seems reasonable considering the measured majority spectrum of Co_2FeAl. The increase of sp-related density of states is reflected in the majority photoelectron yield towards lower energies. In addition a maximum of the majority electron yield is observed at $E - \varepsilon_F = -0.85$ eV which might be related either to the band bottom of the majority sp-bands or to the localized majority d-bands.

On the minority side, as the flat parts of the band structure of Co_2FeAl are well connected to final states, we would expect a broad range with large negative spin polarization starting from the Fermi level (LDA+U) or slightly below (GGA). Instead, the minority photoelectron yield is even weaker than on the majority side, indicating only sp-type contributions. Such bands are present only in the GGA calculations. Additionally, at $E - \varepsilon_F = -0.62$ eV a strong minority peak is observed that might be related to bulk interband transitions. Again, this energy fits better to the GGA band structure. Thus we conclude that GGA calculations reproduce the spin-resolved photoelectron spectra of Co_2FeAl better than the LDA+U calculations. No surface states are needed to explain the measured spin-resolved photoelectron spectra of Co_2FeAl.

Unfortunately we do not have a calculated band structure of CCFA at hand. However, from the spin-resolved partial densities of states as given by Wurmehl and coworkers [37] we can see that the localized majority bands at the Fermi energy are related to Cr and thus should persist to some extent in CCFA. However, also Fe induced sp-bands should be present at the Fermi energy, though only close to the Brillouin zone boundaries. The measured intensity close to the Fermi level is connected to a sharp peak appearing slightly shifted in both spin channels. The minority peak ($E - \varepsilon_F = -0.19$ eV) has the higher energy than the majority peak ($E - \varepsilon_F = -0.23$ eV) and is cut off by the Fermi distribution, leading to a slightly negative spin polarization at the Fermi energy. The origin of this minority peak might be connected either to bulk transitions or to a surface state. Both majority and minority peaks are rather narrow in energy, indicating the localized character of the corresponding transition. Since the Fe-related sp-bands did not lead to such peaks in Co_2FeAl, we exclude that the band bottom of the sp-bands is a possible origin. Thus we restrict our search to resonant bulk transitions from localized states with d-character. Such states can be found in the Co_2CrAl majority band structure, indicating a bulk origin of the majority peak in the photoelectron spectrum. In the minority channel calculations, bulk transitions from localized bands might exist as well. However, such transitions have not shown up at all in the Co_2FeAl minority photoelectron spectra. Thus we assume that the minority state is indeed a surface state. This conclusion is supported by a positive curvature in the dispersion, moving the peak above the Fermi energy for larger electron momentum. This dispersion was found in angle-resolved measurements (not shown here). Since a true surface state is possible only in a band gap we conclude that the bulk band structure of CCFA might indeed be half metallic.

12.4 Conclusions

In this report we have presented the surface electronic structure of $Co_2Cr_{1-x}Fe_xAl$ as obtained by low-energy spin-resolved photoemission. We showed that due to its surface sensitivity this technique is well suited to investigate possible surface states that may alter significantly the performance of real devices. We found that the surface spin polarization of Co_2CrAl is canceled out completely, most probably due to bulk Co-Cr disorder. In contrast, the measured Co_2FeAl spectra could be explained solely on the base of GGA band structure calculations. For $Co_2Cr_{0.6}Fe_{0.4}Al$, a negative surface spin polarization is found at the Fermi energy which is attributed to the presence of a surface state. This surface state supports the picture of $Co_2Cr_{0.6}Fe_{0.4}Al$ being half metallic in the bulk.

Acknowledgements This work was supported by the Research Center OPTIMAS and the DFG Forschergruppe through DFG FG559 "New materials with high spin polarization" (project P10). We would like to thank C. Herbort and M. Jourdan for preparation of the samples and G.H. Fecher for supporting us with the band structure data. B. Krenzer is acknowledged for discussion about the interpretation of LEED patterns; R. Fetzer and J. Fischer are acknowledged for experimental assistance.

References

1. Balke B, Wurmehl S, Fecher G, Felser C, Kübler J (2008) Sci Technol Adv Mater 9:014102
2. Galanakis I, Mavropoulos P, Dederichs PH (2006) J Phys D, Appl Phys 39:765
3. Picozzi S, Continenza A, Freeman AJ (2004) Phys Rev B 69:094423
4. Miura Y, Nagao K, Shirai M (2004) Phys Rev B 69:144413
5. Felser C, Elmers HJ, Fecher G (2005) The properties of $Co_2Cr_{1-x}Fe_xAl$ Heusler compounds, vol 676. Springer, Berlin, p 113
6. Kudryavtsev YV, Uvarov VN, Oksenenko VA, Lee YP, Kim JB, Hyun YH, Kim KW, Rhee JY, Dubowik J (2008) Phys Rev B 77:195104
7. Hoshino T, Fujima N, Asato M, Tatsuoka H (2010) J Alloys Compd 504:S531
8. Miura Y, Uchida H, Oba Y, Abe K, Shirai M (2008) Phys Rev B 78:064416
9. Inomata K, Wojcik M, Jedryka E, Ikeda N, Tezuka N (2008) Phys Rev B 77:214425
10. Herbort C, Arbelo E, Jourdan M (2009) J Phys D, Appl Phys 42:084006
11. Herbort C, Arbelo Jorge E, Jourdan M (2009) Appl Phys Lett 94:142504
12. Zarei S, Hashemifar S, Akbarzadeh H, Hafari Z (2009) J Phys Condens Matter 21:055002
13. Fecher G, Kandpal H, Wurmehl S, Morais J, Lin HJ, Elmers HJ, Schönhense G, Felser C (2005) J Phys Condens Matter 17:7237
14. Fecher G, Felser C (2007) J Phys D, Appl Phys 40:1582
15. Karthik SV, Rajanikanth A, Takahashi YK, Okhubo T, Hono K (2006) Appl Phys Lett 89:052505
16. Wurmehl S, Kohlhepp JT, Swagten HJM, Koopmans B (2008) J Phys D, Appl Phys 41:115007
17. Inomata K, Okamura S, Miyazaki A, Kikuchi M, Tezuka N, Wojcik M, Jedryka E (2006) J Phys D, Appl Phys 39:816
18. Xu X, Zhang D, Wang W, Wu Y, Wang Y, Jiang Y (2010) J Magn Magn Mater 322:3351
19. Jullière M (1975) Phys Lett A 54:225
20. Kallmayer M, Klaer P, Schneider H, Jorge EA, Herbort C, Jakob G, Jourdan M, Elmers HJ (2009) Phys Rev B 80:020406(R)

21. Klaer P, Kallmayer M, Blum CGF, Graf T, Barth J, Balke B, Fecher GH, Felser C, Elmers HJ (2009) Phys Rev B 80:144405
22. Klaer P, Arbelo Jorge E, Jourdan M, Wang WH, Sukegawa H, Inomata K, Elmers HJ (2010) Phys Rev B 82:024418
23. Irkhin V, Katsnelson M, Lichtenstein A (2007) J Phys Condens Matter 19:315201
24. Chadov S, Fecher G, Felser C, Minàr J, Braun J, Ebert H (2009) J Phys D, Appl Phys 42:084002
25. Hashemifar SJ, Kratzer P, Scheffler M (2005) Phys Rev Lett 94:096402
26. Mathias S, Miaja-Avila L, Murnane MM, Kapteyn H, Aeschlimann M, Bauer M (2007) Rev Sci Instrum 78:083105
27. Kirschner J (1985) Polarized electrons at surfaces. Springer, Berlin
28. Wang W, Przybylski M, Kuch W, Chelaru L, Wang J, Lu Y, Barthel J, Meyerheim H, Kirschner J (2005) Phys Rev B 71:144416
29. Cinchetti M, Wüstenberg J-P, Sanchez Albaneda M, Steeb F, Conca A, Jourdan M, Aeschlimann M (2007) J Phys D, Appl Phys 40:1544
30. Wüstenberg J-P, Cinchetti M, Albaneda MS, Bauer M, Aeschlimann M (2007) J Magn Magn Mater 316:E411
31. Zangwill A (1988) Physics at surfaces. Cambridge University Press, Cambridge
32. Wüstenberg J-P et al. (2012) Phys Rev B 85:064407
33. Aeschlimann M, Bauer M, Pawlik S, Weber W, Burgermeister R, Oberli D, Siegmann HC (1997) Phys Rev Lett 79:5158
34. Gaier O, Hamrle J, Trudel S, Parra AC, Hillebrands B, Arbelo E, Herbort C, Jourdan M (2009) J Phys D, Appl Phys 42:084004
35. Jourdan M, Jorge EA, Herbort C, Kallmayer M, Klaer P, Elmers HJ (2009) Appl Phys Lett 95:172504
36. Wüstenberg J-P, Fischer J, Herbort C, Jourdan M, Aeschlimann M, Cinchetti M (2009) J Phys D, Appl Phys 42:084016
37. Wurmehl S, Fecher G, Kroth K, Kronast F, Dürr H, Takeda Y, Saitoh Y, Kobayashi K, Lin HJ, Schönhense G, Felser C (2006) J Phys D, Appl Phys 39:803

Chapter 13
Magneto-Optical Investigations and Ion Beam-Induced Modification of Heusler Compounds

Jaroslav Hamrle, Oksana Gaier, Simon Trudel, Georg Wolf, and Burkard Hillebrands

Abstract Half-metallic Co-based Heusler compounds are attracting attention due to their anticipated use as high-performance materials for spintronics applications, such as spin-source or spin-detector. In order to use these materials in applications, their structural and magnetic properties must be well understood. The important phenomena in those materials are, amongst others, exchange and spin–orbit coupling, the latter one giving rise to effect such as magnetic anisotropy and magneto-optical Kerr effect. In this chapter, we present our investigations of magnetic exchange stiffness, magnetic anisotropy, magnetization reversal, and magneto-optical Kerr effect in Co-based Heusler compound thin films. Furthermore, we have also investigated the modification of the compounds under He^{+} and Ga^{+} ion beam irradiation with the aim of improving and tailoring structural properties.

J. Hamrle (✉)
VSB – Technical University of Ostrava, Ostrava, Czech Republic
e-mail: jaroslav.hamrle@vsb.cz

J. Hamrle · O. Gaier · S. Trudel · G. Wolf · B. Hillebrands
Technische Universität Kaiserslautern and Research Center OPTIMAS, Kaiserslautern 67663, Germany

G. Wolf
e-mail: gwolf@physik.uni-kl.de

B. Hillebrands
e-mail: hilleb@physik.uni-kl.de

O. Gaier
Institut Néel, Grenoble, France
e-mail: oksana.gaier@grenoble.cnrs.fr

S. Trudel
University of Calgary, Calgary, Canada
e-mail: trudels@ucalgary.ca

C. Felser, G.H. Fecher (eds.), *Spintronics*, DOI 10.1007/978-90-481-3832-6_13,

13.1 Introduction

Co-based Heusler compounds, with the general composition Co_2MZ, M and Z being a transition metal element and a main group element, respectively, are currently attracting great interest due to their possible use in the novel field of spin-dependent devices, also known as spintronics or magnetoelectronics. This interest in Heusler compounds stems from the half-metallic character of their spin-split band structure, i.e. metallic behavior for one spin component (majority spins), and insulating behavior for the other one (minority spins), as predicted by ab initio calculations for many compounds of this material class [1, 2]. As such, these materials may exhibit a 100 % spin polarization at the Fermi level, which would make them ideal candidates for spin polarizers or spin detectors as the electrodes of tunneling magnetic junctions, amongst other applications.

In addition to half-metallicity, Heusler compounds exhibit several other features which make them suitable candidates for technological applications. One of the most attractive features of Heusler compounds is the possibility to tune their magnetic properties, such as Curie temperature [2], saturation magnetization [3, 4], as well as the width and position of the insulating gap with respect to the Fermi level [5, 6], by changing their composition. Moreover, Co-based Heusler compounds possess relatively high Curie temperatures [7, 8], which is a prerequisite for the stability of the performance of devices incorporating ferromagnetic materials. An indispensable precondition for a successful implementation of Heusler compounds into real devices, however, is a good understanding of their electronic and magnetic properties.

Exchange and spin–orbit coupling are important microscopic phenomena which determine key magnetic properties. Strong exchange interaction is not only a precondition of high Curie temperature, but is also important for maintaining 100 % spin-polarization at room temperature, as well as effectual spin injection into non-ferromagnetic materials. The systematic study of exchange stiffness in Co-based Heusler compounds is presented in Sect. 13.2.

The spin–orbit coupling is responsible for the coupling of the spin to the orbital angular momentum. Hence, it has a strong impact on the depolarization of highly polarized currents required for spin-dependent devices. A possible way to get insight into the spin–orbit interaction is the investigation of the magneto-crystalline anisotropies by magneto-optical techniques, an example of which is magneto-optical Kerr effect (MOKE) magnetometry. The anisotropy constants determined by such magneto-optical techniques are a direct measure for the strength of the spin–orbit interaction (Sect. 13.3). Another approach to studying spin–orbit coupling is the investigation of the MOKE effect itself, as this effect also originates from spin–orbit coupling. In addition, higher orders of the spin–orbit coupling can be indirectly studied through quadratic MOKE (QMOKE) (Sect. 13.4).

Given the strong dependence of the exchange interaction and magnetic anisotropies on the structural order [9], we have also explored the structural modification of Heusler films using both light (He^+) and heavy (Ga^+) ion irradiation (Sect. 13.5). The use of He^+ irradiation was motivated by the possibility of improving the crystalline structure of thin films, without resorting high temperature annealing, which is often undesirable in semiconductor manufacturing processes [10, 11]. Focused Ga^+ ion beam irradiation was proven to provide a spatially resolved approach to modifying the structure [12] and hence has a potential for micro-patterning applications of Heusler films.

13.2 Exchange Stiffness

The exchange stiffness describes the strength of the exchange interaction between individual atomic magnetic momenta in a ferromagnetic material. It is a crucial parameter in the context of dynamic magnetization processes, such as magnetization reversal or spinwaves, as well as for micromagnetic simulations. Moreover, one of the pressing issues regarding Heusler compounds is the reduction of the idealized 100 % spin polarization at finite temperature [9, 13–15]. Some of the proposed mechanisms of this loss of spin polarization include inelastic electron–magnon interactions, which create states near the Fermi level in the minority spin channel band gap [14, 16], and the decrease of spin polarization due to thermally excited magnons [15]. As such, the investigation of magnons, and hence the exchange interaction, is an important issue in order to understand the strong temperature dependence of the spin polarization in Heusler compounds [9].

Our efforts towards the determination of the exchange stiffness in Heusler compounds have concentrated on using Brillouin light scattering (BLS) spectroscopy to determine the exchange stiffnesses of a variety of Co-based Heusler compounds.

Most of the published values were determined by us, namely: Co_2FeSi ($L2_1$) [17], Co_2MnSi ($L2_1$) [18], Co_2FeAl (B2) and $Co_2Cr_{0.6}Fe_{0.4}Al$(CCFA) (B2) [19], Co_2MnAl (B2) [20]. Belmeguenai et al. recently reported on the exchange stiffness of Co_2MnGe [21]. However, the crystallographic ordering of the investigated Co_2MnGe films was not specified. All experiments were done on Heusler films of typical thickness 30–80 nm using BLS spectroscopy, at room temperature (RT). The collected values of the exchange (expressed as both exchange stiffness D and exchange constant A), saturation magnetizations M_s, bulk lattice constant a and bulk Curie temperature T_C are presented in Table 13.1.

Heusler compounds are well known to be systematized by their number of valence electrons N_V. Striking examples are their magnetic moment described by the Slater-Pauling rule [25–27], their Curie temperature T_C [2], or their Fermi level position [5, 6] which are generally proportional to N_V. Here we show that the exchange stiffness D in Heusler compounds scales with N_V as well. The dependence of D on N_V is presented in Fig. 13.1(a). The salient features are: (i) A very large change in D is observed between Co_2FeSi ($D = 7.15$ meV nm^2) and CCFA or Co_2MnAl (both ~2.0 meV nm^2). Such a large change is associated with the introduction of only 0.5 valence electron per atom. (ii) The experimental points are segregated into two branches related to B2- and $L2_1$-ordered compounds. In both branches, D is monotonously increasing with increasing N_V. Although the trends appear to be linear, the investigation of further compounds would clarify the exact functional form

Table 13.1 Magnetic and crystallographic properties of Co-based Heusler compounds

Compound	N_V	a^{bulk} [Å]	M_S^{bulk} [kA/m]	M_S^{bulk} [μ_B/fu]	T_C^{bulk} [K]	thin film order	M_S^{BLS} [kA/m]	M_S^{BLS} [μ_B/fu]	D^{BLS} [meV nm^2]	A^{BLS} [pJ/m]
Co_2FeSi[a]	7.5	5.640	1230	5.95	1100	$L2_1$	1019(10)	4.93	7.2(2)	31.5(5)
Co_2MnSi[b]	7.25	5.654	1020	4.97	985	$L2_1$	970(8)	4.73	5.8(2)	23.5(1)
Co_2MnGe[c]	7.25	5.743	1040	5.31	905	–	743	3.79	3.8(4)	12.0(1.6)
Co_2FeAl[d]	7.25	5.730	978	4.96	≈1000	B2	1027±10	5.21	3.7(1)	15.5(5)
Co_2MnAl[e]	7	5.756	786	4.04	693	B2	560	2.88	1.9	4.8
CCFA[f]	6.95	5.737	668	3.40	750	B2	520(20)	2.65	2.03(16)	4.8(4)

[a][3, 17]; [b][6, 7, 18]; [c][7, 21]; [d][19, 22, 23]; [e][7, 20, 22]; [f][19, 24]

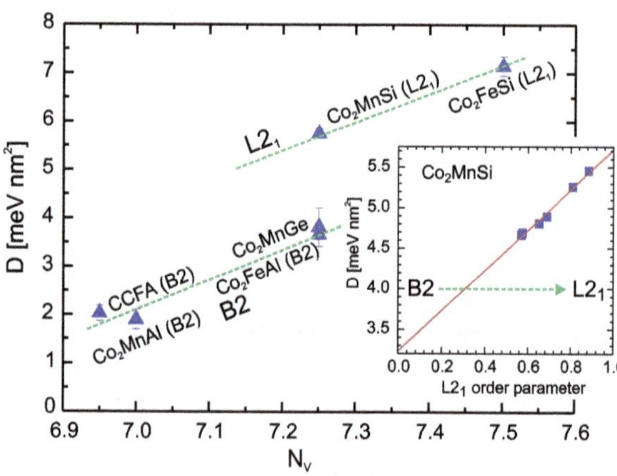

Fig. 13.1 (*Full triangles*) Experimental exchange stiffness D of various Co_2MZ compounds. The *straight dashed lines* are guide to the eye for $L2_1$ and B2 ordered compounds. (*Inset*) Dependence of D on $L2_1$ order for Co_2MnSi. The *straight line* shows a linear fit. Data from [18, 28]

of this dependence. Even though the ordering of the Co_2MnGe sample was not provided [21], it appears to fall on the B2 branch of the data. (iii) The B2-ordered Co_2MnAl and CCFA have similar N_V's (7 and 6.95 valence electrons per atom) and very similar D values.

The observations (i)–(iii) suggest that D is a function of both N_V and the crystallographic ordering. The latter observation is consistent with our study of series of Co_2MnSi samples with a varying degree of $L2_1$ order [28]. The determined values of D are shown in the inset of Fig. 13.1 and linearly increase with increasing degree of $L2_1$ ordering. Extrapolating these data, it can be inferred that D for a perfectly B2-ordered Co_2MnSi sample would be reduced to ~56 % (i.e. to 3.24 meV nm^2) compared to an $L2_1$ ordered sample of the same composition (5.75 meV nm^2). The extrapolated B2 value of D for Co_2MnSi compares favorably with the experimental value for D of B2-ordered Co_2FeAl (3.70 meV nm^2) (both Co_2MnSi and Co_2FeAl have $N_V = 7.25\ e^-$/atom). It again suggests D to be a function of N_V and the crystallographic ordering.

In the following, we qualitatively discuss three possible contributions which may give rise to the observed dependence between the composition (and N_V) and the

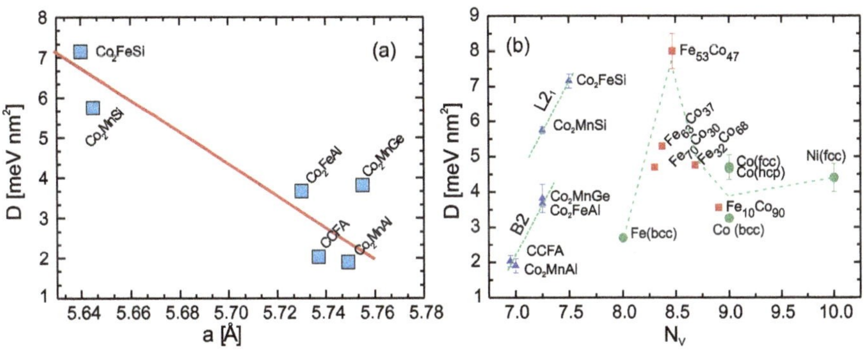

Fig. 13.2 (**a**) Exchange stiffness D as a function of the bulk lattice constant a. The *solid line* is a guide to the eye. (**b**) Exchange stiffness D as a function of number of valence electrons N_V for Co_2MZ Heusler compounds, the ferromagnetic $3d$-metals, and bcc-Fe_xCo_{100-x} compounds [9]. *Lines* are guides to the eye

Heisenberg exchange integral J, which is related to the exchange stiffness D. (i) The increase of N_V adds electrons to the electronic structure. The additional electrons will be primarily found in the t_{2g} orbitals of the transition metal M in the Co_2MZ Heusler compound [29]. Thus, increasing N_V increases the electron density that will participate in exchange. Additionally, ab-initio atom-resolved calculations of the density of states in Co_2MZ (M = Mn, Fe; Z = Al, Si) [2, 30, 31] show that for a given non-magnetic element Z, substituting Mn atoms by the more electronegative Fe atoms results in a lowering of the energy of the M-based bands. This results in an improved alignment of the band energy between M and Co atoms, providing a stronger overlap of the electronic functions, and hence a stronger exchange interaction.

(ii) The difference in electronegativity between Co and Si (−0.02 on the Pauling electronegativity scale) is much less than between Co and Al (0.27). As such, Co-Si bonding is of more covalent character than Co–Al bonding [29]. While this will impact the electronic properties, it is not clear how this affects the exchange interactions. However, this appears to be correlated to an enhanced exchange stiffness.

(iii) Finally, increasing N_V is related with smaller atomic diameter of the constituent elements, leading to a contraction of the unit cell of the Heusler compounds.

Hence, a stronger exchange interaction is expected due to the better overlap of orbitals, a result of the closer proximity between the magnetic elements [32, 33]. Indeed, a larger exchange stiffness is generally observed for a smaller (bulk) lattice constant a, as is shown in Fig. 13.2(a).

Figure 13.2(b) compares the experimental exchange stiffness D associated to a variety of Co-based Heusler compounds, the ferromagnetic $3d$-metals, and Fe-rich bcc-Fe_xCo_{100-x} intermetallic compounds, as a function of N_V (for detailed references see [9]). As discussed above, Co-based Heusler compounds appear to follow a roughly linear dependence of D on N_V, reaching a maximum value with Co_2FeSi ($D = 7.15 \pm 0.20$ meV nm^2) [17]. The Fe_xCo_{100-x} compounds also provide a roughly linear dependence between D and N_V in the x range of 50–100, reaching a maximum value of $D = 8.0 \pm 0.5$ meV nm^2 with $Fe_{53}Co_{47}$ [34] which is (to the best of our knowledge), the largest exchange stiffness value ever reported. This shows that the exchange values of the $L2_1$-ordered Co_2MnSi and Co_2FeSi compounds are extraordinarily large. They are larger than D of the pure ferromagnetic 3d metals, and nearly as large as the maximum value of D obtained for the Fe_xCo_{100-x} series. Higher D values are expected to lead to more stable spin polarization at finite temperature, thereby improving their room temperature spin-dependent transport properties [9].

13.3 Magnetization Reversal and Magnetization Anisotropy

A good understanding of the processes of magnetization reversal, and the involved anisotropy of magnetic thin films, is technologically important. The anisotropy may, for example, be due to magnetocrystalline anisotropy or it may be strain-induced within the sample [35]. The former is a measure of the anisotropy of the spin–orbit interaction due to the directionality of bonding within the unit cell, and is an intrinsic material parameter. The latter is an extrinsic property, and depends on the sample fabrication details. Here, examples of both will be provided.

We have used two magneto-optical methods to probe the anisotropy of Heusler compound thin films, namely BLS spectroscopy and MOKE magnetometry. When BLS is used, the frequency of the observed spin waves will depend on the relative orientation between easy (or hard) axes of magnetization (providing differing effective fields [36]) and the sample's magnetization. Figure 13.3(a) shows the periodic oscillation of the spin wave frequencies observed as a Co_2MnSi film is rotated about the film normal with respect to the direction of the applied magnetic field. The expected 4-fold symmetry of a cubic material is clearly seen in this figure. The solid line is a fit to the data, which allows a quantitative determination of the cubic anisotropy constant, which was found to be 9 kJ/m^3 (90 kerg/cm^3) [28]. Using this method, the systematic decrease of the anisotropy constant and increase of $L2_1$-order was revealed as samples are annealed at higher temperatures (Fig. 13.3(b)). This method was also used to determine the anisotropy constant of $Co_2Cr_{0.6}Fe_{0.4}Al$ [19].

MOKE magnetometry is a tool that is conducive to the study of magnetization reversal. However, the interpretation of Kerr rotation as a function of applied field is not as straightforward, as the observed behavior is a combination of magnetocrystalline anisotropy, as well as other superimposed anisotropies [35]. As such, results varying from the expected four-fold symmetry for a cubic material to uniaxial anisotropy have been observed, including intermediate behavior. Figure 13.4 clearly demonstrates this. In this figure, polar plots of the coercive fields as a function of in-plane sample orientation are shown for $Co_2Cr_{0.6}Fe_{0.4}Al$/Cr/MgO [37], $Co_2FeAl_{0.5}Si_{0.5}$/MgO [38], and non-stoichiometric $Co_2Mn_{0.77}Ge_{0.42}$/MgO [39]. The three samples will be discussed in more detail below. In all three cases the films were epitaxially grown and are single-crystalline.

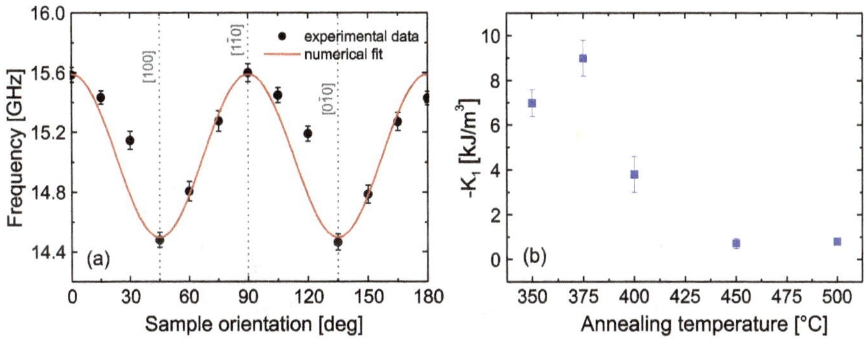

Fig. 13.3 (**a**) Frequency of the Damon–Eshbach (DE) spin-wave mode as a function of the angle between the external magnetic field $H = 30$ mT and the in-plane [110] easy axis direction for a $Co_2MnSi(001)$ film annealed at 375 °C. (**b**) Cubic volume anisotropy constant K_1 for $Co_2MnSi(001)$ films annealed at different temperatures. (Reprinted with permission from O. Gaier, J. Hamrle, S.J. Hermsdoerfer, H. Schultheiß, B. Hillebrands, Y. Sakuraba, M. Oogane, Y. Ando, *J. Appl. Phys.* 103(10), 103910 [28]. Copyright 2008, American Institute of Physics)

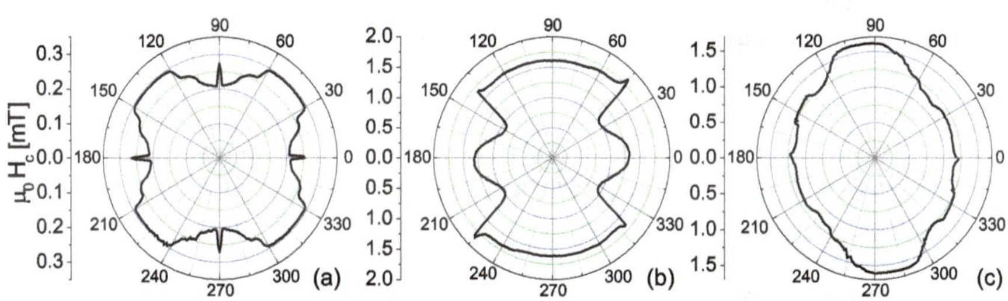

Fig. 13.4 Polar plots of the coercive field for (**a**) $Co_2Cr_{0.6}Fe_{0.4}Al$ [37], (**b**) $Co_2FeAl_{0.5}Si_{0.5}$ [38], and (**c**) non-stoichiometric $Co_2Mn_{0.77}Ge_{0.42}$ [39] epitaxially grown thin films on buffered MgO(001) substrates

The first sample presented in Fig. 13.4(a), an 80 nm thick $Co_2Cr_{0.6}Fe_{0.4}Al$ film grown on a Cr-buffered MgO(001) substrate [37], shows the characteristic 4-fold symmetry expected for a cubic material. The easy axes of magnetization are clearly identified as the $\langle 110 \rangle$ directions from the higher coercive fields. Sharp peaks can also be seen along the hard $\langle 100 \rangle$ directions. The origin of these peaks was found using MOKE microscopy, which enables the observation of magnetic domains during the magnetization reversal [35]. It was shown that the peaks in coercivity arose due to a checkerboard domain configuration with domains separated by 90° domain walls at zero magnetization, which occurs due to magnetic frustration. When the magnetization reversal was examined $\sim$3° from the hard direction, this checker board configuration was not observed.

The complicated polar plot shown in Fig. 13.4(b), corresponding to a 30 nm thick $Co_2FeAl_{0.5}Si_{0.5}$ film grown on MgO-buffered MgO, is believed to arise due to combination of two anisotropies [38]. The first one has four-fold symmetry, and arises from the magnetocrystalline anisotropy of the cubic Heusler compound. The second, uniaxial anisotropy is coincident with one of the cubic axes. The evidence for this superposition of anisotropies lies in the observation of multiple magnetization jumps in the Kerr rotation loops. As is explained in detail in Ref. [9], such behavior has been often observed for samples grown on GaAs and MgO. While for samples grown on GaAs film anisotropy may be due to the directionality of dangling bonds at the GaAs(001) surface [40], no good explanation has yet been put forward to explain this anisotropy of Heusler compounds films grown on cubic MgO substrates.

The final example in Fig. 13.4(c) consists of a 50 nm thick non-stoichiometric $Co_2Mn_{0.77}Ge_{0.42}$ film grown on MgO-buffered MgO [39]. The anisotropy is clearly seen to be uniaxial, despite the epitaxial relationship to the substrate. In this case, the origin of the anisotropy could not be clearly identified. It was tentatively suggested that the non-stoichiometric composition leads to an anisotropic structure. MOKE microscopy could shed some light onto this specific configuration by revealing how the magnetization reversal occurs. However, it is not excluded that this uniaxial anisotropy has the same origin as for the $Co_2FeAl_{0.5}Si_{0.5}$ sample discussed above.

It is worth noting that even though the anisotropy of the magnetization reversal, as seen by MOKE, may deviate from the expected four-fold symmetry of a cubic material, the underlying electronic structure of the material is still well described by a cubic symmetry, as is confirmed by QMOKE measurements (see next section) [39, 41]. As such, the two methods are complementary.

13.4 Quadratic Magneto-Optical Kerr Effect in Heusler Compounds

The linear magneto-optical Kerr effect (MOKE) is often used to determine the magnetization reversal properties, because the Kerr rotation is proportional to the magnetization component in the direction of the propagating light. But in some cases, such as epitaxial Fe thin films [42, 43], one observes an additional MOKE contribution which is proportional to the square of the magnetization. It was recently discovered that some of the Co-based Heusler compounds show a strong quadratic MOKE (QMOKE) contribution. The Co_2FeSi compound showed one of the largest quadratic effects ever reported [41]. But it has also been observed for Co_2MnSi [28], Co_2MnGe [39, 44] as well as for $Co_2FeAl_{0.5}Si_{0.5}$ [45].

The microscopic origin of QMOKE is described in [46]. The authors point out that the QMOKE arises from electronic states which are not only split by first order spin–orbit coupling, $E \sim \xi \mathbf{LS}$, but also by the spin–orbit coupling of the next order, $\sim \xi^2$ (ξ is the spin–orbit-coupling parameter). In reality the excitation spectra are rather complex and depend strongly on the details of the electronic band structure of the material. The Heusler compounds are an interesting system to investigate the QMOKE due to both a large QMOKE signal as well as due to the tunable character of the electronic structure by tuning the crystalline ordering or varying the composition.

In the following a phenomenological description of the QMOKE effect will be discussed. The permittivity tensor ε_{ij} can be expressed as a series expansion in the magnetization up to the second order:

$$\varepsilon_{ij} = \varepsilon_{ij} + \frac{\partial \varepsilon_{ij}}{\partial M_k} M_k + \frac{1}{2} \frac{\partial^2 \varepsilon_{ij}}{\partial M_k \partial M_l} M_k M_l \tag{13.1}$$

where $\frac{1}{2}\frac{\partial^2 \varepsilon_{ij}}{\partial M_k \partial M_l} = G_{ijkl}$ are the magneto-optic coefficients of the second order in the magnetization. For Heusler compounds one can assume a cubic symmetry, which leads to a reduction of the non-zero elements of the magneto-optic tensor. Since all investigated samples are epitaxial thin films, the magnetization is forced into the film plane due to the strong demagnetizing field. Thus the out-of-plane component of the magnetization is not taken into account in the following. By solving the wave equations for a given magnetic layer system one can obtain the complex Kerr angle as a function of the parameters of the optical structure and magnetization components (for details see [47, 48]). Then, the complex Kerr angle for a single magnetic layer with cubic symmetry and a (001)-oriented interface will be of the following form:

$$\begin{aligned} \Theta_{\text{Kerr}} = {} & BKM_L \\ & + A\left[\left(2G_{44} + \frac{K^2}{\varepsilon}\right) - \frac{\Delta G}{2}\cos(4\alpha)\right] M_L M_T \\ & - A\frac{\Delta G}{4}\sin(4\alpha)\left(M_L^2 - M_T^2\right), \end{aligned} \tag{13.2}$$

where A, B are complex optical weighting factors depending on the angle of incidence and on the refraction indexes and thicknesses of each layer in the multilayer structure. K is the linear magneto-optic constant, ε is the dielectric constant and $G_{44} = G_{1212} = G_{1313} = G_{2323}$. ΔG is given by the relation: $\Delta G = (G_{11} - G_{12} - G_{44})/2$, where $G_{11} = G_{iiii}$ and $G_{12} = G_{iijj}$ are the remaining elements of the magneto-optical tensor, of the second order in magnetization, for a cubic system. The angle α describes the relative orientation of the [100] crystallographic direction with respect to the plane of the incident light.

In Eq. 13.2 there are three different contributions to the total Kerr angle: (1) The regular linear or longitudinal MOKE (LMOKE) contribution, proportional to the magnetization component in the direction of the incident light M_L. The linear MOKE contribution does not depend on the sample orientation. (2) A contribution proportional to the product of the magnetization component parallel (M_L) and perpendicular (M_T) to the incident light. This contribution varies with $\cos(4\alpha)$ on the sample orientation and its amplitude depends on ΔG. This term also contains an offset part that is independent of the sample orientation. (3) A contribution which is proportional to the difference of the square of the magnetization components, $M_L^2 - M_T^2$. This contribution depends on in-plane orientation by a $\sin(4\alpha)$ and its amplitude is determined by ΔG as well.

During a magnetization reversal process the total Kerr signal shows a significant difference compared to the pure linear MOKE. While the linear MOKE contribution is odd in field reversal, the QMOKE is even. The influence of the higher order terms on the reversal curve strongly depends on the sample orientation, not only because

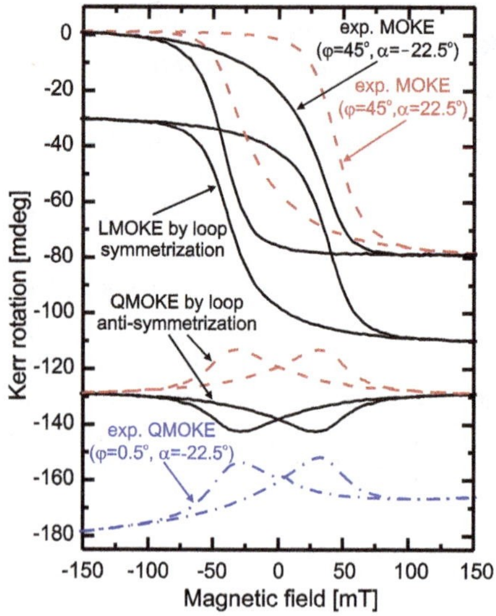

Fig. 13.5 Magnetization reversal loops for a 21 nm thick Co_2FeSi film. Experimentally recorded loops at an angle of incidence $\phi = 45°$ for a sample orientation of $\pm 22.5°$ (*top most solid* and *dashed line*). Symmetrized and antisymmetrized loops show the separated LMOKE and QMOKE signal (*middle curves*, see text inside the figure). The *antisymmetrized curves* are recorded for sample orientation $\pm 22.5°$, indicated by a *solid* and a *dashed line*. The *bottom curve* is recorded at an angle of incidence of $\varphi = 0.5°$ representing only the QMOKE signal (*dash-dot-line*) [41]

the effects depend on $\cos(4\alpha)$ and $\sin(4\alpha)$, but also because the magnitude of M_T depends on the sample orientation due to the magnetic anisotropy of the sample. This will lead to an asymmetric shape of the reversal loop. The reversal loop can be split into a symmetric and a antisymmetric part, containing the linear MOKE and the quadratic MOKE, by

$$\theta_{\text{sym/asym}} = \left[\theta_{\text{inc}}(H) \mp \theta_{\text{dec}}(H)\right]/2 \tag{13.3}$$

where $\theta_{\text{inc/dec}}$ are the Kerr angles from the increasing/decreasing field branch.

Experimental results obtained on a Co_2FeSi thin film with a thickness of 21 nm deposited on an MgO substrate are presented in Fig. 13.5. The Kerr signal was measured at an angle of incidence of 45° and at sample orientations of $\pm 22.5°$. The measured signal shows an asymmetry. The symmetrized signal corresponds to the pure linear MOKE signal and is the same for the $\pm 22.5°$ orientations. The antisymmetrized signal corresponds to the QMOKE and is of opposite sign for the different orientations. To date, the QMOKE we measured for Co_2FeSi is the largest ever reported for any thin-film system [41].

We have designed and built a dual-beam MOKE magnetometer dedicated to the study of QMOKE in thin films [45, 49]. By applying saturation fields along specific directions, where one of the two QMOKE contributions vanishes, it is possible to separate the contributions and determine their relative strengths. The method is

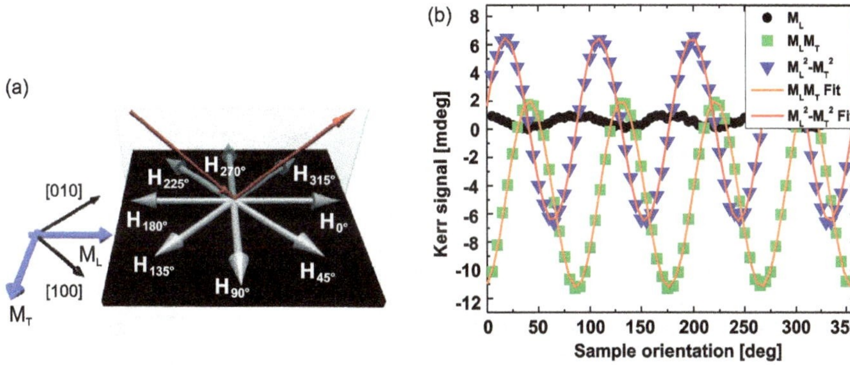

Fig. 13.6 (**a**) Scheme of the 8-directional method. The plane of incident and reflected light is indicated. (**b**) QMOKE obtained with the 8-field method on a $Co_2FeAl_{0.5}Si_{0.5}$ thin film annealed at 575 °C as a function of in-plane sample orientation α [38]. Squares correspond to the $M_L M_T$ contribution, *triangles* to the $M_L^2 M_T^2$ contribution and *dots* to the linear MOKE contribution. The signal was measured at perpendicular incidence. *Solid lines* are fits using Eq. 13.2

sometimes referred to as the eight field or eight direction method [9]. The directions of the saturation field are indicated in Fig. 13.6(a). For magnetic fields in the 0°, 90°, 180° and 270° directions the product term in Eq. 13.2 vanishes and the $M_L^2 - M_T^2$ contribution remains. For the 45°, 135°, 225° and 315° directions, the M_L and M_T components are equal, so the squared difference terms vanish, leaving only the $M_L M_T$ contribution. Changing the sample orientation with respect to the incident light, the different contributions can be extracted as a function of the sample orientation. An example for such a measurement is given in Fig. 13.6(b) for a $Co_2FeAl_{0.5}Si_{0.5}$ thin film with a layer thickness of 30 nm deposited on a MgO substrate with a 20 nm thick MgO seed layer and annealed at a temperature of 575 °C after deposition [38]. The measured data can be fit with a cosine or sine function, to determine a quantitative value for the QMOKE strength, according to Eq. 13.2. In Fig. 13.6(b) it is obvious that the measured data fits very well the predicted curve, indicating a perfect cubic symmetry.

In the previous sections it was already mentioned that the ordering of the Heusler compounds strongly depends on the post-deposition annealing temperature. The change in crystallographic ordering is accompanied by a change in the electronic band structure of the material, and this should also affect the QMOKE. In Fig. 13.7 the amplitude and the offset of the QMOKE contributions as a function of the post deposition annealing temperature is shown. Figure 13.7(a) gives the results obtained for $Co_2FeAl_{0.5}Si_{0.5}$ [38] and in Fig. 13.7(b) for Co_2MnSi [50]. For both compounds, an increasing $L2_1$ ordering with the annealing temperature is reported [28, 51].

The as-prepared (no annealing) $Co_2FeAl_{0.5}Si_{0.5}$ shows nearly no quadratic MOKE, and so does Co_2MnSi for the low annealing temperatures. With the higher annealing temperatures the QMOKE increases, and the largest value is found for the samples with the high annealing temperatures (600 °C for $Co_2FeAl_{0.5}Si_{0.5}$ and

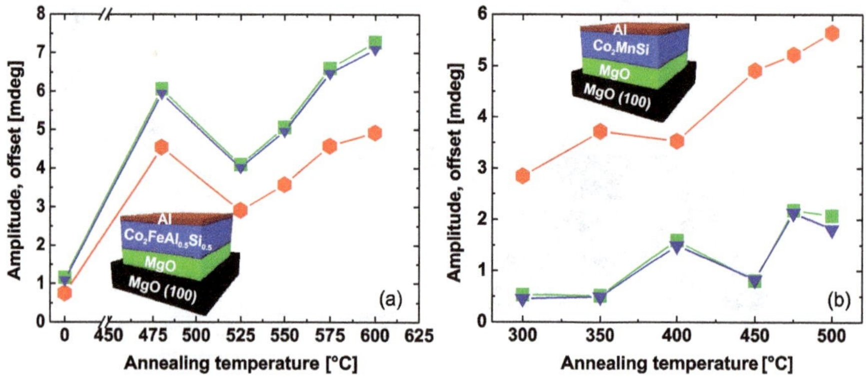

Fig. 13.7 Amplitude and offset of the QMOKE contributions as a function of the annealing temperature for (**a**) the $Co_2FeAl_{0.5}Si_{0.5}$ [38] compound and (**b**) Co_2MnSi [50] compound. *Green squares*: cosinus amplitude of $M_L M_T$ contribution; *blue triangles*: sinus amplitude of $M_L^2 - M_T^2$ contribution; *red hexagon*: offset of $M_L M_T$ contribution. The *solid lines* are only guides to the eye

500 °C for Co_2MnSi) and thus the highest degree of $L2_1$ ordering. In the intermediate temperature range the QMOKE varies, which might be attributed to the mixed ordering.

It is worth mentioning that the absolute values of the QMOKE should not be compared amongst different samples. Even though the layer stacks are fairly similar, the absolute Kerr signal depends also on the index of refraction. Furthermore, the ratio of the offset to the amplitude of the $M_L M_T$ component is not the same for the different compounds. Whereas in the case of $Co_2FeAl_{0.5}Si_{0.5}$ the offset is always smaller than the amplitude, it is always larger for Co_2MnSi, because here also the refraction index and the linear magneto-optic constant play a role, which are different for the two compounds. But the offset and the amplitude follow the same trend, which is a clear indication that the ΔG factor is changed with the annealing. These results show that the strength of the QMOKE depends on the structural ordering of the compound. The highest $L2_1$-ordered systems tend to show a stronger QMOKE.

13.5 Modification of Heusler Compounds Using Ion Irradiation

Ion irradiation, and in particular irradiation with light He^+ ions, has been demonstrated to be an excellent tool for tailoring the magnetic properties of technologically relevant thin films and multilayers [10, 11, 52–57]. In the field of heavy ion irradiation, applications involving Ga^+ ions, which are a standard ion source in focused ion beam (FIB) systems, are predominant. Most of these applications have in common that the modification of magnetic properties results from ion-induced disorder and interfacial mixing [58]. However, Ravelosona et al. have demonstrated that He^+ irradiation can also be employed to enhance the chemical order in magnetic alloys (FePt/Pd) when performed in combination with a mild thermal anneal

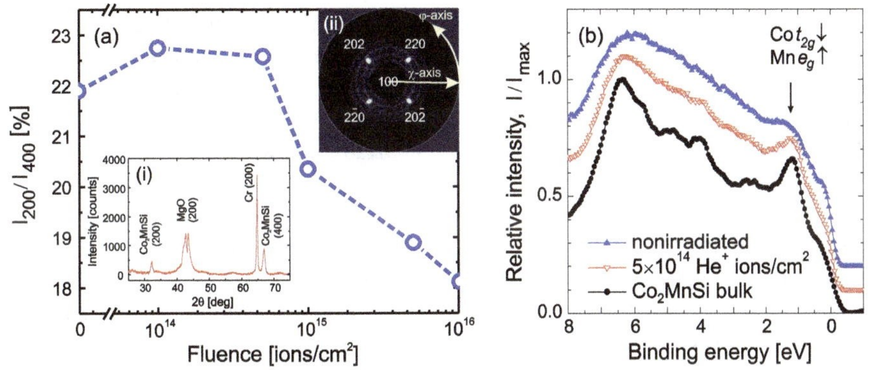

Fig. 13.8 (**a**) Ratio of the (200) and (400) integrated XRD intensities as a function of the applied ion irradiation fluence. The *insets* demonstrate (i) the θ–2θ-scan of the as prepared MgO/Cr(40 nm)/Co_2MnSi(30 nm)/Al(1.3 nm) film and (ii) the corresponding pole figure of the fundamental (220) reflections. (**b**) Hard X-ray photoemission spectroscopy (HAXPES) ($h\nu = 7.94$ keV) spectra of Co_2MnSi films irradiated with 30 keV He^+ ions in comparison to a nonirradiated sample and a cleaved Co_2MnSi bulk sample. The enhanced intensity of Co and Mn d states at about 1.3 eV, characteristic for bulk Co_2MnSi, is marked by an arrow [59]. (Reprinted with permission from O. Gaier, J. Hamrle, B. Hillebrands, M. Kallmayer, P. Pörsch, G. Schönhense, H.J. Elmers, J. Fassbender, A. Gloskovskii, C.A. Jenkins, C. Felser, E. Ikenaga, Y. Sakuraba, S. Tsunegi, M. Oogane, Y. Ando, *Appl. Phys. Lett.* 94(15), 152508. Copyright [2009], American Institute of Physics)

[10, 11]. Based on this work, the present authors investigated the effect of He^+ irradiation on Heusler films utilizing Co_2MnSi as a representative member of this family of compounds [59]. The main goal of these experiments was to explore the applicability of light-ion irradiation technique as an alternative technique to the conventional high-temperature annealing for the promotion of a partially disordered Heusler structure to the $L2_1$ phase. In addition, the effect of Ga^+ irradiation on the properties of Co_2FeSi thin films was also studied [60], which was motivated by the potential micro-patterning application for Heusler films. In the following, the results of these studies will be summarized and briefly discussed.

13.5.1 Irradiation of Co_2MnSi Thin Films with He^+ Ions

The investigated samples are an epitaxial MgO(100)/Cr(40 nm)/Co_2MnSi(30 nm) structure covered by either Al or Ta capping layers. The epitaxial growth is confirmed by the fourfold symmetry of the (220) equivalent reflections observed in the XRD pole figure scans (Fig. 13.8, inset (ii)). The crystallographic structure of the Co_2MnSi layers prior to irradiation is found to be a mix of the B2 and A2 phases, which is deduced from the analysis of the X-ray θ–2θ scans (Fig. 13.8, inset (i)) and the absence of (111) reflections indicative for the $L2_1$ phase in the corresponding pole figure scans. For the irradiation experiments, He^+ ions of two different energies are used, namely 30 and 130 keV. Based on simulations performed with the

SRIM software [61, 62], the ion fluences are chosen in the range between 10^{14} and 10^{16} ions/cm^2. This guarantees that most of the impinging He^+ ions are stopped in the MgO substrate and that the defect densities caused by the ion bombardment are kept low. In case of the 30 keV ions, the irradiation is carried out at RT as well as at elevated temperatures of 150 and 250 °C. The irradiation with 130 keV He^+ ions is performed at RT only.

XRD characterization of the Co_2MnSi films exposed to 30 and 130 keV He^+ ions at RT shows that the crystallinity and epitaxy of the Co_2MnSi layer are largely preserved after the irradiation. In the (111) pole figures measured after the irradiation, the $L2_1$ superstructure reflections are still absent, clearly showing that no transition to the $L2_1$ structure has been induced by the He^+ bombardment under our experimental conditions. However, a detailed analysis of θ–2θ scans recorded from the samples irradiated with 30 keV ions at RT yields a slight increase of the ratio of (200) and (400) integrated intensities at the fluences of 1×10^{14} and 5×10^{14} ions/cm^2 (Fig. 13.8(a)). This result, when compared with simulated intensities (PowderCell software [63]), is an indication of an improvement of the B2 structure, and possibly a local appearance of $L2_1$-ordered regions in the irradiated films. Characterization of the films by various other techniques supports this conclusion [59]. In particular, the investigation of the Co_2MnSi electronic structure by means of hard X-ray photoemission spectroscopy (HAXPES) reveals that the valence band spectrum of the thin film resembles much more closely that of the $L2_1$-ordered bulk material after the irradiation with He^+ ions at 5×10^{14} ions/cm^2 (Fig. 13.8(b)).

As discussed by Bernas et al. for the FePt(Pd) alloy [64], a transition from a disordered to an ordered structure requires ordered nuclei from which to grow, as well as a high enough mobility of the atoms. Therefore, if the 30 keV He^+ irradiation carried out at RT indeed resulted in the nucleation of local regions with $L2_1$ order, performing the irradiation in combination with a mild anneal could be expected to result in the formation of the $L2_1$ structure on the long-range scale. However, no evidence of $L2_1$ phase transition is found for the Co_2MnSi films irradiated with 30 keV He^+ ions at 150 °C and 250 °C.

13.5.2 Irradiation of Co_2FeSi Thin Films with Ga^+ Ions

The $Co_2FeSi(100)$ film used for the irradiation with Ga^+ ions consists of an epitaxial MgO(100)/Co_2FeSi(11 nm)/Al(4 nm) structure, where Co_2FeSi is $L2_1$ ordered. The crystallographic structure is determined by XRD measurements similar to those described in Sect. 13.5.1. The irradiation is carried out at RT under ion fluences ranging between 3×10^{14} and 9×10^{16} ions/cm^2. After ion bombardment, the film is investigated by MOKE magnetometry.

The magnetic and magneto-optical properties of Co_2FeSi are found to be affected in several ways by Ga^+ irradiation. On the one hand, both the coercivity H_C and the LMOKE signal at saturation decrease with increasing ion fluence (Fig. 13.9). This decreasing trend is rather slow up to a fluence of $\approx 6\times10^{15}$ ions/cm^2 and

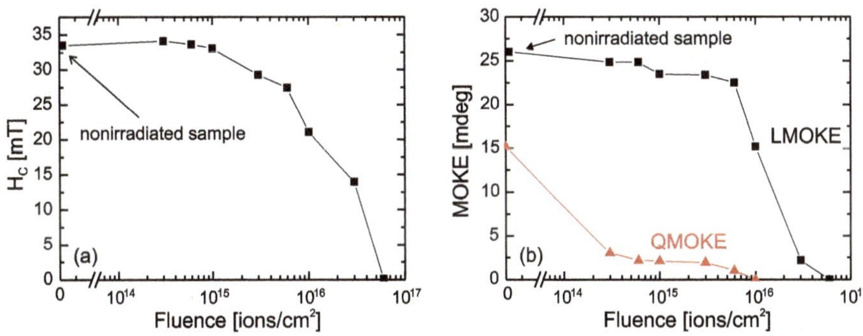

Fig. 13.9 Dependence of (**a**) H_C and (**b**) the amplitudes of the LMOKE and QMOKE for Co_2FeSi samples submitted to various Ga^+ ion fluences. For the determination of the QMOKE amplitude the 8-directional method was employed [60]

exhibits a notable drop for higher fluences. At the highest applied fluence of 9×10^{16} ions/cm^2, both H_C and the LMOKE amplitude become zero, suggesting that the ferromagnetic property of Co_2FeSi is completely suppressed. On the other hand, the initially large QMOKE contribution to the Kerr rotation is drastically reduced even at the lowest fluence values (Fig. 13.9(b)).

A tentative explanation for the behavior of LMOKE and H_C found in the lower range of applied fluences might be the formation of the quaternary compound $Co_2FeSi_{1-x}Ga_x$, whose electronic and magnetic properties are very similar to those of Co_2FeSi in the compositional range $x \leq 0.5$ [65]. The breakdown of the ferromagnetic order at higher fluences is probably a result of an increasing damage impact of Ga^+ irradiation. The QMOKE however appears to be much more sensitive to both the crystallographic order and chemical composition. As such QMOKE may be used as a suitable probing technique for the quality of Heusler films.

13.6 Conclusion

In conclusion, we have investigated the exchange stiffness, magnetization reversal, magnetic anisotropy and quadratic magneto-optical Kerr effect in high spin polarization Co-based Heusler compounds, as well as the modification of Heusler compounds using He^+ and Ga^+ ion irradiation. We have found that the exchange stiffness increases with an increasing number of valence electrons, as well as with increasing $L2_1$ crystallographic order parameter. Samples routinely present both two-fold and four-fold magnetization-reversal anisotropy. Whereas the four-fold anisotropy is obviously related to the cubic symmetry of the crystal, the occasional appearance of 2-fold anisotropy contributions is yet to be clearly understood. The cubic anisotropy constant is reduced with increasing crystallographic $L2_1$ order. Some of Heusler compounds (such as Co_2FeSi, Co_2MnSi, $Co_2FeAl_{0.5}Si_{0.5}$) also present a strong quadratic magneto-optical Kerr effect (QMOKE), suggesting the presence of spin–orbit coupling of higher order. Furthermore, QMOKE was demonstrated to increase with increasing crystallographic $L2_1$ order. Investigation of ion irradiation of Heusler compounds was motivated by a possible tailoring of its magnetic and structural properties. Namely, it was demonstrated that He^+ irradiation can improve crystallographic ordering and the electronic structure of the Co_2MnSi compound.

Acknowledgements We would like to thank all our collaborators. Without their contributions this work would be impossible. Namely (and their respective groups) Martin Aeschlimann, Yasuo Ando, Stanislav Chadov, Mirko Cincetti, Hans-Joachim Elmers, Jürgen Fassbender, Gerhard H. Fecher, Claudia Felser, Andrii Hloskovskii, Koichiro Inomata, Martin Jourdan, Gerhard Jakob, Jürgen Kübler, Takahide Kubota, Mikihiko Oogane, Kamil Postava, Yuya Sakuraba, Horst Schneider, and Masafumi Yamamoto. We gratefully acknowledge DFG FG 559 (project P9) for financing. ST acknowledges the Alexander von Humboldt foundation for a postdoctoral fellowship.

References

1. de Groot RA, Mueller FM, Engen PGv, Buschow KHJ (1983) Phys Rev Lett 50:2024
2. Kübler J, Fecher GH, Felser C (2007) Phys Rev B 76:024414
3. Wurmehl S, Fecher GH, Kandpal HC, Ksenofontov V, Felser C, Lin HJ, Morais J (2005) Phys Rev B 72:184434
4. Galanakis I, Dederichs PH, Papanikolaou N (2002) Phys Rev B 66:174429
5. Fecher GH, Felser C (2007) J Phys D, Appl Phys 40:1582
6. Balke B, Fecher GH, Kandpal HC, Felser C, Kobayashi K, Ikenaga E, Kim JJ, Ueda S (2006) Phys Rev B 74(10):104405
7. Webster PJ (1971) J Phys Chem Solids 32:1221
8. Wurmehl S, Fecher GH, Kroth K, Kronast F, Dürr HA, Takeda Y, Saitoh Y, Kobayashi K, Lin HJ, Schönhense G, Felser C (2006) J Phys D, Appl Phys 39:803
9. Trudel S, Gaier O, Hamrle J, Hillebrands B (2010) J Phys D, Appl Phys 43(19):193001
10. Ravelosona D, Chappert C, Mathet V (2000) Appl Phys Lett 76:236
11. Ravelosona D, Chappert C, Bernas H, Halley D, Samson Y, Marty A (2002) J Appl Phys 91:8082
12. Melngailis J (1987) J Vac Sci Technol B 5(2):469
13. Sakuraba Y, Hattori M, Oogane M, Ando Y, Kato H, Sakuma A, Miyazaki T, Kubota H (2006) Appl Phys Lett 88(19):192508
14. Rajanikanth A, Takahashi YK, Hono K (2009) J Appl Phys 105:063916 and references therein
15. Shan R, Sukegawa H, Wang WH, Kodzuka M, Furubayashi T, Ohkubo T, Mitani S, Inomata K, Hono K (2009) Phys Rev Lett 102:246601
16. Dowben PA, Skomski R (2004) J Appl Phys 95:7453
17. Gaier O, Hamrle J, Trudel S, Hillebrands B, Schneider H, Jakob G (2009) J Phys D, Appl Phys 42(23):232001
18. Hamrle J, Gaier O, Min SG, Hillebrands B, Sakuraba Y, Ando Y (2009) J Phys D, Appl Phys 42:084005
19. Gaier O, Hamrle J, Trudel S, Parra AC, Hillebrands B, Arbelo E, Herbort C, Jourdan M (2009) J Phys D, Appl Phys 42(8):084004
20. Kubota T, Hamrle J, Sakuraba Y, Gaier O, Oogane M, Sakuma A, Hillebrands B, Takanashi K, Ando Y (2009) J Appl Phys 106(11):113907
21. Belmeguenai M, Zighem F, Roussigné Y, Chérif SM, Moch P, Westerholt K, Woltersdorf G, Bayreuther G (2009) Phys Rev B 79:024419
22. Buschow KHJ, van Engen PG, Jongebreur R (1983) J Magn Magn Mater 38:1

23. Fecher G (2008). Private communication
24. Block T, Felser C, Jakob G, Ensling J, Mühling B, Gütlich P, Cava R (2003) J Solid State Chem 176:646
25. Galanakis I, Mavropoulos P, Dederichs PH (2006) J Phys D, Appl Phys 39:765
26. Fecher GH, Kandpal HC, Wurmehl S, Felser C, Schönhense G (2006) J Appl Phys 99:08J106
27. Kübler J (1984) Physica B and C 127:257
28. Gaier O, Hamrle J, Hermsdoerfer SJ, Schultheiß H, Hillebrands B, Sakuraba Y, Oogane M, Ando Y (2008) J Appl Phys 103(10):103910
29. Kandpal HC, Fecher GH, Felser C (2007) J Phys D, Appl Phys 40:1507
30. Özdoğan K, Aktaş B, Galanakis I, Şaşioğlu E (2007) J Appl Phys 101:073910
31. Özdoğan K, Şaşioğlu E, Aktaş B, Galanakis I (2006) Phys Rev B 74:172412
32. Şaşıoğlu E, Sandratskii LM, Bruno P, Galanakis I (2005) Phys Rev B 72:184415.
33. Kurtulus Y, Dronskowski R, Samolyuk GD, Antropov VP (2005) Phys Rev B 71:014425
34. Liu X, Sooryakumar R, Gutierrez CJ, Prinz GA (1994) J Appl Phys 75(10):7021
35. Hubert A, Schäfer R (2009) Magnetic domains: the analysis of magnetic microstructures. Springer, Berlin
36. Hillebrands B (2005) In: Zhu Y (ed) Modern techniques for characterizing magnetic materials. Springer, Berlin, p 543
37. Hamrle J, Blomeier S, Gaier O, Hillebrands B, Schäfer R, Jourdan M (2006) J Appl Phys 100(10):103904
38. Trudel S, Wolf G, Hamrle J, Hillebrands B, Klaer P, Kallmayer M, Elmers HJ, Sukegawa H, Wang W, Inomata K (2011) Phys Rev B 83:104412
39. Trudel S, Hamrle J, Hillebrands B, Taira T, Yamamoto M (2010) J Appl Phys 107:043912
40. Krebs JJ, Jonker BT, Prinz GA (1987) J Appl Phys 61:2596
41. Hamrle J, Blomeier S, Gaier O, Hillebrands B, Schneider H, Jakob G, Postava K, Felser C (2007) J Phys D, Appl Phys 40(6):1563
42. Postava K, Hrabovsky D, Pistora J, Fert AR, Visnovsky S, Yamaguchi T (2002) J Appl Phys 91(10):7293
43. Buchmeier M, Schreiber R, Bürgler DE, Schneider CM (2009) Phys Rev B 79:064402
44. Muduli PK, Rice WC, He L, Collins BA, Chu YS, Tsui F (2009) J Phys Condens Matter 21:296005
45. Trudel S, Wolf G, Schultheiss H, Hamrle J, Hillebrands B, Kubota T, Ando Y (2010) Rev Sci Instrum 81:026105
46. Osgood RM III, Bader SD, Clemens BM, White RL, Matsuyama H (1998) J Magn Magn Mater 182(3):297
47. Visnovsky S (1986) Czechoslov J Phys 36:625
48. Visnovsky S, Lopusnik R, Bauer M, Bok J, Fassbender J, Hillebrands B (2001) Opt Express 9(3):121
49. Trudel S, Wolf G, Schultheiss H, Hamrle J, Hillebrands B (2010) J Phys Conf Ser 200(11):112010
50. Wolf G, Hamrle J, Hillebrands B (2011) J Appl Phys 110:043904
51. Tezuka N, Ikeda N, Miyazaki A, Sugimoto S, Kikuchi M, Inomata K (2006) Appl Phys Lett 89:4
52. Chappert C, Bernas H, Ferré J, Kottler V, Jamet JP, Chen Y, Cambril E, Devolder T, Rousseaux F, Mathet V, Launois H (1998) Science 280:1919
53. Devolder T, Ferré J, Chappert C, Bernas H, Jamet JP, Mathet V (2001) Phys Rev B 64:064415
54. Mougin A, Mewes T, Lopusnik R, Jung M, Engel D, Ehresmann A, Schmoranzer H, Fassbender J, Hillebrands B (2000) IEEE Trans Magn 36:2647
55. McCord J, Gemming T, Schultz L, Fassbender J, Liedke MO, Frommberger M, Quandt E (2005) Appl Phys Lett 86:162502
56. Engel D, Schmoranzer H, Ehresmann A, Mertins HC, Abramsohn D, Gudat W (2004) Physica B 345:185
57. Fassbender J, McCord J (2008) J Magn Magn Mater 320:579

58. Nastasi M, Mayer JW, Hirvonen JK (1996) Ion-solid interactions: fundamentals and applications. Cambridge University Press, Cambridge
59. Gaier O, Hamrle J, Hillebrands B, Kallmayer M, Pörsch P, Schönhense G, Elmers HJ, Fassbender J, Gloskovskii A, Jenkins CA, Felser C, Ikenaga E, Sakuraba Y, Tsunegi S, Oogane M, Ando Y (2009) Appl Phys Lett 94(15):152508
60. Hamrle J, Blomeier S, Gaier O, Hillebrands B, Schneider H, Jakob G, Reuscher B, Brodyanski A, Kopnarski M, Postava K, Felser C (2007) J Phys D, Appl Phys 40(6):1558
61. Ziegler JF, Biersack JF, Littmark JP (1985) The stopping and range of ions in solids. Pergamon, Elmsford
62. Ziegler JF (2006) SRIM manual. www.srim.org
63. http://www.bam.de/de/service/publikationen/powder_cell.htm
64. Bernas H, Attané JP, Heinig KH, Halley D, Ravelososn D, Marty A, Auric P, Chappert C, Samson Y (2003) Phys Rev Lett 91:077203
65. Gercsi Z, Hono K (2007) J Phys Condens Matter 19:326216

Chapter 14
$Co_2Fe(Al_{1-x}Si_x)$ Heusler Alloys and Their Applications to Spintronics

Kouichiro Inomata and Hiroaki Sukegawa

Abstract This work reports on the structural and magnetic properties of the Heusler $Co_2FeAl_xSi_{1-x}$ epitaxial thin films and their applications to magnetic tunnel junctions (MTJs), giant magnetoresistive (GMR) devices and spin transfer magnetization switching. It is shown for $Co_2FeAl_xSi_{1-x}$ that the Fermi level position can be tuned by the composition. The temperature dependence of the tunneling magneto resistance (TMR) in epitaxial MTJs using $B2$-$Co_2FeAl_{0.5}Si_{0.5}$ (CFAS) electrode is fitted well by a spin wave excitation model for tunneling spin polarization. Half-metallic $B2$-CFAS provided a large GMR up to 34 % at room temperature. Magnetization switching was observed in the resistance-current curves and exhibited a two-step switching process originating from the interplay between the magnetocrystalline anisotropy of the CFAS layers and the spin-transfer torque. A small average intrinsic switching current density is obtained by analyzing the data using the thermal activation model. The results show that the use of the Heusler alloy CFAS with high spin polarization is an effective way to reduce the switching current density even in MTJs.

14.1 Introduction

The performance of spintronics depends on the spin polarization of the current. Therefore, a highly spin-polarized current source is strongly desired in spintronics. A promising method for creating highly spin-polarized current is the use of half-metallic ferromagnets (HMFs) [1], which exhibit a semiconductor behavior with a band gap at the Fermi level (E_F) for one spin direction, while the other spin band shows typical metallic behavior and thus have 100 % spin polarization P at E_F. HMFs can lead to infinitely large tunneling magnetoresistance (TMR) in magnetic tunneling junctions (MTJs) and a large current-perpendicular-to-plane GMR (CPP-GMR), which are indispensable for nonvolatile magnetoresistive random access

K. Inomata (✉) · H. Sukegawa
National Institute for Materials Science, 1-2-1 Sengen, Tsukuba 305-0047, Japan
e-mail: INOMATA.Kouichiro@nims.go.jp

H. Sukegawa
e-mail: SUKEGAWA.Hiroaki@nims.go.jp

C. Felser, G.H. Fecher (eds.), *Spintronics*, DOI 10.1007/978-90-481-3832-6_14,

memory (MRAM) and hard disk drive (HDD), respectively. A low current spin transfer torque magnetization switching is also expected in spintronics devices using HMFs such as MTJs and CPP-GMR devices due to their high spin polarization. Moreover, HMFs enable spin injection into a semiconductor with a high efficiency, because only the spin-up electrons should be injected into the semiconductor, allowing the creation of spin-dependent devices such as spin transistors with superior performance. Among some kinds of HMF, Co-based full-Heusler alloys with $L2_1$ structure have been increasingly investigated since many full-Heusler alloys have been predicted to become half-metallic [2] after the first theoretical prediction of half-metallicity of Co_2MnGe and Co_2MnSi [3]. The half-metallicity of the full-Heusler alloys depends on the atomic site disorder. In particular, Co antisite destroys the half-metallicity in the full-Heusler alloys [4]. The Co_2CrAl compound preserves the nearly half-metallicity at the surface whereas in other compounds the surface states kill the spin polarization at the Fermi level [2]. However, some of the interfaces have been found to almost retain the half-metallicity [5, 6]. TMR using a full-Heusler alloy was observed first in a MTJ using $B2$-structured $Co_2Cr_{0.6}Fe_{0.4}Al$ (CCFA) electrodes at room temperature (RT) [7]. After that MTJs using full-Heusler alloy electrodes have been extensively studied using $Co_2(Cr_{1-x}Fe_x)Al$ [8, 9], Co_2FeAl [10], $Co_2FeAl_{1-x}Si_x$ [11, 12], Co_2MnSi [13–15], $Co_2Mn_{1-x}Fe_xSi$ [16] and $Co_2MnAl_xSi_{1-x}$ [17]. In this chapter we describe the structural and magnetic properties of $Co_2FeAl_{1-x}Si_x$ full-Heusler thin films and their applications to MTJs, CPP-GMR devices and spin transfer magnetization switching, and demonstrate the excellent performance of $Co_2FeAl_{1-x}Si_x$ full-Heusler alloys.

14.2 Experimental Method

Full-Heusler $Co_2FeAl_{1-x}Si_x$ alloy films were deposited on Cr- or MgO-buffered MgO(001) substrates using an ultrahigh vacuum magnetron sputtering system with the base pressure of below 8×10^{-8} Pa from stoichiometric compositions of the $Co_2FeAl_xSi_{1-x}$ targets at an ambient substrate temperature followed by post annealing or at heated substrate temperatures. Exchange-biased MTJs with an AlO_x or an MgO barrier using $Ir_{22}Mn_{78}$ antiferromagnet were sputter-deposited on Cr-buffered MgO(001) substrates at an ambient substrate temperature. The bottom Heusler electrode was annealed at T_a after deposition in the same chamber to obtain $B2$ or $L2_1$ structure and smooth its surface and to induce a perfectly flat bottom Heusler/barrier interface. The Ar pressure during sputtering is 0.1 Pa and the typical sputtering rate was 0.03 nm/s for a Heusler alloy, which yields the best smooth surface morphology. The MgO barrier was deposited by electron beam (EB) evaporation from MgO grains or sputtering of the MgO target. The pressure during evaporation is under 5×10^{-6} Pa and the typical evaporation rate is 0.01 nm/s. The upper Heusler or CoFe electrode was sputtered on the barrier, followed by the deposition of antiferromagnetic $Ir_{22}Mn_{78}$ at room temperature (RT), which is used to exchange bias the upper ferromagnetic layer. Finally a Ru cap layer is deposited at RT. MTJs were patterned into $10 \times 10\ \mu m^2$ by the conventional photolithography and Ar ion etching. The patterned MTJs were annealed at temperatures around 350–500 °C in a high vacuum by applying a magnetic field. The pseudo spin valve multilayers of MgO(001)/Cr(10)/Ag(100)/CFAS(t_F)/Ag spacer(5)/CFAS(t_F)/Ag(5)/Ru(8) was fabricated by magnetron sputtering for investigating CPP-GMR and spin transfer magnetization switching. The multilayer was patterned into a 200 nm scale ellipsoidal shape by using e-beam lithography and Ar ion milling.

The crystalline structure was investigated by X-ray diffraction (XRD, Rigaku ATX-G) with a Cu-Kα source for both θ–2θ (out-of-plane) and ϕ–$2\phi_x$ (in-plane) scans. The film roughness was evaluated by atomic force microscopy (AFM). Magnetic properties were measured using a vibrating sample magnetometer (VSM) from 5 K to RT. Nuclear magnetic resonance (NMR) measurement was carried out on ^{59}Co nuclei in Heusler alloy films on Cr- or MgO-buffered MgO(001) by the spin echo method at a low temperature in order to investigate the local structure around the Co atoms [18]. The magnetoresistance measurement of the MTJs was performed by a conventional dc four-point technique.

14.3 Density of States of $Co_2FeAl_xSi_{1-x}$

Ab initio calculations using the OPENMX ab initio free software package [19] were carried out to investigate the role of Al addition and the effect of structural disorder on both the density of states (DOS) and the magnetic properties in $Co_2FeAl_xSi_{1-x}$. The DOS calculations have been performed on the basis of the density-functional theory (DFT) within the general gradient approximation (GGA). We also performed LDA+U calculations to account for the on-site electron–electron Coulomb interaction of the d electrons of Fe and Co elements. Pregenerated fully relativistic pseudo potentials and the pseudo atomic orbitals with a cutoff radius of 6.5 a.u. were downloaded from the OPENMX website [19]. Basis orbitals were fixed to $2s2p1d$ for each element with an energy cutoff of 180 Ry for the numerical integrations; moreover, the energy convergence criterion for the fitting procedure was set to 10^{-7} hartree. The spin–orbit interaction was not considered in the calculations. These conditions were tested on other previously reported full-Heusler alloys, and excellent reproducibility was confirmed. Initially, 16 atoms in the $L2_1$ unit cell composed of eight Co, four Fe, and four Si atoms were taken for the self-consistency calculations considering the bulk Co_2FeSi alloy. The Al addition was carried out by a gradual replacement of those four Si atoms by Al atoms. In such a way, one of the four Si atoms replaced by an Al atom results in $x = 0.25$, while the two Al atom addition in place of the two Si atoms results in $x = 0.50$, etc. in the $Co_2FeAl_xSi_{1-x}$ formula. To model the effect of disorder, the position of Si and Al atoms was swapped with Fe atoms for the $B2$-type disorder and the position of Co by Fe or Si atoms for the $A2$-type disorder. Therefore, the smallest amount of disorder that can be introduced was 12.5 % for the $A2$ and 25 % for the $B2$-type disorder. Finally, note that experimentally obtained lattice parameters were used in the calculations without geometry optimization (relaxation).

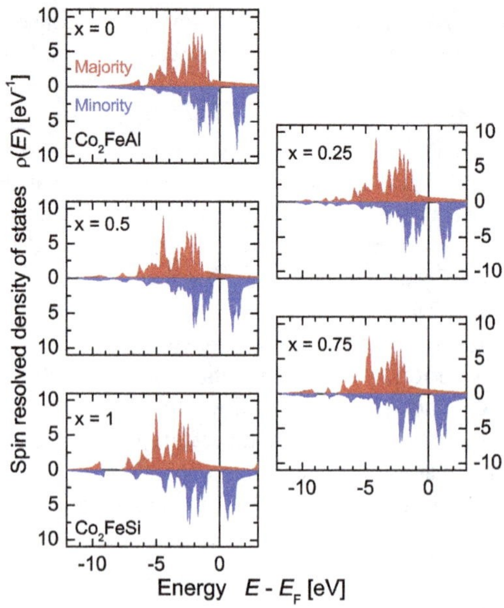

Fig. 14.1 Spin-resolved DOS for $Co_2FeAl_{1x}Si_x$

Figure 14.1 shows the spin-resolved DOS for the series of alloys [20]. All the compositions exhibit a half-metal gap, which is independent of x. Fermi-level position, however, can be tuned by the composition x. When $x = 0$, E_F lies close to the upper edge of the gap, near the minimum energy of the conduction band. In such a case, the half-metallicity can be spoiled by the increase of temperature and structural disorder because some types of disorder form additional states in the minority gap. As a consequence of the Al addition, the Fermi level is shifted deeper into the minority band gap and situates around the middle of the gap within the range of $x = 0.25$–0.5. When the Al concentration is increased to $x = 0.75$ and to $x = 1$, the maximum valence electron energy band becomes close to E_F, predicting lower resistance of the half-metallicity against disorder. Fecher and Felser also found a similar tendency [21]. Figure 14.2 summarizes the effect of various disorders on the magnetization and on the ground-state spin polarization in $Co_2FeAl_xSi_{1-x}$ [20]. Since the exact value of the predicted spin polarization depends on the considered Coulomb interaction strength in the GGA+U calculations, and its exact value is unknown, we only distinguished the completely spin polarized structures ($P = 1$) using half-filled circles from the ones where $P < 1$ (in full squares).

14.4 Experimental Results

14.4.1 Co_2FeSi

Co_2FeSi with the $L2_1$ structure has E_F near top edge of the valence band as shown in Fig. 14.1. Experimentally it has been verified that Co_2FeSi crystallizes in the

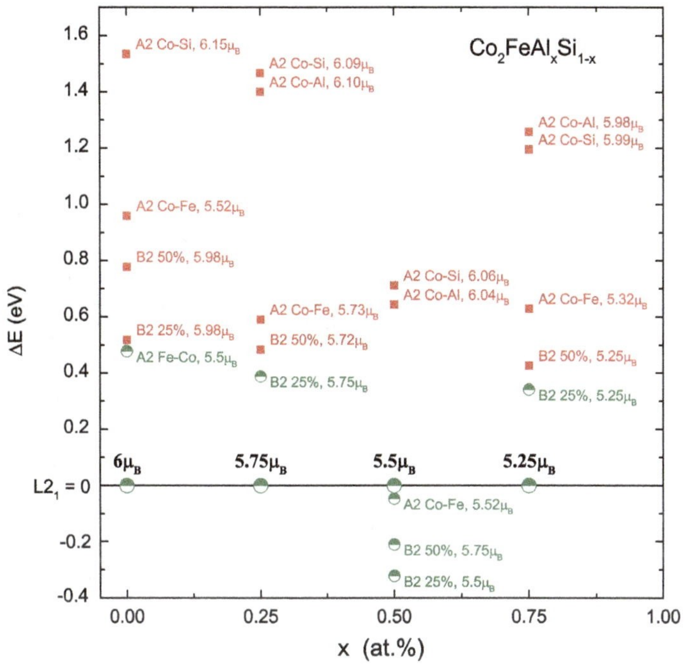

Fig. 14.2 Effect of various disorders on the magnetization and on the ground-state spin polarization in $Co_2FeAl_xSi_{1-x}$. The *half-filled circles* represent $P = 1$, while the *full squares* indicate decreased spin polarization ($P < 1$) as a result of theoretical calculations

$L2_1$ structure in the bulk by quenching after annealing at a high temperature [22] or in the film deposited on a GaAs substrate [23]. We deposited Co_2FeSi films on thermally oxidized Si (SiO_2) and MgO(001) substrates without a buffer layer with substrate heating or annealing after the deposition at RT. All the films on the SiO_2 substrate exhibited the A2 structure, while the films on a MgO(001) substrate had different structures depending on the heat treatment conditions. The $L2_1$ structure can be realized by adopting substrate heating above 473 K or post deposition annealing above 773 K, which suggests that the $L2_1$ structure of the Co_2FeSi film is easily fabricated. The lattice constant of the Co_2FeSi films is 0.567 nm, which is nearly the same as that of the bulk, 0.564 nm [22]. The surface roughness of the Co_2FeSi film investigated by the X-ray reflectivity analysis is larger than that of the Co_2FeAl film, where both films are deposited on the MgO(001) substrate. This may be due to the larger lattice misfit between Co_2FeSi and MgO compared with that between Co_2FeAl and MgO, where the lattice constant of Co_2FeAl is 0.573 nm. Figure 14.3 shows the in-plane (200) rocking curves for Co_2FeSi and Co_2FeAl thin films deposited on a MgO(100) substrate at 773 K, which demonstrates the larger strain for Co_2FeSi with a full width at half-maximum (FWHM) $\Delta\phi$ of 1.2 degree, while it is 0.7 degree for Co_2FeAl [8]. We investigated the TMR for the exchange-biased MTJs with Co_2FeSi electrodes and an AlO_x barrier deposited on a MgO(001) substrate without a buffer layer. The maximum TMR obtained is 44 % at RT and 68 %

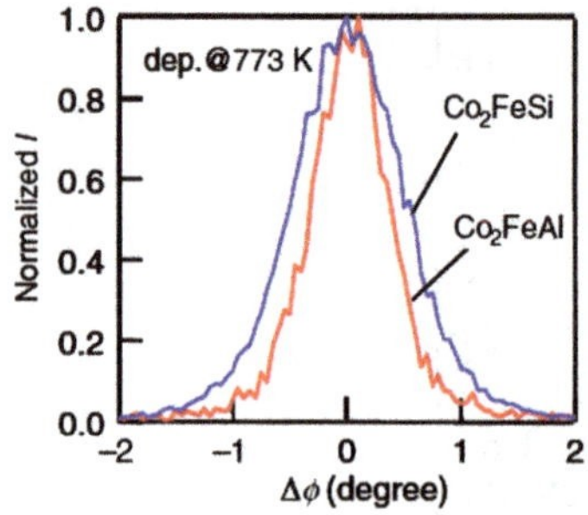

Material	*a* (nm)	Δφ (°)	Mismatch
Co_2FeSi	0.564	0.7	5.7%
Co_2FeAl	0.573	1.2	4.0%

Fig. 14.3 In-plane (200) rocking curves for Co_2FeSi and Co_2FeAl thin films deposited on a MgO(100) substrate at 773 K

at 5 K, which was annealed at 573 K after the deposition at 473 K [8]. The TMR is significantly smaller than the expectation from the theoretical prediction of high spin polarization for Co_2FeSi.

14.4.2 $Co_2FeAl_{0.5}Si_{0.5}$

The E_F position in the minority spin band gap is very important for achieving a large TMR using half-metallic ferromagnets. When the E_F is near the gap edge, spin-flipped inelastic tunneling easily occurs, leading to the reduction of TMR. In $Co_2FeAl_{0.5}Si_{0.5}$ (CFAS) E_F is at the midpoint in the gap as shown in Fig. 14.1. We first investigated the structure of CFAS films annealed at different temperatures. Figure 14.4(a) shows the out-of plane XRD spectra for 30-nm-thick CFAS films deposited on a Cr-buffered MgO substrate with various post-annealing temperatures T_a. All the films exhibit epitaxial growth of CFAS with (100) orientation. We have also investigated the (111) orientation of the films by (111) in-plane $\tilde{\phi}$ scan, as shown in Fig. 14.4(b). The strong (111) intensity is observed in the film with T_a above 500 °C, but not in the films with T_a below 400 °C. Thus, the $L2_1$ structure is formed by post-annealing above 500 °C, while the annealing below 400 °C leads to the $B2$ structure. Figure 14.5 shows the surface roughness of the CFAS films deposited on a MgO(001) substrate with and without the Cr-buffer layer as a function of T_a. The AFM image for the film with for $T_a = 400$ °C is also shown in the inset. The roughness first decreases with increasing T_a and then increases steeply with T_a above 500 °C, while it is significantly improved by using the Cr-buffer layer, which may be due to the improvement of the lattice misfit by the Cr metal layer. In order to verify the half-metallicity of CFAS we have fabricated a MTJ consisting of MgO(100)/Cr(40)/CFAS(80)/(Mg(0.7) + Al(1.3))O_x/CoFe(3)/$Ir_{22}Mn_{78}$(12)/Ru(7) with CoFe for one electrode, because the spin polarization P of CoFe is known to be about 0.5 [12], where the numbers are the layer thicknesses in nanometer and CoFe means $Co_{75}Fe_{25}$. CFAS layer was annealed at 430 °C to obtain a per-

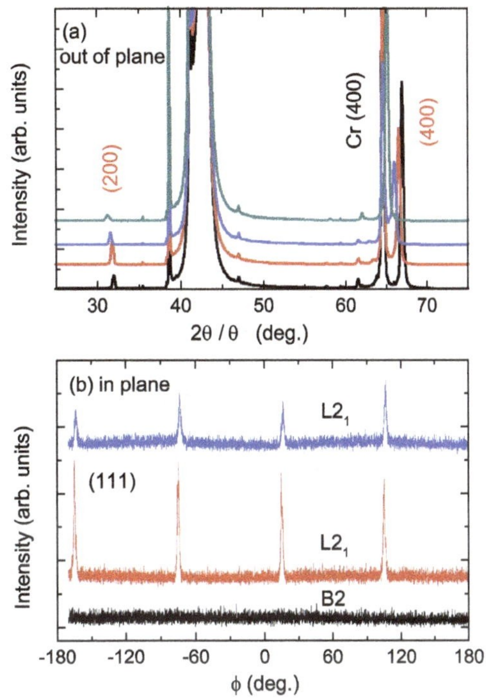

Fig. 14.4 XRD patterns for (**a**) out-of-plane and (**b**) ϕ scans for (111)

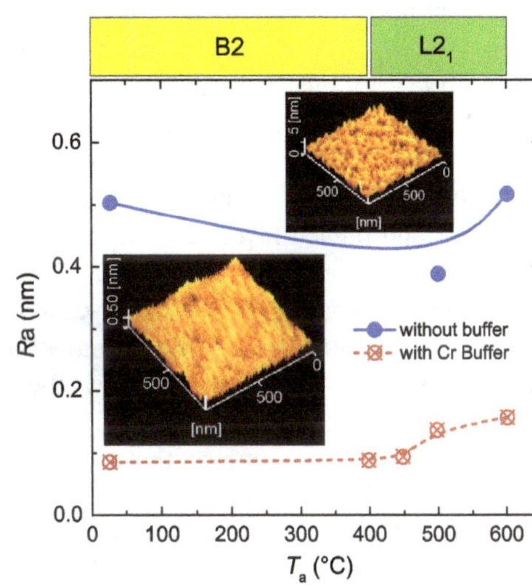

Fig. 14.5 Surface roughness measured by AFM for CFAS films with various annealing temperatures

fect $B2$ structure. The $(Mg(0.7) + Al(1.3))O_x$ tunnel barrier was formed by inductively coupled plasma treatment for 120 s in a mixed $Ar + O_2$ atmosphere (7.0 Pa). The whole sample was annealed at 290 °C for 10 min ex situ and cooled down to RT in vacuum (1×10^{-4} Pa) under a magnetic field of 3 kOe.

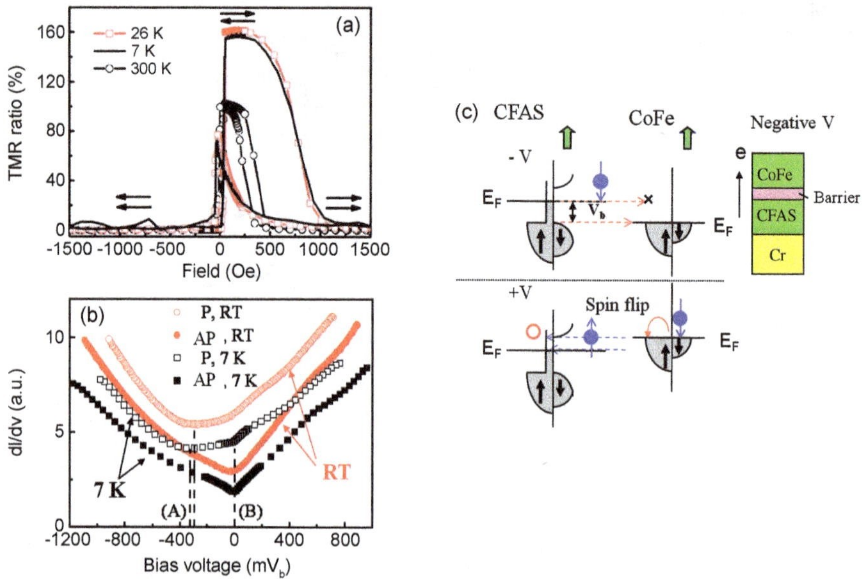

Fig. 14.6 (**a**) Typical resistance versus field curves measured at 7, 26 and 300 K. (**b**) The dependence of differential conductance on bias voltage ($\mathrm{d}I/\mathrm{d}V$) of the CFAS/(Mg-Al)O_x/CoFe at 7 and 300 K. (**c**) schematic tunneling for parallel configuration in a CFAS/(Mg-Al)O_x/CoFe MTJ

Typical resistance versus field curves measured at 7, 26, and 300 K are shown in Fig. 14.6(a). A bias voltage of 1.2 mV was applied for measurement at 300 K, while 2.5 mV for measurements at 26 and 7 K. A TMR ratio of 102.3 % at RT was achieved. The TMR ratio reaches the maximum (162 %) at 26 K and then decreases slightly with lowering measurement temperature. The reduction of TMR ratio is probably ascribed to the incomplete antiparallel magnetization state between the free layer and pinned layer resulting from the coercivity enhancement of the free CFAS layer at a very low temperature. The maximum TMR of 162 % corresponds to the spin polarization $P = 0.92$ for CFAS, suggesting half-metallicity of CFAS, which was estimated using Julliere's model for TMR ratio and $P = 0.5$ for CoFe at a low temperature. Figure 14.6(b) shows the dependence of differential conductance on bias voltage ($\mathrm{d}I/\mathrm{d}V - V$) of the CFAS/(Mg-Al)$O_x$/CoFe at 7 and 300 K. Here, a negative bias voltage means that electrons are tunneling from bottom CFAS to top CoFe. Obvious asymmetry of $G = \mathrm{d}I/\mathrm{d}V$ curves for the parallel configuration $[G_P(V)]$ with respect to the polarity of bias voltage was observed at both 7 and 300 K. The existence of a plateau in $\mathrm{d}I/\mathrm{d}V$ in the negative bias voltage regime also suggests half-metallicity of CFAS. The critical voltages (V_c) corresponding to the minimum $G_P(V)$ at 7 and 300 K are -320 and -290 mV, respectively. When the bias voltage $V < -|V_c|$, both channels of the majority spins and the minority spins in the MTJs will open, and thus $G_P(T)$ increases rapidly. In view of the contribution of thermal fluctuation, the difference of $|V_c|$ between 0 and 300 K should be equal to $|k_B T/e| \sim 26$ mV, where k_B is the Boltzmann constant and e is the charge of elec-

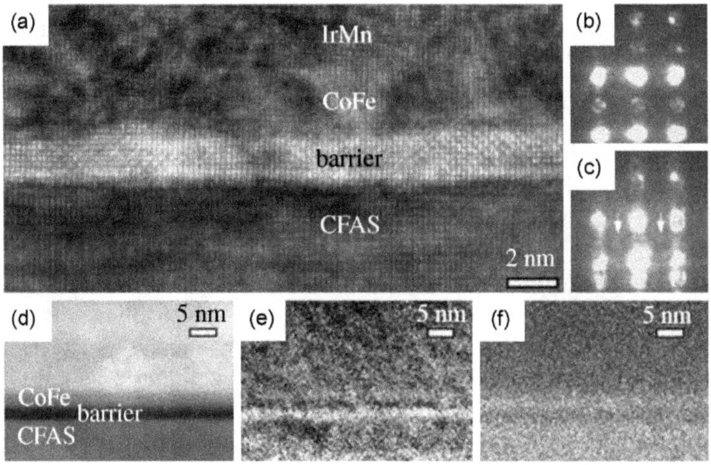

Fig. 14.7 (**a**) High resolution TEM image of an epitaxial $MgAl_2O_4$ barrier on $B2$-CFAS. $MgAl_2O_4$ [100] axis is parallel with CFAS [110] axis in horizontal direction because of the small lattice mismatch between them. Both [001] axes of CFAS and $MgAl_2O_4$ are along the vertical direction. (**b**) Diffraction pattern taken in CFAS. (**c**) Diffraction pattern taken in barrier, 022 planes of spinel structure are indicated by arrows. (**d**) HAADF-STEM image. (**e**) EELS map of Al. (**f**) EELS map of Co. Mg map is absent because of the undistinguishing Mg region in the image

tron. Here, our data result ($\sim$30 meV) is highly consistent with the theoretical analysis and reveals the large energy separation between the Fermi level and the top of the valence band in CFAS. The plateau is not observed in a positive bias voltage regime. This is understood by considering not half-metallic nature of CoFe. When a positive bias was applied the down-spins in CoFe flipped into the spin-up band, which can tunnel into the spin-up band of CFAS as shown in Fig. 14.6(c) schematically.

We found that (Mg-Al)O_x barrier has crystallized $MgAl_2O_4$ spinel structure. Cross-sectional high resolution transmission electron microscope (HRTEM) images of CFAS/(Mg-Al)O_x/CoFe multilayers are shown in Fig. 14.7(a). (Mg-Al)O_x barrier clearly shows the crystalline feature, unlike the amorphous structure report by Sakuraba et al. [14] using a similar (Mg-Al)O_x barrier. The nanobeam diffraction pattern shown in Fig. 14.7(b) clearly indicates that the bottom CFAS is $B2$ ordered, and that the (Mg-Al)O_x barrier (Fig. 14.7(c)) is consistent with $MgAl_2O_4$ spinel. The high-angle annular dark field (HAADF) image in Fig. 14.7(d) shows uniform thickness of each layer with dark contrast from the oxide barrier. The energy filter elemental maps for Al and Co (Figs. 14.7(e) and 14.7(f)) show that Al is uniformly distributed within the barrier layer sandwiched with CFAS and CoFe, suggesting intermixing of Mg and Al during the oxidation process of the barrier, which retarded the formation of separate MgO and Al_2O_3 layers. The crystallized $MgAl_2O_4$ brought an ideal barrier for spin dependent tunneling, where spin independent conductance $G_{SI} = 0$, and a low RA of $1.7 \times 10^4\ \Omega\,\mu m^2$ for CFAS/$MgAl_2O_4$/CoFe MTJs. Although the barrier with crystalline structure is formed, we do not believe that coherent tunneling occurs in our junctions at this stage because of the following reasons. First, the maximum TMR ratio of the MTJs with inserted Mg layer is the same as that of the MTJs fabricated under identical conditions except for a barrier without inserted Mg layer (amorphous AlO_x barrier) in [24]. Second, the behavior of $G_P(V)$ in our junction is quite different from that of MTJs with the contribution of coherent tunneling [25], while it is greatly in accord with the character of half-metal with large band gap. A large TMR of 835 % at 5 K (360 % at RT) has been observed in a CFAS/MgO/CFAS [26], which is also consisting with half-metallic nature of CFAS.

To reveal the band structure of CFAS more clearly, spin wave excitation model was adopted in this study. The original definition of spin polarization is $P_0 = (D_M - D_m)/(D_M + D_m)$, where D_M and D_m stand for the DOS at the Fermi level for the majority and minority spins, respectively. In a half-metal, if the Fermi level position is very close to the edge of the valence band or the conduction band, a phenomenon called "positive feedback" may occur under the influence of thermal activity, leading to a collapse of the half-metallic property [27]. For simplicity, we describe P_0 as a temperature-related $P_0(T)$ with an integral of DOS from $E_F - k_B T$ to $E_F + k_B T$. According to this description, a material with a narrow band gap or a narrow energy separation between the Fermi level and the conduction or valence band edge smaller than the thermal activation energy $k_B T$ is not a half-metal at temperature T. Finally, the decaying of P_{eff} could be written as $P_{\text{eff}} = P_0(T)(1 - \alpha T^{3/2})$, where the decay factor α is a material-dependent constant and may be affected by the circumstance of interface.

In general, the fitting of TMR ratio should use a modified Julliere's model which includes the contribution from the spin-independent conductance (G_{SI}) of MTJs [28]. However, we found that G_{SI} was negligibly small because of a crystallized barrier in our study. Attributing to Fermi level tuning, $P_0(T)$ of CFAS should be a constant from 0 K to RT in our junctions. This inference is confirmed by the perfect agreement between experimental data and the fitting curve of temperature dependence of TMR ratio for CFAS/$MgAl_2O_4$/CoFe MTJs, as shown in Fig. 14.8(a). A CoFe/$MgAl_2O_4$/CoFe MTJs is employed to acquire $P_0(T)$ of CoFe pinned layer. The good fitting result for CoFe/$MgAl_2O_4$/CoFe MTJs is also plotted in Fig. 14.8(a). The fittings give values of $P_0(T)$ for CoFe and CFAS in MTJs at 300 K being 0.493 and 0.92, respectively. Compared with the poor performance of Co_2FeSi, the high $P_0(T)$ of CFAS at room temperature clearly substantiates the success of Fermi level tuning in CFAS.

On the other hand, the values of α for CoFe (3 nm) film with exchange bias and CFAS (30 nm) film obtained from the magnetization versus temperature curves, such as the inset shown in Fig. 14.8(a), are only 3.5×10^{-6} and 4.3×10^{-6}, respectively, while those for CoFe and CFAS obtained by TMR effect are 2×10^{-5} and 3.2×10^{-5}, respectively. Such high values of α obtained by TMR effect are because TMR effect relies heavily on the magnetic properties of the upmost monolayer of electrodes at the interface [27], where the magnetization usually decays very fast [29]. Additionally, the large values of α than those reported elsewhere are also due to other researchers neglecting the temperature dependence of G_T, which is the prefactor for direct spin-dependent elastic tunneling with a function of

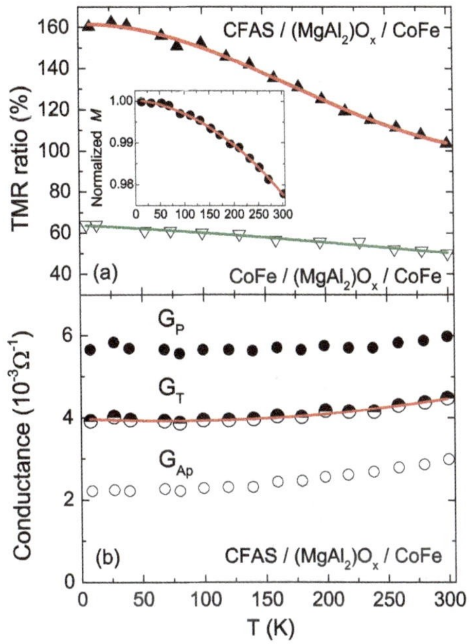

Fig. 14.8 (**a**) Temperature dependence of TMR ratio for CFAS/$MgAl_2O_4$/CoFe MTJs and CoFe/$MgAl_2O_4$/CoFe MTJs and the fitting curves by Bloch $T^{3/2}$ law. The inset shows the dependence of normalized magnetization on temperature from 10 to 300 K for CFAS (30 nm) single layer and the fitting curve by Bloch $T^{3/2}$ law. (**b**) Temperature dependence of G_P, G_{AP} and G_T for CFAS/$MgAl_2O_4$/CoFe MTJs and the fitting curve of G_T, where G_P and G_{AP} are the conductance of MTJs for parallel and antiparallel configurations, respectively; G_T is the prefactor for direct spin-dependent elastic tunneling

$G_T = (G_P + G_{AP})/2$ in the case that $G_{\text{SI}} = 0$, where G_P and G_{AP} are the conductance of magnetic tunneling junctions for parallel and antiparallel configurations, respectively. The unreasonable assumption of G_T will bring a reduced α and a weaker temperature dependence of spin polarization. Figure 14.8(b) shows the temperature dependence of G_P, G_{AP}, G_T for CFAS/$MgAl_2O_4$/CoFe MTJs, and the fitting curve of G_T using the formula (2) given in Ref. [27]. The fitting yields the barrier height $\phi = 0.82$ eV with an effective thickness of $MgAl_2O_4$ barrier being $t_{\text{eff}} = 2$ nm, while 0.34 eV with $t_{\text{eff}} = 1.3$ nm. To sum up, our work strongly suggests that α is as important as $P_0(T)$ to realize higher P_{eff}.

14.4.3 Coherent Tunneling Using B2-Co_2FeAl

Coherent tunneling using an MgO barrier can provide a large TMR due to half-metallic Δ_1 band in certain bcc metals and alloys [30, 31] such as bcc Co, Fe [32, 33], CoFe [34] and CoFeB [35]. In earlier studies of TMR in MTJs consisting of Heusler alloy electrodes, main concern is to put on the half-metallicity of whole conduction electrons of Heusler alloys. Nevertheless, two contributions, i.e., their total band half-metallicity and coherent tunneling (specific band half-metallicity) were mixed in the obtained TMR. Here we report a giant TMR of 360 % at RT and its "remarkable" oscillation due to coherent tunneling in a MTJ with a $B2$-Co_2FeAl Heusler alloy "without total band half-metallicity". This observation is further investigated and confirmed by first-principles electronic band calculations. Our results clearly and experimentally show that besides bcc Co, Fe and CoFe(B), Heusler alloys have promising potentials for giant TMR due to coherent tunneling as well as their tunable properties (such as magnetization and furthermore magnetic damping) based on a large variety constituent elements. In addition, Co_2FeAl has the lowest damping constant in Heusler alloys [36], which is a substantial factor in spin dynamics such as spin injection magnetization switching with a low current. Our finding thus suggests that the Co_2FeAl-based MTJs may play a key role in future spintronics devices.

Thin films for spin-valve-type MTJs were prepared on Cr-buffered MgO(001) substrates. The bottom Co_2FeAl electrode was deposited at RT and subsequently annealed at 480 °C for 15 min in order to smooth its surface and to induce a perfectly flat bottom Co_2FeAl/MgO interface. The structure of the annealed Co_2FeAl layer has been found to be $B2$ structure by using XRD. To further investigate the local environments around Co atoms, we have also measured ^{59}Co NMR. Figure 14.9(a) and (b) show ^{59}Co NMR spectra for Co_2FeAl annealed at 480 °C after deposition on Cr- and MgO-buffered MgO(100) substrates and bulk Co_2FeAl with $L2_1$ structure as a reference, respectively. The peak intensity around 193 MHz in the spectrum in Fig. 14.9(b) corresponds to the nearest neighbor configuration of 4 Fe + 4 Al around a Co atom in $L2_1$ structure. The satellite peaks with lower intensities suggest the existence of $B2$-type disordering, in which Fe and Al atoms swap. The two spectra in Fig. 14.9(a) are quite different from the spectrum in Fig. 14.9(b) and show a perturbed $B2$ structure. The closer inspection of the spectra in Fig. 14.9(a) indicates that the structure corresponding to the Co_2FeAl film grown on a Cr buffer is less resolved in high frequency side than that of a film grown on a MgO-buffer layer, suggesting a less ordered $B2$ structure in the case of Cr buffer. The presence of a high number of resolved resonance lines in Fig. 14.9(a) suggests that in addition to the $B2$-type disorder some admixture of the $A2$-type disordering is also present in the Co_2FeAl films. This may be attributed to the Co-rich composition of $Co_{52.8}Fe_{25.4}Al_{21.8}$ for the Co_2FeAl film, which was found in the inductively coupled plasma (ICP) analysis.

The MgO tunnel barrier was formed by rf sputtering directly from a sintered MgO target under an Ar pressure of 10 mTorr in this study. As shown in Fig. 14.10(a), we designed a wedge-shaped MgO tunnel barrier layer with the MgO thickness (t_{MgO}) ranging from 1.0 nm to 2.5 nm on each 2×2 cm^2 substrate. The top CoFe electrode was then grown on the MgO tunnel barrier at RT. A 10 nm-thick IrMn antiferromagnetic layer was subsequently deposited on the top electrode in order to establish an exchange bias field with the electrode. Lastly, the sample is capped with a 7 nm Ru layer. The structure of the fabricated MTJs was investigated by cross-sectional high resolution transmission electron microscopy

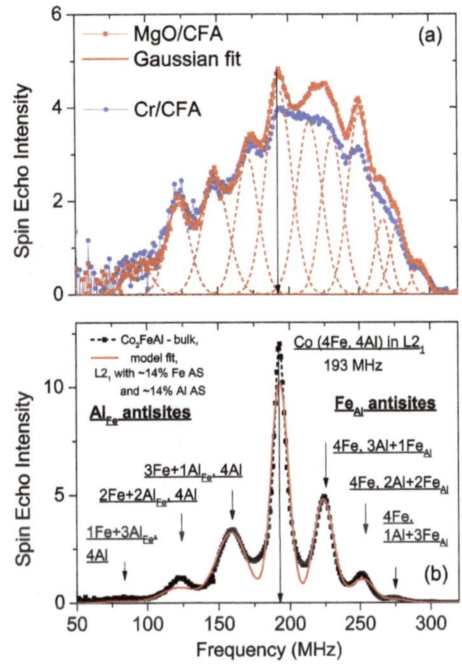

Fig. 14.9 (**a**) ^{59}Co NMR spectra at 4.2 K for 30 nm-thick Co_2FeAl thin films in situ annealed at 480 °C after deposition on a Cr-buffered (*blue*) and MgO-buffered (*red*) MgO(100) substrates. The *red solid line* is Gaussian fit for each resonance peaks. (**b**) Spectrum for bulk Co_2FeAl with $L2_1$ structure. The peak intensity at 193 MHz corresponds to the nearest-neighbor atoms of 4 Fe + 4 Al around a Co atom in $L2_1$ structure. The satellite structure extending symmetrically on both frequency sides of the main line reveals $B2$-type disordering, in which Al and Fe atoms are swapped

(HRTEM). A typical HRTEM image is shown in Fig. 14.10(b) for an MTJ structure with $t_{MgO} = 2.0$ nm. This image clearly shows that all the layers from the Co_2FeAl lower electrode to the CoFe upper electrode were grown epitaxially and were single crystalline. Although atomic steps can be seen at the lower interface, no appreciable lattice defects were observed inside the MgO tunnel barrier, which may be attributed to the small lattice misfit between Co_2FeAl and MgO [37]. We should point out that the quality of the sputtered-MgO barrier is comparable with that of MgO barrier prepared by electron-beam method [11, 25, 38].

Magnetotransport properties of the MTJ (junction area: $10 \times 10\ \mu m^2$) have been measured in a temperature range of 10–290 K, where a magnetic field was applied along Co_2FeAl [110] (‖MgO [100]). In Fig. 14.10(c), we show the t_{MgO}-dependence of observed TMR ratios. The TMR of the MTJ is modest for thin MgO barrier but dramatically increases with increasing t_{MgO}. This result is consistent with the coherent tunneling effect rather than the half-metallicity mechanism, in which the TMR is almost independent of t_{MgO} [39]. Notably, the highest TMR ratio of 330 % for Co_2FeAl/MgO/CoFe MTJ is much higher than that of $Co_2FeAl_{0.5}Si_{0.5}$/MgO/CoFe MTJs (109 % at RT) [37] and Co_2MnSi/MgO/CoFe

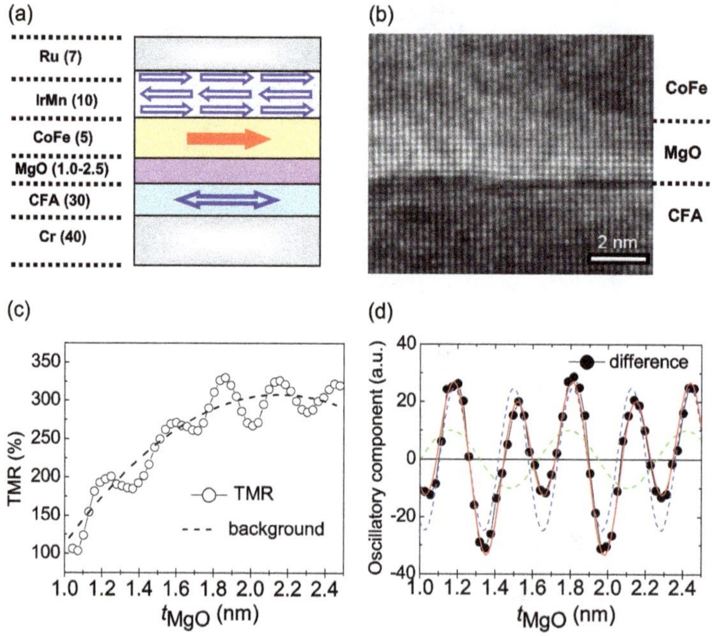

Fig. 14.10 (**a**) Schematic of the magnetic tunnel junction structure. (**b**) High-resolution cross-section transmission electron microscopy images of an MTJ with $t_{MgO} = 2.0$ nm. The *vertical* and *horizontal directions*, respectively, correspond to the MgO(001) and MgO(100) axis. The *arrow* in the micrograph shows the atomic step at the interface. (**c**) The MgO barrier thickness (t_{MgO}) dependence of TMR ratio at RT. Strong oscillation of TMR as a function of t_{MgO} is observed. *Dashed green line* represents background curve, which was fitted to quadratic function. (**d**) Oscillatory component of TMR, i.e., after subtracting the background as a function of t_{MgO}. The oscillatory component of TMR is fitted relatively well by superposing two cosine curves (*solid red line*) with different periods (λ): a short-period oscillation with $\lambda = 0.32$ nm (*dashed blue line*) and a long-period oscillation with $\lambda = 0.63$ nm (*dashed green line*)

MTJs (217 % at RT) [25], suggesting that the Co_2FeAl electrode is more effective in enhancing the TMR effect. The most interesting result in Fig. 14.10(c) is that the TMR ratios of MTJ exhibit an unexpectedly oscillatory behavior as a function of t_{MgO}. The result is remarkable because such an oscillation of TMR is normally observed only in Fe/MgO/Fe MTJs prepared by molecular-beam-epitaxy (MBE) [32, 40], while it is absent in sputter-deposited CoFe/MgO/CoFe [34] and CoFeB/MgO/CoFeB [35] MTJs. To get further insight into the physical origin of the result, as shown in Fig. 14.10(d), we replot the oscillatory component of TMR (i.e., after subtracting the background curve) as a function of t_{MgO}. We found that the oscillatory component of TMR is fitted relatively well by superposing two kinds of oscillations function (cosine curve) with different periods (λ): one of which is a short-period oscillation with $\lambda = 0.32$ nm (dashed blue line) and the other one is a long-period oscillation with $\lambda = 0.63$ nm (dashed green line). Moreover, the amplitude of the short-period is twice larger than that of the long-period one. It should be pointed out that the TMR oscillation observed in the Fe/MgO/Fe MTJ

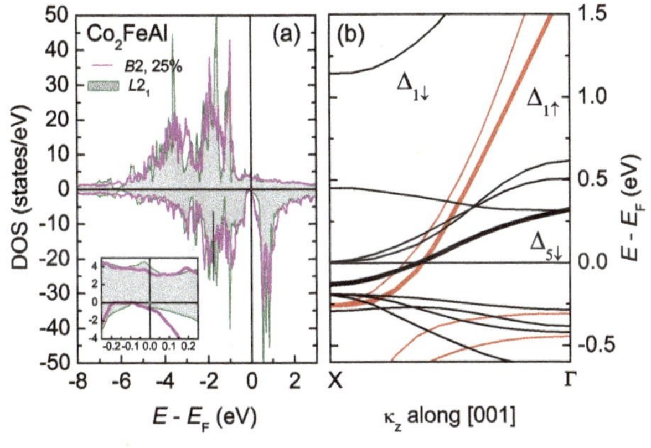

Fig. 14.11 (**a**) Density of states of Co_2FeAl with the ordered $L2_1$ structure (*green line*) and with the disordered $B2$ structure (*red line*). *Inset* in (**a**) shows the magnification of DOS near the Fermi level (E_F). (**b**) Band dispersion of the disordered $B2$-Co_2FeAl along the [001](Γ–X) direction. The *red and black lines* represent majority-spin and minority-spin bands, respectively. Overall three bands composed of two in the majority and one in the minority channel cross E_F. The *thick red and thick black lines* represent majority-spin Δ_1 and minority-spin Δ_5 bands, respectively

was also found to be a superposition of the short- ($\lambda = 0.32$ nm) and long-period ($\lambda = 0.99$ nm) oscillations [40]. These results strongly indicate that the period of the oscillations is independent of electrodes, and thus the origin of the oscillatory should only relate to the MgO tunnel barrier. Butler et al. proposed a model of interference between tunneling states for the TMR oscillation in epitaxial Fe/MgO/Fe MTJs [30]. An interference in the evanescent states at E_F in the MgO barrier between Δ_1 and Δ_5 states at $\mathbf{k}_{\parallel} = 0$ can cause an oscillation of tunneling transmittance as a function of t_{MgO}, where $\mathbf{k}_{\parallel} = 0$ is parallel components of the tunneling electrons wave vector. The tunneling transmittance for a given $\mathbf{k}_{\parallel}$ oscillates as a function of t_{MgO} with a period proportional to $1/(k_1 - k_2)$, where k_1 and k_2 are real components of Δ_1 and Δ_5 states. Thus, we suggest that the large TMR and its "remarkable" oscillation in Co_2FeAl/MgO/CoFe MTJ may be originated from the coherent tunneling effect, in which both the electrons' spins and their Bloch states symmetry are conserved.

To support the interpretation of our experimental results we have studied the band structure of disordered $B2$-Co_2FeAl along the Γ–X ($\mathbf{k}_{\parallel} = 0$) direction in the Brillouin zone. The calculations were performed using the pseudopotential method with plane-wave basis set. The electronic exchange-correlation energy was treated under the local-density approximation (LDA) for the experimental lattice constant of Co_2FeAl, 0.572 nm. We used 120 κ points in the irreducible wedge of the Brillouin zone of the primitive cell and a plane-wave cutoff energy of 500 eV. These parameters ensure good convergences for total energy. Figure 14.11(a) shows the spin-resolved densities of states (DOS) of the ordered $L2_1$ (green line) and the disordered $B2$ (red line) Co_2FeAl structure, respectively. In agreement with previous theoretical band predictions [2, 41], we found that: (i) there is no peak in the majority-spin DOS at the Fermi level (E_F), both in the ordered $L2_1$ and the disordered $B2$ structure; (ii) a few additional states are found in the minority-spin DOS

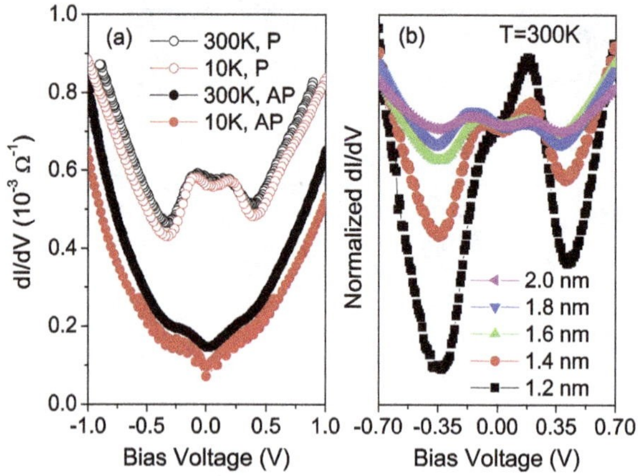

Fig. 14.12 (**a**) Bias voltage dependence of differential conductance ($G = \mathrm{d}I/\mathrm{d}V$) spectra at 300 and 10 K for Co_2FeAl/MgO/CoFe MTJ ($t_{\mathrm{MgO}} = 1.8$ nm). Here, the negative bias voltage corresponds to the tunneling of electrons from bottom Co_2FeAl to top CoFe. P and AP stand for parallel and antiparallel alignment of the two electrodes. (**b**) t_{MgO} dependence of $\mathrm{d}I/\mathrm{d}V$ spectra in the parallel alignment $[G_P(V)]$ measured at 300 K. Each spectrum is normalized by the value at zero bias

at E_F [see inset of Fig. 14.11(a)]. This result has been explained by Galanakis [2], due to the strong hybridization between the Fe 3d and Co 3d, which will cause charge transfer from the Co minority-spin states to the Fe minority-spin states, and give rise to additional states at E_F and Co_2FeAl loses half-metallicity. Furthermore, within our calculation we found that the mixing of 25 % Fe atom with Al atom (disordered $B2$) perturbs the symmetry of the Co and Fe sites creating new states at the conduction end of the band gap, which narrows the pseudo gap considerably.

Figure 14.11(b) shows the band structure of the disordered $B2$ Co_2FeAl along the Γ–X ($\mathbf{k}_{\parallel} = 0$) direction. Overall three bands composed of two in the majority (solid lines) and one in the minority (dotted lines) channel cross E_F. From the symmetry analysis, surprisingly, we found that one of the majority-spin conduction bands (thick red line) has a Δ_1 (spd-like state) symmetry, while the minority-spin conduction band (thick black line) has a Δ_5 (pd state) symmetry. It is clear that the state with Δ_1 symmetry has the smallest decay rate across the MgO barrier, followed by the Δ_5 state owing to the symmetry compatibility with MgO complex bands within the energy gap. One can therefore immediately confirm that the Co_2FeAl along the [001] direction behaves as a half-metal in terms of the Δ_1 symmetry. Note that the band structure of the disordered $B2$ Co_2FeAl is quite similar to that of bcc Co [42], but contrasts sharply with bcc Fe for which in addition to the Δ_1 band, there are other bands (Δ_2 and Δ_5) in the majority channel that cross E_F at $\mathbf{k}_{\parallel} = 0$. This effect is, we speculate, the reason for the much larger TMR ratio for Co_2FeAl/MgO/CoFe and Co/MgO/Co [32] than Fe/MgO/Fe [31] MTJs.

Figure 14.12(a) shows the dependence of differential conductance on bias voltage ($\mathrm{d}I/\mathrm{d}V - V$) curves of Co_2FeAl/MgO/CoFe MTJs at 10 and 300 K. The dip around zero bias observed in G = $\mathrm{d}I/\mathrm{d}V$ curve for antiparallel configuration $[G_{AP}(V)]$ at

10 K is so-called zero-bias conductance anomaly as usually observed in typical MTJs due to magnon excitations [43], which disappears at 300 K. Slightly asymmetrical G curves for the parallel configuration $[G_P(V)]$ with respect to the polarity of bias voltage was observed at both 10 and 300 K. The small asymmetrical feature probably reflects a slight different in the quality of the lower-Co_2FeAl/MgO and the upper-MgO/CoFe interfaces in our MTJs as confirmed by HRTEM (see Fig. 14.10(b)). It is interesting to note that, as marked by the arrows in Fig. 14.12(a), $G_P(V)$ curves present a local minimum at about -0.35 and $+0.38$ eV for the negative and positive bias, respectively. This behavior is remarkable. To the best of our knowledge, such a conductance anomaly has not been observed in other Heusler-based MTJs. Here, we explain the local minimum by evoking the band dispersion of the disordered $B2$-Co_2FeAl in the following way. As shown in Fig. 14.12(b), in fact the top of minority-spin 5 band, which dominates the down spin conduction, lies at $\sim$0.348 eV above E_F. When the energy of the conduction electrons arriving across the barrier overcomes the top of this Δ_5 band, the conduction channel associated with this state disappears. Therefore, the amplitude of the relative variation of $G_P(V)$ curve from zero bias to the local minimum quantifies the contribution of the five electrons in P channel conductivity. This interpretation remains in agreement with the results of Fig. 14.12(b) which show $G_P(V)$ curves measured at 300 K for Co_2FeAl/MgO/CoFe MTJs with various MgO thicknesses. We clearly found the amplitude of the local minimum increases pronouncedly with decreasing the MgO thickness. This important feature strongly indicates that the contribution of Δ_5 electrons to the tunneling becomes more significant in the thin MgO thickness regime.

We analyzed the temperature dependence of TMR measured for various epitaxial MTJs with a MgO barrier and Co_2FeAl as an electrode using Julliere's model for TMR and the spin wave excitation model, $P_{\mathrm{eff}} = P_0(1 - \alpha T^{3/2})$ as described in 14.4.2. It is demonstrated that the P_{eff} values of the electrodes are significantly different between two electrodes facing a MgO barrier; a top electrode on a MgO barrier has significantly smaller P and larger α than those of a bottom electrode underneath the barrier. The analysis exhibits that the bottom Co_2FeAl has almost $P_0 \sim 1$ and a small α of 0.75×10^{-5}, while top Co_2FeAl has $P_0 = 0.68$ and $\alpha = 3.54 \times 10^{-5}$ [44]. The MTJs fabricated are Co_2FeAl(30)/MgO(1.8)/Co_2FeAl(5)/$Ir_{80}Mn_{20}$(12)/Ru(7) (unit: nm) and Co_2FeAl(30)/MgO(1.8)/CoFe(5)/$Ir_{80}Mn_{20}$(12)/Ru(7) on a Cr(40)- or MgO(10)-buffered MgO(001) substrate. The MTJs were patterned into an area of 10×10 μm^2 using conventional photolithography and Ar ion milling process. Patterned MTJs were annealed at T_a from 300 to 475 °C for 1 hour in high vacuum by applying a 5 kOe magnetic field. Figure 14.13 shows the temperature dependence of TMR for three different MTJs consisting of (a) Co_2FeAl/MgO/CoFe and (b) Co_2FeAl/MgO/Co_2FeAl, deposited on a Cr-buffered MgO(100) substrate, respectively and (c) CoFe/MgO/CoFe deposited on a MgO-buffered MgO(100) substrate as a reference. All the MTJs are epitaxially grown and exchange biased using an IrMn antiferromagnet. The TMR at 10 K (RT) is 700 % (330 %), 415 % (223 %) and 268 % (140 %) for (a), (b) and (c), respectively. The comparison between (a) and (c) suggests that Co_2FeAl has much larger

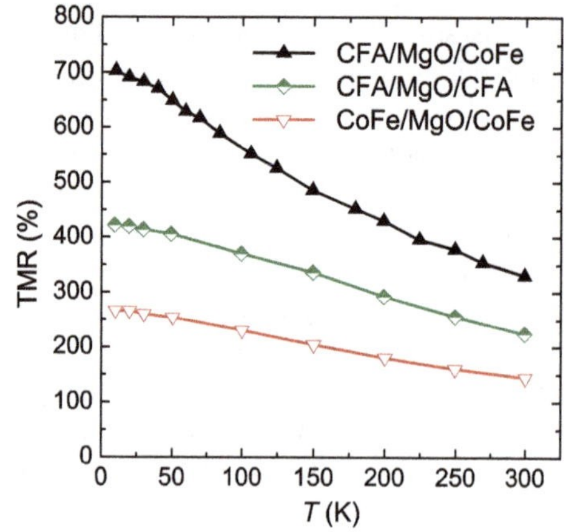

Fig. 14.13 Temperature dependence of TMR ratio for the three MTJs. The TMR ratios increase as the temperature is reduced attaining a value of 700 %, 415 %, and 268 % at 10 K for $Co_2FeAl/MgO/CoFe$, $Co_2FeAl/Mg/Co_2FeAl$ and CoFe/MgO/CoFe MTJs, respectively

P than that of CoFe. However, the TMR in (b) is significantly smaller than that in (a) in contradiction to the prediction of higher P_0 for Co_2FeAl than CoFe. This apparent inconsistency can be understood by considering different P_0 values between the bottom and top Co_2FeAl electrodes facing a MgO barrier due to the different interface structures. In fact we have obtained higher TMR of 785 % and 360 % at 10 K and RT, respectively for a $Cr/Co_2FeAl/MgO/CoFe(0.5)/Co_2FeAl$ MTJ with 0.5 nm thick CoFe insertion between MgO barrier and top Co_2FeAl, which improves the Co_2FeAl interface [45]. In order to clarify the different P_0 values between the bottom and top electrodes we have analyzed the temperature dependence of TMR using Julliere's model for TMR ratio and spin wave excitation model for P of electrodes in the MTJs. We have neglected the spin-independent contribution to the tunneling as in reference [12] and assumed P_0 to be the same as the value at 10 K. An excellent fitting curve for the temperature dependence of TMR in (c) was obtained using $P_0 = 0.76$ and $\alpha = 3.2 \times 10^{-5}$ for both CoFe, assuming the same P_0 and α for top and bottom CoFe electrodes facing a MgO barrier. By using these values for CoFe we successfully fitted the experimental result for $Cr/Co_2FeAl/MgO/CoFe$ MTJ as shown with a solid line in Fig. 14.14(a), which gives $P_0 = 0.99$ and $\alpha = 7.5 \times 10^{-6}$ for the bottom Co_2FeAl on a Cr buffer. It is noticeable that the P_0 value implies almost half-metallicity of the bottom Co_2FeAl on a Cr buffer due to the coherent tunneling effect. This is consistent with the calculated band dispersion of $B2$-Co_2FeAl along [001] direction, in which Δ_1 band exists for the majority spin but not for the minority spin at E_F. The small $\alpha = 7.5 \times 10^{-6}$ may be originated from the very flat interface between the bottom CAF and a MgO barrier. The small α of 7.5×10^{-6} and the large P_0 of 0.99 bring a very large tunneling spin polarization at RT, P_{eff} (RT) = 0.96 for $Cr/Co_2FeAl/MgO$ structure.

The temperature dependent TMR in $Cr/Co_2FeAl/MgO/Co_2FeAl$ MTJ can also be fitted well using the same values of P_0 and α estimated for bottom Co_2FeAl as shown in Fig. 14.14(b), which leads to $P_0 = 0.68$ and $\alpha = 3.5 \times 10^{-5}$ for the MgO barrier/top Co_2FeAl structure. The P_0 and α of the top Co_2FeAl on a MgO barrier are significantly smaller and larger compared with those of the bottom Co_2FeAl, respectively although the α is comparable to the value of CoFe, suggesting that the top

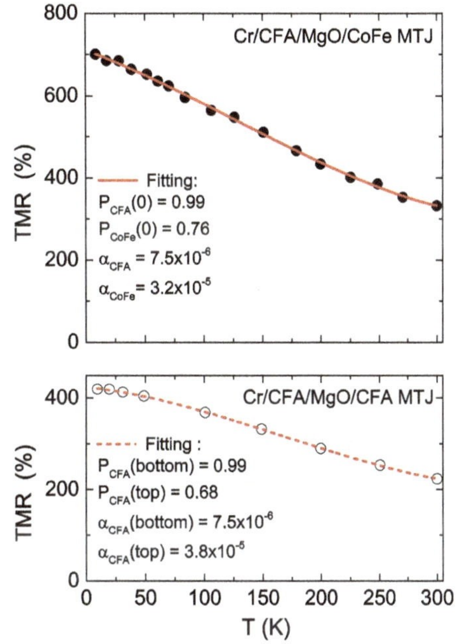

Fig. 14.14 Fitting of the temperature dependence of TMR using Julliere's model for TMR ratio and spin wave excitation model for tunneling spin polarization P for (**a**) Cr/Co_2FeAl/MgO/CoFe and (**b**) Cr/Co_2FeAl/MgO/Co_2FeAl MTJs (*solid and dashed red lines*). *Circles* are experimental data

Co_2FeAl on a MgO barrier has a degraded interface with disordered structure and larger roughness compared with the bottom interface due to the thinner MgO barrier and top Co_2FeAl thicknesses. Bottom Co_2FeAl on a Cr buffer is in-situ annealed at 480 °C before depositing a MgO barrier, which provides highly $B2$-ordered and better interface of the bottom Co_2FeAl. The comparable α between top Co_2FeAl and CoFe may be based on nearly the same magnitude of T_c and similar top interface structure with each other. By applying $P_0 = 0.76$ and $\alpha = 3.2 \times 10^{-5}$ for CoFe to an exchange-biased MgO buffer(20)/Co_2FeAl(30)/MgO(1.8)/CoFe(5)/IrMn(10)/Ru MTJ fabricated, we have successfully fitted the temperature dependence of TMR for the MTJ using $P_0 = 0.90$ and $\alpha = 3.2 \times 10^{-6}$ for bottom Co_2FeAl on a MgO buffer. The $P_0 = 0.90$ is slightly lower than 0.99 of the Co_2FeAl on a Cr buffer. This may be caused by the difference of the disordering in Co_2FeAl between on a Cr buffer and a MgO buffer, which was revealed by ^{59}Co NMR measurements as shown in Fig. 14.9. The values of P_0 and estimated above are listed in Table 14.1 for Co_2FeAl and CoFe on different buffers and a MgO barrier. The significantly small P_0 of Co_2FeAl on a MgO barrier (0.68) compared with that on a MgO buffer (0.90) suggests the interface and bulk structure differences of Co_2FeAl between on the barrier and buffer, because MgO barrier thickness (1.8 nm) is much thinner than that of the MgO buffer (20 nm). The structure of Co_2FeAl on a thin MgO barrier may be more disordered at the interface and the bulk. We can calculate TMR ratio for MgO buffer/Co_2FeAl/MgO/Co_2FeAl MTJ using P_0 and estimated for top and bottom Co_2FeAl, which provides 310 % and 183 % at a low temperature and RT, respectively. The 183 % at RT is in excellent agreement with the experimentally obtained TMR of 188 % at RT. This result indicates that the P_0 and α values listed in Table 14.1 for bottom and top electrodes facing a MgO barrier are acceptable.

Table 14.1 *P* and α values for bottom and top Co_2FeAl and CoFe estimated from the temperature dependence of TMR ratio

	P		$\alpha(\times 10^{-5})$
	P (10 K)	*P* (RT)	
Cr buffer/Co_2FeAl/MgO	0.990	0.960	0.75 (b-Co_2FeAl)
MgO buffer/Co_2FeAl/MgO	0.902	0.867	0.75 (b-Co_2FeAl)
MgO/t-Co_2FeAl	0.680	0.551	3.54 (t-Co_2FeAl)
MgO buffer/CoFe/MgO	0.753	0.628	3.2 (b-CoFe)
MgO/t-CoFe	0.760	0.634	3.2 (t-CoFe)

b and t mean bottom and top layers adjacent to MgO barrier, respectively

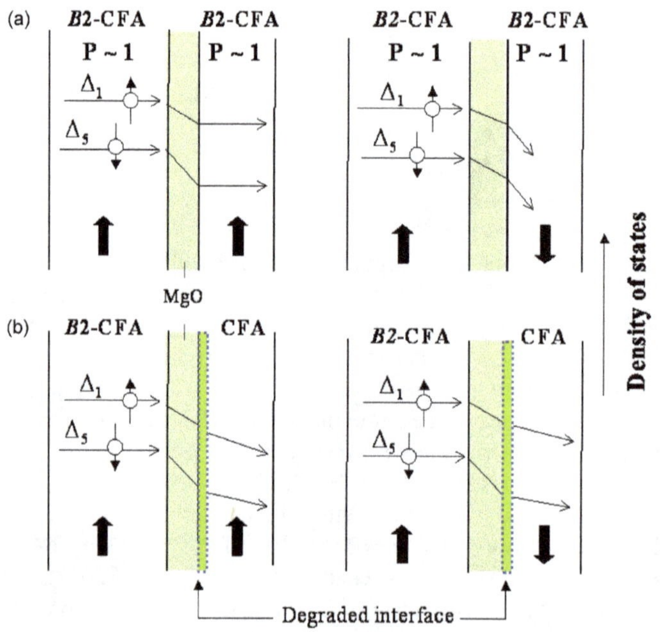

Fig. 14.15 Schematic tunneling mechanism based on the coherent tunneling effect for Cr/Co_2FeAl/MgO/Co_2FeAl MTJs with perfect (**a**) and disordered (**b**) interfaces between MgO barrier and top Co_2FeAl, where the bottom interface is assumed to be perfect. Interface wave function matching and tunneling probability for the disordered top interface (**b**) are changed, which leads to TMR reduction

The tunneling mechanism based on the coherent tunneling effect is illustrated schematically in Fig. 14.15 for Cr/Co_2FeAl/MgO/Co_2FeAl MTJs with perfect (a) and disordered (b) interfaces between MgO barrier and top Co_2FeAl, where the bottom interface is assumed to be perfect because we have obtained $P_0 \sim 1$ for Cr/Co_2FeAl/MgO barrier structure. The changes of interface wave function matching and tunneling probability for the disordered interface will decrease and increase

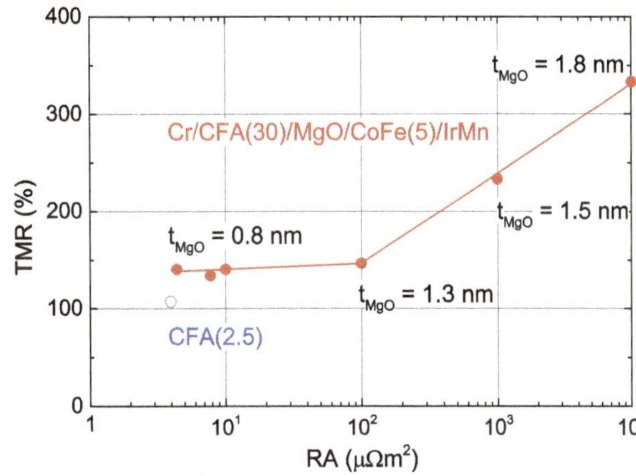

Fig. 14.16 TMR ratio at RT as a function of resistance (R) area (A) product for Cr(40)/ Co_2FeAl(30)/MgO(tMgO)/ CoFe(5)/IrMn(10)/Ru MTJs with different MgO barrier thicknesses

Δ_1 and Δ_5 band contributions to tunneling in (a) and (b), respectively, resulting in the TMR reduction and the reduced tunneling spin polarization. If we could realize the same interface for MgO barrier/top Co_2FeAl as that of the MgO buffer/top Co_2FeAl with $P_0 = 0.9$, the TMR of 1800 % at RT is predicted to be achieved for a Cr/Co_2FeAl/MgO/Co_2FeAl MTJ, since bottom Co_2FeAl has $P_0 \sim 1$ and both Co_2FeAl electrodes have $\alpha = 0.75 \times 10^{-6}$. This will be a challengeable issue.

Finally resistance (R) × area (A) product dependence of TMR at RT with different MgO barrier thicknesses t_{MgO} is shown in Fig. 14.16 for Cr(40)/Co_2FeAl(30)/ MgO(t_{MgO})/CoFe(5)/IrMn(10)/Ru MTJs. The TMR for $t_{Co_2FeAl} = 2.5$ nm is also plotted for a MTJ with $t_{MgO} = 0.8$ nm. The TMR decreases with decreasing MgO thickness in accordance with a feature of coherent tunneling effect. Surprisingly, however, TMR keeps over 100 % at RT for $RA = 5\ \Omega\,\mu m^2$ and even for $t_{Co_2FeAl} = 2.5$ nm. This large TMR may be caused by that the coherent tunneling effect works effectively for a MTJ with small RA and thin Co_2FeAl thickness, which may be due to the good interface structure between bottom Co_2FeAl and a MgO barrier. MTJs with a thin free layer accompanied with a large TMR and a small RA are indispensable for magnetization switching by spin transfer torque which is a key technology in high density MRAMs and spin logic etc. Thus, Co_2FeAl will be a promising material for practical applications in spintronics.

14.4.4 CPP-GMR Using $Co_2FeAl_{0.5}Si_{0.5}$ [46, 47]

The layer structure of MgO(001)/Cr(10)/Ag(100)/CFAS(t_F)/Ag spacer(5)/ CFAS(t_F)/Ag(5)/Ru(8) was deposited by magnetron sputtering. The thicknesses of both the bottom and top CFAS layers (t_F) were varied from 2.5 to 22 nm. All the sputtering processes were performed at RT. The Ag underlayer (100 nm) was deposited as the bottom electrode. The Cr/Ag underlayers were in situ annealed at 300 °C for 30 min to improve the surface flatness and subsequently the following

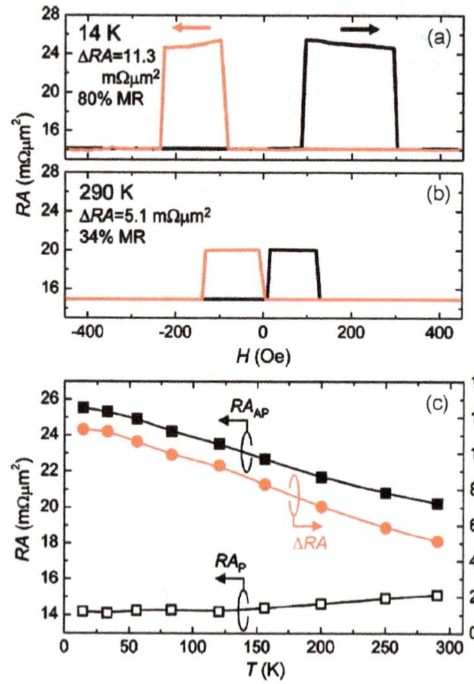

Fig. 14.17 MR curves of the CFAS(2.5)/Ag(5)/CFAS(2.5) PSV at (**a**) 14 K and (**b**) 290 K. (**c**) Temperature dependence of RA for the parallel state RA_P (open square) and for the antiparallel state RAAP (*filled square*), and ΔRA (*filled circle*). The lead resistance was subtracted at each temperature

layers were deposited. After the deposition of all the layers, the films were annealed at 500 °C for 30 min under a magnetic field of 5 kOe. Pseudo spin valves (PSVs) were fabricated by patterning the films into pillars designed to elliptical shapes from $0.07 \times 0.14\ \mu m^2$ to $0.20 \times 0.40\ \mu m^2$. The pillars were Ar ion etched through the CFAS/Ag/CFAS trilayer. The actual pillar size was determined by scanning electron microscopy. The CFAS films were found to have the $B2$-ordered structure by 4-axis XRD. The MR was measured by the four probe method in a temperature range of 14–290 K. Figures 14.17(a) and 14.17(b) show the MR curves of the PSV device with $t_F = 2.5$ nm at 14 K and 290 K, respectively [47]. For obtaining the intrinsic value of resistance-area product (RA) of the CFAS/Ag/CFAS trilayer, the lead resistance given as a segment of R versus $1/A$ plot (0.13 Ω at 14 K and 0.3 Ω at 290 K) was subtracted. The ΔRA value and the MR ratio were 11.3 $m\Omega\,\mu m^2$ and 80 % at 14 K, and 5.1 $m\Omega\,\mu m^2$ and 34 % at 290 K, respectively. RA for parallel magnetization configuration (RA_P) showed a small increase with increasing temperature as shown in Fig. 14.17(c). The ratio of RA_P at 14 K to that of 290 K was 0.94, which is in good agreement with the value expected from the temperature dependence of the resistivity of the CFAS film, $\rho_F = 62.4$ and 71.0 $\mu\Omega$ cm at 14 K and 290 K, respectively. On the other hand, RA for the antiparallel magnetization state (RA_{AP}) decreases with increasing temperature. Consequently, ΔRA decreased monotonically. These behaviors of RA and ΔRA were common features of all of our devices with various CFAS thicknesses.

Figure 14.18(a) shows the ΔRA values for various t_F measured at 14 and 290 K [47]. At 14 K, ΔRA increased up to $t_F = 8.0$ nm, while it showed an almost con-

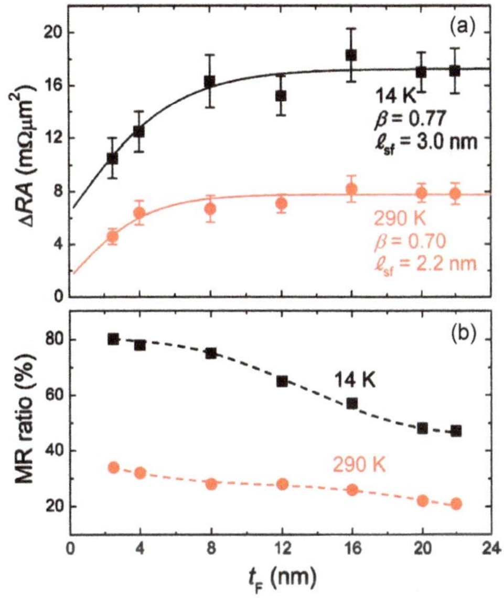

Fig. 14.18 Variations of (**a**) ΔRA and (**b**) MR ratio of the CFAS/Ag/CFAS PSV for the CFAS layer thickness $t_F = 2.5$–22 nm at 14 K (*square*) and 290 K (*circle*). The *fitted curves* in (**a**) are by the Valet–Fert model

stant value of ~17 mΩ μm² for $t_F > 8.0$ nm. At 290 K, ΔRA increased only for the thickness from $t_F = 2.5$ nm to 4.0 nm. Figure 14.18(b) shows the dependence of MR ratio on t_F. The MR ratio was almost constant while ΔRA increases with increasing t_F, but decreased for the larger t_F due to the constant ΔRA but increasing RA_P along with t_F. If the spin diffusion length of the ferromagnetic layer (lsf) is much longer than the layer thickness (t_F), the entire ferromagnetic layer should contribute to the spin-dependent transport. Thus, the ΔRA value should increase in proportion to t_F. On the other hand, when l_{sf} is comparable to t_F, the ΔRA value should be saturated at a certain thickness. The present data for the CFAS/Ag/CFAS PSV indicate that the spin diffusion length of the CFAS alloy is comparable to or shorter than the layer thickness. In order to evaluate the spin asymmetry and the spin diffusion length, the general Valet–Fert model [48] was used, which gives a short spin diffusion length $l_{sf} \sim 3.0$ nm and a relatively high bulk spin asymmetry $\beta \sim 0.77$ at 14 K. The short spin diffusion length effectively limits the contribution of the bulk scattering, suggesting that the interfacial scattering plays an important role in the relatively large CPP-GMR effect of the CFAS/Ag/CFAS devices. The origin of the high CPP-GMR is attributed to high majority-spin conductance because of the good Fermi surface matching between full-Heusler and Ag [49].

14.4.5 Spin Transfer Magnetization Switching in a Spin-Valve Nanopillar with $Co_2FeAl_{0.5}Si_{0.5}$ [50]

GMR multilayer film consisting of Cr(20)/Ag(50)/CFAS(20)/Ag(4)/CFAS(2.5)/Ru(15) (in nm) was prepared on a single-crystal MgO(001) substrate using a mag-

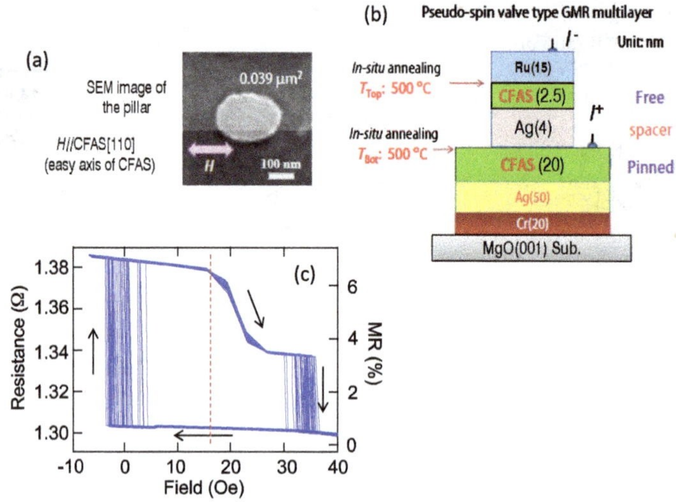

Fig. 14.19 (**a**) Scanning electron microscopy image of pillar after milling. (**b**) Schematic illustration of $Co_2FeAl_{0.5}Si_{0.5}$ (CFAS)/Ag/CFAS CPP-GMR nanopillar. (**c**) Resistance-magnetic field (R–H) curves measured at room temperature; ten loops are shown. H_{offset} indicates hysteresis offset field

netron sputtering system. All the layers were deposited at RT. Both the bottom and top CFAS layers were post annealed at 500 °C for 15 min in situ to improve $B2$ ordering. The multilayer was patterned into a 200 nm scale ellipsoidal shape [long and short axes were 250 and 190 nm, respectively; see Fig. 14.19(a)] by using e-beam lithography and Ar ion milling. A pillar was made by etching the multilayer down to the surface of the bottom CFAS layer, as shown schematically in Fig. 14.19(b). The longitudinal direction of the pillar was aligned along CFAS[110], corresponding to the easy magnetization axis of unpatterned CFAS layers. The electrical properties were characterized by measuring the resistance-magnetic field (R–H) and resistance-current (R–I). Figure 14.19(c) represents the R–H curves of the nanopillar measured with a constant current of 0.1 mA (corresponding to 2.5×10^5 A/cm^2 in current density). A large MR ratio of 7.5 % was obtained with an RA of 5 mΩ μm^2 and a ΔRA of 3.7 mΩ μm^2. Note that the maximum MR ratio of the nanopillar reached 9.0 % even though the thickness of the top CFAS layer was reduced for the purpose of spin-transfer switching. The hysteresis of the R–H curves corresponded to the magnetization switching of the top CFAS layer. When the magnetic field reached around zero, the resistance abruptly changed from the low to the high state, indicating transition of the magnetization configuration from parallel (P) to antiparallel (AP). When the field was increased from zero, an intermediate (I) state was observed in addition to the P and AP states. The I state may reflect fourfold in-plane magnetocrystalline anisotropy of CFAS layers, which can stabilize the moment at an angle of 90° with respect to an applied magnetic field direction. To discuss quantitatively, we determined the magnetocrystalline anisotropy constant K_1 of a 2.5 nm thick CFAS layer to be -1.0×10^5 erg/cm^3 by measuring magnetization curves

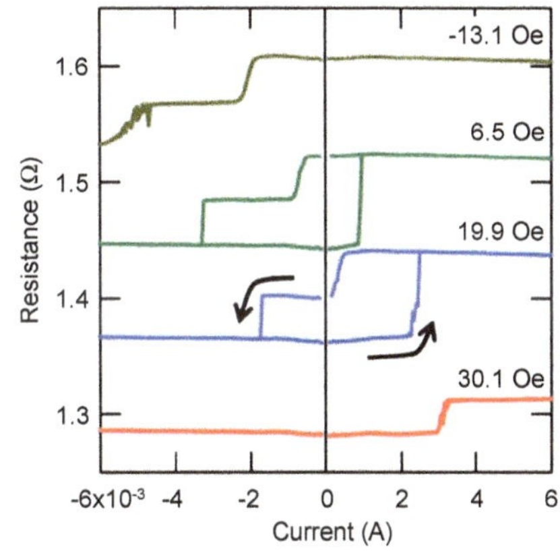

Fig. 14.20 Magnetic field dependence of resistance-current (R–I) curves of CFAS/Ag/CFAS CPP-GMR nanopillar measured at room temperature. Before measuring each curve, the magnetization configuration was reset by applying +250 Oe. For clarity, the *curves* are vertically offset by 0.08 Ω each. Current sweep directions are displayed with *arrows*

along [110] easy axes and [100] hard axes for CFAS at RT. The absolute value of K_1 is larger enough to overcome the demagnetization energy (around 4×10^4 erg/cm^3, estimated from the dimension of the nanopillar). Consequently, the presence of the I state presumably shows that the moment of the top CFAS layer was trapped at an angle of 90° with respect to that of the bottom CFAS layer. This two-step switching behavior has also been reported in a fully epitaxial Fe/Ag/Fe(001) nanopillar and is attributed to the fourfold magnetocrystalline anisotropy of the epitaxial Fe(001) layers [51]. The hysteresis offset field (H_{offset}) and the coercive field (H_c) were determined from the R–H curves to be 16 and 18 Oe, respectively.

Figure 14.20 shows R–I curves under different magnetic fields, where a positive current indicates that electrons flow from the top electrode to the bottom one. Spin transfer magnetization switching due to a dc current was clearly observed. The switching current density for dc current (J_c) is of the order of 10^6 A/cm^2. In addition, three states were evident in the R–I curves, corresponding to the P, AP, and I states. The magnetization switched from P to AP when a current was applied in the positive direction under the magnetic field between −10 and 20 Oe; the switching process included the I state when a current was applied in the negative direction. The switching from AP to I is gradual while the switching from I to P is abrupt in the negative current direction. This suggests that the detailed mechanism of these two switching processes is different. However, when the applied magnetic field increased in the field range from −10 to 20 Oe, the shapes of the R–I curves did not change significantly, simply shifting in the positive direction, as expected from the typical behavior of spin transfer switching under a magnetic field. The P state was no longer stable and non-static states appeared in the negative current region below −10 Oe, whereas AP state was not observed above 20 Oe. These behavior in the R–I curves shows that the spin transfer torque dominates all the transition observed.

To evaluate the thermal stability factor ($\Delta = KV/k_BT$), where K is the magnetic anisotropy constant, V is the volume of the free layer, and T is the absolute temperature, and intrinsic critical current density (J_{c0}), i.e., J_c in the short pulse current limit, we used the thermal activation model [52–54] for switching currents deduced

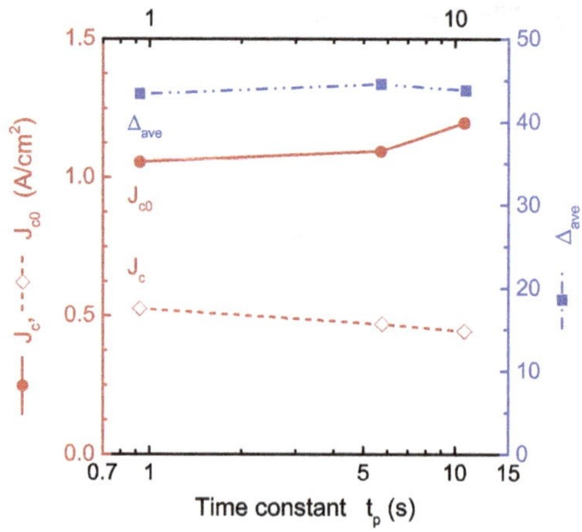

Fig. 14.21 t_p dependence of a dc critical current density J_c, intrinsic critical current density J_{c0} and thermal stability factor Δ_{ave} under a magnetic field of 14 Oe. All measurements were carried out at room temperature

from the R–I curves under application of offset magnetic field H_{offset} (= 16 Oe). We defined switching current I_c as the current at which the resistance switches completely from the P (I) state to the AP (P) state. For simplicity, a single effective barrier ($U \sim KV$) was assumed. The I_c distribution was measured by repeating the R–I curves 200 times, from which J_{co} can be obtained [52–54]. We obtained $J_{c0}^{+(-)} = 6.6 \times 10^6$ A/cm^2 (-1.2×10^7 A/cm^2) and $\Delta_l^{+(-)} = 47(36)$ [49]. Consequently, the average critical current density ($J_{c0} = (J_{c0}^+ + J_{c0}^-)/2$) was 9.3×10^6 A/cm^2. J_{c0} is predicted to be proportional to $\alpha M_s^2 t/\eta = cM_s^2 t$, where η is the spin transfer efficiency, M_s is the saturation magnetization, t is the thickness of switching layer, and c is defined as $c = \alpha/\eta$. The small J_{c0} observed in the CFAS/Ag/CFAS nanopillar is caused by low c in the CFAS layers, because CFAS should have high P and low α. Braganca et al. [55] reported small J_{c0} of 1.1×10^7 A/cm^2 using 4.5 nm thick Py with a low M_s (560 emu/cm^3). From the M_s of the 2.5 nm thick CFAS at RT (960 emu/cm^3) and the relationship of $J_{c0} \propto cM_s^2 t$, we roughly obtain the ratio of c between CFAS and Py, c(CFAS)/c(Py) ~ 0.5. This apparently shows that the use of CFAS layers is effective for the reduction of J_{c0}. Estimated J_{c0} and Δ are almost independent of the duration time of current, $t_p(I)$, indicating the thermal effect is negligibly small (see Fig. 14.21).

14.5 Summary

We have investigated the structural and magnetic properties of full-Heusler $Co_2FeAl_xSi_{1-x}$ epitaxial thin films prepared on MgO(100) substrates using an ultrahigh-vacuum magnetron sputtering system and their applications to MTJs, CPP-GMR devices and spin transfer magnetization switching. We showed that Fermi level position can be tuned by the composition x in $Co_2FeAl_xSi_{1-x}$. Co_2FeSi (CFS) with $L2_1$ structure has E_F near top edge of the valence band, thus a large TMR characteristic to a half-metal is difficult to realize in MTJs using a CFS electrode. $B2$-

$Co_2FeAl_{0.5}Si_{0.5}$ (CFAS) exhibits a half-metallic band gap theoretically and experimentally even at RT and a large spin polarization over 0.9. While $B2$-Co_2FeAl, which is not half-metallic, is exploited to be a new ferromagnetic electrode for pronounced coherent tunneling inducing giant TMR such as 360 % at RT in an exchange-biased MTJ with an MgO barrier. Temperature dependence of TMR in epitaxial MTJs using $B2$-CFAS or $B2$-Co_2FeAl electrode is fitted well by spin wave excitation model for tunneling spin polarization. Half-metallic $B2$-CFAS provided a large CPP-GMR up to 34 % at RT in a CFAS/Ag/CFAS PSV, of which mechanism is attributed to high majority-spin conductance because of the good Fermi surface matching between full-Heusler and Ag. Magnetization switching was observed in the R–I curves and exhibited a two-step switching process originating from the interplay between the magnetocrystalline anisotropy of the CFAS layers and the spin-transfer torque. A small average intrinsic switching current density (J_{c0}) of 9.3×10^6 A/cm^2 was obtained by analyzing the data using the thermal activation model. The results show that the use of the full-Heusler CFAS alloy with high spin polarization is an effective way to reduce J_{c0} even in MTJs.

Acknowledgements This paper summarizes the works collaborated with R. Shan, W.H. Wang, T. Furubayashi, T. Nakatani, Z. Gercsi, S. Kasai, S. Mitani and K. Hono (NIMS), and M. Wojcik and E. Jedryka (Polish Academy of Science), supported partly by the NEDO, CREST and JST-DFG (FE633/6-1).

References

1. de Groot RA et al. (1983) Phys Rev Lett 50:2024
2. Galanakis I, Dederiches PH, Papanikolaou N (2002) Phys Rev B 66:174429
3. Ishida S, Fujii D, Kashiwagi S, Asano S (1995) J Phys Soc Jpn 64:2152
4. Picozzi S, Continenza A, Freeman AJ (2002) Phys Rev B 66:094421
5. Nagao, Shirai S, Miura Y (2004) J Phys Condens Matter 16:S5725
6. Galanakis I (2004) J Phys Condens Matter 16:8007
7. Inomata K, Okamura S, Goto R, Tezuka N (2003) Jpn J Appl Phys 42:L419
8. Inomata K et al. (2006) J Phys D, Appl Phys 39:816
9. Marukame T et al. (2006) Appl Phys Lett 88:262503
10. Okamura S et al. (2005) Appl Phys Lett 86:232503
11. Tezuka N, Ikeda N, Sugimoto S, Inomata K (2006) Appl Phys Lett 89:252508
12. Shan R et al. (2009) Phys Rev Lett 102:246601
13. Kammerer S et al. (2004) Appl Phys Lett 85:79
14. Sakuraba Y et al. (2006) Appl Phys Lett 88:192508
15. Ishikawa T et al. (2009) J Appl Phys 105:07B110
16. Kubota T et al. (2009) Appl Phys Lett 94:122504
17. Sakuraba Y et al. (2010) Phys Rev B 81:14422
18. Inomata K et al. (2008) Phys Rev B 77:214425
19. http://staff.aist.go.jp/t-ozaki/
20. Nakatani TM et al. (2007) J Appl Phys 102:033916
21. Fecher GH, Felser C (2007) J Phys D 40:1582
22. Wurmehl S et al. (2006) Appl Phys Lett 88:032503
23. Hashimoto M, Herfort J, Schonherr HP, Ploog KH (2005) Appl Phys Lett 87:102506
24. Tezuka N, Ikeda N, Sugimoto S, Inomata K (2006) Appl Phys Lett 89:112514

25. Tsunegi S et al. (2008) Appl Phys Lett 93:112506
26. Tezuka N et al. (2009) Appl Phys Lett 94:162504
27. Attema JJ, de Wijs GA, de Groot RA (2007) J Phys Condens Matter 19:315212
28. Shang CH, Nowak J, Jansen R, Moodera JS (1998) Phys Rev B 58:R2917
29. Wang WH et al. (2005) Phys Rev B 71:144416
30. Butler WH, Zhang XG, Schulthess TC, MacLaren JM (2001) Phys Rev B 63:054416
31. Mathon J, Umerski A (2001) Phys Rev B 63:220403(R)
32. Yuasa S et al. (2004) Nat Mater 3:868
33. Yuasa S et al. (2006) Appl Phys Lett 89:042505
34. Parkin SSP et al. (2004) Nat Mater 3:862
35. Djayaprawira D (2005) Appl Phys Lett 86:092502
36. Mizukami S et al. (2009) J Appl Phys 105:07D306
37. Wang WH et al. (2008) Appl Phys Lett 92:221912
38. Marukame et al. (2007) Appl Phys Lett 90:012508
39. Sukegawa H et al. (2009) Phys Rev B 79:184418
40. Matsumoto R et al. (2007) Appl Phys Lett 90:252506
41. Miura Y, Nagao K, Shirai M (2004) Phys Rev B 69:144413
42. Bagayoko D, Ziegler A, Callaway J (1983) Phys Rev B 27:7046
43. Moodera JS, Nowak J, van de Veerdonk RJM (1998) Phys Rev Lett 80:2941
44. Wang WH et al. (2010) Phys Rev B 81:140402(R)
45. Wang WH et al. (2010) Phys Rev B 82:092404
46. Furubayashi T et al. (2008) Appl Phys Lett 93:122507
47. Nakatani T et al. (2010) Appl Phys Lett 96:212501
48. Valet T, Fert A (1993) Phys Rev B 48:7099
49. Sakuraba Y et al. (2010) Phys Rev B 82:094444
50. Sukegawa H et al. (2010) Appl Phys Lett 96:042508
51. Lehndorff R et al. (2007) Phys Rev B 76:214420
52. Sharrock MP (1999) IEEE Trans Magn 35:4414
53. Pakala M et al. (2005) J Appl Phys 98:056107
54. Morota M et al. (2008) J Appl Phys 103:07A707
55. Braganca et al. (2005) Appl Phys Lett 87:112507

Chapter 15
Transport Properties of $Co_2(Mn, Fe)Si$ Thin Films

Horst Schneider, Enrique Vilanova Vidal, and Gerhard Jakob

Abstract Thin Heusler films with the composition $Co_2Mn_{1-x}Fe_xSi$ were grown by both sputter and pulsed laser deposition. The samples show a high degree of structural order and very good magnetic properties. The availability of thin film samples on dielectric substrates allowed the systematic investigation of their electronic properties by transport experiments. The normal Hall effect shows a transition from a hole-like charge transport in Co_2MnSi to an electron-like transport in Co_2FeSi. This is in agreement with calculations, which predict that the substitution of Mn by Fe leads to a band filling and a shift of the Fermi energy. Furthermore, the behavior of the anomalous Hall effect was studied. It is the sum of two opposing mechanisms: an intrinsic contribution, caused by the topology of the Fermi surface and a temperature dependent impurity scattering.

15.1 Introduction

A major problem of magnetic tunneling junctions with Heusler electrodes is the strong temperature dependence of the magnetoresistive effect [1]. For Co_2MnSi and Co_2FeSi electrodes this could be caused by the position of the Fermi energy E_F close to the edge of the half-metallic gap [2]. This allows thermally induced breaking of the half-metallicity [1]. The situation might be improved by the use of an intermediate compound with the stoichiometry $Co_2Mn_{1-x}Fe_xSi$ (CFMS). Doping of Co_2MnSi with Fe causes a shift of E_F across the gap, while the electronic band structure remains almost unchanged. For $Co_2Mn_{0.5}Fe_{0.5}Si$, the Fermi level is expected in the center of the gap, which promises the best temperature stability [2].

The alloy series CMFS is also interesting from a fundamental point of view. Bulk Co_2FeSi samples with a saturation magnetization of 6 μ_B/fu have been reported [3]. However, ab initio band structure calculations fail to reproduce this value [4]. Instead, an LDA+U formalism is required [3]. On the other hand, for the description of Co_2MnSi no Coulomb repulsion is required. This raises the question of the im-

H. Schneider · E. Vilanova Vidal · G. Jakob (✉)
Institut für Physik, Johannes Gutenberg – Universität, 55099 Mainz, Germany
e-mail: jakob@uni-mainz.de

C. Felser, G.H. Fecher (eds.), *Spintronics*, DOI 10.1007/978-90-481-3832-6_15,

portance of electron correlations in Heusler alloys and of the validity of the various band structure calculations.

The fabrication of tunneling junctions is an intricate process and unwanted extrinsic contributions can easily be introduced [5, 6]. Hall experiments on the other hand are comparatively simple and have already been applied to Heusler alloys [7–9]. In CMFS, the increase of the Fe content results in a filling of the majority states at the half-metallic gap. The accompanying increase of E_F results in a decrease of their density of states at the Fermi edge [2]. This signifies a change of the Fermi surface topology, which will affect the normal Hall effect [10].

In order to perform a quantitative analysis based on semi-classical transport theory the anisotropic scattering time, the Fermi velocity, and the effective mass tensor have to be known. Parameters obtained from band structure calculations [11] or experimental data [12] could reproduce the experimental Hall coefficients. The anomalous Hall voltage depends on the sign of the charge carriers as well. A corresponding change is therefore expected. In this work, the experimental data were compared with calculated band structures. They were obtained within the LSDA+DMFT framework, which takes into account dynamical correlations in addition to the static correlations modeled by LSDA+U [13].

Since the Hall voltage is inversely proportional to the thickness of the measured sample, transport measurements were performed on thin films with a thickness below 100 nm. The deposition processes as well as structural and magnetic properties are illuminated in the following sections. Afterwards, the transport data are presented and discussed.

15.2 Deposition and Structure

One set of Co_2FeSi samples was deposited by RF magnetron sputtering from stoichiometric targets directly on $Al_2O_3(11\bar{2}0)$ and MgO (100) substrates [14]. For both substrates, best results with respect to crystallinity were obtained at a deposition temperature of 700 °C. After deposition, the films were covered with an Al protection layer.

As a second growth technique, pulsed laser deposition was employed [15]. Its advantages are superior vacuum conditions as well as a higher flexibility. The latter aspect was put to use in the deposition of CMFS [16]. The ternary compounds Co_2MnSi and Co_2FeSi were deposited from stoichiometric targets. Films with quaternary stoichiometry were grown by ablation from a target consisting of one half-disk of Co_2MnSi and Co_2FeSi each. The laser pulses and the stepper motors of the target carousel were synchronized to achieve the desired composition. The high deposition temperature ensures a high interdiffusion between the two target materials [17]. XMCD spectra of such films point toward complete intermixing [18]. The greatest disadvantage of pulsed laser deposition is the formation of melt droplets. Fortunately, this is of little consequence for transport measurements: a notable influence of the laser droplets requires a surface coverage close to the percolation limit, whereas for the investigated films droplets it is about 1 %.

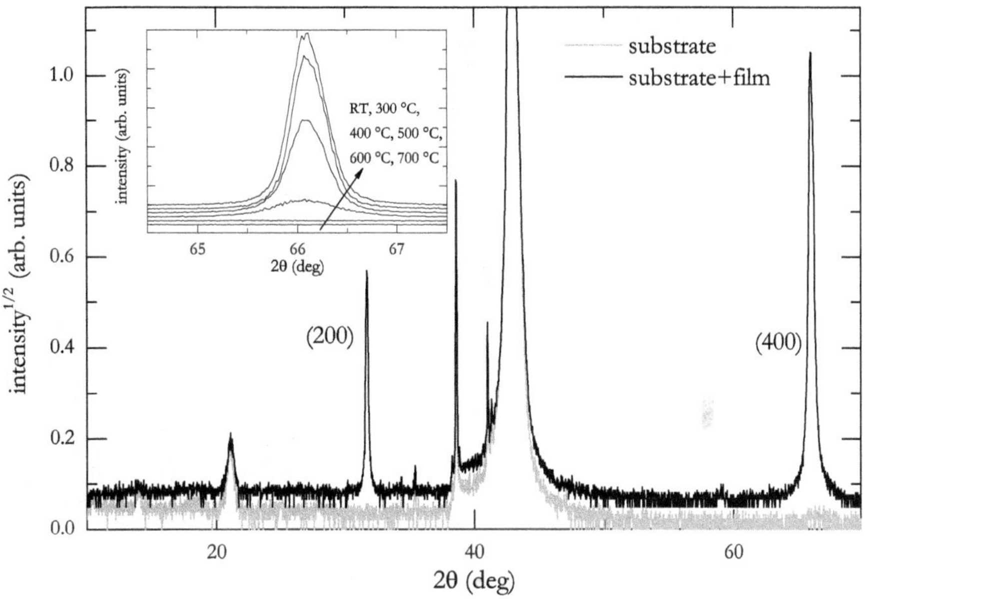

Fig. 15.1 Bragg scans of a Co_2FeSi/MgO sample sputter deposited at a substrate temperature of 700 °C and an uncoated substrate. The *inset* shows θ–2θ scans of the (400) reflection of films deposited at various substrate temperatures. The graphs are shifted for better visibility

The structure of the films was analyzed by X-ray and electron diffraction. Figure 15.1 shows a θ–2θ scan of a Co_2FeSi film sputter deposited on MgO at 700 °C. It shows (100)-oriented growth, no impurity phases are detectable. The corresponding ω-scan of the (400) reflection has a FWHM of 0.3°, which evidences a very good out-of-plane growth. Comparative scans of the in plane film and substrate reflections reveal epitaxial growth with $(010)_{\text{film}} \parallel (011)_{\text{sub}}$. Despite the rather large lattice mismatch between substrate and film, no epitaxial strain could be detected. Transmission electron micrographs show that the crystal structure relaxes within a few monolayers [19]. It can be seen in the inset of Fig. 15.1 that a decrease of the substrate temperatures causes a broadening of the scattered reflections. This structural change coincides with a change of the surface morphology. The surface of high-temperature films features deep trenches. At reduced deposition temperatures the surface flattens. This indicates that low temperature deposition favors nanocrystalline rather than single crystalline growth. For films deposited on Al_2O_3, (110) oriented growth is observed. In these films, several epitaxial domains are present. The preferred orientation is $(1\bar{1}0)_{\text{film}} \parallel (0001)_{\text{sub}}$. It accounts for more than 90 % of the scattered intensity.

The presence of the {111} and {311} reflections reveals film growth in the ordered $L2_1$ structure. Similar to standard Rietveld refinements of powder patterns, the investigation of the relative peak intensities allows an estimation of the atomic disorder. Measured integrated intensities were compared with calculated values. The experimental intensities are compatible with a Co–Si disorder or a surplus of Fe on

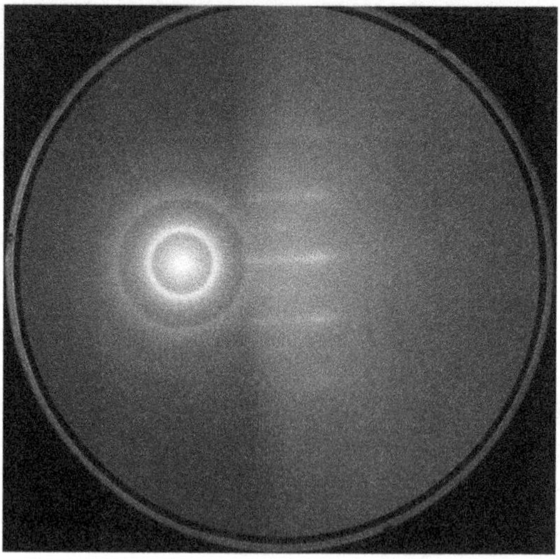

Fig. 15.2 RHEED pattern of a $Co_2FeSi/Cr/MgO$ film after deposition at room temperature

Si sites of the order of 10 %. The latter conjecture was confirmed with NMR spectroscopy [20]. Due to the similar form factors of Fe and Co, disorder between these sites cannot be assessed in standard X-ray analysis.

Similar results were obtained for laser deposited samples over the whole CMFS series: epitaxial films with a rough surface at high substrate temperatures and smooth nanocrystalline samples at room temperature. If an additional Cr or Fe buffer is deposited, epitaxial, single-phase growth can be realized also by room temperature deposition. This is evidenced by RHEED patterns as presented in Fig. 15.2. However, these samples do not grow in the $L2_1$, but the B2 structure. After annealing above 300 °C, the superstructure reflections can be detected. Their maximum relative intensity is reached between 400 and 500 °C. Above this temperature, the scattered intensity of all reflections diminishes. This decrease is the result of interdiffusion between film and buffer layer [21, 22].

15.3 Magnetism

The bulk magnetic properties were investigated with SQUID and vibrating sample magnetometry. The results are presented in Figs. 15.3 and 15.4. The saturation magnetization of sputtered Co_2FeSi films can be extrapolated to 5.0 μ_B/f.u. at 0 K. Hysteresis loops for different film orientations with respect to the magnetic field reveal an in plane uniaxial anisotropy for the (110)-oriented Co_2FeSi/Al_2O_3 samples. The difference in coercitivity of Co_2FeSi/MgO on the other hand is smaller than 1 % (see inset of Fig. 15.3). This unusual result was confirmed by Kerr magnetometry [23]. The dependence of the magnetic moment on sample temperature is shown in Fig. 15.3. It exhibits a $T^{3/2}$ behavior, in agreement with the traditional spin wave description.

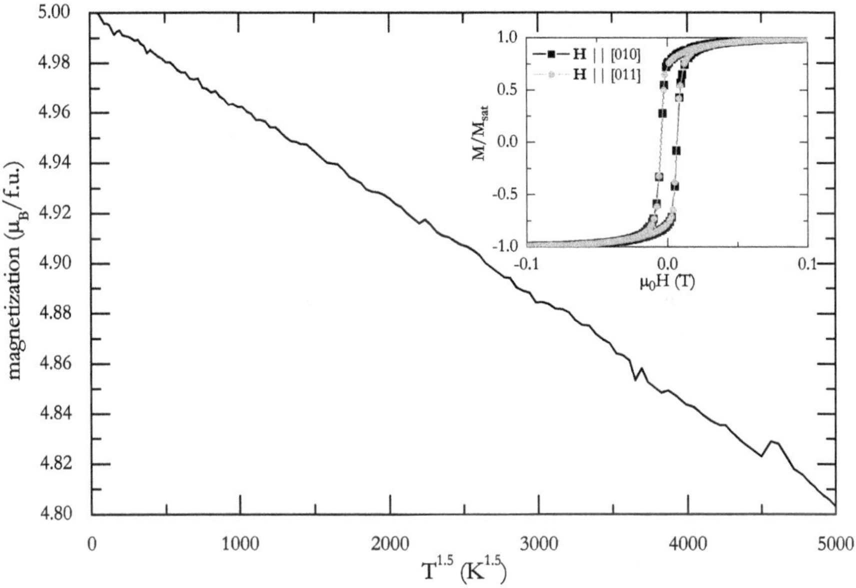

Fig. 15.3 Temperature dependence of the saturation magnetization of a sputter deposited Co_2FeSi/MgO film. The *inset* shows the region of interest for different film orientations with respect to the external field

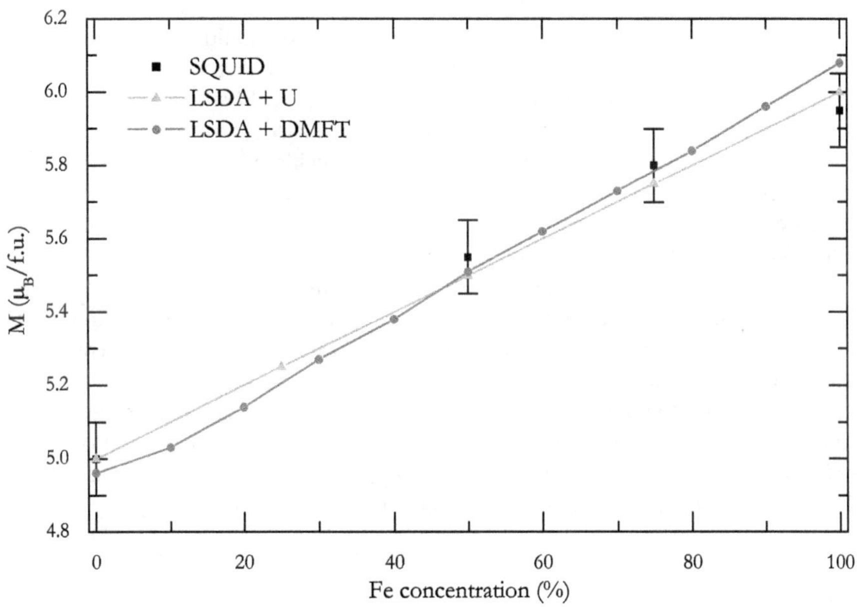

Fig. 15.4 Comparison of CMFS thin film magnetometry data obtained at 5 K and theoretical values. The LSDA+U values were taken from Ref. [2] and the LDSA+DMFT points from Ref. [13]

A higher atomic order of the laser deposited films has a clear effect on the sample magnetization. Figure 15.4 shows that the magnetization of Co_2FeSi lies close to the theoretically expected value of 6 μ_B/f.u. Also for the other CMFS samples the dependence of the saturation magnetization on the Fe/Mn ratio is in good agreement with the general Slater–Pauling rule for the half-metallic Heusler compounds as well as with particular band structure calculations taking into account electron correlations [2, 4, 13]. For samples deposited on metallic buffer layers, it can be seen that their magnetization is correlated with their crystalline quality. For as-deposited films, the saturation magnetization is reduced by about 10–15 % compared to the films annealed at 400–500 °C. For these temperatures, the magnetization assumes also values consistent with the Slater–Pauling rule. The drop of crystal quality at higher annealing temperatures is mirrored in a rapid drop of the magnetization [19].

15.4 Transport Properties

Despite the rough topology of the epitaxial films, they are electrically conducting down to a thickness of 5 nm. Photolithographic sample patterning was used for the creation of well-defined transport geometries. Subsequently, Ohmic and Hall resistivities were measured using standard DC techniques. To rule out effects from either sample oxidation or a metallic Al cap, transport properties of films with and without protection layer were compared. They showed no significant difference.

The inset of Fig. 15.5 demonstrates that over the accessible temperature range the films show metallic behavior. The data can be fit with a $T^{7/2}$ law below 70 K, at higher temperatures a $T^{1.65}$ dependence is found. These exponents can be the result of more than one scattering mechanism each and should therefore regarded as effective values. For conventional ferromagnets a T^2 behavior from one-magnon processes is expected. Incoherent [24] or s–d scattering [25] should yield different exponents. Because of the absence of minority charge carriers in half-metallic ferromagnets, other scattering mechanisms have to be considered. In a rigid band model a $T^{9/2}$ dependence has been proposed [26]. For a non-rigid band, a T^3 dependence was predicted [27]. However, the above models do rely on simplified assumptions of the Fermi surface, which are not valid in CMFS.

Figure 15.5 shows the magnetoresistive behavior at a sample temperature of 4 K. An anisotropic magnetoresistance effect with respect to the current direction is observed, which implies the presence of spin–orbit coupling in our films. The spontaneous resistivity anisotropy has a small negative value of -0.8 %. This effect is dominating the low field magnetoresistive response. Above magnetic saturation, the resistivity shows a linear dependence on the applied field. In conventional ferromagnets this decrease is attributed to the suppression of spin-disorder scattering [28]. This signifies an incomplete spin polarization in these films.

15.5 Hall Effect

Figure 15.6 shows Hall resistivities for CMFS measured at various temperatures. The data show the behavior typical of ferromagnets: The increase of the sample

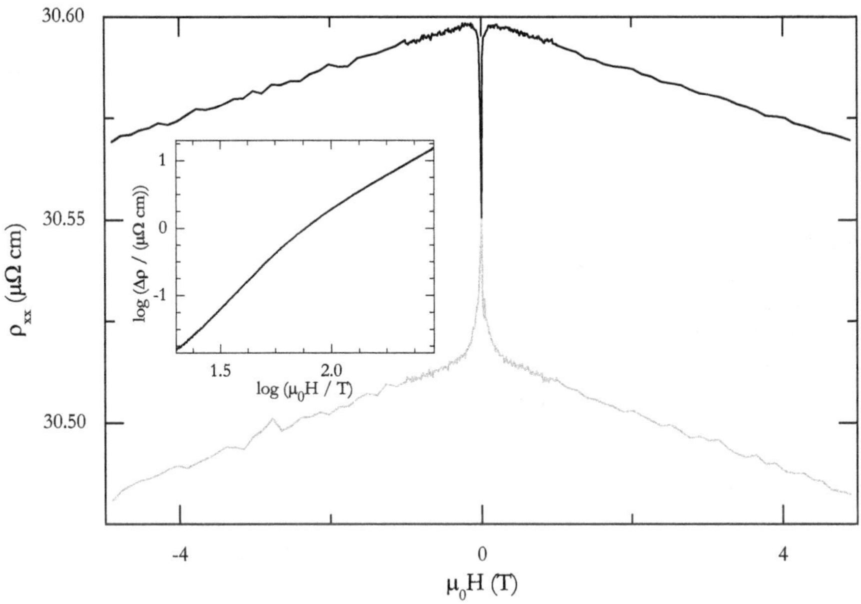

Fig. 15.5 Magnetoresistance of Co_2FeSi at a temperature of 4 K. The *inset* shows the temperature dependent zero-field resistivity

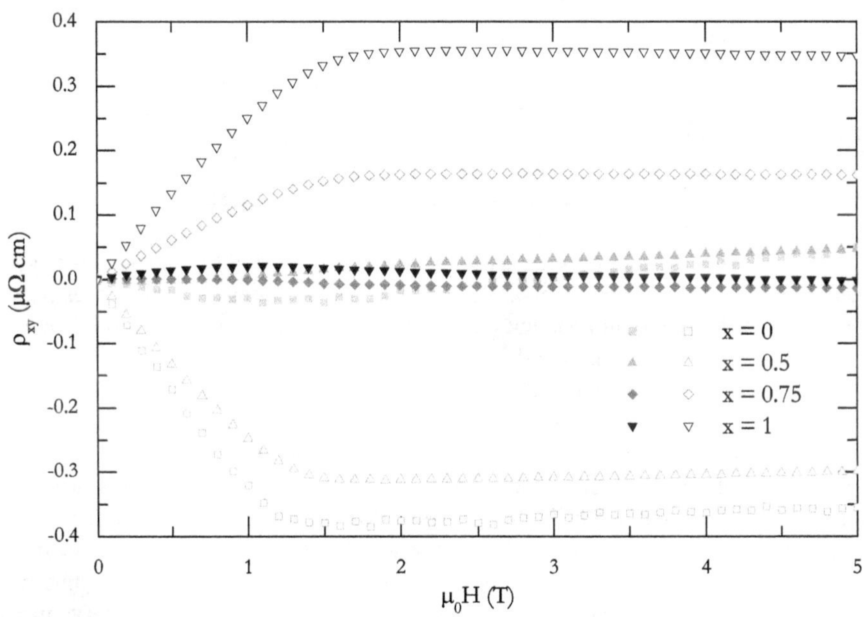

Fig. 15.6 Hall resistivity of laser ablated CMFS Heusler films. *Open* (*filled*) *symbols* represent data measured at 300 K (5 K) for the respective Fe concentrations *x*

magnetization causes a steep change due to the anomalous contribution. At higher field values the normal Hall effect determines the field dependence. This is summed up by the formula $\rho_{xy}(B, M) = R_0 B + R_s \mu_0 M$ with the normal and anomalous Hall coefficients R_0 and R_s. At a sample temperature of 5 K, the normal effect gives rise to a slope of $+5.6 \cdot 10^{-3}$ μΩ cm/T in Co_2MnSi. Increasing the Fe fraction causes a sign reversal of this value. It reaches $-1.1 \cdot 10^{-3}$ μΩ cm/T for Co_2FeSi.

In a simple Fermi sphere model these values correspond to 5 holes/f.u. for Co_2MnSi and 25 electrons/f.u. for Co_2FeSi. These high apparent charge carrier densities are a sign of a partially compensated Fermi surface. In order to obtain more meaningful results the multi-sheeted structure of the Fermi surface has to be taken into account [29].

For this purpose the electronic structures presented in Fig. 15.7 were computed. They were generated with the relativistic full-potential Green function (SPR-KKR) method [13, 30]. As the anisotropic scattering time is generally not known, a full a priori treatment is not easily possible. Furthermore, the calculation of the effective mass tensor requires a much higher density of k-space points than used in the calculations here. Due to computational limits, the density cannot be improved at present. Therefore, in the following the respective Fermi surface properties are discussed qualitatively.

Figure 15.7(a) shows the calculations for Co_2MnSi. It can be seen that the majority bands are unoccupied near Γ. With increasing k-values they cross E_F and are occupied near M. Therefore, the majority states are hole-like with respect to their Fermi surface topology. Looking at the minority states, it is eye-catching that according to this calculation there exists a non-vanishing minority contribution to the Fermi surface, because the respective valence band states are not completely filled. Therefore, Co_2MnSi is not half-metallic considering transport. But since the total number of holes in the pockets at the Γ point is small, the deviation from the Slater–Pauling rule is negligible. For the discussion of the Hall effect the important point is that also the minority charge carriers are hole-like in their topology. In total, contributions to the Hall effect in a multiband model should result in a positive sign of the Hall effect as observed.

The electronic structure of Co_2FeSi presented in Fig. 15.7(c) shows two obvious differences to Co_2MnSi. First, the minority carriers contribute no longer near the center of the Brillouin zone but close to the M point. The Fermi energy is now located at the bottom of the conduction bands of the minority states. Thus, small electron pockets are formed at the M point. And second, the shape of the Fermi surface of the majority carriers has changed. As pointed out in the introduction, the increased electron density of Co_2FeSi is mainly accounted for by the majority carriers, and the hole-like Fermi surface shrinks correspondingly. This reduction process is not caused by a uniform filling of all bands. Mainly two bands take up the electrons and form a circular hole-like Fermi surface. The third majority carrier sheet is more structured. Topologically, the contribution of this band is also hole-like, but locally the curvature shows strong variations between negative and positive curvature. The effective mass thus reverses sign correspondingly. Since both the anisotropic scattering time and the exact values of the effective mass tensor are unknown, this

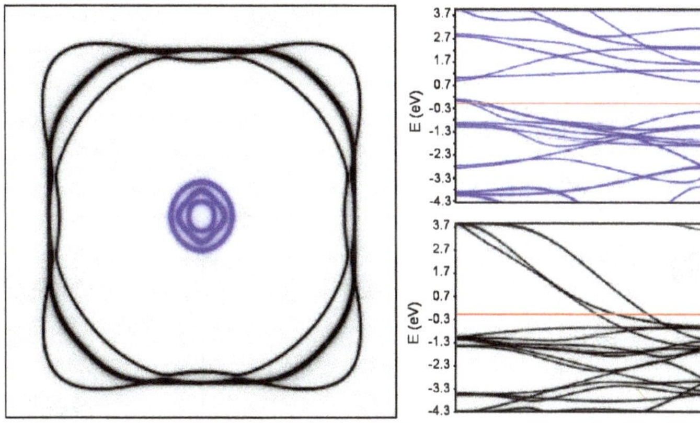

(a) Co_2MnSi

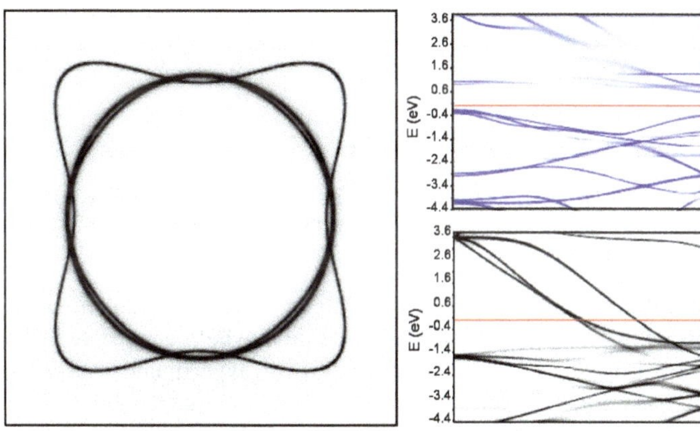

(b) $Co_2Mn_{0.5}Fe_{0.5}Si$

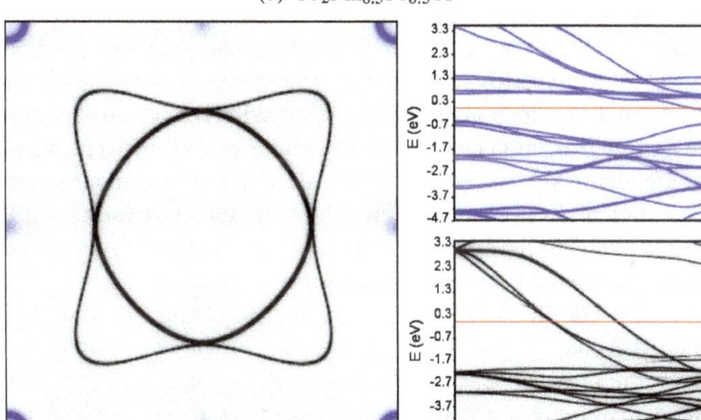

(c) Co_2FeSi

Fig. 15.7 Spin-resolved band structure of $Co_2Mn_{1-x}Fe_xSi$. *Left*: Intersections of the Fermi surface with the $(k_x, k_y, 0)$ plane. *Right*: Dispersion relations in the [110] direction. *Blue lines* show minority states, *black lines* refer to majority states

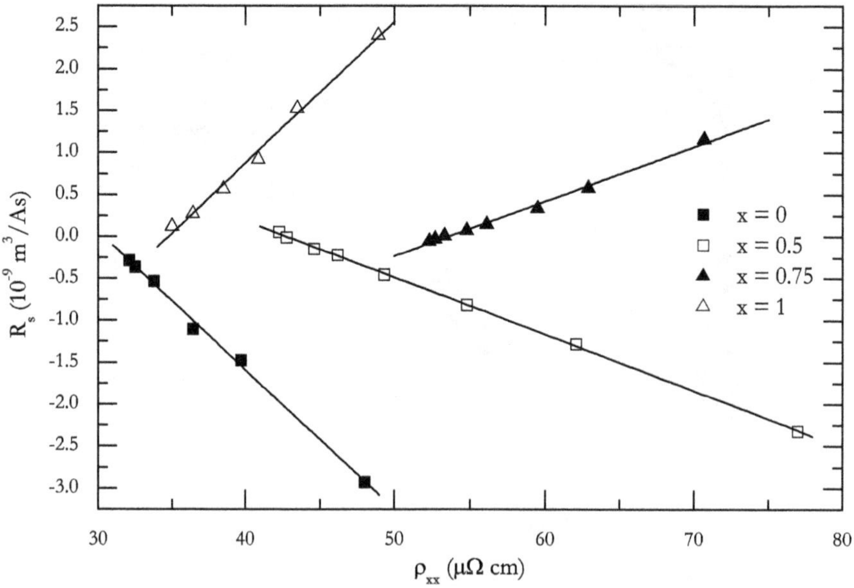

Fig. 15.8 Anomalous Hall coefficients of laser ablated $Co_2Fe_xMn_{1-x}Si$ films for the respective Fe concentrations x. The *lines* represent linear fits of the data

contribution could be electron-like in the Hall measurements. The minority carriers are electrons anyway, therefore the Hall effect in Co_2FeSi is expected to differ significantly from Co_2MnSi. This is confirmed by the experimental observation of an electron-like behavior.

$Co_2Fe_{0.5}Mn_{0.5}Si$ is located in the center of the half-metallic gap and no minority spin contributions are present in Fig. 15.7(b). The absence of minority carriers and the intermediate shape of the Fermi surface should lead to an intermediate normal Hall effect as is found experimentally in Fig. 15.6.

The sign of the anomalous Hall effect is reversed correspondingly. In principle, this is to be expected from symmetry considerations, if all the relevant scattering mechanisms are identical. To understand the nature of the anomalous scattering, the anomalous coefficient R_s is plotted against the longitudinal resistivity in Fig. 15.8. A linear relationship is found for all samples, indicating skew scattering as the relevant temperature dependent scattering mechanism [31].

Apart from this contribution, a non-vanishing low temperature contribution is present, which also changes sign from negative in Co_2MnSi to positive in Co_2FeSi. As in the temperature dependence only skew scattering is present, the low temperature anomalous Hall effect is possibly intrinsic in character. Such a contribution was shown to occur if spin–orbit coupling lifts the degeneracy of band crossings [32, 33]. On such positions in k-space large Berry curvatures occur and nonperturbative contributions to the Hall conductivity arise. Such band crossings are present at the Fermi edge in Fig. 15.7. In the paper by Onoda et al. [33] band crossings between a spin up and a spin down band are discussed. In a half-metal there should exist only one spin direction and therefore no intrinsic Hall effect should be visible. However, the spin–orbit coupling present in a fully relativistic treatment causes a small mixing of spin up and spin down channels. The resulting breaking of the time reversal symmetry can lead to a nonzero intrinsic Hall conductivity also in $Co_2Mn_{0.5}Fe_{0.5}Si$ [32].

15.6 Summary

$Co_2Mn_{1-x}Fe_xSi$ films have successfully been fabricated by sputtering and pulsed laser deposition. Large fractions of these samples grow in the ordered $L2_1$ structure. An iron surplus is present especially in the sputter deposited samples. It causes a reduction of the saturation magnetization. The laser deposited films show magnetization values compatible with the Slater–Pauling rule. The magnetoresistive data reveal the presence of an anisotropic magnetoresistive effect as well as the presence of spin-flip scattering in zero field even at low temperatures.

The flexibility of pulsed laser deposition allowed the systematic variation of the band structure of the investigated Heusler thin films by varying the Fe/Mn ratio. To interpret the experimental data, high accuracy band structure calculations were performed. A shift of the Fermi energy relative to the minority spin conduction band and an evolution from a hole-like transport to electron-like charge carriers was predicted. An observed sign reversal of the normal Hall coefficient is consistent with these calculations. In addition, the anomalous Hall effect shows corresponding changes and was found to be a superposition of an extrinsic skew scattering mechanism and an intrinsic contribution.

Acknowledgements This work was financially supported by the Deutsche Forschungsgemeinschaft (project P2 in research unit FOR 559).

References

1. Sakuraba Y, Hattori M, Oogane M, Ando Y, Kato H, Sakuma A, Miyazaki T, Kubota H (2006) Appl Phys Lett 88:192508
2. Balke B, Fecher GH, Kandpal HC, Felser C, Kobayashi K, Ikenaga E, Kim JJ, Ueda S (2006) Phys Rev B 74:104405
3. Wurmehl S, Fecher GH, Kandpal HC, Ksenofontov V, Felser C, Lin HJ, Morais J (2005) Phys Rev B 72:184434
4. Galanakis I, Dederichs PH, Papanikolaou N (2002) Phys Rev B 66:174429
5. Moodera JS, Kinder LR, Wong TM, Meservey R (1995) Phys Rev Lett 74:3273
6. Song JO, Lee SR, Shin HJ (2005) Curr Appl Phys 7:18
7. Hordequin C, Risouiu D, Ranna L, Pierre J (2000) Eur Phys J B 16:287
8. Husmann A, Singh LJ (2006) Phys Rev B 73:172417
9. Gofryk K, Kaczorowski D, Plackowski T, Mucha J, Leithe-Jasper A, Schnelle W, Yu G (2007) Phys Rev B 75:224426
10. Hurd CM (1972) The Hall effect in metals and alloys. Plenum, New York
11. Butler WH (1984) Phys Rev B 29:4224

12. Evtushinsky DV, Kordyuk AA, Zabolotnyy VB, Inosov DS, Büchner B, Berger H, Patthey L, Follath R, Borisenko SV (2008) Phys Rev Lett 100:236402
13. Chadov S, Fecher GH, Felser C, Minár J, Braun J, Ebert H (2009) J Phys D, Appl Phys 42:084002
14. Schneider H, Jakob G, Kallmayer M, Elmers HJ, Cinchetti M, Balke B, Wurmehl S, Felser C, Aeschlimann M, Adrian H (2006) Phys Rev B 74:174426
15. Schneider H, Vilanova E, Balke B, Felser C, Jakob G (2009) J Phys D, Appl Phys 42:084012
16. Schneider H, Vilanova Vidal E, Chadov S, Fecher GH, Felser C, Jakob G (2010) J Magn Magn Mater 322:579
17. Kallmayer M, Pörsch P, Eichhorn T, Schneider H, Jenkins CA, Jakob G, Elmers HJ (2009) J Phys D, Appl Phys 42:084008
18. Kallmayer M, Klaer P, Schneider H, Arbelo Jorge E, Herbort C, Jakob G, Jourdan M, Elmers HJ (2009) Phys Rev B 80:020406(R)
19. Schneider H, Herbort C, Jakob G, Adrian H, Wurmehl S, Felser C (2007) J Phys D, Appl Phys 40:1548
20. Wurmehl S, Kohlhepp JT, Swagton HJM, Koopmans B, Blum CGF, Ksenofontov V, Schneider H, Jakob G, Ebke D, Reiss G (2009) J Phys D, Appl Phys 42:084017
21. Tezuka N, Okamura S, Miyazaki A, Kikuchi M, Inomata K (2006) J Appl Phys 99:08T314
22. Jourdan M, Conca A, Herbort C, Kallmayer M, Elmers HJ, Adrian H (2007) J Appl Phys 102:093710
23. Hamrle J, Blomeier S, Gaier O, Hillebrands B, Schneider H, Jakob G, Postava K, Felser C (2007) J Phys D, Appl Phys 40:1563
24. Richter R, Wolf M, Goedsche F (1979) Phys Status Solidi B 95:473
25. Goodings DA (1963) Phys Rev 132:542
26. Kubo K, Ohata N (1972) J Phys Soc Jpn 33:21
27. Furukawa N (2000) J Phys Soc Jpn 69:1954
28. Campbell IA, Fert A (1982) Transport properties of ferromagnets. In: Ferromagnetic materials, vol 3. North-Holland, Amsterdam, p 747
29. Ashcroft NW, Mermin ND (1976) Solid state physics. Harcourt Brace, New York
30. Minár J, Chioncel L, Perlov A, Ebert H, Katsnelson MI, Lichtenstein AI (2005) Phys Rev B 72:045125
31. Smit J (1958) Physica 24:39
32. Fang Z, Nagaosa N, Takahashi KS, Asamitsu A, Ogasawara RMT, Yamada M H, Tokura KY, Terakura K (2003) Science 302:92
33. Onoda S, Sugimoto N, Nagaosa N (2006) Phys Rev Lett 97:126602

Chapter 16
Preparation and Investigation of Interfaces of $Co_2Cr_{1-x}Fe_xAl$ Thin Films

Martin Jourdan

Abstract In the framework of spin polarization investigations of Heusler compounds by the measurement of the magnetoresistance (TMR) of tunneling junctions with AlO_x barrier special emphasis is put on the role of the interfaces.

It is demonstrated how an unsuitable morphology can limit the TMR. The barrier morphology could be improved by inserting a Mg layer at the Heusler/barrier interface. Evidence is given that this very thin Mg layer is acting as a seed layer for improving the Al morphology and not as a barrier for coherent MgO tunneling. Thus the Jullière model can be used for evaluating a relatively large spin polarization of 67 % for the B2 ordered Heusler compound $Co_2Cr_{0.6}Fe_{0.4}Al$, which is close to theoretical predictions.

The magnetic surface and bulk moments of the ferromagnetic Heusler compounds $Co_2Cr_{0.6}Fe_{0.4}Al$ and Co_2CrAl were comparatively investigated by X-ray magnetic circular dichroism measurements (collaboration with Prof. H.-J. Elmers). We provide evidence that the magnetism of the film interface region is fully developed with an interface magnetic ordering temperature which can be higher than in the bulk.

16.1 Introduction

Magnetoresistance is the property of a material or composite to change its resistance in an applied magnetic field. In spin valves or magnetic tunneling junctions the magnetoresistance effect is associated with the electronic band structure of the ferromagnetic electrode materials. A large spin polarization is beneficial for obtaining large tunneling magnetoresistances (TMR). This is required for various application like e.g. magnetic random access memory (MRAM), in which the magnetoresistance effect has to be larger than the variation of the junction resistances on the chip. Conventional ferromagnets like CoFe have a maximum spin polarization of 55 % [1]. Heusler compounds are predicted to show a huge spin polarization [2]. Here the

M. Jourdan (✉)
Institut für Physik, Johannes Gutenberg – Universität, 55099 Mainz, Germany
e-mail: Jourdan@uni-mainz.de

C. Felser, G.H. Fecher (eds.), *Spintronics*, DOI 10.1007/978-90-481-3832-6_16,

Heusler alloy $Co_2Cr_{0.6}Fe_{0.4}Al$ is investigated, which is theoretically predicted to be 100 % spin polarized [3].

At the moment the highest TMR-values are obtained employing epitaxial MgO tunneling barriers [4–6]. In these junctions the magnitude of the tunneling magnetoresistance (TMR) effect is based on the long decay length of one specific state of the conventional ferromagnetic electrodes in the epitaxial MgO tunneling barrier (coherent tunneling) [7–9]. In this case no large total spin polarization is required in order to obtain a large TMR. If instead an amorphous or polycrystalline AlO_x barrier is used, the basic prerequisite for a state selection is not fulfilled, e.g. the symmetry at the barrier is broken and $k_{\parallel}$ is not conserved. In this case the semi-classical Jullière model can be applied [10], which relates the TMR with the total spin polarization of the ferromagnetic electrodes. However, the interface spin polarization can differ from the bulk value in general [11].

Following the latter approach it is expected that Heusler ferromagnets, for which large spin polarizations are predicted (e.g. [3]), produce large TMR-values of junctions with an AlO_x barrier. However, only in the case of $Co_2MnSi/AlO_x/Co_2MnSi$-junctions a huge TMR of 570 % (T = 6 K) was obtained [12]. In combination with epitaxial MgO barriers even larger TMR values were measured employing this Heusler compound [13, 14]. Moderately large TMR values ($\simeq$110 %, 5 K) with AlO_x barrier were obtained with $Co_2FeAl_{0.5}Si_{0.5}$ [15] and Co_2FeSi [16] electrodes. All other Heusler electrodes, including $Co_2Cr_{0.6}Fe_{0.4}Al$ (TMR(5 K) $\simeq$ 72 % with CoFe counter electrode [17]) show smaller magnetoresistances in combinations with AlO_x barriers.

The results presented in this chapter are limited to the compound $Co_2Cr_{0.6}Fe_{0.4}Al$, although Co_2CrAl, Co_2FeAl, and $Co_2FeAl_{0.3}Si_{0.7}$ thin films were prepared and investigated as well (our papers [18, 19]). It is demonstrated how the TMR can be enhanced by an improvement of the AlO_x barrier morphology resulting in a new lower limit of the Jullière spin polarization of $Co_2Cr_{0.6}Fe_{0.4}Al$. The results suggest that more than the few above mentions Heusler compounds have a large spin polarization in agreement with the theoretical predictions, but morphological limitations of the AlO_x barrier hamper the experimental confirmation.

Additionally the influence of atomic disorder and temperature on the TMR of junctions with $Co_2Cr_{0.6}Fe_{0.4}Al$ electrode, as well as results of tunneling spectroscopy are discussed.

16.2 Preparation of $Co_2Cr_{0.6}Fe_{0.4}Al$ Thin Films and Tunneling Junctions

Tunneling junction stacks consist of MgO(100)/MgO(100) buffer /$Co_2Cr_{0.6}Fe_{0.4}$ Al(100)/Mg/AlO_x/$Co_{70}Fe_{30}$/MnFe/Ag. The $Co_2Cr_{0.6}Fe_{0.4}Al$ thin films were radio frequency (rf) sputter deposited at room temperature on MgO(100) substrates with e-beam evaporated MgO(100) buffer layers. The MgO buffer layers are improving the surface quality as compared to the as supplied MgO substrates. Compared to a dc sputtering process rf sputtering results in improved crystallographic order and morphology, probably due to gentle sputter-cleaning of the growing film during deposition. After deposition the thin films were annealed at 550 °C. Very thin rf sputtered Mg films (0.2–0.8 nm) serve as a seed layer for the rf sputtered Al (1.6–2.0 nm), which was subsequently plasma oxidized to form the tunneling barrier. The layer stacks were completed by 4 nm of rf-sputtered Co70Fe30, 15 nm of rf-sputtered MnFe (exchange bias layer), and 17 nm of rf-sputtered Ag (capping layer). Square shaped Mesa structures of size 60 μm × 60 μm were patterned by photolithography and ion beam etching.

16.3 Structural Analysis of $Co_2Cr_{0.6}Fe_{0.4}Al$ Films

The compound $Co_2Cr_{0.6}Fe_{0.4}Al$ belongs to the Heusler-type ($L2_1$ structure) materials for which 100 % spin polarization at the Fermi energy, i.e. half metallic properties are predicted [20–22]. However, considering real samples the effects of impurities and crystal imperfections have to be taken into account [22, 23].

Epitaxial $Co_2Cr_{0.6}Fe_{0.4}Al$ thin films were prepared by rf-sputtering at room temperature on MgO(100) substrates (process described in detail in our paper [24]). After the deposition the samples where annealed in-situ under UHV conditions at temperatures between 400 °C and 700 °C.

Different types of site disorder are possible in $Co_2Cr_{0.6}Fe_{0.4}Al$. Cr/Fe–Al disorder (B2 structure) is most likely due to a very small difference in the total formation energy of the ordered and the disordered structure [23]. However, this type of disorder is predicted to have only a small influence on the spin polarization of $Co_2Cr_{0.6}Fe_{0.4}Al$ [23]. Possible types of disorder which strongly reduce the spin polarization are Co-Cr/Fe disorder (DO_3 structure) or disorder on all sites (A2 structure). The characteristic X-ray reflections for the disorder in $Co_2Cr_{0.6}Fe_{0.4}Al$ are (111) and (200). Both are present in the cases of the fully ordered $L2_1$ structure and for pure Co-Cr/Fe disorder (with slightly different intensity ratio). However, (111) disappears completely if there is full disorder on the Cr/Fe–Al positions, which is always the case for $Co_2Cr_{0.6}Fe_{0.4}Al$. If there is disorder on all positions, (200) disappears as well.

Figure 16.1 (upper panel) shows $\Theta/2\Theta$-scans of $Co_2Cr_{0.6}Fe_{0.4}Al$ thin films annealed at different temperatures obtained in Bragg–Brentano geometry. Scattering at Bragg planes which are parallel to the substrate surface is observed (specular reflections).

With increasing annealing temperature of the samples the intensity ratio $I_{(200)}/I_{(400)}$ is increasing, indicating an increased degree of order on the Co positions (B2 order) as deduced from simulations (*Powdercell* [25]). However, there is almost no contrast between scattering from Co and Cr/Fe and disorder between those elements cannot be detected by standard X-ray diffraction. A continuous structure improvement due to annealing follows from the analysis of the width of the $\Theta/2\Theta$-peaks using the Debye–Scherrer formula. With increasing annealing temperature, the peaks become narrower with a minimum width of the (200)-peak of

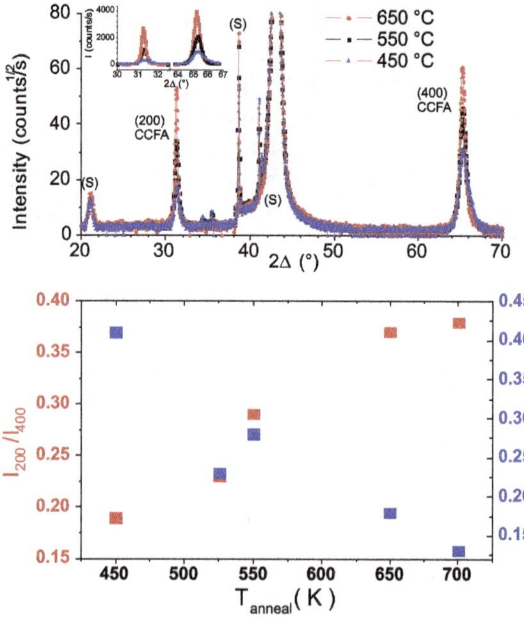

Fig. 16.1 *Upper graph*: X-ray diffraction $\Theta/2\Theta$ scan of three samples annealed at 450 °C (*blue triangles*), 550 °C (*black squares*) and 650 °C (*red dots*). The (200) and (400) peak of $Co_2Cr_{0.6}Fe_{0.4}Al$ and additional peaks (S) due to the substrate are labeled. The *inset* shows the (200) and (400) peaks in a more detailed way. *Lower graph*: Intensity ratio I_{200}/I_{400} (*red*) and width of the X-ray diffraction (200)-peak (*blue*) vs. annealing temperature [26]

0.13° (Fig. 16.1, lower panel). This minimum width corresponds with a correlation length of $\simeq$66 nm, which is a substantial fraction of the thicknesses of the films (80 nm).

The question remains how Co-Cr/Fe disorder (DO_3 structure) can be detected. In the case of $Co_2Cr_{0.6}Fe_{0.4}Al$ ^{57}Fe conversion electron Mößbauer spectroscopy (CEMS) experiments performed by Vadim Ksenofontov on the thin film samples analyzed above by X-ray diffraction proved to be a suitable method. In our paper [26] ^{57}Fe CEMS spectra of $Co_2Cr_{0.6}Fe_{0.4}Al$ thin films annealed at different temperatures are shown.

The spectrum obtained investigating a $Co_2Cr_{0.6}Fe_{0.4}Al$ thin film annealed at 450 °C consists of a clear sextet corresponding to an identical hyperfine field $H_{hf} = 290$Oe at all ^{57} Fe sites. However, upon increasing the annealing temperature a central quasi-paramagnetic peak with increasing fraction appears in the spectra. The site assignment of Fe atoms in Mößbauer spectra is based on the composition of their local surrounding in the B2 structure (Fe/Cr–Al disorder). The magnetic sextet can be attributed to Fe atoms in Fe/Cr and Al positions of the fully ordered $Co_2Cr_{0.6}Fe_{0.4}Al$ Heusler structure with a first coordination sphere comprising eight Co atoms. The central peak stems from Fe in a nonmagnetic surrounding. Such an environment is possible if a ^{57}Fe atom is situated on a Co site which is surrounded by Cr and Al atoms only.

Thus the CEMS experiments in combination with X-ray diffraction demonstrate that annealing of $Co_2Cr_{0.6}Fe_{0.4}Al$ monotonously increases the degree of B2 order in the sense that Al moves to the correct sites and the structural correlation length increases. However, at $T_{annealing} \simeq 550$ °C disorder between Co and Fe (and presumably Cr) atoms start to build up.

Similar conclusions concerning Co–Fe disorder in $Co_2FeAl_{0.5}Si_{0.5}$ thin films were drawn by Inomata et al. based on NMR experiments [27]. Thus it can be assumed that this regarding the spin polarization destructive type of disorder is developing in all Heusler materials with elevated annealing temperatures.

16.4 Interface and Bulk Magnetic Moment of $Co_2Cr_{0.6}Fe_{0.4}Al$ Thin Films

Large tunneling magnetoresistances (TMR) of junctions with ferromagnetic Heusler electrodes and AlO_x or MgO barriers were obtained at low temperatures. However, when increasing the temperature a pronounced reduction of the TMR is always observed (e.g. [12, 15, 16, 28]). This is surprising considering the high magnetic ordering temperatures of the Heusler compounds. A self-evident explanation of the strong temperature dependence of the magnetoresistance of devices based on Heusler thin films is investigated: The possible reduction of the magnetic interface moments compared to the bulk properties of the samples. Such a reduction explains the decrease of the TMR when rising the temperature of junctions with conventional ferromagnetic electrodes [29–31] to 300 K.

We measured the bulk magnetic properties of epitaxial Heusler thin films by SQUID (Superconducting Quantum Interference Device) magnetometry and X-ray Magnetic Circular Dichroism (XMCD, [32, 33]) experiments in transmission mode. The magnetization of the Heusler interface with a thin AlO_x layer was investigated by XMCD experiments in total electron yield (TEY) mode.

In XMCD experiments the sample is irradiated with circular polarized soft X-rays in the direction of the surface normal. It is magnetized first parallel (M^+) then antiparallel (M^-) with respect to this direction in a field of $\mu_0 H = \pm 1.2$ T for each X-ray energy. The XMCD-TM mode is based on the generation of luminescence in the MgO substrate by the X-rays which are transmitted through the Heusler thin film. The relation between luminescence intensity I(E) and X-ray absorption A(E) Heusler film is given by $I \propto e^{-A}$. The XMCD-TEY mode is employed for the evaluation of the $Co_2Cr_{0.6}Fe_{0.4}Al$ interface moments. The measured conversion electron count is proportional to the X-ray absorption. The electron escape depth amounts to 2–3 nm [34, 35].

The XMCD experiments were performed at the German synchrotron light source BESSY II (beamline UE56/1-SGM) in collaboration with and using the experimental set-up of the group of Prof. H.-J. Elmers. $Co_2Cr_{0.6}Fe_{0.4}Al$ and Co_2CrAl thin films annealed at different temperatures were investigated. The X-ray absorption spectra are evaluated by employing the XMCD sum rules [36, 37], which are based on the integration of the $L_{2,3}$ absorption peaks. Parameters of the calculation of the element specific magnetic moments are the numbers n of d-electron holes and a correction factor for the Cr moment due to the overlap in energy of the $L_{2,3}$ peaks [38, 39].

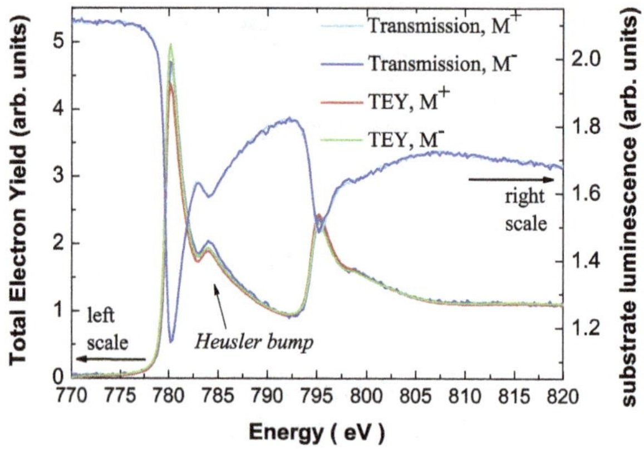

Fig. 16.2 *Left scale*: X-ray energy dependent (Co-edges) XMCD-TEY absorption intensity (*red* and *green curves*) and XMCD-TM absorption intensity (*blue* and *cyan curves*) for up M^+ and down M^- magnetization of Co_2CrAl covered by AlO_x. The shoulder feature labeled *Heusler bump* is characteristic for the Heusler structure. *Right scale*: X-ray energy dependent XMCD-TM luminescence intensity [18]

As an example, in Fig. 16.2 the XMCD-TEY absorption and the XMCD-TM luminescence raw data are shown as well as the derived bulk absorption (at the Co-edges). The most interesting results obtained by XMCD on the Co_2CrAl thin films are an interface magnetic ordering temperature which is higher than the bulk value and an interface magnetization that is larger than the bulk value. This statement is independent of the sum rule evaluation. A clear asymmetry of the X-ray absorption spectra at the Co edge for both magnetization directions of the sample annealed at 550 °C and 700 °C is observed, although the SQUID and XMCD-TM investigations detect the absence of bulk ferromagnetism in these samples at 300 K (Fig. 16.2).

The sum rule derived magnetic moments of the XMCD-TEY and XMCD-TM measurements are shown in Fig. 16.3. The magnitude of the magnetic moments as determined from the XMCD experiments is larger than the total magnetic moment measured by the SQUID magnetometer. This is due to systematic errors in the sum rule evaluation. However, the bulk and interface magnetic moments determined by XMCD can be compared. An interface magnetic moment larger than the bulk value is observed for all Co_2CrAl samples and temperatures. An explanation for this effect is improved interface order on the Co sites induced by an energetically favorable layer termination as motivated by scanning tunneling microscopy investigations (our paper [40]). Alternatively, the interface magnetism could be intrinsically increased as measured for conventional ferromagnets [41, 42].

In the case of $Co_2Cr_{0.6}Fe_{0.4}Al$ a direct comparison of the XMCD-TEY and XMCD-TM analysis reveals, within the experimental errors, identical interface and bulk magnetic moments. This includes all element specific moments (Co, Cr and Fe) as well as the respective fraction of the orbital momentum which is always smaller than 10% of the spin momentum. There is no indication of weakened magnetism at the interface of the Heusler layer.

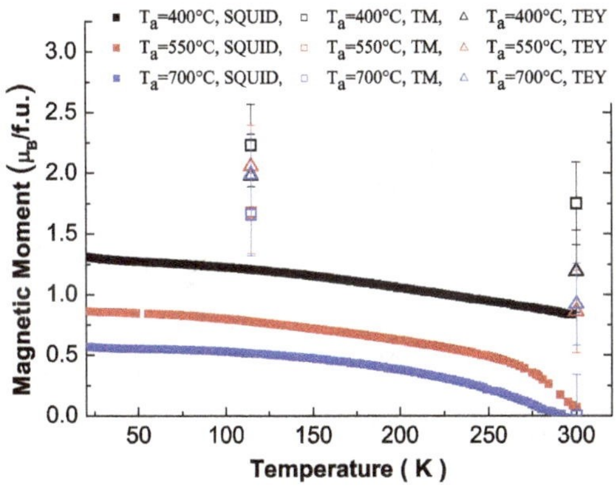

Fig. 16.3 Temperature dependence of the magnetization of Co_2CrAl/Mg(2.5 Å)/AlO_x(18 Å)/ AlO_x(10 Å) thin films annealed at 400 °C (*black symbols*), 550 °C (*red symbols*), and 700 °C (*blue symbols*) measured with a SQUID (*filled symbols*), XMCD-TM (*open squares*), and surface sensitive XMCD-TEY (*open triangles*). The *bars* indicate the maximum error. The sample annealed at 400 °C shows increased roughness resulting in interface oxidation [18]

Thus, evidence is given that the strongly temperature dependent magnetoresistance of tunneling junctions with Heusler electrodes is not induced by weakened interface magnetism in these compounds.

16.5 Structural Order, Morphology and Tunneling Magnetoresistance

An important process parameter in tunneling junction preparation is the annealing temperature of the Heusler base electrode. This electrode has to be deposited at room temperature in order to obtain a smooth surface which is a prerequisite for the preparation of a thin and defect free tunneling barrier. By the annealing process of the $Co_2Cr_{0.6}Fe_{0.4}Al$ thin film the crystallographic order as well as the morphology are changed. The observed dependence is consistent with the change of the bulk crystallographic order of $Co_2Cr_{0.6}Fe_{0.4}Al$ with annealing temperature as discussed in Sect. 16.2. Combining X-ray diffraction and Mößbauer spectroscopy a continuous increase and decrease of the degree of different types of atomic site transposition were identified. However, the abrupt increase of the TMR at annealing temperatures $\simeq$500 °C coincides with a change in the morphology of the Heusler thin films as well. Reflection High Energy Electron Diffraction (RHEED) shows streaks arranged on circular arcs, which is characteristic for scattering at a two-dimensional single crystalline surface. In contrast to these patterns the RHEED images observed at reduced annealing temperatures consist of more and irregularly arranged spots, which indicates scattering at three-dimensional structures (our paper [43]).

Similar morphological changes can be observed investigating the very smooth rf sputtered $Co_2Cr_{0.6}Fe_{0.4}Al$ thin films by in-situ scanning tunneling microscopy.

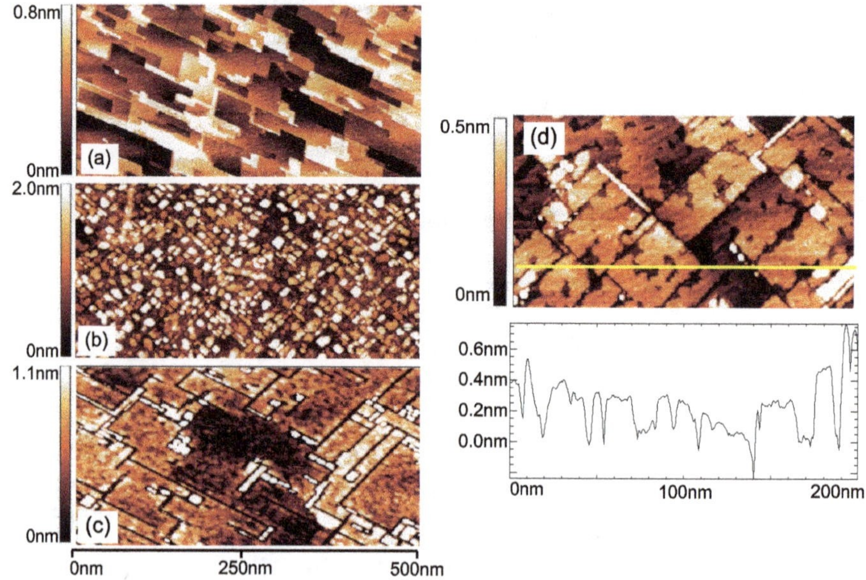

Fig. 16.4 In-situ STM images. **(a)** $Co_2Cr_{0.6}Fe_{0.4}Al$ thin film after annealing; **(b)** 18 Å Al deposited directly on a $Co_2Cr_{0.6}Fe_{0.4}Al$ surface; **(c)** 18 Å of Al on 2.5 Å Mg on a $Co_2Cr_{0.6}Fe_{0.4}Al$ thin film; **(d)** 2.5 Å Mg seed layer on a $Co_2Cr_{0.6}Fe_{0.4}Al$ surface (*height profile below*) [24]

Since the morphology of the Heusler base electrode strongly influences the morphology of the tunneling barrier it has a major influence on the magnetoresistance of the tunneling junction.

16.6 TMR Enhancement and Barrier Optimization

An AlO_x tunneling barrier which is over or under oxidized or has pin holes can strongly reduce the TMR. Thus by applying the Jullière model for the evaluation of the spin polarization from the TMR, only a lower limit can be deduced. For the estimation of how close the deduced value is to the ideal case, a careful characterization of the barrier morphology is necessary.

Figure 16.4(b) shows an Al layer which was deposited directly after the annealing process of a $Co_2Cr_{0.6}Fe_{0.4}Al$ thin film with a morphology as shown in Fig. 16.4(a). Obviously Al has unfavorable wetting properties on the clean Heusler surface resulting in an island morphology which is not suitable for a tunneling barrier.

In order to induce improved wetting properties of the Al layer without degrading the Heusler surface or generating spin flip scattering centers for the tunneling electrons, Mg was selected as a surfactant or seed layer on the $Co_2Cr_{0.6}Fe_{0.4}Al$ surface. Due to its high reactivity this metal is expected to oxidize as well when the Al barrier is oxidized in order to form the tunneling barrier. Figure 16.4(d) shows an in-situ STM image of 2.5 Å of rf sputtered Mg on a $Co_2Cr_{0.6}Fe_{0.4}Al$ surface.

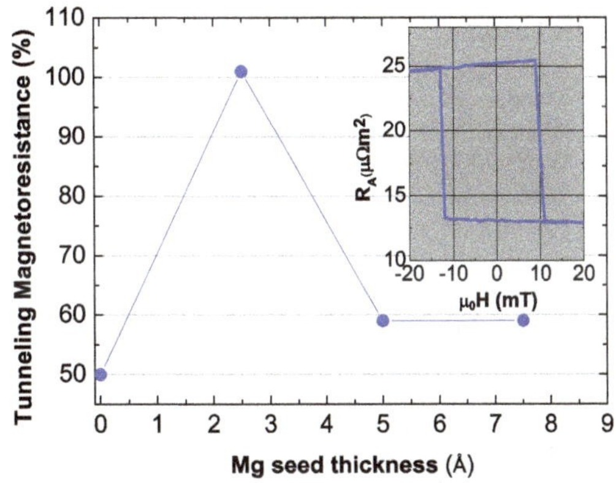

Fig. 16.5 Dependence of the tunneling magnetoresistance (TMR) on the Mg layer thickness of $MgO/Co_2Cr_{0.6}Fe_{0.4}Al/Mg/AlO_x$(18 Å)/CoFe/MnFe/Ag tunneling junctions. The *inset* shows the magnetoresistance loop of the junction with a Mg thickness of 2.5 Å [24]

The morphology shown in the STM image strongly suggests an almost complete wetting of the $Co_2Cr_{0.6}Fe_{0.4}Al$ by the Mg layer. The square-shaped terraces of the Heusler film are covered by an additional layer with curved defects. The depth of the defect structures amounts to 2–3 Å, which is consistent with the thickness of the deposited Mg layer. Figure 16.4(c) shows an in-situ STM image of 18 Å of rf sputtered Al on 2.5 Å of Mg on a $Co_2Cr_{0.6}Fe_{0.4}Al$ thin film. The observed morphology of the Al essentially mirrors the morphology of the Mg layer below. The Al surface is smooth with deep trenches which could be induced by grain boundaries of the Mg seed layer which are related to the morphology of the $Co_2Cr_{0.6}Fe_{0.4}Al$ surface.

The tunneling magnetoresistances at $T = 4$ K of junctions with a 18 Å AlO_x barrier on Mg layers of different thicknesses are shown in Fig. 16.5. The in-situ STM images shown above were obtained from the same junctions. Without Mg seed layer, only junctions with aged (i.e. slightly degraded) $Co_2Cr_{0.6}Fe_{0.4}Al$ layers before Al deposition resulted in a tunneling barrier. Junctions deposited on fresh $Co_2Cr_{0.6}Fe_{0.4}Al$ layers (Fig. 16.4(b)) were electrically shortened.

A clear maximum in the TMR is observed for the thinnest Mg layer, demonstrating that the function of this layer is indeed to act as a seed layer for the Al. It is expected that the Mg is oxidized during the Al oxidation as well. However, the thinnest Mg layer results in the largest TMR effect indicating that the tunneling process is still dominated by the polycrystalline AlO_x barrier and the increase in TMR is due to its improved morphology.

Thus it is reasonable to evaluate the spin polarization of the Heusler compound $Co_2Cr_{0.6}Fe_{0.4}Al$ by applying the Jullière model [10]. Assuming a spin polarization of 50 % for Co70Fe30, the TMR value of 101 % results in a spin polarization of $Co_2Cr_{0.6}Fe_{0.4}Al$ of 67 %, which is larger than all previously reported values for this compound. This value is relatively close to a theoretical prediction of the bulk spin polarization of $\simeq 77$ % for B2 ordered $Co_2Cr_{0.6}Fe_{0.4}Al$ [23].

Acknowledgements Financial support by the German research foundation (DFG-Jo404/4-1), project P11 in research unit 559, is acknowledged.

References

1. Moodera JS, Mathon G (1999) J Magn Magn Mater 200:248
2. Alanakis I (2002) J Phys Condens Matter 14:6329
3. Wurmehl S, Fecher G, Kroth K, Kronast F, Dürr H, Takeda Y, Saitoh Y, Kobayashi K, Lin HJ, Schönhense G, Felser C (2006) J Phys D, Appl Phys 39:803
4. Parkin SSP, Kaiser C, Panchula A, Rice PM, Hughes B, Samant M, Yang SH (2004) Nat Mater 3:862
5. Yuasa S, Nagahama T, Fukushima A, Suzuki Y, Ando K (2004) Nat Mater 3:868
6. Ikeda S, Hayakawa J, Ashizawa Y, Lee YM, Miura K, Hasegawa H, Tsunoda M, Matsukura F, Ohno H (2008) Appl Phys Lett 93:082508
7. Mavropoulos Ph, Papanikolaou N, Dederichs PH (2000) Phys Rev Lett 85:1088
8. Butler WH, Zhang X-G, Schulthess TC, MacLaren JM (2001) Phys Rev B 63:054416
9. Mathon J, Umersky A (2001) Phys Rev B 63:220403R
10. Jullière M (1975) Phys Lett 54A:225
11. Galanakis I (2002) J Phys Condens Matter 14:6329
12. Sakuraba Y, Hattori M, Oogane M, Ando Y, Kato H, Sakuma A, Miyazaki T, Kubota H (2006) Appl Phys Lett 88:192508
13. Ishikawa T, Hakamata S, Matsuda K, Uemura T, Yamamoto M (2008) J Appl Phys 103. 07A919
14. Tsunegi S, Sakuraba Y, Oogane M, Takanashi K, Ando Y (2008) Appl Phys Lett 93:112506
15. Tezuka N, Ikeda N, Miyazaki A, Sugimoto S, Kikuchi M, Inomata K (2006) Appl Phys Lett 89:112514
16. Ebke D, Schmalhorst J, Liu N-N, Thomas A, Reiss G, Hütten A (2006) Appl Phys Lett 89:162506
17. Inomata K, Okamura S, Miyazaki A, Kikuchi M, Tezuka N, Wojcik M, Jedryka E (2006) J Phys D, Appl Phys 39:816
18. Jourdan M, Arbelo Jorge E, Herbort C, Kallmayer M, Klaer P, Elmers HJ (2009) Appl Phys Lett 95:172504
19. Arbelo Jorge E, Jourdan M, Kallmayer M, Klaer P, Elmers HJ (2010) J Phys Conf Ser 200:072006
20. Galanakis I, Dederichs PH, Papanikolaou N (2002) Phys Rev B 66:174429
21. Block T, Felser C, Jakob G, Ensling J, Mühling B, Gütlich P, Cava RJ (2003) J Solid State Chem 176:646
22. Fecher G, Kandpal H, Wurmehl S, Morais J, Lin H-J, Elmers H-J, Schönhense G, Felser C (2005) J Phys Condens Matter 17:7237
23. Miura Y, Nagao K, Shirai M (2004) Phys Rev B 69:144413
24. Herbort C, Arbelo Jorge E, Jourdan M (2009) Appl Phys Lett 94:142504
25. http://www.bam.de/de/service/publikationen/powder_cell.htm or http://www.ccp14.ac.uk/tutorial/powdcell/index.html
26. Ksenofontov V, Herbort C, Jourdan M, Felser C (2008) Appl Phys Lett 92:262501
27. Inomata K, Wojcik M, Jedrika E, Ikeda N, Tezuka T (2008) Phys Rev B 77:214425
28. Marukame T, Ishikawa T, Hakamata S, Matsuda K, Uemura T, Yamamoto M (2007) Appl Phys Lett 90:012508
29. Moodera JS, Nowak J, van de Veerdonk RJM (1998) Phys Rev Lett 80:2941
30. Hagler T, Kinder R, Bayreuter G (2001) J Appl Phys 89:7570
31. Yuan L, Liou SH, Wang D (2006) Phys Rev B 73:134403
32. Schneider K, Schönhense G (2002) Rep Prog Phys 65:R1785
33. Bechtholt PS (2005) In: 36th IFF spring school – magnetism goes nano
34. O'Brien WL, Tonner BP (1994) Phys Rev B 50:2963
35. Regan TJ, Ohldag H, Stamm C, Nolting F, Lüning J, Stöhr J, White RL (2001) Phys Rev B 64:214422
36. Thole BT, Carra P, Sette F, van der Laan G (1992) Phys Rev Lett 68:1943
37. Carra P, Thole BT, Altarelli M, Wang X (1993) Phys Rev Lett 70:694

38. Elmers HJ, Fecher GH, Valdaitsev D, Nepijko SA, Gloskovskii A, Jakob G, Schönhense G, Wurmehl S, Block T, Felser C, Cramm S (2003) Phys Rev B 67:104412
39. Kallmayer M, Conca A, Jourdan M, Schneider H, Jakob G, Balke B, Gloskovskii A, Elmers HJ (2007) J Phys D, Appl Phys 40:1539
40. Herbort C, Arbelo E, Jourdan M (2009) J Phys D, Appl Phys 42:8406
41. Wagner K, Weber N, Elmers H-J, Gradmann U (1997) J Magn Magn Mater 167:21
42. Ney A, Poulopoulos P, Farle M, Baberschke K (2000) Phys Rev B 62:11336
43. Conca A, Jourdan M, Adrian H (2007) J Phys D, Appl Phys 40:1534

Chapter 17
Tunnel Magnetoresistance Effect in Tunnel Junctions with Co_2MnSi Heusler Alloy Electrode and MgO Barrier

Yasuo Ando, Sumito Tsunegi, Yuya Sakuraba, Mikihiko Oogane, Hiroshi Naganuma, and Koki Takanashi

Abstract We demonstrated that a large TMR ratio of 753 % has been observed at 2 K in a MTJ using a Co_2MnSi Heusler alloy electrode and a crystalline MgO tunnel barrier. At room temperature (RT), we also have observed a large TMR ratio of 217 %, which value at RT is much larger than that of MTJs using an amorphous Al-oxide tunnel barrier. However, the temperature dependence of the TMR ratio was still large. In order to improve the interface, we investigated the TMR effect in Co_2MnSi/CoFeB(0–2 nm)/MgO/CoFe MTJs. TMR ratio was enhanced by inserting a thin CoFeB layer at the Co_2MnSi/MgO interface. The MTJ with CoFeB thickness of 0.5 nm exhibited the highest TMR ratio. From the conductance–voltage measurements for the fabricated MTJs, we inferred that the highly spin polarized electron created in Co_2MnSi can conserve the polarization through the 0.5 nm thick FeB layer.

Y. Ando (✉) · S. Tsunegi · M. Oogane · H. Naganuma
Department of Applied Physics, Graduate School of Engineering, Tohoku University, Sendai, Japan
e-mail: ando@mlab.apph.tohoku.ac.jp

M. Oogane
e-mail: oogane@mlab.apph.tohoku.ac.jp

H. Naganuma
e-mail: naganuma@mlab.apph.tohoku.ac.jp

Y. Sakuraba · K. Takanashi
Institute for Materials Research, Tohoku University, Sendai, Japan

Y. Sakuraba
e-mail: y.sakuraba@imr.tohoku.ac.jp

K. Takanashi
e-mail: koki@imr.tohoku.ac.jp

C. Felser, G.H. Fecher (eds.), *Spintronics*, DOI 10.1007/978-90-481-3832-6_17,

17.1 Introduction

Magnetic tunnel junctions (MTJs) are key devices in the spintronics field because MTJs are fundamental devices used for magnetic random access memory (MRAM), magnetic sensors and innovative spintronics devices, such as spin-transistors and microwave generators. High spin polarization materials are necessary to increase tunnel magnetoresistance (TMR) effects and to enhance spintronics device performance. Some full-Heusler alloys (Co_2MnSi, $Co_2Cr_{0.6}Fe_{0.4}Al$, Co_2MnGe, etc.) are the most promising high spin polarization material, because they have been predicted to possess half-metallic electronic structure and have a high Curie temperature [1–3]. Experimentally, some groups have used Heusler alloys as MTJ electrodes with an amorphous Al-oxide barrier [4–8]. We have achieved a TMR ratio of 159 % (at 2 K) in a MTJ with Co_2MnSi/Al-O/CoFe structure [9], and 203 % (at 2 K) in the optimized MTJ with Co_2MnSi/Mg/Al-O/CoFe structure [10]. We also demonstrated a large TMR ratio of 570 % (at 2 K) in a MTJ with Co_2MnSi/Al-O/Co_2MnSi structure [11]. These results proved the half-metallic property of Co_2MnSi experimentally. However, in these MTJs, large temperature dependences of the TMR ratio were a serious problem, resulting in that the TMR ratios at room temperature (RT) reduced below 100 %.

According to Julliere's model [12], for MTJs with an amorphous tunneling barrier the TMR ratio depends only on the density of states (DOS) at the Fermi level of the ferromagnetic electrodes. On the other hand, in MTJs with a (001)-oriented crystalline MgO tunneling barrier and bcc-structured ferromagnetic electrodes, Δ_1-band selective coherent tunneling occurs and enhances the TMR ratio drastically compared with an MTJ with an amorphous barrier [13, 14]. Some groups have reported very large TMR ratios greater than 200 % at RT in the MTJs with Fe, CoFeB electrodes and a MgO barrier [15, 16]. Recently, some groups have attempted to use Heusler alloy electrodes and a crystalline MgO barrier together in the MTJs. Ishikawa et al. observed very large TMR ratio at LT and oscillation of TMR ratio against the MgO barrier thickness [17]; Tezuka et al. observed TMR ratio over 300 % at RT [18, 19]. These results are apparently related to the coherent tunneling process through the crystalline MgO barrier. In addition, Miura et al. suggested theoretically that coherent tunneling can enhance the TMR ratio in a MTJ with Heusler alloy electrodes and an MgO barrier [20].

This chapter presents our recent works on TMR effects in high-quality epitaxially grown Co_2MnSi/MgO/CoFe MTJs. Particularly, we note that the Co_2MnSi/MgO interface structure is important for realizing a large TMR effect, and that the interface structure is sensitive to the electrode fabrication conditions, the preparation method for the MgO barrier, and insertion materials between Co_2MnSi/MgO interface. Furthermore, we discuss the coherent tunneling process and inelastic tunneling process, considering results of the tunneling conductance measurement.

17.2 Tunneling Magnetoresistance in $Co_2MnSi/MgO/Co_{50}Fe_{50}$ Magnetic Tunnel Junctions

17.2.1 Preparation of $Co_2MnSi/MgO/CoFe$ Magnetic Tunnel Junctions

The MTJs with structure of MgO(001)-sub./Cr(40)/Co_2MnSi(30)/MgO(2–2.5)/ $Co_{50}Fe_{50}$(5)/IrMn(10)/Ta(5) (unit : nm) were fabricated [21]. The (001)-oriented epitaxial Co_2MnSi bottom electrode was grown at ambient temperature by inductively coupled plasma (ICP) assisted magnetron sputtering. The deposition pressure was 0.1 Pa and the rate was about 0.03 nm/s. The film was subsequently annealed at 500 °C to reduce site-disorder and create highly oriented film. To obtain the stoichiometric film composition, we used a composition-adjusted Co-Mn-Si alloy sputtering target (Co, 43.7 %; Mn, 27.95 %; Si, 28.35 %). The crystallographic structure and surface morphology of the bottom Co_2MnSi were observed using X-ray diffraction (XRD) with Cu K_α radiation; surface roughness was verified using atomic force microscopy (AFM). We measured magnetization curves using a vibrating sample magnetometer (VSM) and a superconducting quantum interference device (SQUID) magnetometer. Film compositions were examined using inductively coupled plasma (ICP) analysis. The result of the XRD θ–2θ measurement for the bottom electrode showed only the (002) Cr peak and (002) and (004) Co_2MnSi peaks except for the peaks from MgO substrates, indicating perfect (001)-orientation of the Co_2MnSi. Additionally, we confirmed epitaxial growth and highly $L2_1$ order of Co_2MnSi film from the ϕ-scan measurement. The long range order parameters S_{B2} and S_{L2_1} were ca. 1.0 and ca. 0.8, respectively. Results of AFM measurements showed that the surface of the Co_2MnSi was very smooth: R_a and peak–valley $(P–V)$ values were about 0.2 and 2.0 nm, respectively. The magnetization of Co_2MnSi was almost 5 μ_B, as predicted by the Slater–Pauling curve for Heusler alloys. These results show that the prepared Co_2MnSi thin film is suitable for the bottom electrode of the MTJ [22]. The MgO barrier was formed using a direct-magnetron sputtering and electron beam (EB) evaporation system. We describe the MTJs with a sputtered MgO barrier as SP-MTJ and the EB-evaporated MgO barrier as EB-MTJ. The Ar pressure was 0.1 Pa in the MgO sputtering and the deposition rate was about 0.008 nm/s. The pressure during evaporation was about 2×10^{-6} Pa and typical evaporation rate was 0.01 nm/s. The MTJs were patterned into 8×8 to 90×90 μm^2 areas using photolithography and Ar ion etching. Patterned MTJs were annealed in the range of 300 to 485 °C for 1 hour in high vacuum by applying a 1 T magnetic field to improve the crystallinity of the MgO barrier.

17.2.2 Tunneling Magnetoresistance of the $Co_2MnSi/MgO/CoFe$ Magnetic Tunnel Junctions

Magnetoresistance (MR) measurements were performed using the standard dc four-probe method at a bias voltage of 1 mV, For comparison, CoFeB/MgO/CoFeB MTJ

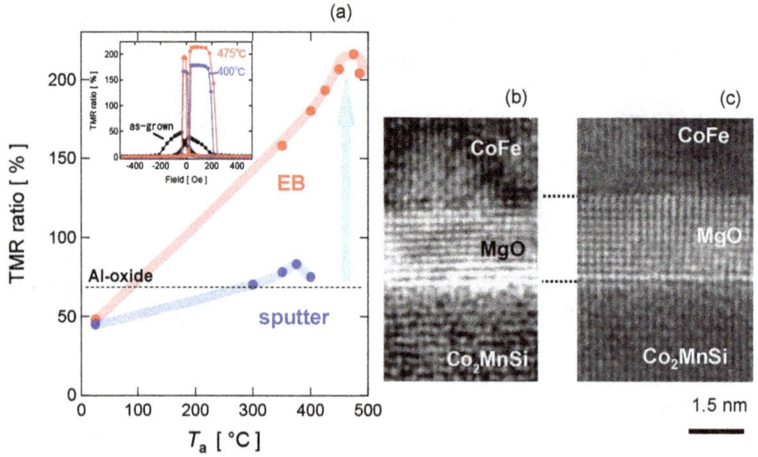

Fig. 17.1 (**a**) Annealing temperature (Ta) dependence of the TMR ratio for the Co_2MnSi/MgO/CoFe-MTJs (*blue*: SP-MTJ, *red*: EB-MTJ). The MR curves of the EB-MTJ as-deposited and annealed at $T_a = 400$ °C and 475 °C are also shown as the *inset*. (**b**) Cross-sectional high-resolution transmission electron micrograph (TEM) images along the $[1\bar{1}0]$ direction of the Co_2MnSi film for the SP-MTJ annealed at 375 °C and (**c**) the EB-MTJ annealed at 475 °C, respectively

was fabricated in the same conditions and both the TMR effect and the differential conductance were investigated. Figure 17.1(a) shows the annealing temperature (T_a) dependence of the TMR ratio at RT for the SP-MTJ and the EB-MTJ. The MR curves of the EB-MTJ are also shown as the inset. The TMR ratio increased with increasing annealing temperature. The improvement of TMR ratio by annealing was explained to result from improvements of both crystallinity and (001)-orientation of the MgO barrier at the electrode/barrier interface [23], while the degradation of TMR by high-temperature annealing was caused by atomic diffusion, especially of Mn atom in the pinning layer [24]. The TMR ratio of the as-deposited EB-MTJ is almost the same as that of the as-deposited SP-MTJ even though the anti-parallel magnetic configuration in the EB-MTJ is poor. This means that the quality of MgO layer in the EB-MTJ should be better than that of the SP-MTJ. For the SP-MTJ, the TMR ratio showed a maximum of 80 % at $T_a = 375$ °C, this TMR ratio was rather smaller than those of the MTJs with CoFeB electrodes and an MgO barrier, reported previously [23]. In addition, the $R \times A$ value is almost 10^8 $\Omega\,\mu m^2$ and larger than the EB-MTJ ($\approx 10^5$ $\Omega\,\mu m^2$) and the previous CoFeB/MgO/CoFeB MTJ ($\approx 10^5$ $\Omega\,\mu m^2$), although the MgO barrier thickness was same [23]. Both the small TMR ratio and the large $R \times A$ value for the SP-MTJ are thought to result from tunneling electron scattering at the MgO barrier interface. On the other hand, the EB-MTJ showed a maximum TMR ratio of 217 % at $T_a = 475$ °C. This TMR ratio is much larger than that of the SP-MTJ; it is comparable to that of the CoFeB/MgO/CoFeB MTJ reported previously.

Figures 17.1(b) and (c) show cross-sectional high-resolution transmission electron micrograph (TEM) images along the $[1\bar{1}0]$ direction of the Co_2MnSi film for

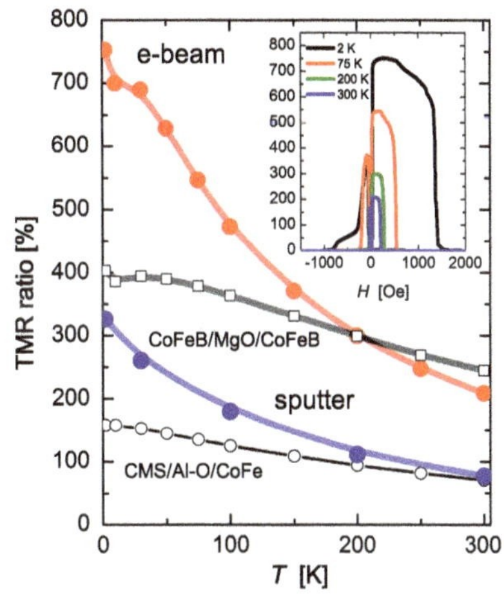

Fig. 17.2 The temperature dependence of the TMR ratio both for the SP-MTJ (*blue circle*) and EB-MTJ (*red circle*) exhibited a maximum TMR ratio at RT. As references, the temperature dependences for both CoFeB/MgO/CoFeB MTJ (*closed black circle*) and Co_2MnSi/Al-oxide/CoFe MTJ (*white circle*) are shown together. The TMR curves measured at various temperatures on the EB-MTJ are also shown as the *inset*

the SP-MTJ annealed at 375 °C and the EB-MTJ annealed at 475 °C, respectively. Both TEM images reveal epitaxial growth from the bottom Cr/Co_2MnSi electrode to the CoFe top electrode and the crystallinity of the MgO barrier is almost same. However, for the SP-MTJ, a disarray structure was observed at the Co_2MnSi/MgO interface. On the other hand, for the EB-MTJ, extremely smooth and abrupt interfaces were formed, as shown in the TEM image. We performed X-ray absorption spectroscopy (XAS) to investigate the Co_2MnSi/MgO interface for both Co_2MnSi/MgO multilayers prepared using direct sputtering and the EB-evaporation. For the Co_2MnSi/MgO layer prepared using sputtering, the shoulder feature just after the main Co L_3 absorption peak that originates from $L2_1$ ordering of the Co_2MnSi [25] was very weak, indicating the disarray pertaining at the Co_2MnSi/MgO interface. Also, the Mn L_3 absorption peak was clearly visible indicating that oxidation of Mn atoms occurred. On the other hand, for the Co_2MnSi/MgO layer prepared using EB-evaporation, there was a clear shoulder of the Co L_3 absorption peak and no additional structure of the Mn absorption spectrum. These indicated that the Co_2MnSi at the MgO interface had an $L2_1$ ordered structure and a clean Co_2MnSi/MgO interface was formed. We note that the XAS results were consistent with the HR-TEM images.

Figure 17.2 shows the temperature dependence of the TMR ratio both for the SP-MTJ and EB-MTJ exhibited a maximum TMR ratio at RT. As references, the temperature dependences for both CoFeB/MgO/CoFeB MTJ and Co_2MnSi/Al-oxide/CoFe MTJ are shown together in Fig. 17.2. The TMR curves measured at various temperatures on the EB-MTJ are also shown as the inset of Fig. 17.2. The TMR ratio drastically increases with decreasing temperature for both MTJs. We observed a large TMR ratio of 753 % for the EB-MTJ at 2 K. The TMR ratio value is larger than those for CoFeB/MgO/CoFeB MTJ (400 %) and for Co_2MnSi/Al-oxide/CoFe MTJ (160 %) prepared under the same fabrication condition. This result reveals that the half-metallicity of the Co_2MnSi can be realized not only in the MTJ with an amorphous Al-oxide barrier, as previously reported, but in the

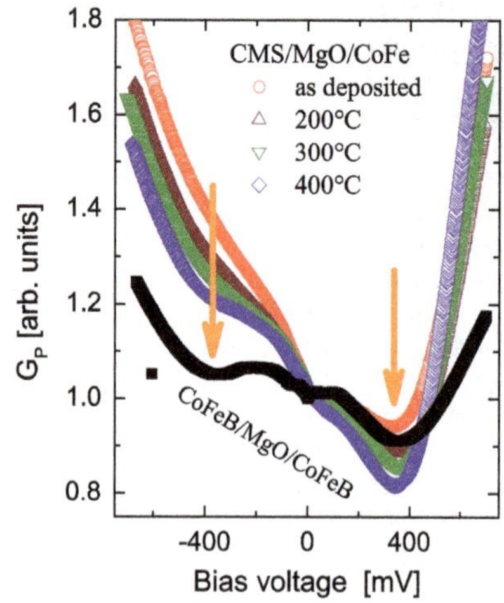

Fig. 17.3 G_P–V curves measured at 10 K for the EB-MTJ annealed at various T_a. For comparison, the G_P curve for the CoFeB/MgO/CoFeB MTJ (*black curve*) is also shown

MTJ with a crystalline MgO barrier. However, temperature dependence of TMR ratio for the $Co_2MnSi/MgO/CoFe$ MTJ is not improved in comparison with that for CoFeB/MgO/CoFeB MTJs. In this experiment, both good crystallinity and high (001)-orientation of the MgO barrier could promote the coherent tunneling process through the Δ_1-band of the Co_2MnSi, and we expected an enhancement of the TMR effect at RT. However, the temperature dependence of TMR even for the MgO barriers was still large.

17.2.3 Tunneling Conductance Characteristics

The bias voltage dependence of the differential tunneling conductance $G(= dI/dV)$ on bias voltage at 10 K and 300 K was measured using an ac lock-in amplifier technique for both parallel (G_P) and anti-parallel (G_{AP}) magnetic configurations to investigate the tunneling process mechanism. A positive bias voltage is defined by electrons tunneling from the lower Co_2MnSi electrode to the upper CoFe electrode.

Figure 17.3 shows the G_P–V curves measured at 10 K for the EB-MTJ annealed at various T_a. For comparison, the G_P curve for the CoFeB/MgO/CoFeB MTJ is also shown. For all MTJs, the shapes of the G_P curves were asymmetric with respect to the sign of bias voltage. The linear slope at the small bias voltage seems to be caused by the difference between the interfacial characteristic of the Co_2MnSi/MgO and MgO/CoFe. The crucial rise of the G_P in the negative bias voltage is suppressed by increasing T_a. In addition, the dip structure observed at ±400 mV is enhanced by increasing T_a; this dip structure resembles that for the CoFeB/Mg/CoFeB MTJ [26, 27] and Fe/MgO/Fe MTJ [28]. We infer that this dip structure is related to the coherent tunneling process through the crystalline MgO barrier and that this coherent tunneling plays an important role in realizing the large TMR effect. The shape of G_P–V curve was almost identical between RT and 10 K.

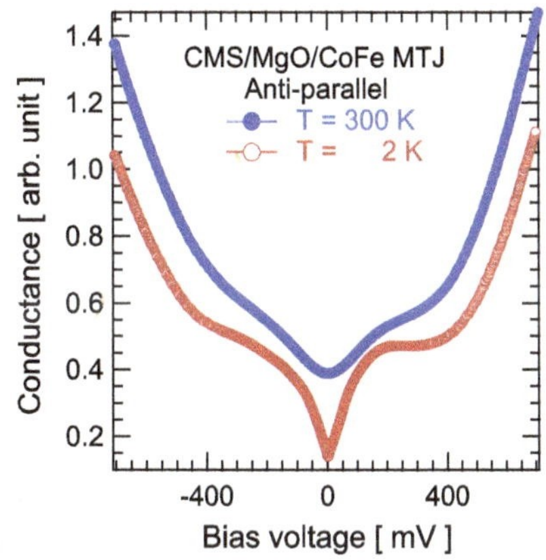

Fig. 17.4 G_{AP}–V curves (at the anti-parallel magnetic configuration) measured at both RT (*blue curve*) and 10 K (*red line*) for the EB-MTJ (Co_2MnSi/MgO/CoFe) annealed at $T_a = 475$ °C

Furthermore, the crucial rise of G_P at the negative bias voltage, which is the possible origin of the large temperature dependence of TMR ratio for the MTJ with Al-oxide barrier [29], seems to be suppressed at both RT and low temperature (LT). Therefore, the dominant origin of the large temperature dependence of TMR ratio for the EB-MTJ is thought to be different from that of the Co_2MnSi/Al-oxide/CoFe MTJ.

Figure 17.4 shows G_{AP}–V curves measured at both RT and 10 K for the EB-MTJ annealed at $T_a = 475$ °C. The G_{AP} at 10 K increased sharply at low bias voltage compared with the G_{AP} at RT. The zero-bias anomaly of G_{AP} originates from inelastic tunneling processes and two possible explanations exist for the large inelastic tunneling probability in the MTJs with Co_2MnSi electrodes; magnon excitations attributable to the low Curie temperature at the Co_2MnSi/MgO interface [30] and magnetic impurity scatterings caused by some Mn and Si oxides [31]. In this MTJ, impurities such as Mn and Si oxides were not found from the XAS measurement [25], this large zero-bias anomaly at 10 K for the EB-MTJ is thought to result mainly from the inelastic tunneling process caused by magnon excitation. Theoretically, the exchange interaction energy of the Co_2MnSi at the Co_2MnSi/MgO interface is small, indicating that the magnon can be excited easily [32]. At RT, the zero-bias anomaly of the G_{AP} almost disappeared because these inelastic tunneling processes are stimulated by thermal excitation energy. Therefore, we suggested that the inelastic tunneling process attributable to the magnon excitation can be the dominant reason for the large temperature dependence of the TMR effect for EB-MTJ.

17.3 Tunneling Magnetoresistance in Co_2MnSi/CoFeB/MgO/$Co_{50}Fe_{50}$ Magnetic Tunnel Junctions

We have demonstrated a giant TMR ratio of 753 % at 2 K in MTJs with Co_2MnSi/MgO/CoFe structure in which the MgO barrier was deposited by EB evaporation system. This result indicates that Heusler alloys as electrodes of MgO-based MTJs have great potential to achieve giant TMR ratio by half-metallicity and coherent tunneling. However, the TMR ratio decreased drastically with increasing temperature. We think that the large temperature dependence originated with the inelastic tunneling process occurred by fluctuation of magnetization at the Co_2MnSi/MgO interface. On the other hand, CoFeB/MgO/CoFeB MTJ has very small temperature dependence on spin polarization. This indicates that the fluctuation of magnetization at the CoFeB/MgO interface is relatively small, compared to that of Co_2MnSi/MgO interface. Then, insertion of a very thin CoFeB layer into the Co_2MnSi/MgO interface can improve temperature dependence of the TMR ratio. In this work, to achieve high TMR ratio at RT, we fabricated MTJs with a structure of Co_2MnSi/CoFeB/MgO/CoFe and investigated their spin-dependent tunneling properties.

17.3.1 Preparation of Co_2MnSi/CoFeB/MgO/CoFe Magnetic Tunnel Junctions

The magnetron sputtering technique was used to prepare all thin films. We fabricated MTJs with a structure of MgO(20 nm)/Co_2MnSi(50)/CoFeB(d_{CoFeB})/MgO(2.3)/ $Co_{50}Fe_{50}$(5)/IrMn(10)/Ta(5) substrates [33] on single crystalline MgO (001). A Co–Mn–Si alloy target (Co, 43.7 %; Mn, 27.95 %; and Si, 28.35 %) was used to prepare Co_2MnSi films, and the deposited film composition was Co, 50.1 %; Mn, 24.0 %; Si, 25.9 % as measured by inductively coupled plasma analysis. The Co_2MnSi bottom electrode was deposited at RT and annealed at 500 °C to obtain good chemical ordering. The CoFeB modifying layer was deposited from 0.3 nm to 2.0 nm at RT. The MgO tunneling barrier was formed by RF-direct sputtering using an MgO target. The Ar pressure was 0.33 Pa; this Ar pressure was optimized to obtain a highly (001)-oriented MgO layer on the CoFeB thin film. All the MTJs were patterned into a 4-terminal structures by photo-lithography and Ar ion milling. TMR effect was measured at 310 K and 10 K. To improve the crystallinity of the MgO barrier and CoFeB, patterned MTJs were annealed in the range of 300–475 °C for 1 hour in high vacuum by applying a magnetic field of 1 T. For comparison, we prepared an MTJ with a structure of MgO(001)-subs./Cr/CoFeB/MgO/CoFe/IrMn/Ta, and measured the TMR effect.

17.3.2 Tunneling Magnetoresistance of the Co_2MnSi/CoFeB/MgO/CoFe Magnetic Tunnel Junctions

TMR ratio increased with increasing annealing temperature, because of crystallization of the CoFeB and the MgO layers, and showed a maximum at 425–450 °C.

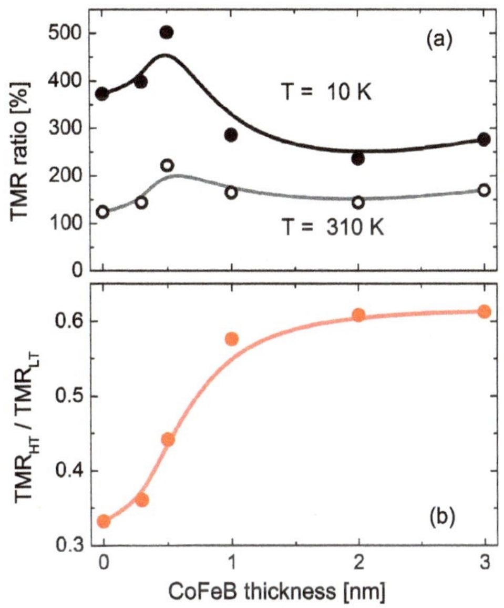

Fig. 17.5 (**a**) CoFeB thickness dependence of TMR ratio at 310 K (*open circles*) and 10 K (*closed circles*) for the optimized MTJ for each CoFeB thickness. (**b**) The temperature dependence of the TMR ratio (TMR_{RT}/TMR_{LT}) with each CoFeB thickness

However, for the MTJ with 2.0 nm thick CoFeB layer, TMR ratio decreased rapidly over 300 °C. We inferred that this reduction in TMR was caused by the large amount of boron diffusion in the Co_2MnSi or MgO layers. Figure 17.5(a) shows CoFeB thickness dependence of TMR ratio at 310 K and 10 K for the optimized MTJ for each CoFeB thickness. The MTJ with CoFeB thickness of 0.5 nm shows the highest TMR ratio, up to 510 % at 10 K and 222 % at 310 K. This TMR ratio is larger than those of the Co_2MnSi/MgO/CoFe MTJ and the CoFeB/MgO/CoFe MTJ. Larger TMR ratio of Co_2MnSi/MgO/CoFe MTJ than that of CoFeB/MgO/CoFe MTJ at low temperature (LT) indicates that spin polarization of Co_2MnSi is larger than that of CoFeB. However, TMR ratio of Co_2MnSi/MgO/CoFe MTJ at RT is smaller than that of CoFeB/MgO/CoFe MTJ. This large temperature dependence of TMR for Co_2MnSi/MgO/CoFe MTJ is due to inelastic tunneling process, as mentioned before. Figure 17.5(b) shows the temperature dependence of the TMR ratio (TMR_{RT}/TMR_{LT}) with each CoFeB thickness. TMR_{RT}/TMR_{LT} ratio tends to saturate beyond a thickness of 1.0 nm. This indicates that the temperature dependence of TMR ratio is improved by CoFeB insertion.

17.3.3 Tunneling Conductance Characteristics

Figures 17.6(a) and (b) show bias voltage dependence of conductance (G) measured at 10 K for parallel magnetic configuration. G is normalized by that at zero bias. The MTJs shown in Fig. 17.6 were annealed at the optimized temperature. For the MTJs with 0.3 and 0.5 nm thick CoFeB, the conductance curves are similar to the MTJ with Co_2MnSi single-layer electrode. The shape of the conductance curve reflects

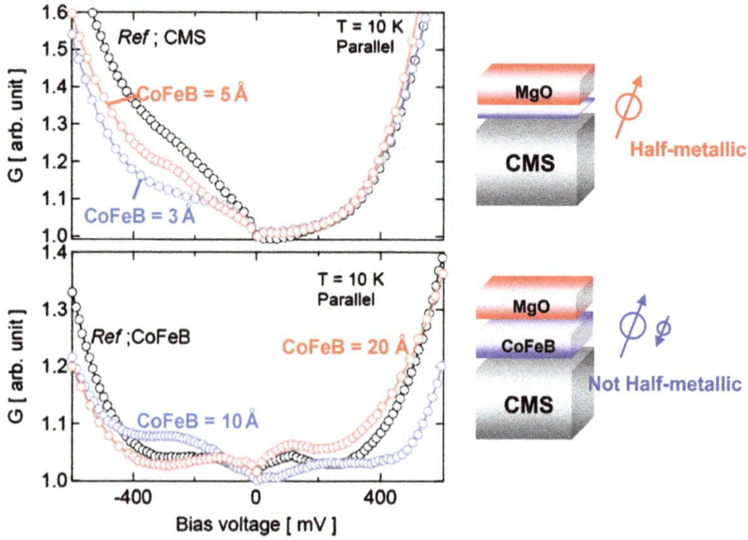

Fig. 17.6 Bias voltage dependence of conductance (G) measured at 10 K for parallel magnetic configuration

the density of states of half-metallic Co_2MnSi, as reported previously [34, 35]. On the other hand, for the MTJ with 1.0 and 2.0 nm thick CoFeB, conductance curves are similar to the MTJ with the CoFeB single-layer electrode [26]. The shape of the conductance curve is characteristic in the MTJs with coherent tunneling through the Δ_1-band of CoFeB [27]. We infer that the highly spin polarized electron in the Co_2MnSi layer can conserve the polarization through the 0.5 nm thick CoFeB layer.

Figure 17.7 shows the bias voltage dependence of G measured at 10 K for anti-parallel magnetic configuration for the MTJs shown in Fig. 17.5. G_{AP} for the Co_2MnSi/MgO/CoFe MTJ increased sharply at low bias voltage. This large zero-bias anomaly of G_{AP} at low bias is caused by the inelastic tunneling processes occurring due to magnetization fluctuation at the Co_2MnSi/MgO interface. On the other hand, for the MTJs with CoFeB insertion layer, zero-bias anomaly seems to be small compared to G_{AP} of Co_2MnSi/MgO/CoFe MTJ. This result indicates that the CoFeB insertion layer suppresses fluctuation of magnetization and improves the temperature dependence of the TMR ratio. From the results of the G–V measurements, we think that the enhancement of TMR ratio by insertion of CoFeB up to 0.5 nm thickness results from suppression of inelastic tunneling process, and the reduction of TMR ratio by insertion of CoFeB over 0.5 nm comes from relatively small spin polarization of CoFeB compared to that of Co_2MnSi.

17.4 Summary

In summary, we have fabricated epitaxially grown Co_2MnSi/MgO/CoFe MTJs with an MgO barrier prepared using both direct sputtering and EB-evaporation tech-

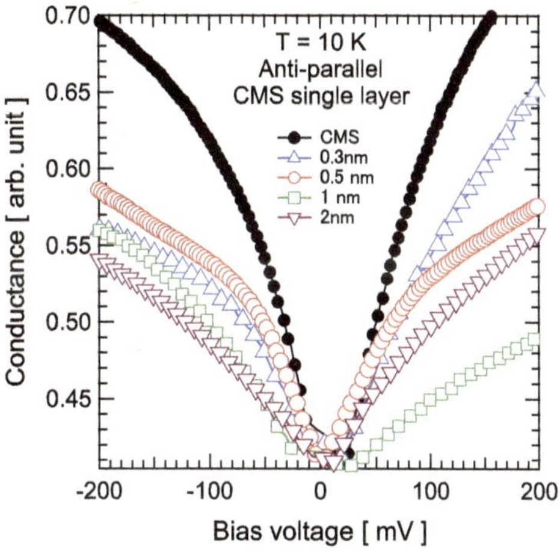

Fig. 17.7 The bias voltage dependence of G measured at 10 K for anti-parallel magnetic configuration for the MTJs shown in Fig. 17.5

niques. For the MTJ with a MgO barrier prepared using sputtering, the TMR ratio was comparable to those of the MTJs using an Al-oxide barrier. From HR-TEM images, XA measurements and conductance curves, we found that the Co_2MnSi/MgO interface has a disarrayed structure and contains oxides. This poorer interface structure is the reason for the low TMR ratio and large resistance for the MTJ with MgO barrier prepared by sputtering. It should be noted that we can obtain higher TMR ratio by further optimization of sputtering conditions of MgO barrier to improve interface structure. On the other hand, for the MTJ with the MgO barrier prepared using EB-evaporation, we observed the very large TMR ratio of 753 % at 2 K. However, unfortunately, the temperature dependence of TMR ratio remained large. We discussed the origin of the large temperature dependence of the TMR ratio. Results of this study suggest that this problem is attributable to inelastic tunneling processes because of magnon excitation at the Co_2MnSi/MgO interface. Also, we can improve the TMR ratio by insertion of very thin CoFeB films. From the G–V measurement, we expect that the highly spin polarized electron in the Co_2MnSi layer can conserve the polarization through the 0.5 nm thick CoFeB layer. Furthermore, the temperature and bias voltage dependence of TMR ratio was improved by inserting CoFeB. This improvement is due to the suppression of the fluctuation of magnetic moment at the Co_2MnSi/MgO interface.

Acknowledgements We would like to thank all our collaborators in the DFG research unit FG 559. This work would be the result of efforts of all members. This study was supported by DFG FG 559 (project P12) and the Strategic Japanese–German Cooperative Program on *Nanoelectronics* by JST as the matching fund. Also, the part of this work was supported by the *High-Performance Low-Power Consumption Spin Devices and Storage Systems* program under Research and Development for Next-Generation Information Technology, and Grand-in-Aid for Scientific Research for Priority Area *Creation and Control of Spin Current* by the Ministry of Education, Culture, Sports, Science and Technology of Japan (MEXT). This research was partly conducted at the Laboratory for Nanoelectronics and Spintronics, Research Institute of Electrical Communication, Tohoku University.

References

1. Galanakis I (2002) J Phys Condens Matter 14:6329
2. Picozzi S, Continenza A, Freeman AJ (2002) Phys Rev B 66:094421
3. Wurmehl S, Fecher GH, Kroth K, Kronast F, Dürr H, Takeda Y, Saitoh Y, Kobayashi K, Lin HJ, Schönhense G, Felser C (2006) J Phys D, Appl Phys 39:803
4. Tanaka CT, Nowak J, Moodera JS (1999) J Appl Phys 86:6239
5. Inomata K, Okamura S, Goto R, Tezuka N (2003) Jpn J Appl Phys 42:L419
6. Kubota H, Nakata J, Oogane M, Ando Y, Sakuma A, Miyazaki T (2004) Jpn J Appl Phys 43:L984
7. Conca A, Falk S, Jakob G, Jourdan M, Adrian H (2004) J Magn Magn Mater 290–291:1127
8. Kämmerer S, Thomas A, Hütten A, Reiss G (2004) Appl Phys Lett 85:79
9. Sakuraba Y, Nakata J, Oogane M, Kubota H, Ando Y, Sakuma A, Miyazaki T (2005) Jpn J Appl Phys 2(44):L1100
10. Sakuraba Y, Hattori M, Oogane M, Kubota H, Ando Y, Sakuma A, Telling PKND, van der Laan G, Arenholz E, Hicken RJ, Miyazaki T (2007) J Magn Soc Jpn 31:209
11. Sakuraba Y, Hattori M, Oogane M, Ando Y, Kato H, Sakuma A, Miyazaki T, Kubota H (2006) Appl Phys Lett 88:192508
12. Julliere M (1975) Phys Lett 54A:225
13. Butler WH, Zhang XG, Schulthess TC, MacLaren JM (2001) Phys Rev B 63:054416
14. Mathon J, Umerski A (2001) Phys Rev B 63:220403
15. Yuasa S, Nagahama T, Fukushima A, Suzuki Y, Ando K (2004) Nat Mater 3:868
16. Ikeda S, Hayakawa J, Ashizawa Y, Lee YM, Miura K, Hasegawa H, Tsunoda M, Matsukura F, Ohno H (2008) Appl Phys Lett 93:082508
17. Ishikawa T, Hakamata S, Matsuda K, Umemura T, Yamamoto M (2008) J Appl Phys 103:07919
18. Tezuka N, Ikeda N, Sugimoto S, Inomata K (2007) Jpn J Appl Phys 2(46):L454
19. Tezuka N, Ikeda N, Mitsuhashi F, Sugimoto S (2009) Appl Phys Lett 94:162504
20. Miura Y, Uchida H, Oba Y, Nagao K, Shirai M (2007) J Phys Condens Matter 19:365228
21. Tsunegi S, Sakuraba Y, Oogane M, Takanashi K, Ando Y (2008) Appl Phys Lett 93:112506
22. Oogane M, Sakuraba Y, Nakata J, Kubota H, Ando Y, Sakuma A, Miyazaki T (2006) J Phys D 39:834
23. Hayakawa J, Ikeda S, Matsukura F, Takahashi H, Ohno H (2005) Jpn J Appl Phys 44:L587
24. Hayakawa J, Ikeda S, Lee YM, Matsumura F, Ohno H (2006) Appl Phys Lett 89:232510
25. Telling ND, Keatley PS, der Laan GV, Hicken RJ, Arenholz E, Sakuraba Y, Oogane M, Ando Y, Miyazaki T (2006) Phys Rev B 74:224439
26. Matsumoto R, Nishioka S, Mizuguchi M, Shiraishi M, Maehara H, Tsunekawa K, Djayaprawira DD, Watanabe N, Otani Y, Nagahama T, Fukushima A, Kubota H, Yuasa S, Suzuki Y (2007) Solid State Commun 143:574
27. Miao G, Chetry KB, Gupta A, Butler WH, Tsunekawa K, Djayaprawira DD, Xiao G (2006) J Appl Phys 99:08T305
28. Ando Y, Miyakoshi T, Oogane M, Miyazaki T, Kubota H, Ando K, Yuasa S (2005) Appl Phys Lett 87:142502
29. Sakuraba Y, Kubota H, Miyakoshi T, Oogane M, Ando Y, Sakuma A, Miyazaki T (2006) Appl Phys Lett 89:052508
30. Zhang S, Levy PM, Marley AC, Parkin SS (1997) Phys Rev Lett 79:3744
31. Takada I, Inoue J, Itoh H (2008) J Magn Soc Jpn 32:338
32. Sakuma A, Toga Y, Tsuchiura H (2009) J Appl Phys 105:07C910
33. Tsunegi S, Sakuraba Y, Oogane M, Naganuma H, Takanashi K, Ando Y (2009) Appl Phys Lett 94:252503
34. Sakuraba Y, Hattori M, Oogane M, Kubota H, Ando Y, Sakuma A, Miyazaki T (2007) J Phys D, Appl Phys 40:1221
35. Tsunegi S, Sakuraba Y, Oogane M, Takanashi K, Ando Y (2009) J Phys D, Appl Phys 42:195004

Index

C. Felser, G.H. Fecher (eds.), *Spintronics*, DOI 10.1007/978-90-481-3832-6,